Innovative Biotic Stress Management Strategies in Crops

Crops suffer from biotic and abiotic stress during their growing period. Biotic stress in crops is caused by living organisms, specifically viruses, bacteria, fungi, nematodes, insects, arachnids, phanerogamic plants, and weeds. They directly deprive their host of its nutrients leading to reduced plant vigor and, in extreme cases, death of the host plant causing up to 40% yield losses and significantly reducing the quality of the produce annually. This has led to increased concern over food safety and nutritional security in India as well as abroad. Climate change and human interventions have altered ecosystems, reducing biodiversity and creating new niches where pests and pathogens can thrive, and new pests and diseases of crops. The integrative strategies for the management of biotic stresses of crops include managing pests by enhancing the resistance of host plants. Biological, cultural, physical, and chemical methods are needed.

In this book, 29 chapters are included covering the general topic and specific crop-based strategies to manage the stresses caused by insect pests and diseases of different crops such as cereals, pulses, oilseeds, vegetables, fruits, cash crops, medicinal and aromatic plants, spices, and plantation crops by using different approaches with the latest information. At the end of each chapter, references are cited which may facilitate the reader to get more detailed information. The book also focuses more on the diagnosis of biotic stresses and management by adding recent information, which has practical utility.

The purpose of this book is to provide recent information in the field of biotic stress management in crops to students, researchers, teachers, extension workers and farmers, and those who deal with crops that are vulnerable to biotic stress.

Innovative Biotic Stress Management Strategies in Crops

Edited by
Dinesh Singh, Rashid Pervez, and Anoop Kumar

CRC Press
Taylor & Francis Group
Boca Raton London New York

CRC Press is an imprint of the
Taylor & Francis Group, an **informa** business

Designed cover image: Dinesh Singh

First edition published 2026
by CRC Press
2385 NW Executive Center Drive, Suite 320, Boca Raton FL 33431

and by CRC Press
4 Park Square, Milton Park, Abingdon, Oxon, OX14 4RN

CRC Press is an imprint of Taylor & Francis Group, LLC

ISBN: 978-1-032-63154-7 (hbk)
ISBN: 978-1-032-63159-2 (pbk)
ISBN: 978-1-032-63157-8 (ebk)

DOI: 10.1201/9781032631578

Typeset in Times New Roman
by Newgen Publishing UK

Contents

Foreword

बिरसा कृषि विश्वविद्यालय
BIRSA AGRICULTURAL UNIVERSITY
राँची, झारखण्ड, भारत
Ranchi, Jharkhand, India

डॉ. सुनील चन्द्र दुबे
कुलपति
Dr. S.C. Dubey
Vice-Chancellor

Foreword

Biotic stress in crop is caused by living organisms, particularly, bacteria, fungi, viruses, viroids, nematodes, insects, mites, and weeds. These causative agents directly and rapidly damage their hosts at any phase of their life cycle, commencing from seedling establishment, plant maturation or grain or fruit setting and after harvest leading to reduce plant vigour and yield. Now a day, worldwide more attention has been given to protect the crop to meet food and nutritional security of human being. The integrative innovative strategies for management of diseases, insect pests and plant nematodes of crops are enhancing resistance of host plants, using biological, cultural, physical and chemical technologies and improving environmental conditions in favour to the crops.

This book entitled **Innovative Biotic Stress Management Strategies in Crops** is a maiden effort to give an overview of plant diseases, insect pests and plant parasitic nematodes for growers, agriculturists, extension workers and scientists from plant protection disciplines dealing with management. The book has 29 chapters covering most of the innovative management tools and technologies developed by distinguished scientists to diagnose and manage the biotic stress in various major crops. The chapters are well referenced and contain colour photographs of disease symptoms, life cycle descriptions, informative tables and figures.

I wish to place on record the valiant effort made by Drs. Dinesh Singh, Rashid Pervez and Anoop Kumar in producing this invaluable book.

SCDubey 23/12/24
(S. C. Dubey)

कुलपति कार्यालय, प्रशासनिक भवन, बिरसा कृषि विश्वविद्यालय, काँके, राँची 834006 झारखण्ड, भारत
The Office of the Vice-chacellor, V.C. Cell, Administrative Building, Birsa Agricultural University, Kanke, Ranchi 834006 Jharkhand, India
www.bauranchi.org 91-651-2450500 bauvcranchi@gmail.com | vc@bauranchi.org

Preface

India has traditionally produced, consumed, and exported a wide range of commodities. Since the agricultural economy's globalization gained popularity, the nation's output and economics have faced significant obstacles. Current challenges are falling exports and high production costs. It is appropriate to state that every effort is required to revitalize, maintain, and grow India's economy. Depending on the crops, environmental conditions, and pests, biotic stress in crops can result in pre- and postharvest losses of approximately 25–40%. These losses can be caused by a variety of plant diseases, including bacteria, fungi, viruses, nematodes, insects, and other pests. The amount of land accessible per person is steadily declining because of the expanding population.

The book "Innovative Biotic Stress Management Strategies in Crops" has 29 chapters covering most of the innovative management tools and technology developed by distinguished scientists to diagnose and manage the biotic stresses in various crops. This book is an outcome of insect, plant parasitic nematodes, and disease management technologies for the benefit of farmers, extension workers, and entrepreneurs. The book is regarded as one of the best resources for information on the state, developments, and methods of plant protection technology for different types of crops. Students, academics, specialists, and politicians will utilize it as a text, since it is unique in offering concepts and practical information about using management strategies for human well-being.

The editors are thankful to the Director, Joint Director (Research), Head, Division of Nematology, ICAR-Indian Agricultural Research Institute, New Delhi; Director, ICAR-NCIPM, New Delhi and Director, Indian Institute of Sugarcane Research, Lucknow, Uttar Pradesh, for encouragement. The editors are also grateful to all the contributors of this book for writing the chapters. We express our sincere thanks to CRC Press, Taylor & Francis Group, for taking on the task of the publication of this book.

Dinesh Singh
Rashid Pervez
Anoop Kumar

Editor Biography

Dinesh Singh

Dinesh Singh (Principal Scientist) did his MSc (Ag) and PhD in the Department of Mycology & Plant Pathology, Institute of Agricultural Sciences, BHU, Varanasi. He joined ARS services in January 1997 at IIVR, Varanasi, and is presently working as Head & Principal Scientist at the Division of Crop Protection, ICAR-Indian Institute of Sugarcane Research, Lucknow, Uttar Pradesh. Dr. Dinesh Singh has made a significant contribution broadly in the area of plant pathology, particularly in the field of plant bacteriology such as biological control, diagnosis, and diversity under 32 in-house projects and 13 externally funded projects. He has guided 9 PhD and 3 MSc students in the field of plant bacteriology. He has published 194 research papers in peer-reviewed reputed journals, 35 books, 118 book chapters, 47 popular articles, 5 technical bulletins, 9 training manuals, and 2 practical manuals. He is editor of 4 journals. He has been honored and awarded various awards such as NESA Fellowship, NABS Fellow, IPS Fellow, SPPS Fellow, Bioved Fellowship Award, Dr. M. M. Alam Medal, Young Scientist Associate Award, Late Shri P. P. Shinghal Memorial Award, J. P. Varma Memorial Lecture Award, Reviewer Excellence Award, Award for Excellent in Research, Distinguished Scientist in Plant Pathology Award, Best Faculty Award, NESA Green Technology Innovative Award, B P Pal Memorial Lecture Award, and MS Pavgi Award, from various professional societies and agencies.

Rashid Pervez

Rashid Pervez is working as a Principal Scientist at the Division of Nematology, ICAR-Indian Agricultural Research Institute (IARI) Pusa, New Delhi. He mainly is working on entomopathogenic nematodes (EPN). Dr. Pervez has described six new species of EPN including two that are heat tolerant. Dr. Rashid has been the Principal Investigator and Co PI of the various projects funded by ICAR, DST and the Kerala govt. He guided several M. Sc. Students and is an examiner of and PhD students. He is serving as chief editor and editor of various scientific journals like *Indian Phytopathology, Pantnagar Journal of Research Annals of Plant Protection Sciences*, *Trends in Biosciences,* and *Advances in Life Sciences*. He has published 112 research papers in reputed international and national journals, 10 books, 37 books/bulletins, 4 manuals, 5 extension folder, and 70 scientifically popular articles. He has also attended several symposia and conferences, and delivered lectures, and presented 96 research papers. He is the recipient of the Dr. Absar Mustafa Khan Memorial Award, Distinguished Scientist of the Year Award 2017, Scientist of the Year Award 2016, ICAR- Ganesh Shankar Vidyarthi Best Book Award 2015, SPPS Meritorious Scientist Award 2015, Dr. Muthuswami Anandaraj Award 2014, Dr. H. S. Mehta Memorial Award 2013, Young Scientist Award 2007, Young Scientist Associate Award 2005, and Fellow, Nematological Society of India (NSI), Society of Plant Protection Sciences (SPPS), Society for Biocontrol Advancement (SBA), Indian Society for Spices (ISS), Agricultural and Environmental Technology Development Society (AETDS).

Anoop Kumar

Anoop Kumar is a senior scientist in Agriculture Entomology at the ICAR-NCIPM, New Delhi. He graduated in Agriculture with a BSc Ag from Sardar Patel University of Agriculture & Technology, Meerut, UP in 2005, an MSc in Agriculture Entomology from Chandra Shekhar Azad University of Agriculture & Technology, Kanpur in 2007, and a PhD in Entomology and Agriculture Zoology from Banaras Hindu University, Varanasi in 2013. He was awarded a vice-chancellor's silver medal at graduation, a Department topper Fellowship of the University and a book prize post graduation, an ICAR Senior Research Fellowship, and an INSPIRE Fellowship during his PhD. He was awarded a 'Life Fellow' of the Entomological Society of India for devotion to the 'Promotion of Research in Entomology'. He implemented 15 research projects as PI and Co-PI. He has published more than 25 research papers in various reputed journals, 10 book chapters, 10 popular articles, and 2 technical bulletins.

Contributors

Adhikari, Sneha
ICAR-Indian Institute of Wheat and Barley Research, Regional Station,
Shimla, Himachal Pradesh, India.

Agarwal, Sneha
Department of Biotechnology,
SET, MRIIRS,
Faridabad, Haryana, India.

Ajithkumar, K.
Main Agricultural Research Station,
University of Agricultural Sciences
Raichur, Karnataka, India.

Ali, Anwar
Division of Plant Pathology,
Faculty of Agriculture, SKUAST-K,
Sopore, Jammu & Kashmir, India.

Aswathanarayana, D. S.
Department of Plant Pathology,
College of Agriculture,
University of Agricultural Sciences
Raichur, Karnataka, India.

Baitha, Arun
Division of Crop Protection,
ICAR-Indian Institute of Sugarcane Research,
Lucknow, Uttar Pradesh, India.

Balodi, Rekha
ICAR-National Research Institute for Integrated Pest Management,
New Delhi, India.

Banoth, Vinesh
ICAR-Indian Institute of Seed Science,
Mau, Uttar Pradesh, India.

Barman, Ashis Roy
Department of Plant Pathology,
Bidhan Chandra Krishi Viswavidyalaya,
Nadia West Bengal, India.

Bhagwati, Bhabesh
Department of Nematology,
AAU,
Jorhat, Assam, India.

Bharathi, V.
Babu Jagjivan Ram Agricultural College, PJTSAU
Sircilla, Telangana, India.

Bhardwaj, Subhash Chander
ICAR-Indian Institute of Wheat and Barley Research, Regional Station,
Shimla, Himachal Pradesh, India.

Bhat, Mudasir
Division of Plant Pathology,
Faculty of Agriculture, SKUAST-K,
Sopore, Jammu & Kashmir, India.

Chandrawat, B. S.
College of Agriculture,
SKNAU,
Jobner, Rajasthan, India.

Chander, S.
ICAR-National Research Institute for Integrated Pest Management,
New Delhi, India.

Charu, Lata
ICAR-Indian Institute of Wheat and Barley Research,
Regional Station,
Shimla, Himachal Pradesh, India.

Chatterjee, Sitesh
Rice Research Station,
Chinsurah, West Bengal, India.

Das, Amrita
Division of Plant Pathology,
ICAR-Indian Agricultural Research Institute,
New Delhi, India.

Devi, Sailabala
Deputy Director of Research,
Bihar Agricultural University,
Bhagalpur, Bihar, India.

Didwania, Nidhi
Manav Rachna Centre for Medicinal Plant Pathology (MR-CMPP),
MRIIRS,
Faridabad, Haryana, India.

Dipasree, Sadhukhan
Department of Plant Pathology,
Bidhan Chandra Krishi Viswavidyalaya,
Nadia, West Bengal, India.

Dixit, Saurabh
Crop Research Station, Masodha,
Acharya Narendra Deva University of Agriculture and Technology,
Ayodhya, Uttar Pradesh, India

Dutta, Subrata
Department of Plant Pathology,
Bidhan Chandra Krishi Viswavidyalaya,
Nadia, West Bengal, India.

Ganesh, H. S.
Department of Plant Pathology,
College of Agriculture,
University of Agricultural Sciences,
Raichur, Karnataka, India.

Gangwar, Om Prakash
ICAR-Indian Institute of Wheat and Barley Research, Regional Station,
Shimla, Himachal Pradesh, India.

Ghosal, Abhijit
Sasya Shyamala Krishi Vigyan Kendra,
Ramakrishna Mission Vivekananda Educational and Research Institute,
West Bengal, India.

Ghosh, Amalendu
Division of Plant Pathology,
ICAR-Indian Agricultural Research Institute,
New Delhi, India.

Gundappa
ICAR-National Bureau of Agricultural Insect Resources,
Bengaluru, Karnataka, India.

Gupta, Shivanshi
Division of Plant Pathology,
Faculty of Agriculture, SKUAST-K,
Sopore, Jammu & Kashmir, India.

Hasan, S. S.
Division of Crop Protection,
Indian Institute of Sugarcane Research,
Lucknow, Uttar Pradesh, India.

Hithesh, G. R.
Division of Entomology,
ICAR-Indian Agricultural Research Institute,
New Delhi, India.

Jan, Roohi
Division of Plant Pathology,
Faculty of Agriculture, SKUAST-K,
Sopore, Jammu & Kashmir, India.

Jash, Subhendu
Department of Plant Pathology,
Bidhan Chandra Krishi Viswavidyalaya,
Nadia, West Bengal, India.

Jatoth, Rajender
Babu Jagjivan Ram Agricultural College, PJTSAU,
Sircilla, Telangana, India.

Junaid, Raja Rashid
Division of Plant Pathology,
Faculty of Agriculture, SKUAST-K,
Sopore, Jammu & Kashmir, India.

Jayant, Bhatt
Department of Plant Pathology,
Jawaharlal Nehru Krishi Vishwa Vidyalaya,
Jabalpur, Madhya Pradesh, India.

Joshi, Ranjana
Department of Plant Pathology,
College of Agriculture,
University of Agricultural Sciences,
Raichur, Karnataka, India.

Kalita, Sasanka
Department of Biotechnology,
SET, MRIIRS,
Faridabad, Haryana, India.

Keerthi, M.C.
Division of Crop Protection,
ICAR-Indian Institute of Horticultural Research,
Bengaluru, Karnataka, India.

Kumar, Ajay
Department of Plant Pathology,
Banda University of Agriculture and Technology,
Banda, Uttar Pradesh, India.

Kumar, Anshu
Department of Plant Pathology,
Bidhan Chandra Krishi Viswavidyalaya,
Nadia, West Bengal, India.

Kumar, Ashok
Department of Plant Pathology,
Banda University of Agriculture and Technology,
Banda, Uttar Pradesh, India.

Kumar, Dharmendra
Department of Plant Pathology,
Banda University of Agriculture and Technology,
Banda, Uttar Pradesh, India.

Kumar, Manish
Division of Plant Pathology,
Indian Agricultural Research Institute,
New Delhi, India.

Kumar, Rakesh
ICAR-National Research Institute for Integrated Pest Management,
New Delhi, India.

Kumar, Sahani Nilesh
Department of Plant Pathology,
Banda University of Agriculture and Technology,
Banda, Uttar Pradesh, India.

Kumar, Sanjeev
Department of Plant Pathology,
Jawaharlal Nehru Krishi Vishwa Vidyalaya,
Jabalpur, Madhya Pradesh, India.

Kumar, Santosh
Department of Mycology & Plant Pathology,
Institute of Agricultural Sciences,
Banaras Hindu University
Varanasi, Uttar Pradesh, India.

Kumar, Subodh
ICAR-Indian Institute of Wheat and Barley Research, Regional Station
Shimla, Himachal Pradesh, India.

Kumar, Tribhuwan
Department of Molecular Biology and Genetic Engineering,
Bihar Agricultural University,
Bhagalpur, Bihar, India.

Kumar, Vishal
Department of Biotechnology,
SET, MRIIRS,
Faridabad, Haryana, India.

Kundu, Krishnendu
Department of Plant Pathology,
Bidhan Chandra Krishi Viswavidyalaya,
Nadia, West Bengal, India.

Kurulkar, Uday
Department of Nematology,
AAU,
Jorhat, Assam, India.

Lal, Kishan
Department of Plant Pathology,
National P.G. College,
Gorakhpur, Uttar Pradesh, India.

Lokesh Babu, P.
Division of Plant Pathology,
ICAR-Indian Agricultural Research Institute,
New Delhi, India.

Maurya, Manish Kumar
Crop Research Station, Masodha,
Acharya Narendra Deva University of Agriculture and Technology,
Ayodhya, Uttar Pradesh, India.

Meena, Narottam Kumar
ICAR-National Research Centre on Seed Spices,
Ajmer, Rajasthan, India.

Mehta, Neelam
ICAR-National Research Institute for Integrated Pest Management,
New Delhi, India.

Mishra, Raj K.
Division of Crop Protection,
ICAR-Indian Institute of Pulses Research,
Kanpur, Uttar Pradesh, India.

Mishra, Shubham
Department of Plant Pathology,
Jawaharlal Nehru Krishi Vishwa Vidyalaya,
Jabalpur, Madhya Pradesh, India.

Mukhopadhyay, Sabyasachi
Department of Plant Pathology,
Bidhan Chandra Krishi Viswavidyalaya,
Nadia, West Bengal, India.

Neog, Pranjal Pratim
Department of Nematology,
BN College of Agriculture, AAU,
Charial, Assam, India.

Omprakash, S.
Regional Agricultural Research Station, PJTSAU,
Warangal, Telangana, India.

Pandey, Sakshi
Department of Plant Pathology,
Banda University of Agriculture and Technology,
Banda, Uttar Pradesh, India.

Pandey, Sonika
Division of Crop Protection,
ICAR-Indian Institute of Pulses Research,
Kanpur, Uttar Pradesh India.

Paramanandham, Latha
Department of Plant Pathology,
TNAU,
Coimbatore, Tamil Nadu, India.

Patel, Ashwani Kumar
Department of Mycology and Plant Pathology,
Institute of Agricultural Sciences,
Banaras Hindu University,
Varanasi, Uttar Pradesh, India.

Patar, Subhash
Department of Plant Pathology,
Bidhan Chandra Krishi Viswavidyalaya,
Nadia, West Bengal, India.

Patel, Shiv Shanker
Department of Mycology and Plant Pathology,
Institute of Agricultural Sciences,
Banaras Hindu University,
Varanasi, Uttar Pradesh, India.

Pathak, Satyam
Crop Research Station, Masodha,
Acharya Narendra Deva University of Agriculture and Technology,
Ayodhya, Uttar Pradesh, India.

Pati, Susmita
Department of Plant Pathology,
Bidhan Chandra Krishi Viswavidyalaya,
Nadia, West Bengal, India.

Patwari, Anindita
Department of Plant Pathology,
Bidhan Chandra Krishi Viswavidyalaya,
Nadia, West Bengal, India.

Pavitra
Department of Plant Pathology,
College of Agriculture,
University of Agricultural Sciences,
Raichur, Karnataka, India.

Prasad, Pramod
ICAR-Indian Institute of Wheat and Barley Research, Regional Station,
Shimla, Himachal Pradesh, India.

Prasad, Vindeshwari
Crop Research Station, Masodha,
Acharya Narendra Deva University of Agriculture and Technology,
Ayodhya, Uttar Pradesh, India.

Priyadarshini, G.
Babu Jagjivan Ram Agricultural College, PJTSAU,
Sircilla, Telangana, India.

Raghuvanshi, Richa
Department of Soil Science and Agricultural Chemistry,
Banda University of Agriculture and Technology,
Banda, Uttar Pradesh, India.

Raghavendra, K.V.
ICAR-National Research Institute for Integrated Pest Management,
New Delhi, India.

Rajaswaminathan, Vairavan
Department of Plant Pathology,
TNAU,
Coimbatore, Tamil Nadu, India.

Rajkumar
Division of Crop Protection,
ICAR-CPCRI,
Kasaragod, Kerala, India.

Rajpoot, S. K. S.
Crop Research Station, Masodha,
Acharya Narendra Deva University of Agriculture and Technology,
Ayodhya, Uttar Pradesh, India.

Ramesh, K. B.
Division of Entomology,
ICAR-Indian Agricultural Research Institute,
New Delhi, India.

Rashid, Pervez
Division of Nematology,
ICAR-Indian Agricultural Research Institute,
New Delhi, India.

Ray, Sujit Kumar
Department of Plant Pathology,
Bidhan Chandra Krishi Viswavidyalaya,
Nadia, West Bengal, India.

Reddy, C.R.K.
Division of Crop Protection,
ICAR-Indian Institute of Oil Seed Research,
Hyderabad, Telangana, India.

Renuka, M.
Main Agricultural Research Station,
University of Agricultural Sciences,
Raichur, Karnataka, India.

Sah, S. B.
Department of Entomology,
Bihar Agricultural University,
Bhagalpur, Bihar, India.

Saha, Nilanjana
Department of Biotechnology,
SET, MRIIRS,
Faridabad, Haryana, India.

Saha, Tamoghna
Department of Entomology,
Bihar Agricultural University,
Bhagalpur, Bihar, India.

Sahu, Vandana
Department of Mycology and Plant Pathology,
Institute of Agricultural Sciences,
Banaras Hindu University,
Varanasi, Uttar Pradesh, India.

Saini, Paridhi
Department of Biotechnology,
SET, MRIIRS,
Faridabad, Haryana, India.

Sarkar, Ujjwal
Department of Plant Pathology,
Bidhan Chandra Krishi Viswavidyalaya,
Nadia, West Bengal, India.

Savitha, A. S.
College of Agriculture,
University of Agricultural Sciences,
Raichur, Karnataka, India.

Sehgal, M.
ICAR-National Research Institute for Integrated Pest Management,
New Delhi, India.

Shukla, P. K.
ICAR-Central Institute for Subtropical Horticulture,
Lucknow, Uttar Pradesh, India.

Singh, Abhay Pratap
Department of Plant Pathology,
Banda University of Agriculture and Technology,
Banda, Uttar Pradesh, India.

Singh, Dhirendra Kumar
Department of Mycology & Plant Pathology,
Institute of Agricultural Sciences,
Varanasi, Uttar Pradesh, India.

Singh, Dinesh
Division of Crop Protection,
ICAR-Indian Institute of Sugarcane Research,
Lucknow, Uttar Pradesh, India.

Singh, M. R.
Division of Crop Protection,
Indian Institute of Sugarcane Research,
Lucknow, Uttar Pradesh, India.

Singh, S. K.
ICAR-National Research Institute for Integrated Pest Management,
New Delhi, India.

Suroshe, Sachin S.
Division of Entomology,
ICAR-Indian Agricultural Research Institute,
New Delhi, India.

Singh, Shivam
Narayan Institute of Agricultural Science,
Gopal Narayan Singh University,
Jamuhar, Sasaram, Bihar, India.

Srinivas, K.
Division of Crop Protection,
Indian Institute of Sugarcane Research,
Lucknow, Uttar Pradesh, India.

Suresh, J.
Dean (Horticulture),
TNAU,
Coimbatore, Tamil Nadu, India.

Tiwari, Abhishek
Division of Crop Protection,
ICAR-Indian Institute of Pulses Research,
Kanpur, Uttar Pradesh, India.

Vaish, S. S.
Department of Mycology and Plant Pathology,
Institute of Agricultural Sciences,
Banaras Hindu University,
Varanasi, Uttar Pradesh, India.

Upadhyay, Vinod
Regional Agricultural Research Station,
Assam Agricultural University,
Kokrajhar, Assam, India.

Verma, Kritik
Department of Biotechnology,
SET, MRIIRS,
Faridabad, Haryana, India.

Yadav, M. S.
ICAR-National Research Institute for Integrated Pest Management,
New Delhi, India.

Yele, Yogesh
ICAR-National Institute of Biotic Stress Management,
Raipur, Chhattisgarh, India.

Yenjerappa, S. T.
College of Agriculture,
University of Agricultural Sciences,
Raichur, Karnataka, India.

Yogendra
Department of Plant Pathology,
Banda University of Agriculture and Technology,
Banda, Uttar Pradesh, India.

Yogendra, Kumar
Department of Plant Pathology,
Banda University of Agriculture and Technology,
Banda, Uttar Pradesh, India.

Zakir, Amin
Division of Plant Pathology,
Faculty of Agriculture, SKUAST-K,
Sopore, Jammu & Kashmir, India.

Zohra, Shabir
Division of Plant Pathology,
Faculty of Agriculture, SKUAST-K,
Sopore, Jammu & Kashmir, India.

1 Ecological Engineering for Insect Pest Management

Sachin S. Suroshe, Keerthi M.C., Hithesh G.R., Yogesh Yele, Rakesh Kumar and Chander S.

1.1 INTRODUCTION

The term Ecological Engineering (EE) was coined by Howard T. Odum in the 1960s, and it has since become widely used in North America, Europe, and China. It is defined as "the energy supplied by man is small relative to the natural sources, but sufficient to produce large effects in the resulting patterns and processes" (Odum 1962). Odum et al. (1963) further propounded that "environmental manipulation by man using small amounts of supplementary energy to control systems in which the main energy drives still come from natural sources". EE takes into consideration the ecology and engineering concepts to forecast, plan, restore, and manage ecosystems that integrate "human society with its natural environment for the benefit of both" (Figure 1.1). The principles of EE have been applied to a wide range of natural resources and environmental issues, including fisheries and agriculture, as well as wastewater control and coastline conservation (Mitsch and Jørgensen 2003).

To meet the nutritional requirements of the ever-increasing population across the globe, farmers are forced to adopt modern production technologies (Swaminathan and Bhavani 2013). It involves the adoption of high-yielding varieties, intensive and extensive cultivation, improved agronomic practices, and application of synthetic fertilizers and pesticides. Due to the ease of application and immediate knockdown effect, farmers primarily relied on synthetic insecticides to manage dreaded insect pests (Aktar et al. 2009). However, the indiscriminate usage of chemical pesticides led to several problems *viz.,* the development of resistance, resurgence, residues, and secondary pest attacks (Damalas and Koutroubas 2016). Further, the high costs involved in pesticidal control and the risks posed to human health and the environment mean that the use of synthetic pesticides as the sole measure is unsustainable in the long run.

EE for pest management has recently been developed as a paradigm for addressing pest management approaches that focus on using cultural strategies to manipulate habitat and improve biological pest control. EE is also called Habitat Manipulation (HM). HM is another form of Conservation Biological Control (CBC), and an ecologically oriented concept that favors NEs to enhance biological control in agricultural systems (Gurr et al. 2004; Rusch et al. 2010). It is an alternative to the strategy of unilateral intervention with synthetic chemicals. Although HM and CBC provide favorable living conditions for NEs in the agro-ecosystems (principle of 'enemy hypotheses'), the methods used differ (Figure 1.2). The CBC mainly focuses on reducing the use of pesticides by resorting to selective pesticides, careful timing and correct placement (Torres et al. 2018).

In contrast, HM aims to create a suitable ecological infrastructure within the agricultural landscape to provide resources such as food, alternative prey or hosts, and shelter to NEs (Landis et al. 2000). The EE involves mainly the creation and restoration of ecosystems. All applications within this subset of EE have the commonality of using ecosystems for the treatment of human wastes

DOI: 10.1201/9781032631578-1

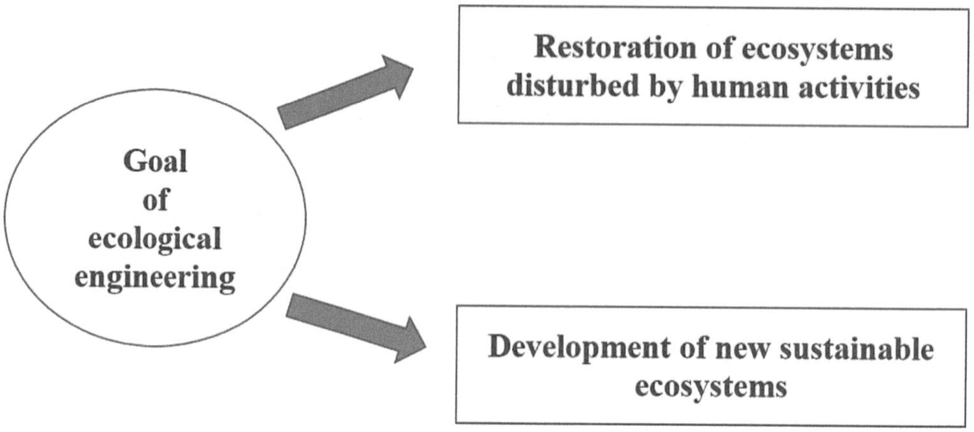

FIGURE 1.1 Goals of ecological engineering.

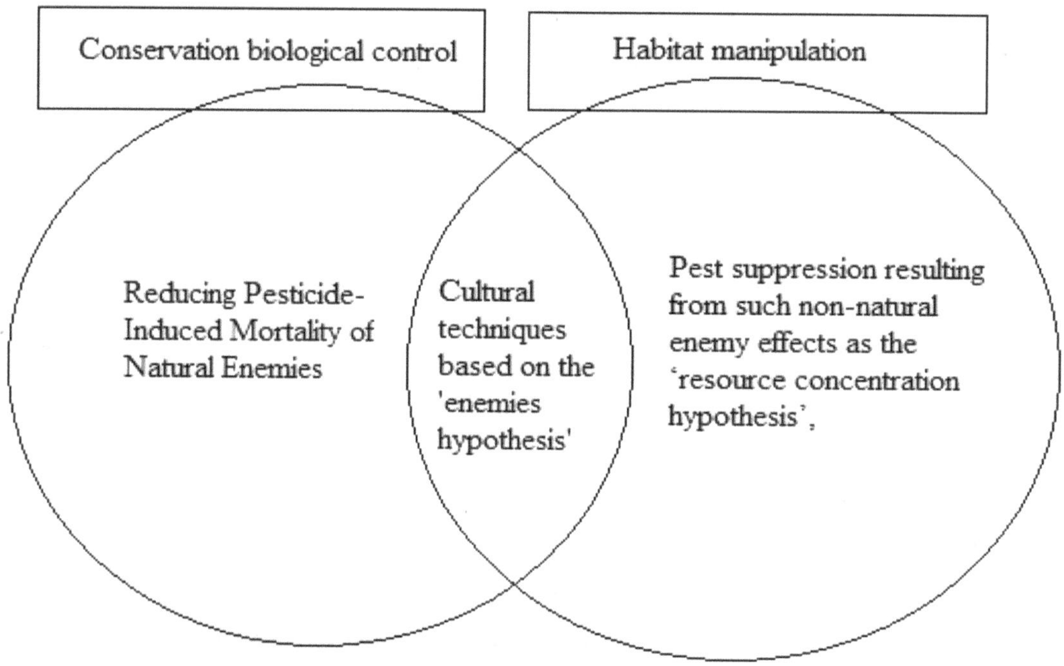

FIGURE 1.2 Habitat manipulation versus conservation biological control. (Adapted from Gurr et al. 2000.)

with an emphasis on genuinely solving problems with an ecological system rather than shifting to another.

Ecological knowledge informs about cultural practices rather than approaches such as synthetic insecticides and genetically engineered crops. A new technique is being developed to improve biological control in an agro-system by preserving or increasing plant diversity or by providing appropriate refugia for pest NEs. It could be argued that all pest management strategies are forms of EE, regardless of whether they act on the physical environment (e.g., through tillage), the chemical environment (e.g., through pesticide use), or the biotic environment (e.g., through the use of novel crop varieties and improved activity of NEs). The employment of cultural practices to manipulate habitat and improve biological control fits the idea of EE (Gurr et al. 2004).

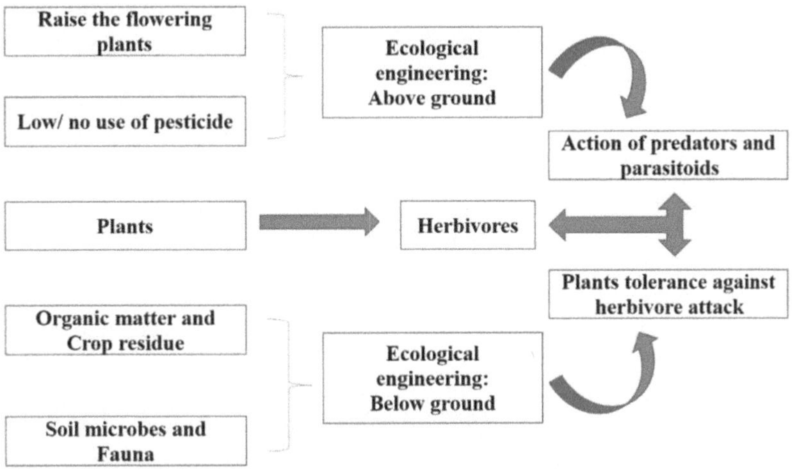

FIGURE 1.3 Approaches of ecological engineering in agro-ecosystem.

The EE in an agricultural ecosystem works based on two approaches above ground and below ground (Figure 1.3). The above-ground approaches include planting flowering plants/suitable cash crops along the orchard border with shorter plants near the main crop and higher plants toward the border to attract NEs and prevent insects from migrating. Further, these approaches do not remove weed plants that offer nectar and pollen to NEs and avoid broad-spectrum pesticide application in crop ecosystems. The below-ground approaches include the balanced application of bio-fertilizers and plant growth-promoting rhizobacteria (PGPR), application of organic matter in the form of farmyard manure (FYM), and vermicompost, crop residue, which enhance the below-ground biodiversity. They also reduce tillage intensity and cover the soil with mulches providing a refuge for soil-dwelling or hibernating NEs (MANAGE 2019).

A detailed understanding of the target pest's life cycle and its ecological and behavioral interactions with its surroundings and natural regulating factors is vital for effective pest management. EE must emphasize why agro-ecosystems are vulnerable and how to make them more pest-resilient. EE redesigned the agro-ecosystem by moving away from linear, one-to-one interactions between target pests and a specific management approach to the web of relationships between insect pests, associated NEs, and crop diversification schemes. Crop diversity has significantly dropped in developed countries, and in the developing world as well, over the last half-century, owing to export-oriented competition for growing only a few cash crops. The decline in crop diversification has led to simpler landscapes. The expansion of monocropping has decreased the abundance and activity of NEs due to the removal of critical food resources and overwintering sites. Such extremely simplistic cropping systems are less conducive to NEs since they require more than prey and hosts to complete their life cycles successfully. Several adults of parasitoids need pollen and nectar to survive and increase longevity. Monocultures, generally, do not provide favorable conditions for NEs. Such simple agricultural systems lack many resources that natural adversaries require to feed, reproduce, and prosper, such as refuge locations, pollen and nectar, and alternative prey or hosts.

1.2 REFUGE LOCATIONS

Perennial crop systems may be more amenable to CBC than ephemeral or short-lived annual crop systems, because of being disturbed less. In contrast, annual crop ecosystems frequently get disrupted by field operations such as plowing, harvesting, and other crop management practices;

that's why resident NEs persist more efficiently on perennial crops (Thorbek and Bilde 2004). Perennial crops like alfalfa, witness a common practice of harvesting entire fields, which disrupts the resident arthropod fauna (*Medicago sativa* L.) (van den Bosch et al. 1982). However, alternative harvesting also known as strip harvesting in alfalfa, according to Hossain et al. (2000) provides a safe haven for the brown lacewing *Micromus tasmaniae* (Walker) and a variety of coccinellid and hemipteran predators. More recently, Clem and Harmon-Threatt (2021) demonstrated that the field borders containing grasses mixed with wildflowers provide winter refuge for NEs *viz.*, beetles, wasps, spiders, and other arthropods in organic soybean farms. Winter refuge is crucial for the survival of NEs. Without these habitat types, NEs are less likely to be present early in the season, which can be vital for controlling insect pests before it's become unmanageable (Geiger et al. 2009). Crop leftovers, plants along field boundaries, and crevices under and beneath the bark of living and dead trees are examples of physical refugia in agro-ecosystems. Keerthi et al. (2020) documented that the cauliflower intercropped with cineraria flower attracted NEs, *i.e.*, coccinellids, spiders and syrphids which then devoured aphid colonies. In addition, the authors also reported that the foliage acts as a shelter site for predators during their resting period and pupation.

Quispe et al. (2017) chose 15 flowering plant species and placed them in three refuge patches in maize fields randomly. The flower crops were divided into three groups: A, B, and C, each with five flowering plants. The specified refuge areas were planted close to the maize plants. The most effective refuges were A and B, which had the highest diversity and abundance of predators and parasitoids, including several insect species known to attack maize pests (*Spodoptera frugiperda* and *Rhopalosiphum maidis*), as well as other species that serve as alternative hosts for these NEs. Group A contain *Gossypium barbadense*, *Aster* sp., *Foeniculum vulgare, Coriandrum sativum* and *Lavandula officinalis*; Group B contain *Helianthus annus, Nicandra physaloides, Salvia officinalis, Bidens pilosa* and *Artemisia absinthium* and Group C contain *Malva parviflora, Rosmarinus officinalis, Phaseolus vulgaris, Galinsoga parviflora* and *Sorghum halepense.*

1.3 POLLEN AND NECTAR

Plant-derived foods such as nectar and pollen play a critical role in the life history of many NEs. While certain parasitoids can receive nutrients from their hosts (Jervis and Kidd 1986), others require non-host foods. Planting a cover crop such as buckwheat provides pollen and nectar for *Dolichogenidea tasmanica*, a major parasitoid of *Epiphyas postvittana* and an important leaf roller pest. This parasitoid oviposits in the early-instar larvae of leaf roller, and its efficacy can be boosted by providing adult foods. The majority of NEs are omnivore that feeds on plants, hosts and prey. Feeding on nectar improves longevity, flying activity, and oviposition of natural enemies. The mirid predatory bug *Macrolophus pygmaeus* completes its development only when it feeds on plant sap. Pollen is an excellent source of reproduction for phytoseiid mites and anthocorid bugs as it is laden with proteins (Lundgren 2009). During periods of prey shortage in the cabbage crop, alyssum with flowers provides resource subsidies for the predatory bugs *Orius laevigatus* and *Orius majusculus*. Pollen-producing plants, such as *Ricinus communis*, supply more pollen to predatory mites. Selecting varieties that produce high nectar levels or extra-floral nectar of a specific composition may help some NEs to establish (Koptur 2005).

Floral and extra-floral nectar are high in carbohydrates, amino acids, and other substances, which can boost NEs' fecundity and also provide enough calories for them to forage (Amoros-Jimenez et al. 2014), and increase search and prey handling, longevity enhancement, and predation resistance. Honeydew is consumed by many parasitoids and other NEs. The mere presence of flowering plants in the field does not guarantee biological control of insect pests. It depends on several criteria like blossom color, external morphology (Vattala et al. 2006), nectar quality and content. So, careful plant selection which could provide pollen and nectar for EE is required.

1.4 ALTERNATIVE PREY OR HOSTS

Crop pest NEs look for alternative hosts and prey populations in non-crop habitats. NEs require alternate prey or hosts when their primary food is not available to support biological control of harmful insect pests. If there is enough alternative prey, generalist predator populations can establish themselves within a crop before the arrival of concerned pests (van Emden 1990). Generalist predators, which rely on a variety of prey species, are more reliant on alternative prey or hosts than specialist predators; this is true for parasitoids as well. Natural pest control by NEs is improved by feeding on alternative hosts and prey during off seasons when hosts and prey density in fields is low. Bianchi and van der Werf (2004) used modeling to show that if pest aphid infestation of wheat is delayed, generalist predators, *Coccinella septempunctata* populations, thrive on aphid populations in non-crop areas. Similarly, if the pea aphids as alternative hosts were present, levels of alfalfa weevil (*Hypera postica*) parasitism by the ichneumonid wasp, *Bathyplectes curculionis* were higher. Access to pea aphid honeydew significantly boosts the wasp's fertility and adult life span.

1.5 APPROACHES TOWARDS CONSERVATION OF NES

All over the world, many cropping systems are annuals; NEs can only inhabit this transient home for a limited time. On the other hand, NEs must locate other nutrients/resources required for survival before sowing and after harvest of specific crops either within or outside the field. The importance of these resources has already been discussed earlier. Any cultural strategy that provides the said resources to the NEs for its enhanced performance suits the concept of EE. The specific approaches have been discussed here.

1. Farm scaping: This is a method of increasing and preserving beneficial insect colonies to improve pest management processes. The goal is to optimize the benefits of agricultural landscape diversification by enhancing biodiversity for increased stability and ecosystem services *viz.,* erosion control, pesticide runoff reduction, aesthetic benefits, higher revenue, nutrient management, pollination services, soil health, and pest suppression (Gurr et al. 2012). There are two primary approaches in terms of pest management: the bottom-up and the top-down. Intercropping, trap crops, companion plantings, mulches, and living mulches are all examples of bottom-up techniques. These methods are intended to "mask" or "disguise" the cash crop or repel pest insects, thus protecting it. These activities may provide additional ecological services, such as nitrogen fixation, erosion prevention, weed suppression, or delivering nutrients to NEs. Top-down techniques are intended to boost natural enemy populations, resulting in better pest control. Insectary plantings, beetle banks, and hedgerows are classic examples of increasing NE populations for pest suppression (Wratten et al. 2000).

2. Trap crop and insectary plants: Trap crops are species that can attract both pests and their NEs simultaneously. For example, borage, *Borago officinalis* (Boraginaceae) has been found attractive to aphids and its parasitoid *Aphidius colemani* and chrysopid predators (Zhu et al. 2005). Binita et al. (2018) investigated how insectary plants increase the abundance of NEs on trap crop plants in an organic cabbage agro-ecosystem. They used two insectary plant species (sweet alyssum, *Lobularia maritima*, and buckwheat, *Fagopyrum esculentum*) alone or in combination, as well as a trap crop mixture of Indian mustard (*Brassica juncea*), red Russian kale (*B. oleracea* var. *acephala*), glossy collards (*B. oleracea* var. *italica*), and cabbage crop (*B. oleracea* var. *capitata*). According to the findings, buckwheat alone attracted the maximum number of NEs followed by a mixture of buckwheat and sweet alyssum. The sweet alyssum alone attracted the fewest NEs. Coccinellid predator *Coleomegilla maculata* and syrphid flies were the most abundant NEs observed visiting insectary plants. In an organic cabbage system, trap crops kept pests away from the cash crop, while insectary plants increased the population of essential NEs. As a result, it has been demonstrated that insectary plants will benefit NEs.

3. Beetle banks: These are essentially 'Grassy ridges' in the field that provide proximal overwintering habitat and faster predator colonization (Macleod et al. 2004). They can be built by raising a ridge (0.52.0 m; HW) by properly conducting two-directional plowing. A mixture of perennial grass seeds and dense shelter seeds of perennial flowers is sown in the ridges to attract and offer shelter for flying NEs such as syrphids, parasitoids, and ladybird beetles. During a 7-year study, Macleold et al. (2004) discovered substantial differences in the overwintering numbers of predatory carabid and staphylinid beetles and spiders in the habitat of a grassy bank consisting of *Agrostis stolonifera*, *Dactylis glomerata*, *Holcus lanatus*, and *Lolium perenne*. Beetle banks were thus proven to help biodiversity conservation in an agro-ecosystem. Beetle banks are simple to construct and can quickly grow in the ecosystem, allowing for the preservation of diverse flora and fauna. In the United Kingdom, the usage of beetle banks is actively encouraged, with government funding available to support their construction. The current payment rate for a 2 m wide bank is £12 per 100 m (DEFRA 2002). Beetle banks' principal role is to increase aphid predation by epigeal polyphagous predators. Beetle banks can serve as a home for birds and a source of food for young birds and other species (Thomas et al. 2001).

4. Hedgerows and insectary plants: These are intended to supply nectar and pollen to beneficial insects such as pollinators and NE, hence boosting agricultural pollination and biological pest suppression in surrounding crops (Morandin et al. 2011). Hedgerows are densely planted rows of shrubs and perennial grasses that encircle farm areas. An effective hedgerow is a single row of shrubs or small trees (alternating large and small shrub/tree species) flanked on both sides by strips of perennial grasses. The width of the hedgerows should be between 1.5 to 3.0 m, with the length determined by the size of the field.

5. Selective use of pesticides and use of transgenic in EE: Pesticides must be used selectively to conserve NEs, with active chemicals that have the least non-target activity. Pesticides' unintended side effects on NEs could be brought down further by adjusting the time, place and mechanism of application. Insecticides that are extremely toxic (Red label) should be avoided, and the use of insecticides with selective modes of action or bio-pesticides (Green label) should be encouraged in crop ecosystems (Table 1.1). The use of biological toxins or extracts of the insect bacterial toxins of *Bacillus thuringiensis* Berliner (*Bt*) has a specific effect only against a small group of lepidopterous larvae like cabbage loppers, bollworms, gypsy moths, etc. It is being widely used as a safer insecticide and the most common bio-pesticide in the cotton, pigeon pea and maize ecosystems.

Moreover, a few commercial crops like cotton and maize are being genetically transformed to express insecticidal cry toxins from the *Bt* bacterium. There is substantial disagreement over the use of transgenic crops in the EE because increased landscape homogenization due to transgenic crops will defeat the purpose of EE and would lead us again towards monocropping. Unquestionably,

TABLE 1.1
List of harmless (0) or variable harmful (1) insecticides on NEs

Family	Insecticide	Reference
Phytoseiidae	Oxadiazine, Spinosad, benzoyl ureas (0)	Cloyd 2012
Anthocoridae	Azadirachtin, carbamates, coragen (0), abamectin benzoate (1)	Bosco et al. 2008
Chrysopidae	Azadirachtin (0)	Cloyd 2012
Coccinellidae	Azadirachtin (0), OP (1)	Cloyd 2012
Braconidae	Juvenoids (0)	Cloyd 2012
Trichogrammatidae	Azadirachtin (0), abamectin (1)	El-Wakeil et al. 2013

further adoption of this technology in developing countries might prove a disaster, particularly in tropical areas where biodiversity hotspots may be threatened. Manachini and Lozzia (2002) emphasized the need for a longer-term risk assessment before using transgenic in EE. Further, baculoviruses (NPV, GV) are a safer management practice for managing the lepidopteran pest in cotton, pigeon pea and coconut agro-ecosystem.

Now, we assess the EE techniques and practices that have been adopted in agricultural crops for the conservation of NEs to suppress insect pests for sustainable crop production.

1.6 ECOLOGICAL ENGINEERING IN RICE

Rice is the primary staple food for more than one-third of the world's population. Over 90 percent of the world's rice is produced and consumed in Asia, with rice accounting for 40–46 percent of all irrigated acreage in the region (Fukagawa and Ziska 2019). For several years experts have linked insect outbreaks in rice fields, notably planthopper outbreaks, to the overuse of chemical insecticides. The overuse of insecticide reduces the diversity, abundance and efficiency of the natural enemy component of the rice ecosystem. There is a huge potential for the use of the EE technique in rice pest management to reduce dependence on pesticides and slow the breakdown of varietal resistance (Yele et al. 2022; Yele et al. 2023). The EE concept is still in its infancy in terms of pest management, particularly in rice. In this section, a few of the new concepts, emphasizing their potential benefits to food security and adaptability for small-holder farmsteads have been discussed.

1.6.1 High Diversity Vegetation Patches (HDVP)

HDVPs are enlarged platforms along the rice bunds for vegetable cultivation. The HDVP could be planted with a variety of crop plants to improve floral diversity and provide resources for a wide range of NEs and pollinators. Platforms could be built at 10 m intervals along the rice field bunds. The platforms could be constructed by piling paddy dirt into a mound, which could then be flattened and molded into a rectangle (30 to 50 cm tall). A layer of cow dung, vermicompost, or any other fertilizer could be used to fertilize the HDVP. It is documented that HDVPs could attract large numbers of pollinators and were used as perches and foraging sites by insectivorous birds like shrikes (*Lanius* spp.) and trillers (*Lalage nigra*).

1.6.2 Diverse Vegetation Patches (DVP)

DVP is like the HDVP which is described earlier. However, to minimize problems associated with patches providing resources for certain pests and to facilitate harvesting and monitoring, DVP segregated the crop plants and went for a single crop species on each bund. That's how the DVP became easier to use than the HDVP method as it allowed better monitoring of NEs and pollinators.

1.6.3 Planting of Floral and Vegetation Strips on Bunds

Rice fields are surrounded by high earthen structures called bunds, which are ideal for planting strips of 'upland' vegetation that does not compete for nutrients with the rice plants. A pioneering attempt of EE to manipulate biodiversity in rice-based ecosystems was initiated by the International Rice Research Institute (IRRI), in collaboration with research institutes in China in 2008. The strategy was to increase and maintain grasses and weeds around the rice fields, planting of sesame on bunds as a source of nectar, intercropping zizania in some fields, planting of vetiver grass alongside the roads and irrigation canals and release of *Trichogramma* spp. The control fields were sprayed several times with chemical pesticides. Follow-up surveys showed that the numbers of egg parasitoids and invertebrate predators were significantly increased. Predators like frogs were also more abundant, and their population density was well over 413 individuals per hectare in EE fields, while frog numbers in the control fields were only 107 (Kong et al. 2016). In the EE fields insecticides were not used at all for the entire period from 2009 to 2014 (Liu et al. 2014). Insecticide use was reduced by 75 percent in ecologically engineered fields (Chen et al. 2016). The survey of EE fields spanning over five years indicated an increase in aquatic predators like Odonata (Zhu et al. 2017b). Furthermore, the parasitoids of leaf folders were significantly higher (Zhu et al. 2017a). This study shows that EE technology checked the pest population below ETL levels and yield losses were minimized. The cost of cultivation has also been reduced, together with a reduction in labor costs and a 75 percent reduction in insecticide use. Similar positive results have also been reported from other parts of China.

Monoculture and the intensification of agriculture have reduced the farmland biodiversity and flowering plant diversity on which NEs depend (Lu et al. 2014). Egg parasitoids *Anagrus* spp. and *A. nilaparvatae* play an important role in the management of leaf and plant hoppers in Asia (Gurr et al. 2011). Laboratory screening experiments proved that volatiles of *Sesamum indicum*, *Impatiens balsamena*, *Emilia sonchifolia*, *Hibiscus coccineus*, *Trida procumbens* and *H. esculentus* attracted and enhanced the performance of *Anagrus* spp. parasitoids (Zhu et al. 2013). Of these, *S. indicum*, *I. balsamena* and *E. sonchifolia* were also attractive to *A. nilaparvatae* and *S. indicum* flowers specifically enhanced the life span of *A. nilaparvatae* and *A. optabilis*. Zhu et al. (2014) studied the influence of various plant species on the performance of the predatory mirid bug *Cyrtorhinus lividipennis*, a key NE of rice planthopper. The presence of flowering plants like *Tagetes erecta, Tridax procumbens, Emilia sonchifolia* and *Sesamum indicum* around the rice plots increased the abundance and survival of *C. lividipennis*. Predation efficiency and consumption on *Nilaparvata lugens* by *C. lividipennis* were increased in flower treatment plots. Among all flower plants, *S. indicum* was the most favorable and strongly promoted predation by *C. lividipennis*. These studies suggest that *S. indicum* is well suited for use as an EE plant on bunds of rice crops.

1.7 ECOLOGICAL ENGINEERING IN MAIZE AND SORGHUM

Stem borers, comprising the immature forms, *i.e.*, the larvae, of a group of lepidopterous insects, cause significant yield losses in subsistence cereal production throughout sub-Saharan Africa and tropical Asian countries. More stem borers were known to cause damage to sorghum and maize, with the *Chilo partellus* being the most prevalent borer pest (Mashwani et al. 2015). Striga weed species (*Striga* spp.) are also some of the greatest biological threats and cause yield losses in sorghum (Spallek et al. 2013). A push-pull technique for managing cereal stem borers and weeds in Africa was developed by scientists led by Dr. ZR Khan of the International Centre of Insect Physiology and Ecology (ICIPE) in Kenya and Rothamsted Research in the United Kingdom, in partnership with other research institutions in eastern Africa. The "push-pull" technique manipulates the distribution and abundance of insect pests and/or NEs by combining a variety of behavior-modifying stimuli. Pests are repelled or deterred away from the main crop (maize or sorghum) in this method by using stimuli (molasses grass and silver leaf, *Desmodium uncinatum*as, Push component) to repel or deter the insect pests. The pests are simultaneously lured (pulled) to other border areas where they are concentrated by using trap crops (Napier grass, *Pennisetum purpureum*), which leads to their management. The exudates of desmodium, a legume intercrop, inhibit the germination of striga, a parasitic weed on sorghum that is an added advantage for weed management through the push-pull concept type EE (Khan et al. 2001).

FIGURE 1.4 Maize intercropped with coriander to study the effect on FAW incidence. (Photograph: Keerthi 2020.)

Several authors also reported that planting an intercrop in maize reduced insect damage and increased the incidence of NEs (Keerthi et al. 2023). Intercropping of maize with coriander increased the abundance of pollinators and also acted as a pollen source for ladybird beetles (Figure 1.4).

1.8 ECOLOGICAL ENGINEERING IN VEGETABLE CROPS

Fruits and vegetables make up to 30 percent of global food consumption. Vegetable crops are plagued by various insect pests right from the nursery to harvesting, leading to massive crop losses. Pesticides are used to safeguard around one-third of agricultural products. However, due to the ill effects of pesticides on humans and the environment, new eco-friendly procedures are the need of the hour. Furthermore, fruits and vegetables are typically consumed fresh or semi-processed; they are likely to have higher levels of pesticide residues than other plant-based foods. EE offers a cost-effective and safe pest management solution in vegetables as it provides higher market value, and the cost of habitat manipulation does not burden the farmers financially.

Field margins are an important tool for increasing the abundance of predators and parasitoids in intensive vegetable cultivation. Pfiffner et al. (2003) studied the parasitism rates of lepidopteran cabbage pests in response to the presence or absence of species-rich wildflower strips. Caterpillar parasitism rates ranged from less than 10 to 64 percent. The parasitized *Pieris rapae* and *Mamestra brassicae* were detected in significantly larger numbers near the wildflower strip than in the control plot. Furthermore, the diversity and richness of parasitic Hymenoptera in sweep-net catches in wildflower strips were substantially higher than in cabbage fields. Keerthi et al. (2020) reported that the cauliflower intercropped with cineraria flower reduced aphid colonies and attracted their Nes, *i.e.*, coccinellids, spiders and syrphids (Figure 1.5). In addition, the authors reported that the foliage acts as a shelter site for predators during pupation (Figure 1.6). Similarly, White et al. (1995) also reported that planting the annual, *Phacelia tanacetifolia* in the vicinity of the cabbage crop helps in the management of *Brevicoryne brassicae* an important sucking pest through increased predation by syrphids.

In addition to offering food and shelter, the intercrop also helps in distracting the pests. For example, Pålsson et al. (2021) presented an "attract and reward" technique in which an apple

FIGURE 1.5 Cauliflower intercropped with winter annual flower crops. (Source of Photograph: Keerthi 2020.)

FIGURE 1.6 The flowering plants provide a shelter for the pupae of coccinellid predators. (Source of Photograph: Keerthi 06 March 2018.)

orchard was intercropped with bean plants infested with aphids (*Aphis fabae*) to disrupt the ant-aphid mutualism and to distract ants away from rosy apple aphids *Dysaphis plantaginea*, to allow NEs to be more active and effective against rosy apple aphid.

1.9 CONCLUSION

The population of natural enemies is subjected to continuous downfall, especially in modern agricultural systems characterized by the removal of plants after harvesting. Conservation biological control is the protection of NEs against the adverse effects of pesticides and incompatible cultural practices, and it improves their efficiency *via* providing food sources. During non-crop seasons, NEs need pollen and nectar from different floral resources for their survival. Conservation of NEs could be achieved by providing alternative habitats and floral resources. Hence, conservation biological control would certainly contribute to the enhancement of the faunal biodiversity in the agricultural environment by providing viable options for sustainable pest management and a safer environment.

REFERENCES

Aktar MW, Sengupta D, Chowdhury A (2009) Impact of pesticides use in agriculture: their benefits and hazards. *Interdiscip Toxicol* 2(1): 1. doi: 10.2478/v10102-009-0001-7

Amoros-Jimenez R, Pineda A, Fereres A, Marcos-Garcia MA (2014) Feeding preferences of the aphidophagous hoverfly *Sphaerophoria rueppellii* affect the performance of its offspring. *Bio Control* 59: 427–435.

Bianchi F, van der Werf W (2004) Model evaluation of the function of prey in non-crop habitats for biological control by ladybeetles in agricultural landscapes. *Ecol Model* 171: 177–193.

Binita S, Finke DL, Piñero JC (2018) The 'Botanical triad': the presence of insectary plants enhances natural enemy abundance on trap crop plants in an organic cabbage agro-ecosystem. *Insects* 10: 181.

Bosco L, Giacometto E, Tavella L (2008) Colonization and predation of thrips (Thysanoptera: Thripidae) by *Orius* spp. (Heteroptera: Anthocoridae) in sweet pepper greenhouses in Northwest Italy. *Biol Control* 44: 331–340.

Chen GH, Zhu PY, Zheng XS, Yao XM, Zhang FC, Sheng XQ, Xu HX, Lu ZX (2016) The practice of ecological engineering on the control of rice insect pests in Jinhua. *China Plant Protec* 36: 31–36.

Clem CS, Harmon-Threatt AN (2021) Field borders provide winter refuge for beneficial predators and parasitoids: a case study on organic farms. *J Insect Sci* 21(3): 2.

Cloyd AR (2012) Indirect effects of pesticides on natural enemies. In: Soundararajan RP (ed.) *Pesticides–advances in chemical and botanical* pesticides. InTech Open, Croatia, pp. 127–150.

Damalas CA, Koutroubas SD (2016) Farmers' exposure to pesticides: toxicity types and ways of prevention. *Toxics* 4(1): 1. doi: 10.3390/toxics4010001

DEFRA (2002) *The Countryside Stewardship scheme 2003: payment summary sheet.* Department for Environment, Food and Rural Affairs, London.

El-Wakeil N, Gaafar N, Sallam A Volkmar C (2013) Side effects of insecticides on natural enemies and possibility of their integration in plant protection strategies. In: Stanislav Trdan (Ed.), *Agricultural and biological sciences "insecticides—development of safer and more effective technologies".* Intech, Rijeka, Croatia, pp.1–54.

Fukagawa NK, Ziska LH (2019) Rice: importance for global nutrition. *J Nutr Sci Vitaminol* 65: S2–S3.

Geiger F, Wäckers FL, Bianchi FJ (2009) Hibernation of predatory arthropods in semi-natural habitats. *BioControl* 54(4): 529–535.

Gurr G, Wratten SD, Altieri MA (2004) *Ecological engineering for pest management: advances in habitat manipulation for arthropods.* Csiro Publishing, Collingwood, Australia, pp. 25–38.

Gurr GM, Barlow ND, Memmott J, Wratten SD, Greathead DJ (2000) A history of methodological, theoretical and empirical approaches to biological control. In: Geoff Gurr, Steve Wratten (Eds.), *Biological control: measures of success.* Springer, Dordrecht, pp. 3–37.

Gurr GM, Scarratt SL, Wratten SD, Berndt L, Irvin N (2004) Ecological engineering, habitat manipulation and pest management. In: Gurr G, Wratten SD, Altieri MA (Eds.), *Ecological engineering for pest management: advances in habitat manipulation for arthropods*, pp. 1–12.

Gurr GM, Wratten SD, Snyder WE, Read DMY (2012) *Biodiversity and insect pests: key issues for sustainable management.* Wiley, Chichester, UK.

Gurr GM, Liu J, Read DM, Catindig JLA, Cheng JA, Lan LP, Heong KL (2011) Parasitoids of Asian rice plant hopper (Hemiptera: Delphacidae) pests and prospects for enhancing biological control by ecological engineering. *Ann Appl Biol* 158(2): 149–176.

Hossain Z, Gurr GM, Wratten SD (2000) The potential to manipulate the numbers of insects in Lucerne by strip cutting. *Aust J Entomol* 39(1): 39–41.

Jervis MA, Kidd NA, Fitton MG, Huddleston T, Dawah HA (1993) Flower-visiting by hymenopteran parasitoids. *J Nat Hist* 27 (1): 67–105.

Jervis MA, Kidd NAC (1986) Host feeding strategies in hymenopteran parasitoids. *Biol Rev* 61: 395–434.

Keerthi MC, Sharma RK, Suroshe SS, Sinha SR (2020) Ecological engineering in cauliflower for aphid management. *Indian J Agric Sci* 90(7): 1356–1358.

Keerthi MC, Suroshe SS, Doddachowdappa S, Shivakumara KT, Mahesha HS, Rana VS, Gupta A, Murukesan A, Casini R, Elansary HO, Shakil NA (2023) Bio-intensive tactics for the management of invasive fall armyworm for organic maize production. *Plants* 12(3): 685.

Khan ZR, Pickett JA, Wadhams L, Muyekho F (2001) Habitat management for the control of cereal stem borers in maize in Kenya. *Insect Sci Appl* 21: 375–380.

Kong SS, Zheng SJ, Zhu PY, Zheng XS, Zheng RQ, Lu ZX (2016) Survey of amphibian diversity in the paddy field on controlling rice pests by ecological engineering. *China Plant Protec* 36: 10–14.

Koptur S (2005) Nectar as fuel for plant protectors. In: Wackers FL, Van Rijn PCJ, Bruin J (eds.) *Plant-provided food for carnivorous insects: a protective mutualism and its applications*. Cambridge University Press, New York, pp. 75–108.

Landis DA, Wratten SD, Gurr GM (2000) Habitat management to conserve natural enemies of arthropod pests in agriculture. *Annu Rev Entomol* 45(1): 175–201.

Liu GL, Zhang XM, Zhao LW, Zhang JL, Wang XT, Chen YB, Zheng XS (2014) Application of ecological engineering technology for managing rice pests and the benefit analysis. *J Zhejiang Agric Sci* 55: 1809–1181.

Lu ZX, Zhu PY, Gurr GM, Zheng, XS, Read DM, Heong KL, Yang YJ, Xu HX (2014) Mechanisms for flowering plants to benefit arthropod natural enemies of insect pests: prospects for enhanced use in agriculture. *Insect Sci* 21(1): 1–12.

Lundgren JG (2009) Nutritional aspects of non-prey foods in the life histories of predaceous Coccinellidae. *Biol Control* 51: 294–305.

MacLeod A, Wratten SD, Sotherton NW, Thomas MB (2004) 'Beetle banks' as refuges for beneficial arthropods in farmland: long-term changes in predator communities and habitat. *Agric Forest Entomol* 6: 147–154.

Manachini B, Lozzia GC (2002) First investigations into the effects of Bt corn crop on Nematofauna. *Bollettino di Zoologia agraria e di Bachicoltura II* 34(1): 85–96.

MANAGE (2019) www.manage.gov.in/publications/edigest/jun2019.pdf accessed on 30.04.2022.

Mashwani MA, Ullah F, Ahmad S, Sohail K, Shah SF, Usman M (2015) Infestation of maize stem borer, *Chilo partellus* (Swinhoe) in maize stubbles and stalks. *AJAR* 3(7): 94–98.

Mitsch WJ, Jørgensen SE (2003) Ecological engineering: a field whose time has come. *Ecol Eng* 20(5): 363–377.

Morandin LM, Long RF, Pease C, Kremen C (2011) Hedgerows enhance beneficial insects on farms in California's Central Valley. *Calif Agric* 65: 197–201.

Odum HT (1962) Man in the ecosystem. In: *Proc Lockwood Conf on the Suburban Forest and Ecol. Bull Connecticut Agri Sta Storrs*, Storrs CT, pp. 57–75.

Odum HT, Copeland BJ, Brown RZ (1963) Direct and optical assay of leaf mass of the lower montane rain forest of Puerto Rico. *PNAS USA* 49(4): 429.

Pålsson J, Porcel M, Dekker T, Tasin M (2021) Attract, reward and disrupt: responses of pests and natural enemies to combinations of habitat manipulation and semiochemicals in organic apple. *J Pest Sci* 18: 1–3.

Pfiffner L, Luka H, Schlatter C, Juen A, Traugott M (2009) Impact of wildflower strips on biological control of cabbage lepidopterans. *Agric Ecosyst Environ* 129(1–3): 310–314.

Quispe R, Mazón M, Rodríguez-Berrío A (2017) Do Refuge plants favour natural pest control in maize crops? *Insects* 8(3): 71. doi: 10.3390/insects8030071

Rusch A, Valantin-Morison M, Sarthou JP, Roger-Estrade J (2010) Biological control of insect pests in agroecosystems: effects of crop management, farming systems, and semi natural habitats at the landscape scale: A review. *Adv Agron* 109: 219–259.

Spallek T, Mutuku M, Shirasu K (2013) The genus *Striga*: A witch profile. *Mol Plant Pathol* 14(9): 861–869.

Swaminathan MS, Bhavani RV (2013) Food production & availability-essential prerequisites for sustainable food security. *Indian J Med Res* 138(3): 383.

Thomas SR, Goulson D, Holland JM (2001) Resource provision for farmland gamebirds: The value of beetle banks. *Ann Appl Biol* 139: 111–118.

Thorbek P, Bilde T (2004) Reduced numbers of generalist arthropod predators after crop management. *J Appl Ecol* 41: 526–538.

Torres JB, Bueno AD (2018) Conservation biological control using selective insecticides–a valuable tool for IPM. *Biol Control* 126: 53–64.

van den Bosch R, Messneger PS, Gutierrez AP (1982) *An introduction to biological control*. Plenum, New York, p. 247.

van Emden HF (1990) Plant diversity and natural enemy efficiency in agroecosystems. In: Mackauer M, Ehler LE, Roland J (eds.) *Critical issues in biological control*. Intercept, Andover, MA, pp. 63–80.

Vattala HD, Wratten SD, Phillips CB, Wäckers FL (2006) The influence of flower morphology and nectar quality on the longevity of a parasitoid biological control agent. *Biol Control* 39: 179–185.

White AJ, Wratten SD, Berry NA (1995) Habitat manipulation to enhance biological control of *Brassica* pests by hover flies (Diptera: Syrphidae). *J Econ Entomol* 88(5): 1171–1176.

Yele Y, Chander S, Suroshe SS, Nebapure SM, Arya PS, Prabhulinga T (2022) Effect of ecological engineering on incidence of key rice pests. *Indian J Entomol* 84(3): 503–508.

Yele Y, Chander S, Suroshe SS, Nebapure S, Tenguri P, Sundaran AP (2023) Ecological engineering in low land rice for brown plant hopper, *Nilaparvata lugens* (Stål) management. *Peer J* 11: e15531.

Zhu J, Obrycki JJ, Samuel A, Ochieng, Thomas C Baker, Pickett, JA, Smiley D (2005) Attraction of two lacewing species to volatiles produced by host plants and aphid prey. *Naturwissenschaften* 92: 277–281.

Zhu P, Gurr GM, Lu Z, Heong KL, Chen G, Zheng X, Yang Y (2013) Laboratory screening supports the selection of sesame (*Sesamum indicum*) to enhance *Anagrus* spp. parasitoids (Hymenoptera: Mymaridae) of rice planthoppers. *Biological Control* 64(1): 83–89.

Zhu P, Zheng X, Zhang F, Xu H, Yao X, Yang Y, Chen G (2017b) Aquatic insect populations in paddy fields as affected by management of rice insect pests through ecological engineering technology. *Chinese J Rice Sci* 31(2): 207–215.

Zhu PY, Lu ZX, Heong KL, Chen GH, Zheng XS, Xu HX, Yang YJ, Nicol HI, Gurr GM (2014) Selection of nectar plants for use in ecological engineering to promote biological control of rice pests by the predatory bug, *Cyrtorhinus lividipennis* (Heteroptera: Miridae). *PLoS One* 9(9): 108.

Zhu PY, Zheng XS, Zhang FC, Barrion AT, Xu HX, Yang YJ, Chen GH, Lu ZX (2017a) Natural enemy functional guilds of rice leaf folder (*Cnaphalocrocis mendinalis*) enhanced in paddy field with sustainable pest management by ecological engineering. *Chinese J Biol Control* 33: 351–363.

2 Entomopathogenic Nematodes

Innovative Insect Pest Management

Rashid Pervez

2.1 INTRODUCTION

Similar to other nematodes, entomopathogenic nematodes (EPNs) are thread-like, colorless, slender, unsegmented, bisexual, and bilaterally symmetrical. EPNs *Steinernema*, *Heterorhabditis*, and *Oscheius* are obligatory insect infections in nature. Infectious juveniles (IJs) of the nematode carry the bacterial symbiont in their intestine, and when they enter an insect host, the IJs release the bacterial symbiont into the insect haemocoel (Poinar 1990; Kaya and Gaugler 1993). The insect will die in 24 to 72 hours. The bacterium and nematode symbionts in the infected insect keep multiplying and growing until a new generation of IJs carrying the bacterial symbionts emerges to begin a new infection cycle in another insect.

Since they have a wide host range, are accessible, have a brief life cycle, can be mass multiplied inexpensively, and are environmentally friendly. EPNs have a lot of potential as biological control agents against crop pests (Ali et al. 2005; Pervez et al. 2007, 2008a, 2008b). The varied species and isolates have a wide range of host ranges, reproduction, infectivity, and survival needs, and they can be found in many soil settings. It appears that EPNs prefer sandy loam soils. This is further supported by the finding that while nematode movement is constrained in soils with high clay contents, EPN mobility and survival are improved in soils with high sand concentrations (Pervez and Ali 2007; Pervez et al. 2014a, 2014b).

EPNs are used to control a variety of pest insects that are important to agriculture. A few of the pests that EPNs are used to control globally include fungus gnats in mushrooms, root weevils on ornamentals, strawberries, cranberries, citrus, and bananas, scarabs on turfs, ornamentals, and blueberries, cutworms, webworms, billbugs, and mole crickets, termites in wooden objects and trees, peach borer moth in apples, and carpenter worms in shade trees.

EPNs have gained enormous popularity as biopesticides for pest control over the past 20 years (Pervez 2022). The use of EPNs to manage unwanted insects, mollusks, plant nematodes, and even soil-borne plant diseases is being studied by thousands of researchers around the globe. Due to their success in reducing insect pests in agricultural and horticultural crops, EPNs are now routinely produced in large quantities. The two main drivers of EPNs' commercial expansion are their ease of manufacture and their exemption from registration requirements. As a result, the availability of EPNs against pests previously believed to be insurmountable to handle was prompted by demonstrations of practical application, primarily in Europe and North America, and later in Japan, China, and Australia.

Nematodes can be purchased commercially and used extensively in many different crops. In recent years, they have gained acceptance as biopesticides with economic viability. The use of EPNs has piqued the interest of farmers in numerous countries for several reasons. The reason for this

DOI: 10.1201/9781032631578-2

interest may vary based on the crop, market, or nation. The most obvious justification for a farmer using an EPN product is the lack of a conventionally safe substitute. Farmers' confidence in EPN products has increased as a result of their exposure to the reliable protection they provide.

2.2 THE EPN BACTERIUM COMPLEX

Researchers are examining the EPN bacterium complex in many parts of the world. Consequently, research projects from various nations or locations are highlighted. While research involving EPN bacterial complexes was highlighted in other nations and regions, North America and Europe focused on the status of commercially available nematodes. Japan was the lone exception, where a focus was placed on the development of commercial nematodes. China, Korea, and India focused their research efforts on using the nematode to control insect pests or soil plant infections. Overall, the level of research differs by nation or place. Research in developing countries typically focuses on proving the efficacy of EPNs or the symbiotic bacteria they host against a range of pests. These studies ultimately aim to use them as biological control agents for soil pests.

Many studies are interested in EPNs and the bacteria that they are associated with. This interest was sparked in part by the lack of efficient, environmentally responsible methods to manage pest insects that live in the soil (Adams et al. 2006) and the potential to generate these nematodes monoxenically *in vitro* before scaling up to commercial production levels. Nematodes have mostly been utilized to control soil pests in high-value crops since they are more expensive than chemical pesticides. The related bacteria were found to create toxins and antibiotics in addition to acting as model systems for symbiotic partnerships, according to microbiologists. People are curious to learn more about the nematode-bacterium combo because of these features of the bacteria (Boemare et al. 1993). In the 1980s and 1990s, Australian researchers made significant strides in the mass production of nematodes and the study of the biology of symbiotic bacteria. Research on these nematodes and the bacteria that live in them is still in its infancy in other parts of the world, and it is typically conducted in collaboration with specialists from industrialized nations. There are some exceptions to this generality (Pervez et al. 2015, 2020).

In contrast to the usual repeated microbial destruction of insect cadavers that have died from other reasons, infected insect corpses do not putrefy because of the nematode-bacterium-insect interaction. Therefore, according to Dutky's theory, symbiotic bacteria produce antimicrobial substances that prevent the growth of rival microorganisms and the putrefaction of insect pests infested with nematodes. The antibiotics produced by *Photorhabdus* spp. during *in vitro* cultivation include genistin, stilbene derivatives, anthraquinone derivatives, furan derivatives, and phenol derivatives. This concept was largely accepted when describing the natural situation in the tripartite, nematode-bacterium-insect interaction. However, following 2 and 5 days of nematode infection, the antibiotic 3,5-dihydroxy-4-isopropyl stilbene (ST) was found at roughly 1500 and 4000 g gl moist insect, respectively, in larval *G. mellonella* cadavers (Poinar 1977).

Depending on the species of the host and the stage of development, *O. carolinensis* grows inside the insect and kills it in 1–5 days. *O. carolinensis* was discovered to frequently associate with *Serratia marcescens*, a bacteria species with the potential to cause insects (Torres-Barragan et al. 2011). It has been a long time since symbiotic bacteria associated with EPNs were defined using phenotypic techniques (morphological and biochemical testing). For phenotypic characterization, Bergey's Manual and several other tests are frequently used, including the test for bioluminescence, dye adsorption on Mac Conkey agar or NBTA, colony morphology on nutrient agar, and the diameter of colonies measured after 48 and 72 hours (Boemare and Akhurst 1988). Based on oligonucleotide cataloging and 16S rRNA sequence analysis, the Xenorhabdus/Photorhabdus group has been categorized as belonging to the gamma subdivision of the purple bacteria (Rainey et al. 1995).

There is still a great deal of uncertainty around the position of the genera *Xenorhabdus* and *Photorhabdus* within the Enterobacteriaceae, even though it seems secure. The fatty acid study proved that the differences were enough to exclude the group from the Enterobacteriaceae, although Rainey et al. (1995) questioned whether it was fair to include them within the Enterobacteriaceae family. Farmer et al. (1989) questioned the inclusion of *Xenorhabdus* spp. in the Enterobacteriaceae because they are unable to reduce nitrate and possess the catalase-negative characteristic (Akhurst 1986a, 1986b). Additionally, scientists learned that *Photorhabdus* sp. produces peculiar yellow or red pigments that are not present in other Enterobacteriaceae and are bioluminescent, a characteristic that is unknown to other Enterobacteriaceae.

Although taxonomic challenges have not yet been fully overcome by molecular methods (16S rRNA sequence comparisons), a variety of group-specific probes have been created. 16S rRNA probes for the four species of *Xenorhabdus* sp. and *P. luminescens* were created. The pigment-coding genes of *P. luminescens* do not cross-hybridize with any *Xenorhabdus* strains, which is an unusual strategy. As more genes are extracted and sequenced, cross-hybridization studies between the *Photorhabdus* and *Xenorhabdus* isolates as well as between species isolated from other settings will be of interest (Akhurst and Boemare 1988).

2.3 EPN BIOLOGY

The life cycles of every known EPN are comparable. The only stage that lives outside of the host is the non-feeding, infectious juvenile stage (IJ). The IJ has bacterial symbiont cells in its gut. The IJ invades and penetrates the hemocoel through organic openings when it encounters a weak host (i.e., anus, mouth, or spiracles). The symbiotic bacterium is then released by the IJ, which results in septicemia and the host's death within 48 hours. The bacterium produces antibiotics that prevent other microbes from colonizing the carcass. The bacteria not only gives the nematode food, but it also breaks down the host tissues and releases the nutrients necessary for nematode development and growth (Pervez et al. 2016a, 2016b).

2.4 INSECT PEST MANAGEMENT

The last three decades have seen an increase in the demand for EPN as biopesticides for insect pest control (Table 2.1). Thousands of researchers worldwide are investigating the potential of

TABLE 2.1
Commercial use of EPNs as bio-insecticides

EPN species	Major pest (s) targeted – as recommended by various commercial companies
Steinernema glaseri	White grubs (scarabs, especially Japanese beetle, *Popillia* sp.
Steinernema kraussei	Black vine weevil, *Otiorhynchus sulcatus*
Steinernema carpocapsae	Turfgrass pests- billbugs, cutworms, armyworms, sod webworms, chinch bugs. Orchard, ornamental and vegetable pests – codling moth, cranberry girdler, dogwood borer and other clearwing borer species, black vine weevil, peach tree borer, shore flies (*Scatella* spp.)
Steinernema feltiae	Fungus gnats (*Bradysia* spp.), shore flies, western flower thrips
Steinernema scapterisci	Mole crickets (*Scapteriscus* spp.)
Steinernema riobrave	Citrus root weevils (*Diaprepes* spp.)
Heterorhabditis bacteriophora	White grubs (scarabs), cutworms, black vine weevil, flea beetles, corn root worm
Heterorhabditis megidis	Weevils
Heterorhabditis indica	Fungus gnats, root mealybug, grubs
Heterorhabditis marelatus	White grubs (scarabs), cutworms, black vine weevil

EPNs to control harmful insects. Because of this, the use of EPNs against insect pests that were once regarded to be challenging to control was made possible thanks to demonstrations of practical application, first in Europe and North America and then later in Japan, China, Australia, and India (Georgis 1990).

Miller (1989) created a method for assessing the pathogenicity of commercially available *S. carpocapsae* to determine its virulence. Grewal et al. (1999) created the sand well method for routine quality assessment of the majority of EPN species at low concentrations. According to Gaugler and Campbell (1991), the ambushing kind of host-searching behavior could account for the restricted movement of some EPNs in the soil. Grewal et al. (1994a, 1994b) claim that *S. feltiae* and *S. riobrave* employ a type of foraging behavior that lies somewhere between ambushing and cruising, as evidenced by *S. carpocapsae* and *S. glaseri*, respectively. Abbas and Saleh (1998) discovered that *S. riobrave* and *S. abbasi* are harmful to fully developed insect larvae.

Management of *Spodoptera depravata* using *S. carpocapsae* (GSN), *S. glaseri* (Dongrae), *S. longicaudum* (Nonsane), and *S. monticolum* was investigated in Korea, whereas Colombian *Heterorhabditis* sp. (SL0708) was also characterized ecologically (Mejia-Torres and Saenz 2013). Within 24 hours, *Heterorhabditis* sp. (ELG), *Heterorhabditis indica*, and *Heterorhabditis* sp. (ELB) eliminated *S. littoralis* (Abdel-Razek and Abd-Elgawad 2007).

Six *Steinernema* isolates from various provinces in Vietnam were evaluated by Phan et al. (2005) for their antibacterial efficacy against *G. mellonella* and *S. littoralis*. Because application problems were frequent and compatibility data is necessary for nematode adoption in IPM systems, researchers examined how droplet size, pressure differentials, hydraulic nozzles, contraction flow fields, and agitation affected nematode viability and virulence (Fife et al. 2003, 2004, 2005). EPNs were evaluated and found effective against red flour beetle, *Tribolium castaneum* (Herbst), the Indian meal moth, *Plodia interpunctella* (Hubner), the Mediterranean flour moth, *Ephestia kuehniella* (Zeller), the saw-toothed grain beetle, *Oryzaephilus surinamensis*, and the yellow mealworm, *Tenebrio molitor* (L.) (Athanassiou et al. 2008; Ramos-Rodriguez et al. 2006).

Four EPN species were discovered to be effective against the juvenile and adults of the granary weevil *Sitophilus granaries* (L.) and the saw-toothed grain beetle *O. surinamensis* (L.) (Traden et al. 2006). In field tests, the red flour beetle and the Indian meal moth were employed to energize treatments for empty grain bins. This study found that 49 to 99 percent of EPNs were promising against *Tribolium castaneum* pupae and *P. interpunctella* larvae. The last instar and adult stages of the pulse beetle, *C. chienensis*, were tested by Fayaz and Javed (2009) against seven Pakistani EPN strains belonging to the genera *Steinernema* and *Heterorhabditis*. They found that *H. bacteriophora*, *S. simkayai*, and *S. pakistanense* exhibited the highest virulence.

It was determined that nine lepidopteran insects were susceptible to *S. glaseri*, *S. feltiae*, and *H. indicus* (Karunakar et al. 1999). It was discovered that *H. armigera* was susceptible to EPN, *S. masoodi*, *S. seemae*, and *S. carpocapsae*, which killed the larvae (Ali et al. 2008). The first and second instars of *H. armigera* larvae were incredibly susceptible to *H. indica*, *S. masoodi*, and *S. mushtaqi* (Banu 2001; Pervez 2010; Pervez 2017). The third and fifth instars of *H. armigera* larvae were the most susceptible to *S. glaseri*, *H. indica*, and *H. bacteriophora* (Jothi and Mehta 2003). *S. mushtaqi* was discovered to be pathogenic against *Clavigralla gibbosa*, *Centrococcus somatics*, and *Maruca vitrata* (Pervez and Ali 2011), whereas *Amsacta moorei*, *Spodoptera litura*, *Lampides boeticus*, *Nezara viriduala*, and *Earias vittella* all succumbed to death within 72 hours (Pervez and Ali 2013). There was evidence that *S. masoodi* and *S. mushtaqi* were more infective against *S. litura* than *S. seemae* and *S. carpocapsae*, which caused 89 percent mortality after 72 hours (Pervez and Ali 2009).

EPNs *Heterorhabditis* sp. (IISR-EPN 01), *Steinernema* sp. (IISR-EPN 02), *Oscheius* sp. (IISR-EPN 08), and *O. gingeri* were all found to be pathogenic to shoot borer larvae *Conogethes punctiferalis* (Guen.) and *Euproctis* sp. (Pervez et al. 2014c). *O. gingeri*, *Steinernema* sp. (IISR-EPN 02), and *Oscheius* sp. (IISR-EPN 05), which each killed 67 percent of the shoot borer pupae, were the most virulent against them (Pervez et al. 2012). Lepidopteran insect pests have recently been managed sustainably using EPNs (Pervez and Rao 2018), along with other insect pests like the root grub (*Holotrichia longipennis* Blanch) that infests ginger (Pervez and Rajkumar 2018), white grub (*Leucopholis burmesterie*) infesting areca nut (Rajkumar et al. 2018), and the legume pod-borer, *Maruca vitrata* infesting pigeonpea (Pervez and Rao 2021).

2.5 INSECT MANAGEMENT UNDER FIELD CONDITIONS

EPNs can be applied using the majority of common liquid application techniques designed to provide herbicides, fertilizers, and irrigation (Table 2.2). When selecting an application system, it's important to take into account the volume, agitation system, pressure, recycle time, ambient conditions, and spray dispersion pattern (Navon 1999).

2.6 MANAGEMENT OF BELOW-GROUND INSECT PESTS

EPNs are often applied by spraying them directly onto the soil's surface. Moisture and soil type both have an impact on their survival and motility. In comparison to sandy loam soil, nematode migration and survival are often lower in clay soil. For application, the majority of EPNs prefer soil temperatures between 12 and 28°C. Pre-application irrigation is typically advised if the soil temperature is higher than 28°C to reduce the temperature before applying EPNs. However, the discovery of heat-tolerant EPNs allows for the mitigation of this detrimental force. It is commonly advised that light irrigation be given after the application of EPNs in the soil to help the EPNs permeate deeper into the soil, where they can operate efficiently and away from the roots (Connick 1993).

2.7 MANAGEMENT OF ABOVE-GROUND INSECT PESTS

The size of the droplet and the spread of the spray are important considerations for administering EPNs in a foliar setting. Both low-volume and high-volume sprayers can be used to apply EPNs, but the pressure in the spray tank shouldn't be too high or the EPNs will shred. Over 300 psi is the maximum pressure that EPNs should be exposed to (2070 kPa) (Grewal 2002). Many of the droplet sizes produced by hydraulic nozzles are too tiny to carry pathogenic particles. It has been found that the type of nozzle used has an impact on the EPNs deposit on the surface of the plant (Akhtar et al. 2012). More EPNs are deposited on the leaves when a flat fan or solid cone nozzle is used,

TABLE 2.2
EPN-based commercial products

Formulation	EPNs	Product	Company
Alginate gel	*S. carpocapsae*	Mioplant	Novartis, Vienna, Austria
	S. carpocapsae	Boden- Niitzlinge	Celaflor, Germany
Flowable gel	*S. carpocapsae*	Biosafe	SDS Biotech, Tokyo, Japan
	S. feltiae	Exhibit	Novartis, Switzerland
	S. feltiae	Stealth	Novartis, Chester, UK
Clay	*H. megidis*	Nemasys – H	MicroBio, Cambridge, UK
	S. feltiae	Nemasys	Microbio
	S. scapterisci	Proactant Ss	Biocontrol,Florida, USA
Water dispersible granule	*S. carpocapsae*	Biosafe	Thermotriology, Columbia
	S. carpocapsae	Helix	Novartis, Mississauga, Candada
	S. feltiae	Magnet	Amycel-Spawn mate, Watsonville
	S. riobrave	Biovector	Thermotriology, Columbia

which causes a greater target insect killing (Lello et al. 1996). As the flow rate through the nozzles increases, there are more EPN deposits on the plant's surface. The amount of EPNs deposited on foliage is frequently increased by adding an adjuvant to the spray solution. EPNs' tolerance to extremes of desiccation, temperature, and UV radiation causes them to occasionally produce insufficient results when applied topically (Ahmad et al. 2009a, 2009b). It has been suggested that adding a fluorescent brightener and applying EPNs in the dark can help to some extent with these issues (Ali et al. 2009, 2010).

2.8 FUTURE DIRECTION

The use of EPNs is becoming more significant in the current situation and its emphasis is on organic farming, where no pesticides are sprayed on plants and no chemicals are permitted in food grains, and the commodities are both cost-effective and competitive. Coordination of activities is needed to integrate the EPNs into an IPM program, both in terms of EPN species, insects, and their environment. The goal of isolating native EPN strains is to increase their genetic diversity and find the ones that can withstand high temperatures. For enhanced formulation, it is essential to find better and more effective adjuvants, surfactants, anti-caking agents, anti-desiccants, and UV retardants. Determining the accurate insect pest management application dosage, procedure, timing, and frequency. A modest cottage enterprise for the manufacturing of EPN could compete with traditional pesticides in rural areas. This potential bioagent will be integrated with current IPM modules for different crops using technology for the application of insect pests to the air, the soil, and stored grains. Standardization in a range of formulations is necessary for the quality of commercial EPN products.

REFERENCES

Abbas MST, Saleh MME (1998) Comparative pathogenicity of *Steinernema abbasi* and *S. riobravae* to *Spodoptera littoralis* (Lepidoptera: Noctuidae). *Int J Nematol* 8: 143–145.

Abdel-Razek AS, Abd-Elgawad MM (2007) Investigations on the efficacy of entomopathogenic nematodes against *Spodoptera littoralis* (Biosd.) and *Galleria mellonella* (L.). *Arch Phytopathol Plant Protec* 40(6): 414–422.

Adams BJ, Fodor A, Koppenhofer HS, Stackebrandt E, Stock SP, Klein MG (2006) Biodiversity and systematic of nematode-bacterium entomopathogens. *Biol Control* 37: 32–49.

Ahmad R, Ali SS, Pervez R (2009a) Field efficacy of *Steinernema masoodi* based biopesticide against *Helicoverpa armigera* (Hubner) infesting chickpea. *Trends Biosci* 2(1): 23–24.

Ahmad R, Hussain MA, Ali SS, Pervez R (2009b) Survival of *Steinernema masoodi* and *S. carpocapsae* (Rhabditida: Steinernematidae) on pigeonpea and chickpea after foliar application. *Arch Phytopathol Plant Protect* 42(2): 112–117.

Akhtar MH, Asif M, Ali SS, Pervez R (2012) Performance of different sprayer nozzles on deposition of infective juveniles of three entomopathogenic nematodes species on pigeonpea. *World J Appl Sci Res* 2(1): 28–31.

Akhurst RJ (1986a) *Xenorhabdus nematophilus* subsp. *Beddingii* (*Enterobacteriaceae*): a new subspecies of bacteria mutualistically associated with entomopathogenic nematodes. *Int J Syst Evol Microbiol* 36: 454–457.

Akhurst RJ (1986b) *Xenorhabdus nematophilus* subsp. *poinarii*: its interaction with insect pathogenic nematodes. *Syst Appl Microbiol* 8: 142–147.

Akhurst RJ, Boemare NE (1988) A numerical taxonomic study of the genus *Xenorhabdus* (Enterobacteriaceae) and proposed elevation of the subspecies of *X. nematophilus* to species. *J Gen Microbiol* 134(7): 1835–1845.

Ali SS, Ahmad R, Hussain MA, Pervez R (2005) *Pest management of pulses through entomopathogenic nematodes*. Indian Institute of Pulses Research, Kanpur, Army Press, Lucknow, India, p. 159.

Ali SS, Pervez R, Ahmad R (2009) Efficacy of *Steinernema masoodi* (Ali et al. 2005) against grey weevil *Myllocerus* sp. infesting early pigeonpea. *Trends Biosci* 2(1): 40–41.

Ali SS, Pervez R, Hussain M, Ahmad R (2008) Susceptibility of three lepidopteran pest to five entomopathogenic nematodes and *in vivo* mass production of these nematodes. *Arch Phytopathol Plant Protect* 41(4): 300–304.

Ali SS, Tickle AN, Nema KK, Pervez R, Shaheen A, Ahmad R (2010) Field efficacy of *Steinernema masoodi* based biopesticide against *Helicoverpa armigera* (Hubner) infesting early pigeonpea. *Trends Biosci* 3(1): 63–65.

Athanassiou CG, Palyvos NE, Kokouli-Durate T (2008) Insecticidal effect of *Steinernema feltiae* (Filipjev) (Nematoda: Steinernematidae) against *Tribolium confusum* du val (Coleptera: Tenebrionidae) and *Ephestia kuehniella* (Zeller) (Lepidoptera: Pyralidae) in stored wheat. *J Stored Product Res* 44: 52–57.

Banu JG (2001) *Studies on entomopathogenic nematodes from Kerala*. Dissertation submitted for Ph. D. (Nematology) to Tamil Nadu Agricultural University, Coimbatore, Tamil Nadu, India, p. 231.

Boemare N, Akhurst RJ, Mourant RJ (1993) DNA relatedness between *Xenorhabdus* spp. (Enterobacteriaceae), symbiotic bacteria of Entomopathogenic Nematodes, and a proposal to transfer *Xenorhabdus luminescens* to a new genus, *Photorhabdus* gen. nov. *Int J Bacteriol* 43: 249–255.

Boemare NE, Akhurst RJ (1988) Biochemical and physiological characterization of colony form variants in *Xenorhabdus* spp. (Enterobacteriaceae). *J Gen Microbiol* 134(3): 751–761.

Connick WJ (Jr.), Nickle WR, Vinyard BT (1993) "Pesta", new granular formulation for *Steinernema carpocapsae*. *J Nematol* 25(2): 198–203.

Farmer JJ, Jorgensen JH, Grimont PAD, Akhurst RJ. Poinar GO (Jr.), Ageron E, Pierce GV, Smith JA, Carter GP, Wilson KL, Hickmman-Brenner FW (1989) *Xenorhabdus luminescens* (DNA hybridization group 5) from human clinical specimens. *J Clin Microbiol* 27: 1594–1600.

Fayaz S, Javed S (2009) Laboratory evaluation of seven Pakistani strains of entomopathogenic nematodes against a stored grain insect pest, pulse beetle *Callosobruchus chienensis* (L.). *Nematology* 41: 255–260.

Fife JP, Derksen RC, Ozkan H E, Grewal PS (2003) Effects of pressure differentials on the viability and infectivity of entomopathogenic nematodes. *Biol Control* 27: 65–72.

Fife JP, Derksen RC, Ozkan HE, Grewal PS, Chalmers JJ, Krause CR (2004) Evaluation of a contraction flow field on hydrodynamic damage to entomopathogenic nematodes–a biological pest control agent. *Biotechnol Bioeng* 86: 96–107.

Fife JP, Ozkan HE, Derksen RC, Grewal PS, Krause CR (2005) Viability of a biological pest control agent through hydraulic nozzles. *Trans Am Soc Agri Eng* 48: 45–54.

Gaugler R, Campbell JF (1991) Selection for enhanced host-finding of scarab larvae (Coleoptera: Scarabaeidae) in an entomopathogenic nematode. *Environ Entomol* 20: 700–706.

Georgis R (1990) Formulation and application technology. In: Gaugler R, Kaya HK (eds.), *Entomopathogenic nematodes in biological control*. CRC Press, Boca Raton, FL, USA, pp. 179–191.

Grewal PS (2002) Formulation and application technology. In: Gaugler R (ed.), *Entomopathogenic nematology*. CABI Publishing, Wallingford, UK, pp. 265–287.

Grewal PS, Converse V, Georgis R (1999) Influence of production and bioassay methods on infectivity of two ambush foragers (Nematoda: Steinernematidae). *J Invertebrate Pathol* 73: 40–44.

Grewal PS, Selvan S, Gaugler R (1994a) Thermal adaptation of entomopathogenic nematodes: niche breadth for infection, establishment, and reproduction. *J Thermal Biol* 19: 245–253.

Grewal PS, Lewis EE, Campbell JF, Gaugler R (1994b) Host finding behavior as a predictor of foraging strategy for entomopathogenic nematodes. *Parasitology* 108: 207–215.

Jothi BD, Mehta U (2003) Efficacy of three species of entomopathogenic nematodes with antidesiccants against *Helicoverpa armigera*. *Int J Nematol* 13(2): 292–232.

Karunakar G, Easwaramoorthy S, David H (1999) Susceptibility of nine lepidopteran insects to *Steinernema glaseri, S. feltiae* and *Heterorhabditis indicus* infection. *Int J Nematol* 9: 68–71.

Kaya HK, Gaugler R (1993) Entomopathogenic nematodes. *Ann Rev Entomol* 38: 181–206.

Lello ER, Patel MN, Matthews GA, Wright DJ (1996) Application technology for entomopathogenic nematodes against foliar pests. *Crop Protect* 15(6): 567–574.

Mejia-Torres MC, Saenz A (2013) Ecological characterization of the Colombian entomopathogenic nematode *Heterorhabditis* sp. SL0708. *Brazil J Biol* 73(2): 239–243.

Miller RW (1989) Novel pathogenicity assessment technique for *Steinernema* and *Heterorhabditis* entomopathogenic nematodes. *J Nematol* 21: 574.

Navon A, Keren S, Salame L, Glazer I (1999) An edible-to-insects calcium alginate gel as a carrier for entomopathogenic nematodes. *Biocontrol Sci Technol* 8(3): 429–437.

Pervez R (2010) Biocontrol potential of entomopathogenic nematodes against different instar larvae of gram pod borer, *Helicoverpa armigera* infesting chickpea. *Curr Nematol* 21(2): 17–21.

Pervez R (2017) Management of gram pod borer *Helicoverpa armigera* (Hübner) infesting chickpea using entomopathogenic nematodes under field conditions. *Curr Nematol* 28(1): 29–34.

Pervez R, Ali SS (2007) Natural occurrence of entomopathogenic nematodes associated with chickpea ecosystem. *Curr Nematol* 18(2): 19–22.

Pervez R, Ali SS (2009) Infectivity of *Spodoptera litura* (F.) (Lepidoptera: Noctuidae) by certain native entomopathogenic nematodes and their penetration in test insect and *in vivo* production. *Trends Biosci* 2(2): 70–73.

Pervez R, Ali SS (2011) Infectivity of *Steinernema mushtaqi* (Rhabditida: Steinernematidae) against insect pests and their mass production. *Arch Phytopathol Plant Protect* 44(14): 1352–1355.

Pervez R, Ali SS, Ahmad R (2007a) Efficacy of some entomopathogenic nematodes against mustard saw fly and *in vivo* production of these nematodes. *Int J Nematol* 17(1): 55–58.

Pervez R, Ali SS, Ahmad R (2008a) Effect of temperatures on the emergence of entomopathogenic nematodes. *Int J Nematol* 18(1): 25–28.

Pervez R, Ali SS, Ahmad R (2008b) Efficacy of entomopathogenic nematodes against green bug, *Nezara viridula* (L.) and their *in vivo* mass production. *Trends Biosci* 1(1–2): 49–51.

Pervez R, Ali SS. (2013) Entomopathogenic nematodes as a potent biopesticides against insect pests. In: Behra KK (ed.), *Newer approaches to biotechnology*. Narendra Publishing House Publishers and Distributors, New Delhi, pp. 213–235.

Pervez R, Devasahayam S, Eapen SJ (2014c) Determination of LD_{50} and LT_{50} of entomopathogenic nematodes against shoot borer (*Conogethes punctiferalis* guen.) infesting ginger. *Ann Plant Protect Sci* 22(1): 169–173.

Pervez R, Eapen SJ, Devasahayam S (2014b) Biodiversity of entomopathogenic nematodes. In: Kambasa KB (ed.), *Advances frontier of biotechnology*. Jaya Publishing House, Delhi, pp. 125–143.

Pervez R, Eapen SJ, Devasahayam S, Jacob TK (2012) Efficacy of some entomopathogenic nematodes against insect pests of ginger and their multiplication. *Nematol Mediteranea* 40(1): 39–44.

Pervez R, Eapen SJ, Devasahayam S, Jacob TK (2014a) Natural occurrence of entomopathogenic nematodes associated with ginger (*Zingiber officinale* Rosc.) ecosystem in India. *Indian J Nematol* 42(2): 238–245.

Pervez R, Rajkumar (2018) Management of root grub [*Holotrichia longipennis* (Blanch)] infesting Ginger (*Zingiber officinale* Rosc.) through entomopathogenic nematodes. *Ann Plant Protect Sci* 26(2): 348–351.

Pervez R, Rao U (2018) Eco-friendly management of lepidopteran insect pests through entomopathogenic nematodes. *J Biol Control* 32(3): 172–178.

Pervez R, Rao U (2021) Infectivity of entomopathogenic nematodes against the legume pod-borer, *Maruca vitrata* Fabricius, infesting pigeon pea. *J Helminthol* 95(15): 1–5.

Pervez R (2022) Diversity of entomopathogenic nematodes in western Uttar Pradesh. *Indian J Nematol* 52(1): 59–65.

Pervez R, Eapen SJ, Devasahayam S, Jacob TK, Ali M A, Thyagarajan P (2016a) *Oscheius* spp. as an alternative to *Heterorhabditis* spp. for ecofriendly management of cardamom root grub (*Basilepta fulvicorne* Jacoby) through entomopathogenic nematodes. *Ann Plant Protect Sci* 24(2): 385–391.

Pervez R, Eapen SJ, Devasahayam S, Jacob TK (2016b) Eco-friendly management of shoot borer *Conogethes punctiferalis* (Guenee) infesting ginger (*Zingiber officinale* Rosc.) through entomopathogenic nematodes. *Curr Nematol* 27(1): 23–27.

Pervez R, Lone SA, Pattnaik S (2020) Characterization of symbiotic and associated bacteria from entomopathogenic nematode *Heterorhabditis* sp. (Nematoda: Heterorhabditidae) isolated from India. *Egypt J Biol Pest Control* 30: 1–9.

Pervez R, Revathi J, Eapen SJ, Devasahayam S, Jacob TK (2015) Isolation and identification of symbiotic bacterium associated with the entomopathogenic nematode, *Heterorhabditis* sp. (IISR-EPN 01) from ginger rhizosphere. *Res J Pharm Biol Chem Sci* 6(2): 339–343.

Phan KL, Tirry L, Moens M (2005) Pathogenic potential of six isolates of entomopathogenic nematodes (Rhabditida: Steinernematidae) from Vietnam. *Bio Control* 50(3): 477–491.

Poinar GO (Jr) (1990) Taxonomy and biology of Steinernematidae and Heterorhabditidae. In: Gaugler R, Kaya HK (eds.), *Entomopathogenic nematodes in biological control*. CRC Press, Boca Raton, pp. 23–61.

Poinar GO (Jr), Thomas GM, Hess R (1977) Characteristics of the specific bacterium associated with *Heterorhabditis bacteriophora* (Heterorhabditis: Rhabditida). *Nematologica* 23: 97–102.

Rainey FA, Ehlers RU, Stackebrandt E (1995) Inability of polyphasic approach to systematic to determine the relatedness of the genera *Xenorhabdus* and *Photorhabdus. Int J Syst Bacteriol* 32: 395–398.

Rajkumar, Patil J, Pervez R, Subhaharan K (2018) Management of white grub (*Leucopholis burmesterie*) infesting arecanut through entomopathogenic nematodes under field conditions. *Ann Plant Protect Sci* 27(1): 112–116.

Ramos-Rodriguez O, Campbell JF, Ramaswamy S (2006) Pathogenicity of three species of entomopathogenic nematodes to some major stored product insect pest. *J Stored Product Res* 42: 241–252.

Torres-Barragan A, Lisnawita L, Buhler W, Cardoza YJ (2011) Studies on the entomopathogenicity and bacterial associates of the nematode *Oscheius caroliensis. Biol Control* 59(2): 123–129.

Trdan S, Vidric M, Valic N (2006) Activity of four ebtomopathogenic nematodes against young adult of *Sitophilus granarious* (Coleptera: Curculioniadae) and *Oryzophilus surinamensis* (Coleptera: Silvanidae) under laboratory condition. *Plant Dis Protect* 113: 168–173.

3 Nematode-Trapping Fungi

The Hidden Microbial Army in Soil for Bio-protection of Roots Against Plant-Parasitic Nematodes

Dharmendra Kumar, Sakshi Pandey, Yogendra, Abhay Pratap Singh, Richa Raghuvanshi, Yogendra Kumar, Nilesh Kumar Sahani, Ajay Kumar and Ashok Kumar

3.1 INTRODUCTION

Plant-parasitic nematodes are infamous plant pathogens that can cause direct damage to their host or act as vectors to transmit the plant virus pathogens (Nicol et al. 2011). More than 4,100 species of plant-parasitic nematodes have an impact on global agriculture and horticulture, causing an estimated annual loss of USD 173 billion (Elling 2013). Nematode-trapping fungi (NTF) are a group of microbial army in soil that form loop-like structures called traps to capture, kill, parasitize and digest the nematodes. NTFs are potential biocontrol resources, which have the advantages of low toxicity, high efficiency, and environmental friendliness, so they have gradually become favoured by people in recent years. A large number of NTFs have received more attention in the past due to their possible use in the biological control of plant-and-animal-parasitic nematodes (Larson 2000). Nematode-trapping fungi occur in all sorts of soil environments where these fungi live both saprophytically and as facultative predators by capturing nematodes (Nordbring-Hertz et al. 2006). To switch over to nematophagous lifestyle, these fungi form traps on hyphae or on conidia in which nematodes can be trapped either mechanically or by adhesion. The traps then capture, penetrate, and kill the nematodes. Alternatively, NTFs can also form conidial traps (CT) directly upon spore germination without an intermediate mycelial phase or on the germination hyphae (Lei et al. 2011). Traps can be formed either spontaneously or be induced in response to signals from the environment, including certain amino acids, valyl peptides and Nemin that were secreted by host nematodes (Dijksterhuis et al. 1994). After the formation of traps, nematophagous activities precede a series of events including adhesion or trapping, penetration and immobilization of nematodes. At this stage, the nematodes become paralyzed (killed), and their internal tissues are rapidly colonized by fungal hyphae (Dijksterhuis et al. 1994). Enzymes released by nematode-trapping fungi during parasitism of nematodes, break down the body contents of the nematodes and the nutrients are translocate elsewhere within the hyphal system for the growth and reproduction of conidia (Zhang et al. 2020). Here, we especially focus on the mechanisms of trap formation, genomics and their biocontrol potential to understand the adaptive evolution and lifestyle transition of NT fungi and their role in future agriculture.

DOI: 10.1201/9781032631578-3

3.2 HABITAT AND FUNGAL DIVERSITY

Nematode-trapping fungi have global occurrence, in nearly all types of soil, but they are most predominant in the soil having high organic matter. Species variability in Nematode-trapping fungi have been investigated in several regions such as Canada (Mahoney and Strongman 1994) China (Liu and Zhang 1994; Liu and Zhang 2003; Zhang et al. 2011; Su et al. 2011) Brazil (Vilela et al. 2012; De Almeida et al. 2012; Falbo et al. 2013; Vilela et al. 2013) Europe (Boag and Lopez-Llorca 1989; Kelly et al. 2009; Saxena 2008) India (Saxena and Mukerji 1991; Kumar and Singh 2006) Latin America (Rubner 1994) and South America (Durand et al. 2005). Most frequently NTFs are isolated from agricultural pasture, coastal vegetation and coniferous leaf litter (Gray 1983; Kelly et al 2009; Kumar et al 2011; Liu et al. 2009; Saxena 2008; Shams Ghahfarokhi et al. 2004). NTFs are most frequently found up to 35 cm depth of soil in deciduous wood and upper organic-rich soil (Gray and Bailey 1985). They are found in agricultural soil and in the rhizospheric zone of crop plants e.g. *Arthrobotrys oligospora* (Persmark and Jannson, 1997). NTFs were also isolated and identified from the soil of polluted environments such as metalliferous mines (Mo et al. 2006, 2008). Instead of terrestrial habitats, NTFs were also found in some aquatic and other unique environments (Hao et al. 2005) such as submerged wood in mangroves (Swe et al. 2008). NTFs follow a seasonal variation with the highest densities and number of species in late summer and autumn (Miao and Liu 2003), possibly due to higher soil temperature and increased input of organic debris.

3.3 HISTORY OF NEMATODE-TRAPPING FUNGI

Corda (1839) described *Arthrobotrys* with *A. superba* as the type species, unaware of its facultative predatory activities. The best-investigated species of NTFs is *A. oligospora*, as described by Fresenius (1852). Since the discovery of the nematophagous ability of *Arthrobotrys oligospora* by Zopf in 1888 and the establishment of its connection with *Orbilia* by Pfister (1994), these fungi were characterized on the basis of conidial features like size, septation and type of conidiogenous cells (Subramanian 1963). For the first time, Rubner (1996), with the help of molecular data, classified these fungi on the basis of traps and categorized them into 3 lineages based on their trapping structures. Thereafter in 1999 the concept of an entirely new genus "Orbiliaceae" was advanced by Scholler et al., and by using 18s and ITSr DNA analysis Scholler et al. (1999) classified these nematode-trapping fungi into 4 genera and more importance was given to trapping structures. However, the classification given by Schollar et al. (1999) was not accepted and criticized when Li et al. (2005) again evaluated the classification with the help of 28s, 5.8s and β tubulin analysis. Due to various confusions that have arisen in the past, these nematode-trapping fungi were placed in 3 genera: (1) *Arthrobotrys* = 53 sp., (2) *Drechslerella* – 14 sp. and *Dactylellina* – 28 sp. There are about 700 sp. of fungi that are hostile to nematodes that feed on either their juveniles, adults or their eggs (Jiang et al. 2017).

Nematode-trapping fungi (NTF) in Orbiliomycetes are a highly specialized group of ancient monophyletic fungi that can form various trapping structures to capture nematodes (Jiang et al. 2017; Su et al. 2017; Zhu et al. 2022). Taxonomically, all Orbiliomycetes NTF have been divided into *Arthrobotrys* (catches nematodes with adhesive networks), *Dactylellina* (produces adhesive knobs, adhesive branches, and non-constricting rings), and *Drechslerella* (captures nematodes using constricting rings) genera based on their trapping structures according to modern molecular biology research (Yang et al. 2007; Yang et al. 2012; Ji et al. 2020). This group of fungi has become an ideal and key group for the biological control of parasitic nematodes and *Rhizoctonia solani* (Soliman et al. 2021; Kumar and Gouda 2018).

3.4 TRAPPING STRUCTURES

NTFs form mainly 3 types of trapping devices (A) adhesive knobs, (B) constricting rings and (C) adhesive nets, which can be further divided into 7 parts namely simple adhesive branches;

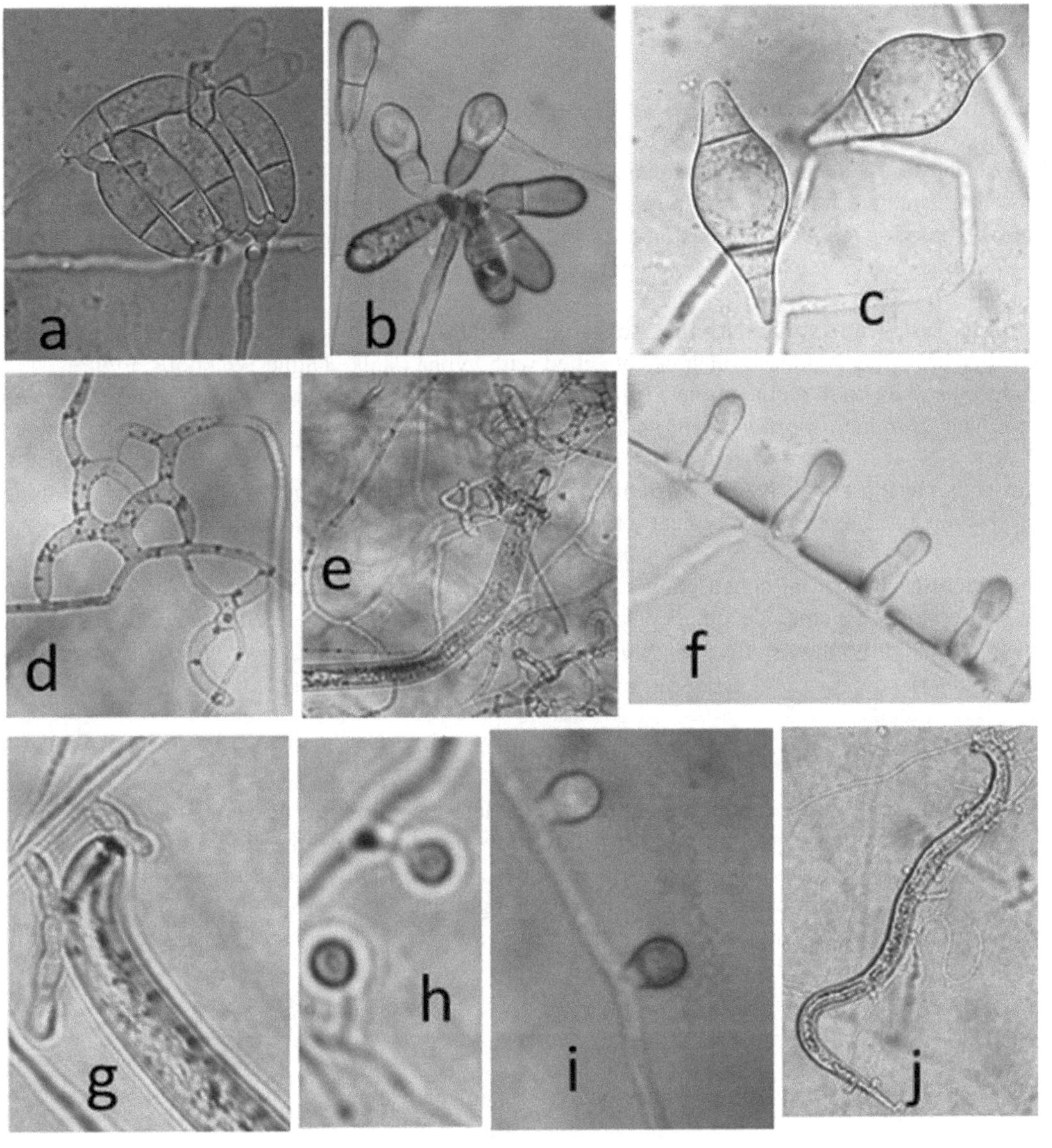

FIGURE 3.1 a. Conidia of *A. musiformis* attached with conidiophore; b. Conidium of *A. conoides* attached with conidiophore; c. Conidia of *Dactylellina. gephyropaga*; d. Adhesive network of *A. oligospora*; e. Trapping of *Heterodera schachtii* by adhesive nets of *A. oligospora*: f. Adhesive branches of *D. cionapagum* ; g. Trapping of *Heterodera schachtii* by adhesive branches of *D. cionapagum*; h. Adhesive stalked knobs of *M. haptotylum*; i. Adhesive sessile knobs of *Dactylellina phymatopaga* (×1350).; j. Trapping of *M. hapla* by *M. haptotylum*.

unstalked adhesive knobs; stalked adhesive knobs; non-constricting rings; constricting rings; 2-D network and 3-D network. Traps are either formed spontaneously or due to certain environmental signals such as compounds secreted by the nematodes (Dijksterhuis et al. 1994; Yang et al. 2011). Different types of traps produced by nematode-trapping fungi are given below:

Adhesive network: Induction of the adhesive network starts with the outgrowth of lateral branches from growing mycelium, which subsequently curve and anastomose to form complex three-dimensional networks. Ex. *Arthrobotrys oligospora*, *Arthrobotrys conoides*, *Arthrobotrys eudermata*, *Arthrobotrys musiformis* (Figure 3.1).

Adhesive hyphal branches: Adhesive hyphal branches consist of one to several elongated cells produced on the hyphae densely side by side (Figure 3.1). When the adhesive branches tend to anastomose, they are called two-dimensional networks. The ability of the hyphal branches to form bridges is used as a main criterion to separate *M. cionopagum* with simple adhesive branches (Figure 3.1) and *M. gephyropaghum* with anastomosed branches (two-dimensional networks, Figure 3.1). Ex. *Monacrosporium cionopagum*, *M. gephyropagum*.

Adhesive sessile and stalked knobs: Adhesive knobs are the globose or subglobose (Su et al. 2017) cells that are either produced on stalks attached with hypha (stalked adhesive knobs, Figure 3.1) or produced sessile knob on the hyphae (unstalked adhesive knobs, Figure 3.1). Ex. *Monacrosporium haptotylum*, *Monacrosporium phymatopagum*.

Non-constricting rings: Non-constricting rings are erect lateral branches from vegetative hypha that thicken and curve to form typically three-celled rings that then fuse to the supporting stalk.

Constricting rings: The constricting rings are three-celled and the most sophisticated trapping device. When the nematode enters into constricting rings, the three-celled ring swells rapidly inwards and tightly captures the prey within 1–2 seconds (Figure 3.2). Ex. *Drechslerella dactyloides*, *D. brochopaga*.

3.5 TRAP FORMATION AND INFECTION MECHANISM

Nematode-trapping fungi capture and parasitize nematodes in a series of events that include adhesion to the nematode surface or trapping, followed by penetration or insertion of fungal hypha into the nematode body, digestion and growth into the nematode body (Dijksterhuis et al. 1994). An essential signal of the transition of nematode-trapping fungus from a saprophytic to a parasitic lifestyle is the formation of traps (Li et al. 2005; Yang et al. 2007). A series of steps is involved in the trapping of nematodes, which are as follows.

Attraction

↓

Adhesion or Mechanical Trapping

↓

Penetration

↓

Immobilization of Nematode

Adhesive traps of NTFs capture nematodes by extracellular polymers that accumulate at the site of infection whereas constricting rings capture the nematodes by rapid swelling of ring cells (Figures 3.2 and 3.3). Following trapping by adhesion or mechanical, NTFs form a penetration tube which pierces the nematode cuticle (Dijksterhuis et al. 1994). In *A. oligospora*, the penetration process is an amalgam of the activity of hydrolytic enzymes, such as proteases, and the mechanical pressure of penetrating fungus. The cuticle of a nematode consists mainly of proteins including collagen (Cox et al. 1981), and various proteases have been extracted and studied from nematophagous fungi that can hydrolyze proteins of the nematode cuticle. The penetration tube swells up to form a large infection bulb in the nematode body. From these swollen and mature

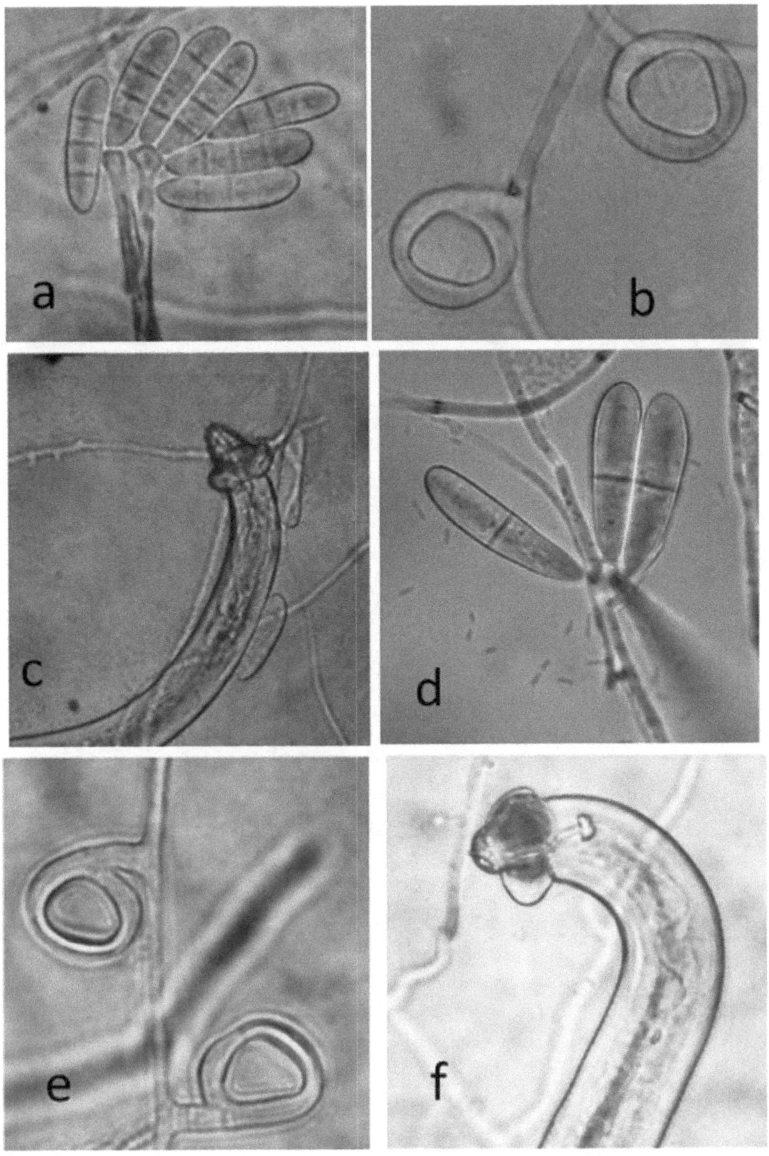

FIGURE 3.2 a. Conidia of *D. brochopaga* attached with conidiophore.; b. Constricting ring of *D. brochopaga*; c. Trapping of *M. incognita* by constricting rings of *D. brochopaga*; d. Conidia of *D. dactyloides* attached with conidiophore; e. Constricting rings of *D. dactyloides*; f. Trapping of nematodes by constricting ring of *D. brochopaga*.

infection bulbs, trophic hyphae develop and allow fungus to grow and utilize the nutrients from the nematodes for further growth.

3.6 TAXONOMY AND PHYLOGENY OF NEMATODE-TRAPPING FUNGI

Corda, in 1839, described the *Arthrobotrys* with the type species *A. superba* Corda and described the conidiophore of the fungus as simple, erect and septate and the conidia as two-celled and an

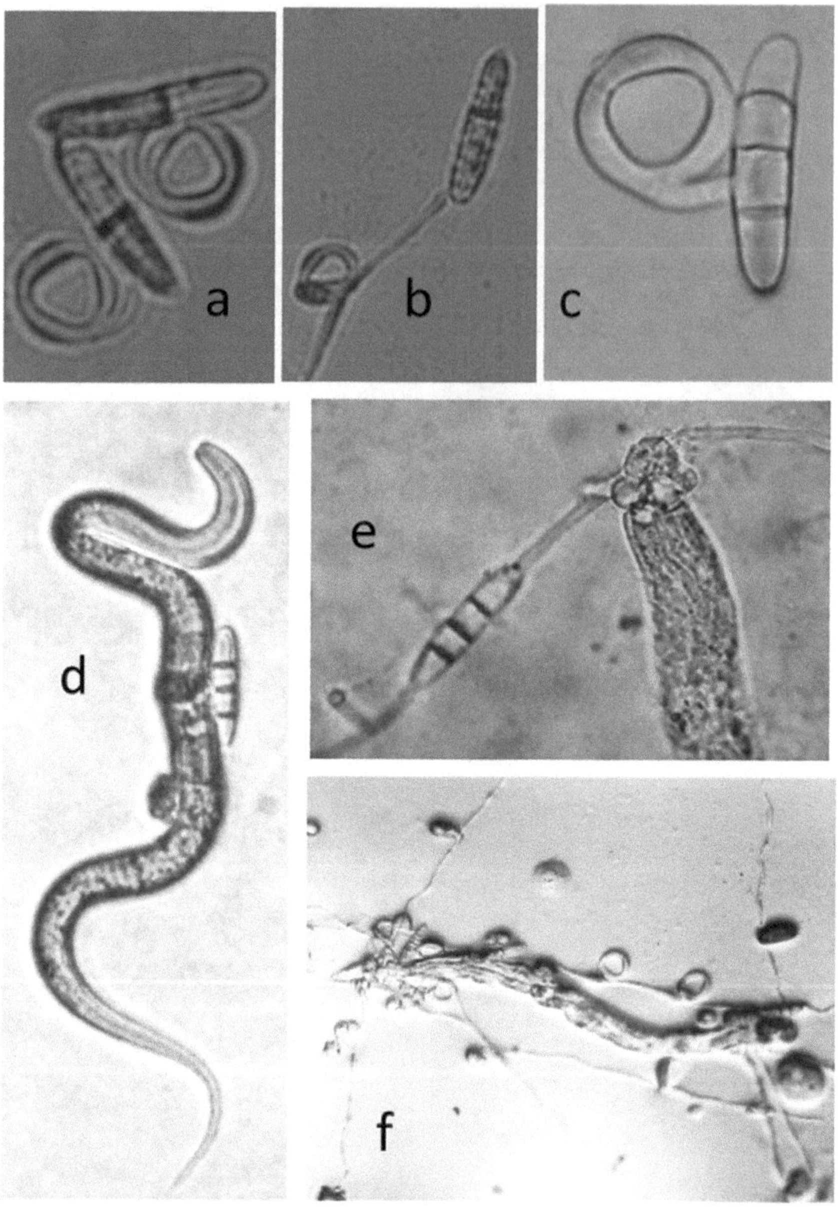

FIGURE 3.3 a. Conidial trap of *Drechslerella dactyloides*; b. Conidial trap on germ tube of *D. dactyloides*; c. Conidial trap of *D. brochopaga*; d. Trapping of nematode by conidial trap of *D. brochopaga*; e. Trapping of nematode by conidial trap of *D. brochopaga* produced on spore germ tube; f. Growth of *D. brochopaga* after parasitism of nematodes in soil.

apiculiform base. The genus *Dactylella* was established by Grove (1884) based on a saprophytic species *D. minuta* Grove. The genus was characterized as vegetative hyphae creeping, sparse, conidiophores erect, simple, septate or non-septate, smooth, hyaline; conidia borne singly at the apex of the conidiophore, ellipsoidal or fusoid or cylindrical, one-celled at first, later 2 – many – septate, hyaline. The genus *Monacrosporium* was proposed by Oudemans (1885) for two species, *M. elegans* Oudemans and *M. subtile* Oudemans. Drechsler introduced the many species of NTFs, from 1937 to 1975, and classified in *Arthrobotrys*, *Dactylaria* and *Dactylella*. Dreshsler classified the didymosporous species to *Arthrobotrys* and the phragmosporous species to *Dactylella* when containing a single conidium at the top of the conidiophore while in *Dactylaria* when more than one conidium is produced in a sympodial branching manner. Drechsler regarded *Monacrosporium* as synonym of *Dactylella*. Subramanium (1963) grouped the nematode-trapping fungi into three genera: *Arthrobotrys* (Corda 1839), *Dactylella* (Grove 1884) and *Monacrosporium* (Oudemans 1885). Cook and Godfrey (1964) provided a key for the identification of nematode-trapping fungi. Subsequently, many genera of nematode-trapping fungi have been created e.g. *Dreshsleromyces* Subram. *Golovinia* Melht, *Monacrosporium* Subram, *Kafiaddina* Mekht, *Candebrella* Rifai and R.C. Cook. *Geniculifera* Rafai, *Nematophagus* Mekht, *Woroninula* Mekht were created (Mekjtieva 1979; Rifai and Cook 1966; Subramanium 1963). Traditional taxonomy had led to a situation where species with various types of trapping devices had been assigned to one genus, while others with similar trapping devices can be found in different genera (Glockling and Dick 1994; Liu and Zhang 1994, 2003; Zhang et al. 1996).

By using molecular tools, some workers concluded that the trapping devices are phylogenetically more informative in delimiting genera and challenged the traditional taxonomy of NTF generally based on the morphological characters of conidia and conidiophores (Liou and Tzean 1997; Ahrean et al. 1998; Scholler et al. 1999; Yang 2007) Rubner (1996) first used trapping structures to try to rationalize the generic concepts of the nematode-trapping fungi in her monographic treatment of the *Dactylella-Monacrosporium* complex. Liou and Tzean (1997) did the phylogeny of 28 species of NTFs of different genera and stated that NTFs were monophyletic and have evolved from a common ancestor. They grouped the morphologically similar but non-predacious NTFs in genetically distinct parts of the tree. Their analysis with more variable internal transcribed spacer ITS 1 and ITS 2 separated the 28 species of NTFs into 3 clades each with a unique trapping device. The phylogenic analysis of Liou and Tzean (1999) indicated that trapping organs may reflect evolutionary relationships and appear more significant than conidiogenous cells and conidia for genus and species delimitation. Ahren et al. (1998) also studied and elaborated the phylogeny of 15 species of NTFs and their closely related non-parasitic species based on the 18 S rDNA and reported that the species within the genera *Arthrobotrys*, *Dactylaria*, *Dactylella*, *Monacrosporium* and *Duddingtonia* formed a monophyletic and isolated clade among an inconclusive cluster of apothecial ascomycetes. They stated that the phylogeny pattern within the clade of nematode-trapping fungi was not concordant with the morphology of the conidia or the conidiophores, but rather with the morphology of trapping structures. Within the clade of NTFs, their data supported the three lineages of NTFs: first, the species which is forming constricting rings, second a non-parasitic species *Dactylella oxyspora* and *Dactylella minuta* and the last species having various adhesive structures. Molecular studies of nematode-trapping fungi made it clear that trapping organs are phylogenetically more informative than the morphology of the dispersive structures on delimiting genera. Hagedorn and Scholler (1999); Scholler et al. (1999) studied and then proposed a new generic concept for the predatory anamorphic Orbiliaceae in which trapping organs are used as the main morphological criterion to delimitate genera. They divided the nematode-trapping fungi according to rDNA sequence data into four genera: *Arthrobotrys* for fungi forming adhesive networks; *Dreshslerella* for fungi forming constricting rings, *Dactylellina* for fungi forming adhesive knobs and *Gamsylella* for forming adhesive columns and unstalked knobs. Li et al. (2005) challenged the concept of genus *Gamsylella* given by Scholler et al. (1999) based on the phylogenies of NTFs inferred from sequence analysis of 28 S rDNA, 5.8S rDNA and B-tublin genes. They transferred the species that produce unstalked adhesive knobs that grow out to form loops from *Gamsylella* to *Dactylellina* and those that produce unstalked adhesive knobs that grow out to form a network from *Gamsylella* to *Arthrobotrys*. *Gamsylella* as currently circumscribed cannot be treated as a valid genus. Three genera concepts given by Li et al. (2005) are given below:

Genera	Trapping devices
Drechslerella	Fungi forming a three-celled constricting ring with a short, strong stalk
Arthrobotrys	Adhesive network only and unstalked adhesive knobs that develops into an adhesive network
Dactylellina	Stalked adhesive knobs, some with non-constricting rings, or unstalked adhesive knobs which grow out to form adhesive branches and loops

3.7 GENOME SEQUENCE AND GENERAL GENOMIC FEATURES OF NEMATODE-TRAPPING FUNGI

The estimated genome sizes of the sequenced genome of nematode-trapping fungi ranged from 35.43 to 43.12 Mb. The number of predicted genes is ranging from 9,927 to 12,107. The first genome sequencing of NTFs was reported of *A. oligospora* Fres. in 2011 (Yang et al. 2011). The 40.07 Mb genome size of *A. oligospora* strain ATCC24927 is similar in size to the ascomycetes fungus *Neurospora crassa* (Galagan et al. 2003). The genome size of *M. haptotylum* was estimated to 40.4 Mb and protein-coding genes to 10,959. The estimated genome size and number of protein-coding genes in *M. haptotylum* are more or less similar to the numbers of genes recorded in *A. oligospora* (Table 3.1). *A. oligospora* shared many more genes with pathogenic fungi than with non-pathogenic fungi. The genes shared between *A. oligospora* and other pathogenic fungi may be functionally related to pathogenicity in these fungi (Yang et al. 2011). Meerupati et al. (2013) reported that 1,161 proteins in *M. haptotylum* and 1,132 proteins in *A. oligospora* are similar to proteins in the pathogen–host interaction protein database (PHI-base database), particularly from plant pathogens by using BLAST analysis. The *D. stenobrocha* genome is smaller (29.02 Mb) than *A. oligospora* (40.07 Mb). Liu et al. (2014) reported that the total estimated genes in *A. oligospora* were 11,479 whereas only 7,781 genes were found in the genome of *D. stenobrocha*. They also stated that more than 1,400 predicted genes were shared strictly between *D. stenobrocha* and *A. oligospora*, showing their close phylogenetic relationship.

3.8 ROLE OF GENE FAMILY EXPANSIONS IN ADAPTATION OF PARASITISM IN NEMATODE-TRAPPING FUNGI

Meerupati et al. (2013) studied the expansions of the gene family and contractions in the genomes of *M. haptotylum* and *A. oligospora* by analysing the gene family evolution by comparing the genome

TABLE 3.1
General features of genome of nematode-trapping fungi

Trapping Devices	Fungi	Genome Size (Mb)	Number of Genes	References
Adhesive network	*Arthrobotrys oligospora* ATCC24927	40.07	11,479	Yang et al. 2011
Adhesive network	*Arthrobotrys oligospora* TWF154	39.62	12,107	Yang et al. 2020
Adhesive network	*Duddingtonia flagrans*	36.64	9,927	Youssar et al. 2019
Adhesive knob	*Monacrosporium haptotylum*	40.40	10,959	Meerupati et al. 2013
Adhesive knob	*D. entomopaga*	38.39	11,130	Ji et al. 2020
Constricting ring	*Drechslerella stenobrocha*	29.02	5,597	Liu et al. 2014
Constricting ring	*Drechslerella brochopaga*	35.43	10,234	Ji et al. 2020
Adhesive column	*D. cionopagum*	43.12	11,284	Ji et al. 2020
No trapping device	*D. cylindrospora*	37.71	10,785	Ji et al. 2020

of 16 fungi. They found a change in the size of 848 gene families of *M. haptotylum* of which 326 (38.4%) gene families were expanded and 522 (61.6%) contracted. Out of 806 gene families that changed in *A. oligospora*, 362 (44.9%) were expanded and 444 (55.1%) were contracted. The lowest proportion of gene families expansion among the ascomyceteous fungi was found in *Neurospora crassa* branch (9%) and the highest in *Metarhizium anisopliae* (52.6%). The expansion of gene families of the nematode-trapping fungi was similar to *Magnaporthe oryzae* (38.0%) inciting sheath blight of rice and higher than *Metarhizium acridum* (26.9%). They found that NTFs share superfluous gene families being expanded between plant-pathogenic fungi as compared to insect- or animal-pathogenic fungi. Twenty-eight expanded gene families were found shared between nematode-trapping fungi and plant pathogens. The nematode-trapping fungi and insect pathogens shared fewer numbers of expanded gene families and the nematode and animal pathogens the least (9 and 2 families, respectively). A comparative analysis of gene families and the expanded Pfam domain families revealed strong similarities. The expansion of gene families in *M. haptotylum* and *A. oligospora* was found to match with all 25 expanded Pfam domains. This indicates that the expanded Pfam domain families are important for the evolution of the two nematode-trapping species (Meerupati et al. 2013).

3.9 IDENTIFICATION OF CORE, LINEAGE- AND SPECIES-SPECIFIC GENES OF *M. HAPTOTYLUM* AND *A. OLIGOSPORA* AND GENE EXPRESSION IN *M. HAPTOTYLUM* DURING INFECTION OF *CAENORHABDITIS BRIGGSAE*

Meerupati et al. (2013) found that ~62% of the genes of *M. haptotylum* and *A. oligospora* were shared by other fungi; ~20% of the genes were shared between *M. haptotylum* and *A. oligospora* and up to 16% of the genes were unique for each fungal genome. A significant enrichment of species-specific genes was found in the cohort of the ten-fold upregulated genes. Of the 10% highly expressed genes during infection of *Cainorhabditus briggsae* by *M. haptotylum* (1,069 genes), 117 were more than ten-fold upregulated. In total, 38 of these 117 infection-regulated genes were specific for the *M. haptotylum* lineage, 15 were unique for the *M. haptotylum* and *A. oligospora* lineage and 64 were common to other fungi. Twenty-eight of the highly infection-regulated genes showed sequence similarities to proteins in the PHI-base, including the members of the DUF3129 Pfam domain family that encode the gas1 proteins of the rice blast fungus *Magnaporthe oryzae* (Xue et al. 2002), an extracellular cutinase (PBC1) from a plant pathogen (Li et al. 2003), a tetraspanin homolog from *Colletotrichum lindemuthianum* (Veneault-Fourrey et al. 2006), and *RBT4* from *Candida albicans* (Braun et al. 2000). Analysis of the sequences of the 10% most highly expressed genes in *M. haptotylum* (1,069 genes) showed that these were slightly enriched with secreted proteins. In the transcriptome of *A. oligospora*, seventy-two secreted proteins were identified among the 10% most highly expressed genes (in total 1,145 genes) of which 10 were SSPs. Genes from expanded Pfam domain families were highly expressed in both fungi, including the subtilisin (peptidase S8), tyrosinase, CFEM, DUF3129 (gas1), WSC and GLEYA families. In total, 2,599 (32%) of the 8,121 ortholog pairs showed significant differences in fold change (q-value<0.01) between the two fungi. Notably, the orthologs displaying the largest differences in expression values, that is, those that were more than 10-fold up- or down-regulated in the pairwise comparison between *M. haptotylum* and *A. oligospora*, were enriched with genes encoding secreted proteins, including many lineage-specific genes and SSPs. Several expanded Pfam domain families were also enriched among the differentially expressed genes (Meerupati et al. 2013). They proposed that two genomic mechanisms are likely to have been of major importance during the evolution of parasitism in nematode-trapping fungi: (1) Formation of novel genes through gene duplication leading to the expansion of gene families resulting in a large number of species-specific genes. Many species-specific genes were highly expressed and upregulated during infection of nematodes. (2) Differential expression of orthologous genes between *M. haptotylum* and *A. oligospora* during the early stage of infection suggests that differential gene expression has been an important mechanism for the evolution of parasitism in nematode-trapping fungi.

3.10 COMMONLY EXPRESSED GENES OF NTFS DURING THE INFECTION CAUSED BY PLANT-PARASITIC NEMATODES

Andersson et al. (2014) reported that commonly expressed genes of *A. dactyloides*, *A. oligospora* and *M. cionopagum* during infection of *Meloidogyne hapla* and *Heterodera schachtii* were subtilisin (peptidase_S8), aspartyl proteases, CFEM, and the carbohydrate-binding WSC domain. Nematode-trapping fungi were also found to express significant levels of domains of proteins and enzymes involved in protein synthesis, amino acid metabolism, and carbon energy. Furthermore, transcripts containing the Atg8 domain (an essential protein in the autophagic pathway) were highly expressed in *A. dactyloides*, *A. oligospora* and *M. cionopagum*. *A. oligospora* has 52 genes which encode the peptidase_S8 domain (Meerupati et al. 2013). However, only one transcript bearing the peptidase_S8 domain was found among the highly expressed transcripts in each library of *A. oligospora* infecting *M. hapla* and *A. oligospora* infecting *H. schachtii* (Andersson et al. 2014).The other cDNA libraries contained also only one transcript with the peptidase_S8 domain among the top 500 expressed genes. Stress proteins were highly expressed in all nematode-trapping fungi. They included heat-shock proteins and chaperones such as DnaJ, HSP70 and HSP90; gluthatione S-transferases; and antioxidant enzymes such as thioredoxin and catalase.

Fungal species have differentially expressed transcripts that encode peptidases (peptidase_M1 and peptidase_M24), lectins (FB_lectin, B_lectin, RicinB_lectin_2), tyrosinase, transcription factors, cell-signaling components, Atg8, various stress-response proteins, proteins containing the WSC domain and the DUF3129 domain (domain of unknown function). In the *Magnaporthe grisea* DUF3129 is found in the gas1 protein that participates in appressorial penetration and lesion formation during infection of rice (Quast et al. 2013). It is very interesting to know that DUF3129 was not expressed at all in *A. dactyloides* while it is highly expressed in *M. cionopagum*. In the library of Top500 most expressed transcripts during the *M. cinopagum* infecting *H. schachtii* DUF3129 was identified in twelve transcripts. During *A. oligospora* infections, this domain was identified in one transcript in *A. oligospora* infecting *M. hapla* and in four transcripts in *A. oligospora* infecting *H. schachtii* among the top 500 most expressed transcripts (Andersson et al. 2014). In nematode-trapping fungi, DUF3129 is a broad gene family and both of the NTFs *M. haptotylum* and *A. oligospora* have 33 genes that encode one of this expanded domain (Meerupati et al. 2013). Accordingly, the DUF3129 domain is highly expressed during infection among the species that form adhesive branches and adhesive knobs. Transcripts encoding the fungal fruit-body lectin (FB_lectin) and the D-mannose binding lectin (B_lectin) were only highly expressed in *A. oligospora* (adhesive net forming fungus) but not in *A. dactyloides* and *M. cionopagum*.

3.11 VIRULENCE ASSOCIATED TRANSCRIPTS

A BLAST search of the top 500 transcripts in each library of NTFs infecting host nematode species was conducted in the pathogen–host interaction protein database (PHI-base). In total, 97 unique PHI-base genes were identified. Among the 97 identified PHI-base genes, 15 were highly expressed by *A. oligospora*, *A. dactyloides* and *M. cionopagum* during infection of plant-parasitic nematodes (Andersson et al. 2014).

3.12 HOST-SPECIFIC GENE EXPRESSION OF NEMATODE-TRAPPING FUNGI

Gene expressions by *A. oligospora* during the infection of *M. hapla vs. H. schachtii* showed that most of the genes had similar expression levels in both of the host cases. It was recorded that 105 genes were at least 5-fold upregulated during the trapping and infection of *M. hapla* compared to the trapping and infection of *H. schachtii* and 65 genes were at least 5-fold upregulated in *A. oligospora* during the infection of *H. schachtii* compared to *M. hapla*. The differentially expressed genes encoding proteins with a predicted secretion signal were peptidases, glucosidases and several gene families that were expanded in the genomes of nematode-trapping fungi such as DUF3129, WSC and tyrosinase (Meerupati et al. 2013).

3.13 CONIDIAL TRAPS AND THEIR ECOLOGICAL SIGNIFICANCE IN SOIL

Some species of NTFs form conidial traps directly upon conidial germination without an intermediate hyphal phase (Persmark and Nordbring Hertz 1997) or on short spore germlings (Kumar et al. 2015) and these conidial traps depend on the capturing and killing of the nematodes for their subsequent growth and survival. These conidial traps are induced in response to cow dung (Dackman and Nordbring-Hertz 1992), rhizosphere soil (Mankau 1962; Pressmark and Nordbring-Hertz 1997) and compounds released by soil nematodes (Kumar 2003). Formation of conidial traps by NTFs may give these fungi an opportunity to capture and extract the abundant nutrition from the nematode body for their growth and development in the soil where they have intense competition with other fungi for nutrients. *Arthrobotrys oligospora* also form frequent conidial traps in close contact with cow dung (Dackman and Nordbring-Hertz 1992). *Drechslerella dactyloides Monacrosporium gephyrophagum* (Parsmark and Nordbring-Hetrz 1997), *Arthrobotrys brochopaga* and *A. dactyloides* frequently form conidial traps in vicinity to the soil and parasitize the nematodes (Kumar et al. 2015; Kumar 2024b)

3.14 GROWTH AND PARASITIC ABILITY OF NEMATODE-TRAPPING FUNGI IN FUNGISTATIC ENVIRONMENT OF SOIL

Fungistatic factors in soil act as inhibitors for the growth of fungi, therefore, the ability of NTFs to overcome the soil fungistasis is of paramount importance for the biocontrol ability of these fungi. Some species of NTF also form conidial traps in response to soil fungistasis, animal dung and nematodes as survival structures to overcome fungistasis (Dackman and Nordbring-Hertz (1992). *Arthtobortys dactyloides*, *Monacrosporium gephyrophagum* (Parsmark and Nordbring-Hetrz 1997) and *Dactylaria brochopaga* frequently form conidial traps in the vicinity of soil and trap the nematodes (Figure 3.3). The ability of *A. dactyloides* and *D. brochopaga* to form conidial traps in the vicinity of soil and trapping of soil nematodes (Figure 3.3) indicates that this fungus has enormous potential to reduce the nematodes in soil (Kumar et al. 2015; Kumar 2024a, 2024b).

3.15 PRODUCTION AND FORMULATIONS OF NEMATODE-TRAPPING FUNGI

Several NTFs are tested that parasitize the nematodes *in vitro* but most importantly, the question arises, how can we deliver them on a large scale? Then there is a need to formulate and deliver them by using different techniques. Formulation of BCAs is the conclusive factor for successful biocontrol measures (Burges 1998) and is used to protect the antagonistic organism from harsh conditions. Formulation of BCAs strengthens the organism aided by its handling and application during production, storage and distribution (Jones and Burges 1998). Formulation of BCAs includes wettable powder, powder, emulsion, granular formulation, oil solutions, blending agents and microcapsules (Liu et al. 2004). Powder and granular formulations are cost-effective because these formulations

TABLE 3.2
General features of formulation of nematode-trapping fungi

Commercial product	Formulation	Fungi	References
Royal 350	Fresh mycelium on organic substrates	*Arthrobotrys irregularis*	Cayrol et al. 1978
Royal 300	Fresh mycelium on organic substrates	*Arthrobotrys robusta*	Cayrol et al. 1983
	Alginate-based formulations	*Monacrosporium ellipsosporum*	Jaffee and Muldoon 1995
	Alginate-based formulations	*Monacrosporium cionopagum*	Jaffee and Muldoon 1995
	Encapsulationin alginate	*Dactylella candida*	Stirling and Mani 1995
	Encapsulationin alginate	*Arthrobotrys dactyloides*	Stirling and Mani 1995

are prepared by combining agricultural or industrial waste materials such as wheat-bran-sand mixture, sawdust-sand-molasses mixture, compost/farm manures, fly-ash, inert charcoal, etc., and their handling is also very easy (Kerry 1988; Khan and Khan 1994; Khan et al. 2005).

Since the nematode-trapping fungus *Arthrobotrys irregularis* developed as a product marketed by the trade name "Royal 300" (Cayrol et al. 1978) followed by a product of *Arthrobotrys robusta* known as "Royal 350" (Cayrol 1981, 1983) the formulation of some other nematode-trapping fungi have been developed. Commercial products (formulations name) of nematode-trapping fungi are given in Table 3.2.

3.16 ROLE OF NEMATODE-TRAPPING FUNGI AS A BIOAGENT

The classical work on biocontrol of nematode populations with the use of nematode-trapping fungi was initiated by Linford (1937) and Linford et al. (1938) by adding chopped green pineapple tops to nematode-infested soil in a pot which increases the nematode-trapping fungi and reduces nematode population. Arndt and Leuprecht (1994) found that *A. conoides* and *A. oligospora* had the greatest impact in lowering the amount of *M. incognita* during the first four weeks of a greenhouse experiment, Whereas *Paecilomyces lilacinus*, *Verticilium chlamydosporium*, and *Trichoderma harzianum* only reached their peak performance eight weeks after seeding. The activity of nematode-trapping fungi *Dactylella candida* and *Arthrobotrys dactyloides* was analyzed following their encapsulation in alginate (Stirling and Mani 1995) and alginate-based formulations of *Monacrosporium ellipsosporum* and *M. cionopagum* have been manufactured on a small scale (Jaffee and Muldoon 1995).

Now lots of nematode-trapping fungi as BCA and various commercial biological nematicides based on nematode-trapping fungi have been developed. Nematode-trapping fungi have the ability not only to reduce the nematode population (Kumar and Singh 2005; Kumar and Singh 2006; Singh et al. 2007; Kumar et al. 2015) but also to enhance the plant growth (Singh et al. 2012) and soil condition. Several NFs were also tested in combined form to manage the nematodes such as root knots, cereal cysts, burrowing nematodes, etc. on different hosts. 62% galls, 94% of *M. javanica* juvenile reduction occurs when treated with a combination of *P. lilacinus* and *Monacrosporium lysipagum* on tomato. *Radopholus similis* can also be managed by either *Monacrosporium lysipagum* individually (86%) or combined with *P. lilacinus* (96%). Application of mass culture of nematode-trapping fungi grown on barley substrates (Kumar et al. 2005) was found to reduce the number of root knots, females, egg masses and second-stage juveniles of root knot nematodes in tomato and rice (Kumar and Singh 2006; Singh et al. 2007, Kumar 2024a, 2024b). Some other nematode-trapping fungi were also reported to parasitize the *R. solani* by hyphal coils (Kumar and Gouda 2018) and manage the sheath blight or rice in pot conditions.

3.17 CONCLUSION

The importance of nematode-trapping fungi arises from their potential use as biological control agents for plant-and-animal-parasitic nematodes. Studies on comparative transcriptomics to the adhesive trap-forming fungus *A. oligospora* showed that the gene duplication and differential regulation of the orthologous genus has been the important mechanism for the evolution of parasitism in nematode-trapping fungi. Comparative transcriptome analysis showed that the divergence in gene expression between the species of nematode-trapping fungi was significantly larger than that related to the plant-parasitic nematode host. A core set of transcripts being highly expressed by some nematode-trapping fungi during infection of plant-parasitic nematodes were also found highly expressed by *A. oligospora* and *M. haptotylum* during infection of *C. briggsae*. Knowledge of the genomic interspecific variation in the trapping mechanisms and host preferences will be useful for researchers in the identification of more potent nematode-trapping fungi for the management of plant-parasitic nematodes. Some species of NTFs were observed to form conidial traps and kill the nematodes in soil which indicates that conidia-based formulations of conidial trap-forming fungi individually, in consortia, can help in higher control of plant parasitic-nematodes in an environmentally and economically beneficial manner.

REFERENCES

Ahren D, Ursing BM, Tunlid A (1998) Phylogeny of nematode-trapping fungi based on 18s rDNA sequences. *FEMS Microbiol Lett* 158: 179–184.

Andersson KM, Kumar D, Bentzer J, Friman E, Ahrén D, Tunlid A (2014) Interspecific and host-related gene expression patterns in nematode-trapping fungi. B*MC Genomic* 15: 968. doi: 10.1186/1471-2164-15-968

Arndt M, Leuprecht B (1994) Trials on alternatives for control of nematodes in vegetable crops. *Gartenbau-Magazine* 3(5): 24–25.

Boag B, Lopez-Llorca LV (1989) Nematodes and nematophagous fungi associated with cereal fields and permanent pasture in Eastern Scotland. *Crop Res* 117(2): 399–406.

Braun BR, Head WS, Wang MX, Johnson AD (2000) Identification and characterization of TUP1-regulated genes in *Candida albicans*. *Genetics* 156: 31–44.

Burges HD (1998) *Formulation of microbial pesticides*. Dordrecht: Kluwer Academic Publishers.

Cayrol JC, Frankowski JP, Laniece A, D'Harde-mare G, Talon JP (1978) Contre les nematodes en champigonniere. Mise an point d'une methode de lutte biologique a l'aide d'un Hyphomycetes predateur: *Arthrobotrys robusta* souche antipolis' (Royal 300). *Pepinieristes Horticulteurs Maraichers Revue Horticole* 184: 23–40.

Cayrol JC (1981) Possibilities of biological control of *Meloidogyne* nematode crop pest with the use of the 'R 350' nematophagous fungus (*Arthrobotrys irregularis*). *La Defense des Vegetaux* 35: 279–282.

Cayrol JC (1983) Lutte biologique contre les *Meloidogyne* au moyen d'*Arthrobotrys irregularis*. *Rev Nématol* 6: 265–273.

Cayrol JC, Frankowski JP, Laniece A, Hardemare GD, Talon JP (1978) Control of nematodes in mushroom cultures. Statement of a method of biological control using a predatory Hyphomycetes: *Arthrobotrys robusta* strain Antipolis (Royal 300). *Pepinieristes Horticulteurs Maraichers Revue Horticole* 184: 23–40.

Cooke RC, Godfrey BES (1964) A key to the nematode destroying fungi. *Trans Br Mycol Soc* 47(1): 61–74.

Corda ACI (1839) *Pracht- Flora Europaischer Schimmelbildungen* (Vol. i–viii, p. 55, 2 Tabs.). Leipzig: Gerhard Fleischer.

Cox GN, Kusch M, Edgar RS (1981) Cuticle of *Caenorhabditis elegans*: its isolation and partial characterization. *J Cell Biol* 90: 7–17.

Dackman C, Nordbring-Hertz B (1992) Conidial traps – A new survival structure of the nematode-trapping fungus *Arthrobotrys oligospora*. *Mycol Res* 96(3): 194–198.

De Almeida GL, Santurio JM, Filho JO, Zonette RA, Camillo G, Flores AG, da Silva JH, de IaR ML (2012) Predatory activity of fungus *Duddingtonia flagrans* in equine strongyle infective larvae on natural pasture in the southern region of Brazil. *Parasitol Res* 110: 657–662.

Dijksterhuis J, Veenhuis M, Harder W, Nordbring-Hertz B (1994) Nematophagous fungi: physiological aspects and structure-function relationships. *Adv Microb Physiol* 36: 111–143.

Drechsler C (1937) Some Hyphomycetes that prey on free living terricolous nematodes. *Mycologia* 29: 447–552.

Durand DT, Boshoff HM, Michael LM, Krecek RC (2005) Survey of nematophagous fungi in South Africa: the onderstepoort. *J Veterinary Res* 72: 185–187.

Elling AA (2013) Major emerging problems with minor *Meloidogyne* species. *Phytopathology* 103: 1092–1102. doi: 10.1094/PHYTO-01-13-0019-RVW

Falbo MK, Soccol VT, Sandini IE, Vicente VA, Robl D, Soccol CR (2013) Isolation and characterization of the nematophagous fungus *Arthrobotrys conoides. Parasitol Res* 112: 177–185.

Fresnius G (1852) Beitrage Zur Mykologie. *Heft* 1(2): 1–80.

Galagan JE, Calvo SE, Borkovich KA, Selker EU, Read ND, et al. (2003) The genome sequence of the filamentous fungus *Neurospora crassa. Nature* 422: 859–868.

Glockling S, Dick M (1994) *Dactylella megalobrocha,* a new species of nematophagous fungus with constricting ring. *Mycol Res* 98: 845–853.

Gray NF (1983) Ecology of nematophagous fungi: distribution and habitat. *Ann Appl Biol* 102: 501–509.

Gray NF, Bailey F (1985) Ecology of nematophagous fungi: vertical distribution in deciduous woodland. *Plant and Soil* 86: 217–223.

Grove WB (1884) New or noteworthy fungi. *Am J Bot* 22: 195–201.

Hagedorn G, Scholler M (1999) A re-evaluation of predatory orbiliaceous fungi. I. Phylogenetic analysis using rDNA sequence data. *Sydowia* 51: 27–48.

Hao Y, Mo M, Su H, Zhang KQ (2005) Ecology of aquatic nematode-trapping hyphomycetes in southwestern China. *Aquatic Microbial Ecol* 40: 175–181.

Jaffee BA, Muldoon AE (1995) Susceptibility of root-knot and cyst nematodes to the nematode-trapping fungi *Monacrosporium ellipsosporum* and *M.* cionopagum. *Soil Biol Biochem* 27: 1083–1090.

Jiang X, Xiang M, Liu X (2017) Nematode-trapping fungi. *Microbiol Spectr* 5: 10. doi: 10.1128/microbiolspec.funk-0022-2016

Jiang X, Xiang M, Liu X (2017) Nematode-trapping fungi. *Microbiol Spectr* 5: 963–974.

Ji X, Yu Z, Yang J, Xu J, Zhang Y, Liu S, Zou C, Li J, Liang L, Zhang KQ (2020) Expansion of adhesion genes drives pathogenic adaptation of nematode-trapping fungi. I. *Science* 23: 101057. doi: 10.1016/j.isci.2020.101057

Ji X, Yu Z, Yang J, Li J, Liang LM, Zhang KQ (2020) Expansion of adhesion genes drives pathogenic adaptation of nematode-trapping fungi. I. *Science* 23: 101057.

Jone KA, Burges HD (1998) Technology of formulation and application. In: *Formulation of microbial pesticides-beneficial microorganisms, nematodes and seed treatments,* Burges HD (Ed.) Dordrecht: Kluwer Academic, pp. 7–30.

Kelly P, Good B, Hanrahan JP, Fitzpatrick R, de Waal T (2009) Screening for the presence of nematophagous fungi collected from Irish sheep pastures. *Veterinary Parasitol* 165: 345–349.

Kerry BR (1988) Two microorganisms for the biological control of plant parasitic nematodes. *Proc Brighton Crop Protect Conf* 2: 603–607.

Khan MR, Khan MW (1994) Single and interactive effects of root knot nematode and coal smoke on okra. *New Phytolog* 126: 337–342.

Khan MR, Khan SM, Mohiddin FA (2005) Root knot problem of some winter ornamental plants and its biomanagement. *J Nematol* 37: 198–206.

Kumar D (2003) Studies on predacity and biocontrol potential of *Arthrobotrys dactyloides*. Ph.D. Thesis. Banaras Hindu University Varanasi India.

Kumar, D. (2024a) Effectiveness of various nematode-trapping fungi for biocontrol of the *Meloidogyne incognita* in tomato (*Lycopersicion esculentum* Mill.). *Rhizosphere* 29: 100845. doi: 10.1016/j.rhisph.2023.100845

Kumar D (2024b) Evaluation of nematophagous and biocontrol efficacy of *Dactylellina phymatopaga* against Meloidogyne graminicola on rice (Oryza sativa L.). *Biol Control* 188: 105425.

Kumar D, Singh KP (2006) Assessment of predacity and efficacy test of *Arthrobotrys dactyloides* against root knot disease of tomato. *J Phypathol* 154: 1–5.

Kumar D, Singh KP, Jaiswal RK (2005) Screening of different media and substrates for cultural variability and mass culture of *Arthrobotrys dactyloides* Drechsler. *Mycobiology* 32(4): 215–222.

Kumar D, Maurya N, Kumar P, Singh H, Addy SK (2015) Assessment of germination and carnivorous activities of a nematode-trapping fungus *Arthrobotrys dactyloides* in fungistatic and fungicidal soil environment. *Biol Control* 82: 76–85.

Kumar D, Gouda SA (2018) Evaluation of mycoparasitic efficacy of nematode-trapping fungi against *Rhizoctonia solani* inciting sheath blight disease in rice (*Oryza sativa* L.). *Biol Control* 122: 31–40.

Kumar N, Singh RK, Singh KP (2011) Occurrence and colonization of nematophagous fungi in different substrates, agricultural soils and root galls. *Arch Phytopathol Plant Protect* 44: 182–1195.

Larsen M (2000) Prospects for controlling animal parasitic nematodes by predacious micro fungi. *Parasitology* 120(7 suppl): 121–131.

Li L, Ma M, Liu Y, Zhou J, Qu Q, Lu K, Fu D, Zhang K. (2011, September) Induction of trap formation in nematode-trapping fungi by a bacterium. *FEMS Microbiol Lett* 322(2): 157–165. doi: 10.1111/j.1574-6968.2011.02351.x

Li D, Ashby AM, Johnstone K (2003) Molecular evidence that the extracellular cutinase Pbc1 is required for pathogenicity of *Pyrenopeziza brassicae* on oilseed rape. *Mol Plant Microbe Interact* 16: 545–552.

Li Y, Hyde KD, Jeewon R, Lei C, Vijaykrishna D, Zhang KQ (2005) Phylogenetics and evolution of nematode trapping fungi (orbilliales) estimated from nuclear and proteins coding genes. *Mycologia* 97: 1034–1046.

Linford MB (1937) Stimulated activity of natural enemies of nematodes. *Science* 85: 123–124.

Linford MB, Yap F (1938) Root knot injury restricted by a nematode trapping fungus. *Phytopathology* 28: 14–15.

Liou GY, Tzean SS (1997) Phylogeny of the genus *Arthrobotrys* and allied nematode-trapping fungi based on rDNA sequences. *Mycologia* 89(6): 876–884.

Liou JY, Shih JY, Tzean SS (1999) Esteya, a new nematophagous genus from Taiwan, attacking the pinewood nematode (*Bursaphelenchus xylophilus*). *Mycol Res* 103(2): 242–248.

Liu K, Zhang W, Lai Y, Xiang M, Wang X, Zhang X, Liu X (2014) *Drechslerella stenobrocha* genome illustrates the mechanism of constricting rings and the origin of nematode predation in fungi. *BMC Genom* 15: 114. doi: 10.1186/1471-2164-15-114

Liu X, Zhang KQ (1994) Nematode trapping species of *Monacrosporium* with special reference to new species. *Mycol Res* 98: 862–868.

Liu X, Zhang KQ (2003) *Dactylella shizishana* sp. Nov. from Shizi mountain China. *Fungal Diversity* 14: 103–107.

Liu X, Zhang M, Che Y (2009) The living strategy of nematophagous fungi. *Myycoscience* 50: 20–25.

Liu XZ, Zhang KQ, Li TF (2004) *Biological control of plant-parasitic nematodes*. Beijing: China Science and Technology Press.

Mahoney CJ, Strongman DB (1994) Nematophagous fungi from the cattle manure in four states of decomposition at three sites in nova scotia Canada. *Mycologia* 86: 371–375.

Mankau R (1962) Soil fungistasis and nematophagous fungi. *Phytopathology* 52(7): 611–615.

Meerupati T, Andersson KM, Friman E, Kumar D, Tunlid A, Ahrén D (2013) Genomic mechanisms accounting for the adaptation to parasitism in nematode-trapping fungi. *PLoS Genet* 9(11): e1003909. doi: 10.1371/journal.pgen.1003909

Mekhtieva NA (1979) *KhishchnyeNematofagovye Griby-Gifomitsety (Predaciuos Nematophagous Hyphomycetes)*. Baku: A Kademiya Nauk Azerbaidzhanskoi SSR.

Miao Z, Liu X (2003) Note on the seasonal fluctuation of nematode trapping fungi population in orchards at the vicinity of Beijing. *Mycosystema* 22: 77–81.

Mo MH, Chen WM, Su HY, Zhang KQ, Duan CQ, He DM (2006) Heavy metals tolerance of nematode trapping fungi in lead polluted soils. *Appl Soil Ecol* 31: 11–19.

Mo MH, Chen WM, Yanf HR, Zhang KQ (2008) Diversity and metals tolerance of nematode trapping fungi in Pb-polluted soils. *J Microbiol* 46: 16–22.

Nicol JM, Turner SJ, Coyne DL, Den Nijs L, Hockland S, Maafi ZT (2011) Current nematode threats to world agriculture. In: *Genomics and molecular genetics of plant-nematode interactions*, Jones J, Gheysen G, Fenoll C (Eds.) The Netherlands: Springer Dordrecht, pp. 21–43.

Nordbring-Hertz B, Jansson HB, Tunlid A (2006) Nematophagous fungi. In: *Encyclopedia of life sciences*. Chichester: John Wiley & Sons, Ltd. , pp. 1–11. http://dx.doi.org/10.1038/npg.els.0004293

Oudemans CA (1885) Aanwinstenvoor de flora mycological van Nederland. *Ned Kruidk Archf Ser* 24: 236–257.

Persmark L, Jansson HB (1997) Nematophagous fungi in rhizosphere of agricultural crops. *FEMS Microbiol Ecol* 22: 303–312.

Persmark L, Nordbring-Hertz B (1997) Conidial trap formation of nematode-trapping fungi in soil and soil extracts. *FEMS-Microbiol Ecol* 22(4): 313–323.

Pfister DH (1994) *Orbillia fimicola,* a nematophagous discomycete and its *Arthrobotrys* anamorph. *Mocologia* 86: 451–453.

Quast C, Pruesse E, Yilmaz P, Gerken J, Schweer T, Yarza P, Peplies J, Glockner FO (2013) The SILVA ribosomal RNA gene database project: improved data processing and web-based tools. *Nucleic Acids Res* 41: D590–D596. doi: 10.1093/nar/gks1219

Rifai MA, Cooke RC (1966) Studies on some didymosporous genera of nematode trapping hyphomycetes. *Trans Br Mycol Soc* 49: 147–168.

Rubner A (1996) Revision of predacious hyphomycetes in the *Dactylella-Monacrosporium* complex. *Studies Mycol* 39: 1–134

Rubner A (1994) Predacious fungi from Ecuador. *Mycotaxon* 51: 143–151.

Saxena G (2008) Observation on occurrence of nematophagous fungi in Scotland. Applied soil ecology: a section of agriculture. *Ecosyst Environ* 39: 352–357.

Saxena G, Mukerji KG (1991) Distribution of nematophagous fungi in Varanasi, India. *Nova Hedwigia* 52: 487–495.

Scholler M, Hagedorn G, Rubner A (1999) A re-evaluation of predatory orbiliaceous fungi. II. A new generic concept. *Sydowia* 51: 89–113.

Shams Ghahfarokhi M, Razzaghi Abyaneh M, Ranjbhar Bahadori S, Eslami A, Zare R, Ebrahimi M (2004) Screening of soil and sheep faecal samples for predacious fungi: isolation and characterization of the nematode trapping fungus *Arthrobotrys oligospora. Iranian Biomedical J* 8: 135–142.

Singh KP, Jaiswal RK, Kumar N, Kumar D (2007) Nematophagous fungi associated with root galls of rice caused by *Meloidogyne graminicola* and its control by *Arhthrobotrys dactyloides* and *Dactylaria brochopaga*. *J Phytopathol* 155: 193–197.

Singh UB, Sahu A, Singh RK, Singh DP, Meena KK, Srivastava JS, Renu Manna MC (2012) Evaluation of biocontrol potential of *Arthrobotrys oligospora* against *Meloidogyne graminicola* and *Rhizoctonia solani* in rice (*Oryza sativa* L.). *Biol Control* 60: 262–270.

Soliman MS, El-Deriny MM, Ibrahim DSS, Zakaria H, Ahmed Y (2021) Suppression of root-knot nematode Meloidogyne incognita on tomato plants using the nematode trapping fungus *Arthrobotrys oligospora* Fresenius. *J Appl Microbiol* 131: 2402–2415.

Stirling GR, Mani A (1995) The activity of nematode trapping fungi following their encapsulation in alginate. *Nematologica* 41(1–5): 240–250.

Su H, Liu S, Li Y, Cao Y, Chen M, Yang X (2011) *Arthrobotrys latispora*, a new nematode trapping fungus from south west China. *Mycotaxon* 117: 29–36.

Su H, Zhao Y, Zhou J, Feng H, Jiang D, Zhang KQ, Yang J (2017) Trapping devices of nematode-trapping fungi: formation evolution and genomic perspectives. *Biol Rev Camb Philos Soc* 92: 357–368.

Subramanian CV (1963) *Dactylella, Monacrosporium* and *Dactylaria*. *J Indian Bot Sci* 42: 291–300.

Swe A, Jeewon R, Pointing SB, Hyde KD (2008) Taxonomy and molecular phylogeny of *Arthrobotrys mangrovispora,* a new marine nematode trapping fungal species. *Bot Marina* 51: 331–338.

Veneault-Fourrey C, Lambou K, Lebrun MH (2006) Fungal Pls1 tetraspanins as key factors of penetration into host plants: A role in re-establishing polarized growth in the appressorium. *FEMS Microbiol Lett* 256: 179–184.

Vilela VL, Feitos TF, Braga FR, de Araujo JV, de Leucena SC, Dantas ES, Athyde ACR, Silwa WW (2013) Efficacy of *Monacrosporium thaumasium* in the control of goat gastrointestinal helmithiasis in a semi-arid region of Brazil. *Parasitol Res* 112(2): 871–877.

Vilela VL, Feitos TF, Braga FR, de Araujo, JV, Souto DV, Santos HE, Silva GL, Athayde AC (2012) Biological control of goat gastrointestinal helminthiasis by *Duddingtonia flagrans* in semi-arid region of north eastern brazil. *Veterinary Parasitol* 188: 127–133.

Xue C, Park G, Choi W, Zheng L (2002) Two novel fungal virulence genes specifically expressed in appressoria of the rice blast fungus. *Plant Cell* 14: 2107–2119.

Yang CT, de Ulzurrun GVD, Goncalves AP, Lin HC, Chang CW, Huang TY, Chen SA, Lai CK, Tsai IJ, Schroeder FC, et al. (2020) Natural diversity in the predatory behavior facilitates the establishment of a robust model strain for nematode-trapping fungi. *Proc Natl Acad Sci* 117: 6762–6770.

Yang J, Wang L, Ji X, Feng Y, Li X, Zou C, Xu J, Ren Y, Mi Q, Wu J, et al. (2011) Genomic and proteomic analyses of the fungus *Arthrobotrys oligospora* provide insights into nematode-trap formation. *PLoS Pathog* 7: e1002179. doi: 10.1371/journal.ppat.1002179

Yang E, Xu L, Yang Y, Zhang XY, Xiang MC, Wang CS, An ZQ, Liu XZ (2012) Origin and evolution of carnivorism in the Ascomycota (fungi). *Proc Natl Acad Sci USA* 109: 10960–10965.

Yang J, Wang L, Ji X, Feng Y, Li X, Zou C, Xu J, Ren Y, Mi Q, Wu J, Liu S (2011) Genomic and proteomic analyses of the fungus *Arthrobotrys oligospora* provide insights into nematode-trap formation. *PLoS Pathog* 7: e1002179. doi: 10.1371/journal.ppat.1002179

Yang, Y, Yang E, An Z, Liu XZ (2007) Evolution of nematode-trapping cells of predatory fungi of the Orbiliaceae based on evidence from rRNA-encoding DNA and multiprotein sequences. *Proc Natl Acad Sci USA* 104: 8379–8384.

Youssar L, Wernet V, Hensel N, Yu X, Hildebrand HG, Schreckenberger B, Kriegler M, Hetzer B, Frankino P, Dillin A, et al. (2019) Intercellular communication is required for trap formation in the nematode-trapping fungus *Duddingtonia flagrans*. *PLoS Genet* 15: e1008029. doi: 10.1371/journal.pgen.1008029

Zhang KQ, Liu XZ, Cao L (1996) Nematophagous species of *Monacrosporium* from China. *Mycol Res* 100(3): 274–276.

Zhang Y, Yu ZF, Xu J, Zhang KQ (2011) Divergence and dispersal of the nematode trapping fungus *Arthrobotrys oligospora* from China. *Environ Microbiol Rep* 3: 763–773.

Zhang Y, Li S, Li H, Wang R, Zhang K-Q, Xu J (2020) Fungi–nematode interactions: diversity, ecology, and biocontrol prospects in agriculture. *J Fungi* 6: 206.

Zhu MC, Li XM, Zhao N, Yang L, Zhang KQ,Yang JK (2022) Regulatory mechanism of trap formation in the nematode-trapping fungi. *J Fungi* 8: 406.

Zopf W (1888) Zur Kenntnis der Infections-Krankheiten niederer Tiere. Nova Acta Leop. Academy of Nature. *Halle* 52: 7.

4 Biological Control of Insect Pests of Crops by Using Entomopathogenic Fungi

Dinesh Singh and Arun Baitha

4.1 INTRODUCTION

Arthropods are invertebrates and have a segmented body and jointed (arthros) legs (podos) with an external cover called an exoskeleton. Insects (springtails, beetles, termites, and ants), crustaceans (sowbugs), arachnids (spiders and mites), myriapods (centipedes and millipedes) and scorpions belong to arthropods. There are different types of arthropod species and several dozen species of arthropods are found in a square mile. A few arthropods cause damage to crop yields. However, many others which are present in all types of soils eat or compete with several root and foliage feeders.

The term "entomopathogenic" refers to those microorganisms that attack insects and use them as hosts to complete their life cycle. They reduce insect pest populations to levels that do not cause economic damage to the crops. They are a means of managing by decreasing disease vectors (Scholte et al. 2004). Earlier, farmers depended on the control of insect pests by using predators, pathogens, and host plant resistance until the discovery of insecticide. In the past, entomopathogenic fungi (EPF) were studied mainly to develop ways to manage the diseases that were destructive to the silkworm industry in the 1800s. In the 19th century, germ theory was first exhibited by Agostino Bassi using silkworms and the muscardine fungus, which was later named *Beauveria bassiana* in his honor. Later, Bassi's studies on the disease in silkworms enabled him to understand his findings in the area of human diseases. The encouragement for the idea to use entomopathogens to control insect pests came mainly from the silkworm disease studies. In 1888, the first groundbreaking field trials with entomopathogenic fungi were commenced by Elie Metchnikoff a Russian microbiologist, who later won the Nobel Prize and named *Metarhizium anisopliae* (recently named *Metarhizium brunneum*) (Vega et al. 2009). Metchnikoff produced mass conidia of fungus on sterilized brewer's mash. The fungal culture was mixed with sand granules to easily spread on field crops. Though the results were inconsistent, however, the work of Metchnikoff created curiosity around the world. Different experiments with "friendly fungi" against insect pests were conducted in Europe and the United States. The first formal and published proposal of microsporidia to control grape phylloxera (*Daktulosphaira vitifoliae* Fitc.) was suggested by John LeConte (1874). The study exhibited visionary ideas for an alternative effective and economical method of insect pest control and showed the way for future scientists to do work on entomopathogenic fungi as listed in Table 4.1.

Entomopathogenic fungi, such as *B. bassiana* and *Lecanicillium* spp. can control insect pests as well as plant diseases (Kim et al. 2008). Fungal bioagents are recognized as generally safe for the environment and non-target species, in comparison with synthetic chemicals. Biological pesticides exhibit a multi-site action like competition for food and place, antibiosis, and induced systemic

 DOI: 10.1201/9781032631578-4

TABLE 4.1
Fungi pathogenic to insect pests of agricultural crops*

Fungus	Crop	Target pest
Akanthomyces attenuatus Zare & Gams	Snap bean and cowpea	*Megalurothrips usitatus* (Bagnall), whitefly, aphid, and thrips
Aschersonia aleyrodis	-	Hemiptera (Aleyrodidae)
Beauveria bassiana PDRL1187	Canola (*Brassica napus* L.)	Mustard aphid, *Lipaphisery simi*, *Aphis craccivora* Koch
B. bassiana	Melon	Whiteflies
B. bassiana	Cabbage	*Myzuspercsicae*
B. brongniartii	-	Coleoptera (Scarabaeidae)
Cladossporium oxysporium	Cowpea	*Aphis craccivora* Koch
Conidiobolus thromboides		Acari Hemiptera, Thysanoptera
Hirsutella thompsonii	Cowpea	*Aphis craccivora* Koch
	-	*Acari*
	Citrus	Rust mite (*Phyllocoptruta oleivora*)
Isaria fumosorosea Belgium	-	Acari, Diptera, Coleoptera, Hemiptera, Thysanoptera
I. fumosorosea	Coconut	Rugose spiraling whitefly (RSW)
Lagenidium giganteum	-	Diptera (Culicidae)
L. longisporum	-	Hemiptera
L. muscarium	-	Acari, Hemiptera, Thysanoptera
Metarhizium anisopliae L6, M440, PDRL711, PDRL526	Cabbage, Canola (*Brassica napus* L.)	Cabbage aphid, mustard aphids, *Lipaphis erysimi*, *Aphis gossypii*, *Aphis craccivora* Koch
M. acridum	-	Orthoptera
Nomuraea rileyi	-	Lepidoptera
Paecilomyces fumosoroseus n32,	Canola (*B. napus* L.), Cabbage	Mustard aphids, Diamondback moth *Lipaphisery simi*, *Plutella xylostella*
P. lilcinus PDRL812	Canola (*B. napus* L.)	Mustard ahpids, *Lipaphis erysimi*
Verticillium lecanii V17, PDRL922	Canola (*B. napus* L.), Cabbage	Cabbage aphid, Mustard Ahpid (*Myzus persicae*, *Lipaphis erysimi*)
V. lecanii	Chili	*Myzus persicae*, *Aphis craccivora* Koch

(*Modified Table 1 from the source Maina et al. 2018.)

resistance in plants are some of the antagonism mechanisms used by fungal bioagents. These fungi are phylogenetically diverse, heterotrophic, eukaryotic, unicellular, or multicellular (filaments) microorganisms. They reproduce via sexual or asexual spores, or both. They have chitinized cells and are generally non-mobile. They are facultative or obligate parasites of insects and have a high capacity for their sporulation and survival. They constitute a group of great interest in the biological control of insects. About 80 percent of the diseases that are found in insects are caused by fungi. Practically, all insects are susceptible to some of the diseases incited by the fungi, which can lead to death. There are fungi that can invade dead insects, called saprophagous, and fungi that infect living insects, called entomophagous (Butt et al. 2006). Of the estimated 1.5 to 5.1 million species of fungi in the world (Hibbett et al. 2007), currently > 750 species of fungi from about >90 genera are known to be pathogenic to insects and mites classified in several phyla namely: Chytridiomycota, Ascomycota, Basidiomycota, and the subphylum Entomophthoromycotina (Ramanujam et al. 2014; Maina et al. 2018; Rajula et al. 2020).

4.2 ADVANTAGES OF ENTOMOPATHOGENIC FUNGI

- Entomopathogenic fungi do not have residues which effect adversely on the environment.
- They are less or non-toxic to non-target beneficial organisms.
- They are mostly specific to a single group or few species with a narrow area of toxic action.
- They may be used in combination with compatible synthetic chemical insecticides.
- They are self-perpetuating under suitable environmental conditions.
- Use of chemical insecticides may be reduced in the control of insect pests.
- They protect biodiversity in managed ecosystem.
- They can reduce the risk of development of pest resistance to mycoinsecticde or slow down due to unique mode of action.

4.3 DISADVANTAGES OF ENTOMOPATHOGENIC FUNGI

- Entomopathogenic fungi need a specific environmental condition to germinate and cause infection to the insect.
- They may be expensive for their produce to use commercially.
- Entomopathogenic fungi have a short shelf life.
- The insect pests must be present before the application of entomopathogenic fungi; hence preventive treatment may become difficult.
- They have a lack of persistence and a low rate of infection under stress environmental conditions.
- They require a high application rate and thorough coverage of spraying due to slow acting.

4.4 CHARACTERISTIC OF ENTOMOPATHOGENIC FUNGAL GENERA

4.4.1 *Akanthomyces* Lebert

The genus *Akanthomyces* is phylogenetically distinct from *Beauveria* and *Cordyceps* (Kepler et al. 2017). *Akanthomyces* as proposed by Lebert, including the type *A. aculeatus*, primarily infects Lepidoptera. Colonies of *Akanthomyces* species reaching 2–2.5 cm diam. after 10 days at 26°C on PDA, fluffy and regular, bottom, circular, compact, white, cream or flesh-colored and reverse cream to pale yellow (rarely yellow). Mycelium is composed of septate, branched, hyaline, and smooth hyphae. Conidiophores are macronematous, mononematous, flexuous, erect, septate, branched, smooth, 13.6–28.4 × 1.8–2.5 μm. Conidiogenous cells are 10–28 × 1.5–2.8 μm in size, phialidic, with a minute collarette, developing at the tip of prostrate or on secondary branches of conidiophores, singly or in verticels, swollen at the base, tapered towards the tip, smooth. Conidia solitary, 2.4–7.2 × 1.7–2.6 μm, variable in size, slimy, ellipsoidal, subcylindrical to cylindrical, base attenuate, apex rounded, 1-celled, smooth, hyaline to subhyaline. White, cream or flesh-colored, cylindrical, attenuated synnemata, covered with a hymenium of phialides. *Akanthomyces* attacks some insect of orders Hemiptera (*Akanthomyces lecanii*), Coleoptera (*A. neocoleopterorum*), Lepidoptera (*A. pistillariiformis*), and Orthoptera (*A. fragilis*) (Mongkolsamrit et al. 2018; Chen et al. 2020). Some *Akanthomyces* species are associated with *Torrubiella* and are pathogens of spiders (e.g. *A. novoguineensis*). Besides its association with *Torrubiella*, *Akanthomyces* has been listed as an asexual state of some *Cordyceps* species, e.g. *Cordyceps tuberculata* and *C. confragosa*, thus having a wider host range (Kepler et al. 2017; Wijayawardene et al. 2017). While revising the phylogenetic affinities of members within the family Cordycipitaceae, Kepler et al. (2017) reported that *Lecanicillium lecanii*, and some other species of *Lecanicillium*, viz., *L. attenuatum*, *L. muscarium*, & *L. sabanense*, fall within *Akanthomyces* and proposed to reject *Lecanicillium*, as the former has priority over *Lecanicillium*.

4.4.2 *Beauveria* (Balsamo) Vuillemin

The genus *Beauveria* (Balsamo) Vuillemin belongs to phylum Ascomycota, order Hypocreales and contains a number of species that are well-known pathogens of a broad host range of arthropods capable of infecting 700 species of hosts, including various species in Acari and Insecta of Arthropoda (Zimmermann 2007). *Beauveria* spp. has a cosmopolitan distribution and occupies diverse habitats, including soils and plant phylloplanes, while some strains are recognized endophytes (Quesada-Moraga et al. 2007; Zimmermann 2007). Earlier, seven taxa have been described at the species level (Rehner et al. 2006 a & b). Several teleomorph ascomycetes stages have been linked, such as *Cordyceps bassiana* (Li et al. 2001; Huang et al. 2002) and *C. scarabaeicola*. *B. bassiana*, *B. brongniartii*, *B. velata*, *B. amorpha*and *B. malawiensis* were originally described as pathogens of insects (Rehner et al. 2006a), while *B. vermiconia* and *B. caledonica* were originally isolated from Chilean volcanic soil (de Hoog and Rao 1975) and Scottish moorland soil (Bisset and Widden 1988), respectively. Glare et al. (1993) demonstrated that *B. vermiconia* was a naturally occurring fungal pathogen of *Hylamorpha elegans (*Coleoptera: Scarabaeidae) in Osorno, Chile. Glare et al. (2008) reported *B. caledonica* as the predominant naturally occurring fungal pathogen isolated from bark beetle cadavers in plantation forests in New Zealand and *Hylobius abietis* (Curculioniodae: Scolytinae) in Scotland.

Beauveria bassiana causes white muscardine disease in insects and grows as a white mold. It produces several dry, powdery conidia in distinctive white spore balls on most common cultural media. Each spore ball contains a cluster of conidiogenous cells. These cells are short and ovoid and terminate in a narrow apical extension called a rachis. The rachis elongates after production of each conidium, resulting in a long zig-zag extension. The conidia are single-celled, haploid, and hydrophobic.

Beauveria bassiana produces many secondary metabolites such as bassianin, bassiacridin, beauvericin, bassianolide, beauverolides, tenellin and oosporein (Quesada-Moraga and Vey 2004). The fungus produces a number of cyclodepsipeptides such as beauvericin, which may be responsible for at least part of its insect toxicity. Hatting et al. (2004) showed that *B. bassiana* could control up to 65% of *Duraphisnoxia* in field conditions. A comparison of the mortality percentage of this chemical demonstrated significant differences and the highest percentage (100%) was observed in *A. annua* extract, *B. bassiana* secondary metabolites (Zibaee and Bandani 2009). Claudio et al. (2016) stated that *B. bassiana* infects an insect comprising numerous complex developmental transitions that require intricate genetic regulation. Specific roles of several regulatory genes in *B. bassiana* have been established. While the roles of others have yet to be cleared.

4.4.3 *Metarhizium* (Metschnikoff) Sorokin

Metarhizium species is also known as green muscardine fungi. They have been recognized for their biological control potential against arthropods. In around 1879, fungi from this genus were being evaluated for control of wheat chafer beetles (*Anisoplia austriaca)*, and sugar beet curculio (*Cleonus punctiventris)*, in Ukraine. *Metarhizium* species are known to attack a wide range of arthropods: about > 200 species in over 50 families. These include many species of agricultural, medical and veterinary importance. Additionally, *Metarhizium* species have been developed in other countries for use against cockchafers, spittlebugs, grubs, borers, and mosquitoes. Genus *Metarhizium* is composed of three species divided into ten clades (or varieties). The most common form is the 28 genetically highly diverse *Metarhizium anisopliae* var. *anisopliae* (Metsch) Sorokin. *Metarhizium anisopliae*, *M. brunneum*, and *M. robertsii* are general types and they infect insect pests from several orders, while *M. acridum* is specific to order Orthoptera (Wang et al. 2016). The *Metarhizium* sp. has septate mycelium and branch conidiophores in the form of a candelabrum, with two or three branches in each septum densely intertwined. Their conidiogenous cells are cylindrical and slightly conical at the apex. Both strains presented conidia of green color and cylindrical shape forming chains aggregated into cylindrical columns. Each cylindrical conidia presented an average length of 4.5–5.1 μm and a width of 2.1–2.6 μm (Gustavo et al. 2021).

Among several species of *Metarhizium*, *M. anisopliae*, is one of the most famous soil-inhabitant entomopathogens, which has a virulence potential on plant and animal insect pests. The soil forms its normal habitat. It does not grow saprophytically in soil but exists as dormant conidia infecting susceptible hosts on contact. *M. anisopliae* causes green and black muscardine disease. The soil-inhabiting larvae of scarab beetles are typical hosts and co-evolution has led to some isolates being specific to one or two genera of scarab. Thus, the most virulent strains are usually those that cause natural epizootics in that host. The host specificity is dose dependent: a high dose will infect a very wide range of hosts. Besides virulence and host specificity, temperature is an important factor in selecting an isolate for development as a mycoinsecticide. Most isolates of *Metarhizium* grow well at 15–30°C. Although, some isolates develop at temperatures as low as 5–10°C and others grow even at high temperatures of 35–40°C. Therefore, the action of the destruxins reducing host mobility would also impair this compartmental defence mechanism. Indeed, *Metarhizium* isolates producing higher quantities of destruxins are found more virulent. Recently, both the virulence and insecticidal activities of the crude extracts of *M. anisopliae* were evaluated against the 2nd-instar larvae of *Spodoptera littoralis* (Boisduval) (Lepidoptera, Noctuidae), a very harmful polyphagous insect pest. Topical application of the crude extracts alone did not cause the mortality of the insect. A combined treatment with fungal suspensions of the *M. anisopliae* and their extracts caused higher mortality rates than the single *M. anisopliae* and extract (Resquín-Romero et al. 2016). Therefore, the potential of certain entomopathogenic fungal isolates for use in an integrated *S. littoralis* management strategy targeting larvae, as well as the potential of the combined use of entomopathogenic fungi and their extracts.

Murad et al. (2006) stated that *M. anisopliae* can control insect pests and are widely applied in biological control. More than 30 products based on *M. anisopliaes* are used to control both foliar and soil insect pests in India, particularly crops like areca nut, coconut, coffee, corn, potato, pigeon pea, sugarcane, and soybean. Formulation of *M. anisopliae* mixed with farmyard manure and applied at root zone @ 4×10^9 conidia/ha reduced 92% of the grub population as compared with chlorpyrifos (85–100%) and carbofuran (76%). Pandey (2013) reported that soil application of *M. anisopliae* mixed with compost (5×10^{11} spores/g) to potatoes in the state of Uttarakhand resulted in significantly lower tuber damage (15.6%) by the cutworm insect as compared to controls (31%).

4.4.4 *Lecanicillium* Gams and Zare (previously *Verticillium lecanii*)

Lecanicillium muscarium Zare and Gams and *L. longisporum* Zare and Gams (both formerly classified as *Verticillium lecanii* (Zimmerman) have attracted some attention as biocontrol agents of Homoptera and spider mites. Conidiophores of *Lacanicillium* spp. are little differentiated from vegetative hyphae. Conidiogenous cells (phialides) are in whorls (verticils) of 2–6, paired, or solitary on hyphae or apically on short side branches. Conidia are hyaline, aseptate, borne in slime droplets or dry chains. The Teleomorph (where known) of *Lecanicillium* is *Torrubiella*. Conidia of *Lecanicillium* are borne in slime balls and rarely found in dry chains, unlike the other Hyprocreales fungi, which produce hydrophobic conidia.

A wheat bran sugarbeet pulp mixture (9:1 w/w) is being promoted as an excellent solid substrate for the fungus. Feng et al. (2000) identified rice bran to be the best medium, which was better than cooked rice. A bran-to-husk ratio of 1:1 was almost as good. Molasses-yeast broth was found the best liquid medium while rice resulted in the highest conidial production (Derakhshan et al. 2008). Optimal temperatures for growth of the fungus and its sporulation in both systems are 20–25°C (Abd El-Ghany 2015). *L. lecanii* Vl-8 (10^8CFU/ml) resulted in the highest control of cowpea aphids (78% reduction) and increased yields by 32% over control plots in Karnataka. More than 60 products based on *L. lecanii* are manufactured in liquid and dry formulations in India for the management of aphids, scales, and other soft-bodied sucking pests on a range of crops.

4.4.5 *Hirsutella* Pat

The genus *Hirsutella* can be cultured on SDAY (Sabouraud dextrose agar + 1% yeast extract) medium. The colony of fungus appears on the medium as white to light gray. The synnemata is produced by fungus measuring 0.8 ± 0.6 μm in length and 50 ± 10.0 μm in diameter) and simple or possessed short lateral branches. Monophiladic conidiogenous cells arose laterally and terminally from the distal portions of the synnemata and an ellipsoid base that tapered at the conidial apex. Each phialide produced a single, smooth-walled, aseptate, fusiform or ellipsoidal conidium.

Hirsutella thompsonii is used for management of mites. Latge et al. (2004) obtained microcycle conidia from submerged fermentation, but from a strain unique in this regard. Kumar et al. (2005) also reported submerged liquid fermentation. They noted that the low concentrations of polyethylene glycol in a dextrose peptone medium allowed the production of mycelial pellets which are competent to produce conidia. A powder formulation, containing both mycelia and conidia, was developed in India during the 2000s, followed by two liquid formulations. However, efforts to commercialize the products were discontinued (Kumar and Singh 2002; Kumar 2010). Undiluted coconut water was equivalent to two standard laboratory liquid media (potato dextrose broth, glucose yeast extract broth) in a surface-mat-on-liquid fermentation, but very inferior to Sabouraud dextrose broth, which produced 5.1×10^5 spores/cm^{-1} fungal mat surface, a yield practical only for artisanal production.

Among Deuteromycetes, infections are known from eriophyids. More than 50 species of *Hirsutella* sp. have been reported, but only some species are pathogenic to eriophyids. *H. athompsonii* Fisher has been tried for commercialization, primarily in the United States in the 1970s and Cuba in the 1980s. Mycohit, a commercial product based on isolate MF(Ag)5 of fungus, was evaluated against the coconut mite (Kumar and Singh 2002). Spraying young coconut palms reduced mite populations by up to 90% and reduced pre-harvest nut damage. Similarly, citrus rust mite (*Phyllocoptruta oleivorus*) was found to be most susceptible in grapefruit with a population of 5000 mites on a single grapefruit. The population of mites suddenly decreased due to *H. thompsonii*. The spores of fungus possess a mucous coat which facilitates adhesion to the host cuticle. The spores germinate and penetrate all body parts in eriophids, but on spider mites, penetration is generally through the legs. *H. thompsonii* and *H. necatrix* are two species that are found effective against spider mites, but variation in pathogenicity was reported between these two species of *Hirsutella*. Hirsutellin A (HtA) protein inhibits cell growth of the host and causes cells to become hypotrophied and disrupt internal organelles and membranes. It caused 100% mortality in citrus rust mite at a concentration of 100 mg/ml.

4.4.6 *Isaria* Persoon in Roemer (*Paecilomyces* Samson)

Isaria spp. cause yellow and red muscardine disease in insects. Genus *Isaria* belongs to phylum Ascomycota, class Sordariomycetes, order Hypocreales and family Cordycipitaceae. *Paecilomyces* was first reported in 1907 as a genus that was closely related to *Penicillium* and contained only one species, *P. variotii* Bainier. Later another species, *P. fumosoroseus* was isolated from most regions of the world and reported to infect several insects belonging to many different orders (Gokce and Kubilay 2005). *P. fumosoroseus* was reported to be one of the most common fungi that attack nymph and adult *Bemisia* in fields and glasshouses and *Trialeurodes vaporariorum* in glasshouses. Several isolates of *Paecilomyces* were evaluated against whiteflies and these isolates were tested successfully in the field and glasshouses. *P. lilacinus* (Thom) Samson is a typical soil-borne fungus and it is common in the tropics and subtropics. *P. eriophytis* attacks on the blackcurrant mite *Cecidophyopsis ribis*, and *Cephalosporium* sp. on *Phytoptus avellanae* (Eriophyoidea) are also found to be effective.

The latter's isolate *C. ribis* also showed a promising result against *Panonychus ulmi*, *Aceria hippocastani*, *Cecidophyes galii*, *A. vaccinii* and *Polyphagotarsonemus latus*.

The entomopathogenic fungi, *Isaria farinosa* and *I. fumosorosea* were known earlier as *Paecilomyces farinosus* and *P. fumosoroseus*, for more than 30 years. Conidiophores of *I. fumosorosea* are usually well developed, synnematous in many species, septate, bearing whorls of divergent branches and conidiogenous cells (phialides), hyaline to pigmented (not black, brown, or olive). Conidiogenous cells are flask-to-oval shaped or subglobose with a distinct neck, borne singly or in whorls. Conidia are 1-celled, ≤3.5–4.0 μm, hyaline to light colored (shades of yellow through red, tan, lavender, or grass-green), in dry divergent chains. Coniophore is smooth and colorless; conidial chains are often long and conidial heads diffuse. *I. farinosa* currently is of minor importance in research and as a biocontrol agent. However, *I. fumosorosea* is considered a species complex. Various strains *I. fumosorosea* are successfully used to control various pest insects, particularly whiteflies. Three *I. fumosorosea* products are sold for control of whiteflies, aphids, thrips, mites, leafminers, plant bugs and several soil pests in food and ornamental crops.

4.4.7 *Nomuraea* Maublanc

Nomuraea riley causes green muscardine disease in insects. The fungus mycelium is septate, white, with flocculent overgrowth. Fungal growth sparses in culture to dense on the insect's body, generally completely covering the host and usually becoming green, or purple gray to purple as sporulation proceeds. Conidiophores are single or rarely synnematous (if synnematous, with a sterile base and distal fertile zone), erect, bearing whorls of short and blocky branches (metulae) with clusters of short phialides on metulae. Conidiogenous cells are short, with blunt apices and little if any distinct neck. Conidia are aseptate, smooth, round to ovoid or elongated and slightly curved, in short, divergent chains, pale to dark green, purple gray to purple, or rarely white color in mass. Conidia are ovoid dull gray-green (seafoam green), rarely white 3.5-4.5 x 1-2 μm.

Natural epizootics of *N. rileyi* have been observed on lepidopteran pests such as *Achaea janata*, *Agrotis ipsilon*, *H. armigera* and *S. litura* in India. Despite the potential for management of several important noctuid pests, to date there are no commercial products based on *N. rileyi* registered in India (Lingappa et al. 2002).

4.4.8 *Cladosporium* Link

Cladosporium oxysporum expands moderately, often floccose at the center of the fungus which consists of woolly tufts, and it can grow up to 650 μm long and 4–5 μm wide. The colony is colored olive to olive-green on top with a velvety surface, and greenish-black at the bottom. The conidiophores are either straight or slightly bent, and the conidia range from oval to lemon-shaped. The fungus produces conidia in unbranched or branched chains arising from cylindrical base cells. After the first spores have formed on the conidiophore, they bud apically to form secondary spores. They have pores connected in very fragile chains that can fall apart at the slightest movement of air.

Cladosporium spp. is also a promising biological control agent against homopterous insects. *Cladosporium herbarum* is associated with controlling cashew whitefly (*Aleurodicus coccis*). It is also found effective as a biocontrol agent against three whitefly species *viz.*, *Bemisia* sp., *Aleurothrixus* sp. and *Dialeurodes* sp. on various plant hosts in Venezuela (Rojas et al. 1998). Moreover, *C. aphids were* recorded as a biocontrol agent against *Alerurochiton aceris* in Finland and *Chionaspis salicis* (L.) (Coccoidea: Diaspidididae). Under field conditions, *C. cladosporioides* caused 20–57% natural mortality of *Hemiberlesia pitysophila* as well as 39% mortality in the laboratory in China (Pan et al. 1989). Whereas, Thumar and Kapadia (1994) mentioned that the species of *Cladosporium* was able to infect *Aleurolobus barodensis* nymphs around the year in India. Han et al. (1997) noticed that the *Cladosporium* spp. naturally occurred and caused epizootic infection to the *Aleurocanthus spiniferus* in China.

There are a number of reports that emphasize the ability of *Cladosporium* spp. to reduce the population of aphids (Vallejo et al. 1996; Han et al. 1997). Abdel-Baky et al. (1998) reported three species of *Cladosporium* (*C. uredinicola*, *C. cladosporioides* and *C. chlorocephalum*) which infect *Bemisia* spp., *Aphis gossypii* and *Empoasca* sp. in Egypt. They described that the fungus *Cladosporium* occurred naturally at a high percentage of incidence (10.0–28.0%) under field conditions. They also reported that *C. uredinicola* was the most dominant one and more virulent in the laboratory tests.

4.4.9 *Pandora* Humber

Pandora is a genus within the order Entomophthorales of the Zygomycota (Elya et al. 2021). *Pandora neoaphidis* acts as an obligate pathogen in various species of aphids. It is a widespread species that is often reported to be the most common fungal insect pathogen in the local aphid community. *P. neoaphidis* (Remaudière et Hennebert) Humber has been often considered the most important natural regulation agent of aphid colonies and considered a potential biological control agent. *P. neoaphidis* is the most common aphid pathogen in agroecosystems with the ability to produce epizootics in host populations. The fungus infects green peach aphid, *Myzus persicae* (Homoptera: Aphididae) which predates spinach (*Spinacea oleracea* L.) in Arkansas, America. It is very difficult for mass production of infectious spores *in vitro* conditions and the viable formulation and storage into an easily applicable commercial product has halted their direct application as a biocontrol (Vega et al. 2012).

There is limited evidence that the ladybird *Harmonia axyridis*, which is invasive in America and Europe, has an advantage over native ladybird species because it feeds more on *Pandora*-infested aphid cadavers (Roy et al. 2008). *Pandora formicae* (Humber & Bałazy) Humber is a rare example of the entomophthoralean fungus that has adapted to exclusively infect social insects, such as wood ants (*Formica polyctena*). The proportion of dead ant bodies with resting spores increased from late summer throughout autumn, which suggests that these fungal spores are the main overwintering fungal structures (Małagocka et al. 2017). *Pandora* sp. nov. inedit. (ARSEF13372) is a recently isolated fungus having a high potential to control psyllid pests.

4.4.10 *Entomophthora* (Cohn) Fresen

Entomophthora muscae causes fatal disease in flies. Genus *Entomophthora* belongs to phylum Zygomycota, order Entomophthorales and family Entomophthoraceae. Vegetative cells of *Entomophthora* are short, rod-like with or without cell walls. Conidiophores are simple. Conidiogenous cells are club shaped. Primary conidia with a prominent apical point and a broad, flat basal papills with 2–12 (up to ca. 40) nuclei, are forcibly discharged by a cannon-like mechanism. The discharged conidia adhere to a substrate in a droplet of discharged cytoplasm. Rhizoids (if present) ca. the diameter of hyphae are numerous, isolated, or fasciculate. Resting spore buds are found laterally from parental hyphae. An unfixed nuclei stain proves to have granular contents in aceto-orcein.

4.5 APPLICATION OF ENTOMOPATHOGENIC FUNGI WITH ANOTHER BIOAGENTS

Naturally, in mixed infections, one entomopathogen may enhance, improve, or suppress the entomopathogenic activity of the other. Several studies have delved into the insecticidal possibilities of the combined presence of entomopathogens. The results mostly point to the synergism between the agents. And while most studies have explored the coupling of fungal entomopathogens with other insecticidal agents such as bacteria (Wraight and Ramos 2005; Mantzoukas et al. 2013), viruses (Pauli et al. 2018), nematodes (Ansari et al. 2006), diatomaceous earth dust (Kavallieratos et al. 2006), etc., the simultaneous presence of fungal entomopathogens has not been sufficiently investigated (Dakhel et al. 2010). Although, fungal entomopathogens are supposed to have the same mode of action, they differ in their virulence depending on abiotic and biotic factors including insect species, fungal strain, secondary metabolites, etc. *I. fumososrosea*, for example, was more virulent against *Sitophilus oryzae* and *B. bassiana* to *S. granarius*, but the effect of their combined action increased host mortality in both hosts (Mantzoukas et al. 2019).

Compatibility between *B. bassiana* at different concentrations with plant leaf extracts from Neem, Chinaberry, Lantana leaves and Mexican sunflower against the *Spodoptera litura* (Lepidoptera: Noctuidae) has been investigated by Afandhi et al. (2020). The combinatorial effect of the fungus with either plant extract was measured in terms of *B. bassiana* colony growth, conidia viability, density, and mortality of insect larvae. Results showed that the fungal colony growth decreased in all combinations while in terms of insect mortality, the most suitable combination was of the fungus with 0.25% Chinaberry whereby *B. bassiana* exhibited the best conidial viability and density as well as enhanced insect mortality (44%).

Mantzoukas et al. (2019) investigated the potential interaction between two potent entomopathogenic fungi, *B. bassiana* and *I. fumosorosea* against two important coleopteran stored product pests, *Sitophilus oryzae* and *S. granarius*. The combined application of the two fungi was additive against *S. granarius* in all nine combinations of the three different conidial suspensions, while it was effective in seven combinations and competitive in two combinations against *S. oryzae*. Moreover, another study showed that the synergistic action of *B. bassiana* and *Trichoderma asperellum* on corn (*Zea mays* L.) against the Asian corn borer *Ostrinia furnacalis* Guenne (Lepidoptera: Crambidae) proved to suppress the immune response of *O. furnacalis* and increased the lethal activity of *Tribolium asperellum*. These studies indicate the advantage of exploring potential synergisms against pests rather than using a single pathogen (Batool et al. 2020). The combined use of *Chromolaena odorata* leaf powder and *B. bassiana* showed a dose-dependent insecticidal effect against *S. oryzae* L. and *T. castaneum* (Kalita et al. 2018). Depieri et al. (2005) tested the compatibility of *B. bassiana* with emulsible neem oil (*Azadirachta indica* Juss) and with neem seeds and leaves aqueous extracts and found that the latter did not impact the viability of the fungal spores although conidia growth and production were reduced. The emulsible neem oil, by contrast, especially at higher concentrations, affected all conidia parameters to a high degree, thus leading to the conclusion of non-compatibility with the fungus.

4.6 MYCOINSECTICIDES

Microbial pesticides occupy around 1.3% of the world's total pesticide market of which, 90% of them are used as insecticides. Fungal biopesticides (mycoinsecticides and mycoacaricides) are predominantly based on the anamorphic phases of the hypocrealean as species of *Beauveria*, *Akantomyces*, *Metarhizium*, *Isaria* (formerly in the genus *Paecilomyces*), *Hirsutella*, *Lecanicillium* spp. (formerly *Verticillium*) and *Nomuraea*, and *Purpureocillium*. Several commercial formulations of entomopathogenic fungi have been developed for crop pest management. Among the 171 products of entomopathogenic fungi developed, products based on *B. bassiana* represent 33.9% of total products, *M. anisopliae* products were 33.9%, and *I. fumosorosea* and *B. brongniartii* products represented 5.8 and 4.1% respectively.

The use of microbial control agents particularly entomopathogenic fungi has been investigated for the control of a wide range of orchard and field crop pests (Nchu et al. 2010; Abaajeh 2014). Boverin, a *Beauveria bassiana*-based mycoinsecticide for the control of Colorado potato beetle and codling moth in the former USSR, was developed in 1965. The first formal and published proposal for microbial control came with John LeConte's suggested use of microsporidia to control grape phylloxera, *Daktulosphaira vitifoliae* (Fitch). The study presented visionary ideas for an effective and economic alternative method of pest control and paved the way for future scientists to study entomopathogenic fungi and many have been formulated into products that are commercially available for use in controlling insects of economic importance. Current commercial entomopathogenic products in the markets being used to control several insects worldwide include, "Cryptogram™" and Bb plus®, Metarhizium 50®, Biogreen® and Green Guard®, BIO 1020® and Green Muscle®. These mycoinsecticides are used to control several insect pests in glasshouse and field crops. The use of mycoinsecticides in pest management particularly in vegetable production has been reported in many countries around the world. (Abaajeh 2014). Although several species of entomopathogenic fungi are effective against many lepidopteran pests in laboratory studies, their field efficacy has not been proven as their success largely depends on weather conditions (temperature and humidity).

Mycoinsecticides form part of biopesticides produced and marketed for pest management. Although biopesticides represent only 3% of the global crop protection business, their growth is at the rate of 10% per year. This growth is an indication that biopesticides will soon play an important role in insect pest management. Of the total global biopesticide market, mycoinsecticide is the second (27%) to *Bacilus thurengiensis* products. Most of these mycoinsecticides are produced in America, Europe and Asia. At least 30 mycoinsecticides were registered in China; among them, *B. bassiana* is the most popular species with up to 14 products for the control of locust, pine moth and diamond back moth. *M. anisopliae* and *P. lilacinus* with eight and seven products respectably registered for application on grubs, corn borer, aphids, and whiteflies. Most of the countries where mycoinsecticdes are produced are in South and North America and Europe. India and South Africa are probably the leading producers of mycoinsecticides in Asia and Africa, respectively. However, mycoinsecticide products are available for use in pest management in many countries of Asia and Africa. Therefore, it is important to note that many countries of Asia and Africa will soon join several countries of Europe and America in the use of mycoinsecticide as a promising alternative to synthetic chemicals. Commercial formulations including aqueous suspensions, suspension concentrates and wettable powders (including talc) are typically applied in the range 10^7–10^8 conidia/ml and 400–750 l/ha, depending on the crops. Various studies conducted under a variety of conditions demonstrated the efficacy of *B. bassiana* in sugarcane internode borer and white grub in sugarcane (Visalakshi et al. 2015). *B. brongniartii*, has emerged as a promising biocontrol agent in sugarcane, due to its long-term persistence in treated fields and activity against another scarab, *Holotrichia serrata*. Application of *B. brongniartii* (1.6×10^{14} spores/ha) that were formulated on sorghum grains, mixed with FYM in furrows, and covered with soil, achieved 46% infection among grubs of *H. serrata* in sugarcane fields in Tamil Nadu (Easwaramoorthy et al. 2004).

4.7 CONCLUSIONS

Chemical pesticides are becoming obsolete, hence microbe-based pesticides have a significant role in reducing crop loss due to insect infestation. More focus has been placed on the use of natural enemies like entomopathogens for pest control in recent years. Entomopathogenic fungi expand the possibilities for insect control and are used as biopesticides and are a specialized group of microbial organisms with traits and modes of action. Soil is important for protection against environmental extremes, for water availability, and, also as a source of insect hosts. Only a few species of *Beauveria*, *Metarhizium*, *Lecanicillium*, *Hirsutella*, *Conidiobolus* and *Isaria* are commercially available, but the recent surge in studies exploring the endophytic attributes of fungi paves the way to further research into their complex ecology. The ecological behavior of the fungi must be known, which will help to develop to more efficient and more diversified applications. A better grasp of their compatibility with other microbial agents, natural arthropod enemies, or plant and/or chemical compounds could generate a richer and stronger pool of available tools for pest control. It is also required to focus on searching for the most appropriate and potential strains or isolates per plant host. Differential rates of endophytism and degrees of fungal colonization make the case for better research into the compatibility of fungal agents with their hosts and the persistence inside the plant. The inoculation method has been shown to influence the success of colonization. Inundative entomopathogenic fungi released as biological control agents against insects and mites are based on the action of the agent and not on successive generations of the fungi. Attention should be predominantly focused on the technical logistics of mass production and formulation of these fungi. Mass production is concerned with creating products that are relatively cost efficient, consistently effective in the field and with a relatively long shelf life.

REFERENCES

Abaajeh AR (2014) Evaluation of entomopathogenic fungi (Ascomycota) for the control of Cydiapomonella (Lepidoptera: Tortricidae). M.Sc. Dissertation, Cape Peninsula University of Technology, South Africa.

Abdel-Baky NF, Arafat SN, Abdel-Salam AH (1998) Three *Cladosporium* spp. as promising biological control candidates for controlling whiteflies (*Bemisia* spp.) in Egypt. *Pak J Biol Sci* 1: 188–195.

Abdelghany TM (2015) Entomopathogenic fungi and their role in biological control. Foster City, CA: OMICS Group eBooks. 731. DOI: 10.4172/978-1-63278-065-2-66

Afandhi A, Pertiwi EP, Purba DP, Widjayanti T, Leksono AS (2020) The diversity of entomopathogenic fungi collected from leaves and rhizospheres of rice implementing integrated pest management. *Biodiversitas* 21(6): 2690–2695.

Ansari MA, Shah FA, Tirry L, Moens M (2006) Field trials against *Hoplia philanthus* (Coleoptera: Scarabaeidae) with a combination of an entomopathogenic nematode and the fungus *Metarhizium anisopliae* CLO 53. *Biol Control* 39: 453–459.

Bainier G (1907) Mycothe 'que de l'e'cole de Pharmacie. XI *Paecilomyces*, genre nouveau de Muce'dine'es. *Bull Soc Mycol Fr* 23: 26–27.

Batool R, Umer M, Wang Y, He K, Zhang T, Bai S, Zhi Y, Chen J, Wang Z (2020) Synergistic effect of *Beauveria bassiana* and *Trichoderma asperellum* to induce maize (*Zea mays* L.) defense against the asian corn borer, *Ostrinia furnacalis* (Lepidoptera, Crambidae) and larval immune response. *Int J Mol Sci* 21: 8215.

Bisset J, Widden P (1988) A new species of *Beauveria* isolated from Scottish moorland soil. *Canadian J Bot* 66: 361–362.

Butt TM, Wang C, Shah FA, Hall R (2006) Degeneration of entomogenous fungi. In: Eilenberg J, Hokka-nen HMT (eds) *An ecological and societal approach to biological control*. Springer, Netherlands, 213–226.

Chen WH, Han YF, Liang JD, Liang ZQ (2020) *Akanthomyces lepidopterorum*, a new lecanicillium-like species. *Phytotaxa* 459(2): 117–123.

Claudio AV, Harm W, Bas JZ, Constantianus JMK, Jan AL, et al. (2016) Genes involved in virulence of the entomopathogenic fungus *Beauveria bassiana*. *J Invertebrate Pathol* 133: 41–49.

Dakhel WH, Latchininsky AV, Jaronski ST (2010) Efficacy of two entomopathogenic fungi, *Metarhizium brunneum*, strain F52 alone and combined with *Paranosema locustae* against the migratory grasshopper, *Melanoplus sanguinipes*, under laboratory and greenhouse conditions. *Insects* 10: 94.

De Hoog GS, Rao V. (1975) Some new hyphomycetes. *Persoonia* 8: 207–212.

Depieri RA, Martinez SS, Menezes AO Jr (2005) Compatibility of the fungus *Beauveria bassiana* (Bals.) Vuill. (Deuteromycetes) with extracts of neem seeds and leaves and the emulsible oil. *Neotrop Entomol* 34: 601–606.

Derakhshan A, Rabindra RJ, Ramanujam B, Rahimi M (2008) Evaluation of different media and methods of cultivation on the production and viability of entomopathogenic fungi, *Verticillium lecanii* (Zimm.) Viegas. *Pak J Biol Sci* 11: 1506–1509.

Easwaramoorthy S, Srikanth J, Santhalakshmi G, Geetha N (2004) Laboratory and field studies on *Beauveria brongniartii* (Sacc.) Petch, against *Holotrichia serrata* F. (Coleoptera: Scarabaeidae) in sugarcane. *Proc Annl Convention Sugar Technol Asso of India* 66: 3–19.

Elya C, De Fine L, Henrik H (2021) The genus *Entomophthora*: bringing the insect destroyers into the twenty-first century. *IMA Fungus* 12(34).

Feng KC, Liu L, Tzeng MY (2000) *Verticillium lecanii* spore production in solid-state and liquid-state fermentations. *Bioprocess Eng* 23: 25–29.

Glare TR, Jackson TA, Ernesto-Cisternas A (1993) *Beauveria vermiconia* (Deuteromycete) is an entomogenous fungus. *Mycol Res* 97: 336–338.

Glare TR, Reay SD, Nelson TL, Moore R (2008) *Beauveria* caledonicais a naturally occurring pathogen of forest beetles. *Mycol Res* 112: 352–360.

Gokce A, Kubilay MER (2005) Pathogenicity of *Paecilomyces* spp. to the glasshouse whitefly, *Trialeurodes vaporariorum*, with some observations on the fungal infection process. *Turk J Agric For* 29: 331–339.

Gustavo AB, María L C Fernando LK Laura LV Pedro DZ (2021) Morphological and molecular identification of entomopathogenic fungi from agricultural and forestry crops. *Floresta Ambient* 28(2): e20180086. doi: 10.1590/2179-8087-FLORAM-2018-0086

Han B, Zhang H, Cui L (1997) Epizootic infection and spatial pattern within epizootic peak period of *Cladosporium* sp. to the population of *Aleurocanthus spiniferus*. *Entomol J East China* 40: 66–70.

Hatting JL, Wraight SP, Miller RM (2004) Efficacy of *Beauveria bassiana* (Hyphomycetes) for control of Russian wheat aphid (Homoptera: Aphididae) on resistant wheat under field conditions. *Biocontrol Sci Technol* 14: 459–473.

Hibbett DS, Binder M, Bischoff JF, Blackwell M, Cannon PF (2007) A higher-level phylogenetic classification of the fungi. *Mycol Res* 111: 509–547.

Huang B, Li C-R, Li Z-G, Fan M-Z, Li Z-Z (2002) Molecular identification of the teleomorph of *Beauveria bassiana*. *Mycotaxon* 81: 229–236.

Inglis GD, Enkerli J, Goettel MS (2012) Laboratory techniques used for entomopathogenic fungi: hypocreales. In: Lacey LL (ed) *Manual of techniques in invertebrate pathology*. Academic Press, Amsterdam, 189–253.

Kalita S, Hazarika LK, Das P, Das K (2018) Efficacy of *Chromolaena odorata* leaf powder and *Beauveria bassiana* against stored grain pests. *Indian J Entomol* 80: 1005.

Kavallieratos NG, Athanassiou CG, Michalaki MP, Batta YA, Rigatos HA, Pashalidou FG, Balotis GN, Tomanovic Z, Vayias BJ (2006) Effect of the combined use of *Metarhizium anisopliae* (Metschinkoff) Sorokin and diatomaceous earth for the control of three stored-product beetle species. *Crop Prot* 25: 1087–1094.

Kepler RM, Luangsa-ard JJ, Hywel-Jones NL, Quandt CA et al (2017) A phylogenetically based nomenclature for Cordycipitaceae (Hypocreales). *IMA Fungus* 8(2): 335–353.

Kim JJ, Goettel MS, Gillespie DR (2008) Evaluation of *Lecanicillium longisporum*, Vertalec for simultaneous suppression of cotton aphid, *Aphis gossypii*, and cucumber powdery mildew, *Sphaerotheca fuliginea*, on potted cucumbers. *Biol Control* 45: 404–409.

Kumar D, Singh KP, Jaiswal RK (2005) Screening of different media and substrates for cultural variability and mass culture of *Arthrobotrys dactyloides* Drechsler. *Mycobiol* 33(4): 215–222.

Kumar PS (2010) *Hirsutella thompsonii* as a mycoacaricide for *Aceria guerreronis* on coconut in India: research, development, and other aspects. In: Sabelis MW, Bruin J (eds) *Trends in Acarology: Proc 12th Int Congress*. doi: 10.1007/978-90-481-9837-5_73. 441-442

Kumar PS, Singh SP (2002) Development of 'Mycohit', the first mycoacaricide based exclusively on *Hirsutella thompsonii*, for suppressing the coconut mite in India. In: Reddy SM, Ram Reddy S, Singarachary MA, Girisham S (eds) *Bioinoculants for Sustainable Agriculture & Forestry*. Scientific Publishers, Jodhpur, India, 209–213.

Latge JP, Soper RS, Madore CD (2004) Media suitable for industrial production of *Entomophthora virulenta* zygospores. *Biotechnol Bioeng* 19: 1269–1284.

LeConte JL (1874) Hints for the promotion of economic entomology. In: *Proceedings of the American Association for the Advancement of Science* 22: 10–22

Li Z, Li C, Huang B, Meizhen F (2001) Discovery and demonstration of the teleomorph of *Beauveria bassiana* (Bals.) Vuill., an important entomogenous fungus. *Chinese Sci Bull* 46: 751–753.

Lingappa S, Patil RK, Rachappa V, Hegde R, Navi SS (2002) Large-scale demonstration of efficacy of *Nomuraearileyi* (F) Samson on soybean. In: Tandon PL (ed) *Biological control of lepidopteran pests. Proceeding of Symptoms. Biological Control of Lepidopteran Pests*, Bangalore, India, 251–253.

Maina UM, Galadima IB, Gambo FM, Zakaria D (2018) A review on the use of entomopathogenic fungi in the management of insect pests of field crops. *J Entomol and Zool Studies* 6(1): 27–32.

Małagocka J, Bruun JA; Eilenberg J (2017) *Pandora formicae*, a specialist ant pathogenic fungus: new insights into biology and taxonomy. *J Invertebr Pathol* 143: 108–114.

Mantzoukas S, Milonas P, Kontodimas D, Angelopoulos K (2013) Interaction between the entomopathogenic bacterium *Bacillus thuringiensis* subsp. *kurstaki* and two entomopathogenic fungi in biocontrol of *Sesamia nonagrioides* (Lefebvre) (Lepidoptera: Noctuidae). *Ann Microbiol* 63: 1083–1091.

Mantzoukas S, Zikou A, Triantafillou V, Lagogiannis I, Eliopoulos PA (2019) Interactions between *Beauveria bassiana* and *Isaria fumosorosea* and Their Hosts *Sitophilus granarius* (L.) and *Sitophilus oryzae* (L.) (Coleoptera: Curculionidae). *Insects* 10: 362.

Mongkolsamrit S, Noisripoom W, Thanakitpipattana D, Wuthikun T, Spatafora JW, Luangsaard JJ (2018) Disentangling cryptic species with isaria-like morphs in Cordycipitaceae. *Mycologia* 110(1): 230–257.

Murad AM, Laumann RA, Lima TA, Sarmento RBC, Noronha EF, Rocha TL, Valadares-Inglis MC, Franco OL (2006) Screening of entomopathogenic *Metarhizium anisopliae* isolates and proteomic analysis of secretion synthesized in response to cowpea weevil (*Callosobruchus maculatus*) exoskeleton. *Comp Biochem Physiol Comp Pharmacol* 142: 365–370.

Nchu F, Maniania NK, Hassanali A, Eloff JN (2010) Performance of a *Metarhiziumanisopliae*-treated semiochemical-baited trap in reducing *Amblyomma variegatum* populations in the field. *Veterinary Parasitol* 169: 367–372.

Pan WY, Chen SL, Lian JH, Qiu HZ, Lan G (1989) A preliminary report on control of *Hemiberlesia pitysophila* using *Cladosporium cladosporioides*. *For Pest Dis* 3: 22–23.

Pandey AK (2013) Field evaluation of *Beauveria bassiana* and *Metarhizium anisopliae* against the Cutworm, *Agrotisipsilon* (Hufnagel) damaging potato in Uttarakhand hills. *J Biol Control* 27(4): 293–297.

Pauli G, Mascarin GM, Eilenberg J, Delalibera I Jr (2018) Within-host competition between two entomopathogenic fungi and a Granulovirus in *Diatraea saccharalis* (Lepidoptera: Crambidae). *Insects* 9: 64.

Quesada-Moraga E, Navas-Cortés JA, Maranhao EAA, Ortiz-Urquiza A, Santiago-Álvarez C (2007) Factors affecting the occurrence and distribution of entomopathogenic fungi in natural and cultivated soils. *Mycol Res* 111: 947–966.

Quesada-Moraga E, Vey A (2004) Bassiacridin, a protein toxic for locusts secreted by the entomopathogenic fungus *Beauveria bassiana*. *Mycol Res* 108: 441–452.

Rajula J, Afroja Rahman, Patcharin K (2020) Entomopathogenic fungi in Southeast Asia and Africa and their possible adoption in biological control. *Biol Control* 151: 104399. https://doi.org/10.1016/j.biocontrol.2020.104399

Ramanujam B, Rangeshwaran R, Sivakmar G, Mohan M, Yandigeri MS (2014) Management of insect pests by microorganisms. *Proc Indian Nat Sci Acad* 80(2): 455–471.

Rehner SA, De Muro MA, Bischoff JF, Rehner AFSA, De Muro MA, Bischoff JF (2006a) Description and phylogenetic placement of *Beauveria malawiensis* sp. nov. (Clavicipitaceae, Hypocreales). *Mycotaxon* 98: 137–145.

Rehner SA, Posada F, Buckley EP, Infante F, Castillo A, Vega FE (2006b) Phylogenetic origins of African and neotropical *Beauveria bassiana*s. l. pathogens of the coffee berry borer, *Hypothenemus* hampei. *J Invert Pathol* 93: 11–21.

Resquín-Romero G, Garrido-Jurado I, Quesada-Moraga E (2016) Combined use of entomopathogenic fungi and their extracts for the control of *Spodoptera littoralis* (Boisduval) (Lepidoptera: Noctuidae). *Biol Control* 92: 101–110.

Rojas T, Pons N, Arnal E (1998) *Cladosporium herbarum* on whiteflies (Homoptera: Aleyrodidae) in Venezuela. *Boletin de Entomologia Venezuela Serie Monografias* 13: 57– 65.

Roy HE, Baverstock J, Ware RL, Clark SJ, Majerus MEN, Baverstock KE, Pell JK (2008) Intraguild predation of the aphid pathogenic fungus *Pandora neoaphidis* by the invasive coccinellid *Harmonia axyridis*. *Ecol Entomol* 33(2): 175–182.

Scholte EJ, Knols BGJ, Samson RA, Takken W (2004) Entomopathogenic fungi for mosquito control: A review. *J Insect Sci* 4: 19–24.

Thumar RK, Kapadia MN (1994) Ovicidal control of *Aleurolobus barodensis* (Maskell) and its suppression by the entomopthogenic fungus. *Indian Sugar* 44: 501–502.

Vallejo LF, Soto SU, Bernal (1996) Identification of pathogenic fungi for *Lutzomyia* sp. (Diptera: Psychodidae), vectors of *Lei hmaniasis*. *Revista Colomb Entomol* 22: 13–17.

Vega FE, Goettel MS, Blackwell M, Chandler D, Jackson MA, Keller S, Koike M, Maniania NK, Monzon A, Ownley BH, Pell JK, Rangel DEN, Roy HE (2009) Fungal entomopathogens: New insights on their ecology. *Fungal Ecol* 4: 149–159.

Vega FE, Meyling NV, Luangsa-ard JJ, Blackwell M (2012) Chapter 6—fungal entomopathogens. In: Vega FE, Kaya HK (eds) *Insect Pathology*, 2nd edn. Academic Press, San Diego, 171–220.

Visalakshi M, Bhavani B, Govinda Rao S (2015) Field evaluation of entomopathogenic fungi against white grub, *Holotrichiacons anguinea* Blanch in sugarcane. *J Biol Control* 29(2): 103–106.

Wang JB, St Leger RJ, Wang C (2016) Advances in genomics of entomopathogenic fungi. In: Lovett B, St. Leger RJ (eds) *Advances in genetics*. Academic Press Inc, 67–105.

Wijayawardene NN, Hyde KD, Tibpromma S, Wanasinghe DN et al. (2017) Towards incorporating asexual fungi in a natural classification: checklist and notes 2012-2016. *Mycosphere* 8: 1457–1555.

Wraight SP, Ramos ME (2005) Synergistic interaction between *Beauveria bassiana* and *Bacillus thuringiensis* tenebrionis-based biopesticides applied against field populations of Colorado potato beetle larvae. *J Invertebr Pathol* 90: 139–150.

Zibaee A, Bandani A (2009) Effect of five different type pesticides on the sunn pest, *Eurygaster integriceps*. *Munis Entomol Zool* 4: 542–550.

Zimmermann G (2007) Review on safety of the entomopathogenic fungi *Beauveria bassiana* and *Beauveria brongniartii*. *Biocontrol Sci Technol* 17: 553–596.

5 Approaches to Managing Insect Vectors in Agriculturally Important Crops

Abhijit Ghosal, Manish Kumar, Sitesh Chatterjee, Amrita Das and Amalendu Ghosh

5.1 INTRODUCTION

Viruses are intracellular obligate parasites and constitute approximately over 50% of the pathogens responsible for plant diseases in cultivated and non-cultivated crops worldwide (Varma and Malathi 2003; Islam et al. 2017). Viral diseases in plants threaten the natural ecosystem and severely affect the cultivation of production of several economic crops (Cooper et al. 2006; Malathi et al. 2017; Kumar et al. 2020). In recent years, the incidences of viral diseases and their damage are reportedly increasing. The frequent appearance of novel viral infections among plants is becoming more widespread because of national and international trade, climate changes, the availability of insect vectors, and the capacity of the virus genome to evolve rapidly (Garcia-Arenal et al. 2001). Plant-infecting viruses have high emergence potential because of their high mutation rate and frequency of undergoing genetic recombination. The exchange of genetic components occurs among the inter- and intra-virus species resulting in high genetic diversity among the viral population. Mixed infection, synergism, and vector-mediated dispersal are some of the major factors that promote virus infection in cultivated and non-cultivated host plants (Chakraborty and Kumar 2021).

Plant virus transmission is predominantly carried out by insect vectors. Insects such as aphids, whiteflies, leafhoppers, and plant hoppers are the known vectors of commonly occurring plant viruses. Interestingly, virus species that belong to the same group depend on similar types of vectors with similar transmission relationships. For example, all the member species of potyviruses (family *Potyviridae*) are transmitted by aphids in a non-persistent and non-circulative manner, whereas ipomovirus and tritimovirus (family *Potyviridae*) are transmitted by whiteflies and eriophyid mites, respectively. The majority of the insect vectors that transmit viruses belong to the order Hemiptera, Thysanoptera, Coleoptera, Orthoptera, and Dermaptera. Among these, hemipteran insects alone are responsible for the transmission of about 90 percent of plant viruses. The best-characterized insect vectors are aphids (transmit potyviruses, cucumoviruses, caulimoviruses, closteroviruses, and luteoviruses, etc.), thrips (transmit tospoviruses, illarvirues, etc.), leafhoppers (transmit phytoreoviruses, and geminiviruses), plant hoppers (transmit nucleorhabdoviruses) and whiteflies (transmit begomoviruses, criniviruses, and ipomoviruses, etc.). Few mite species that belong to the genera *Abacarus* and *Aceria* are well known for transmitting rymoviruses and tritimoviruses (family *Potyviridae*).

Based on the site of virus-vector interaction, the transmission of plant viruses by insect vectors is broadly classified into three categories, viz. non-persistent (stylet borne), semi-persistent (foregut-borne), and persistent (circulates in the insect body). The basis of this classification is the transmission window utilized by the vector to disseminate the virus-infective particles to a new host after feeding on an infected plant, which in turn lasts from a few seconds to minutes,

DOI: 10.1201/9781032631578-5

hours to days, or days to weeks, respectively. In non-persistent or stylet-borne virus transmission, viruses are acquired and retained in the insect stylet and thus lose their infectivity early. Whereas in semi-persistent ways, the infective virus particles are used to bind to the lining of the foregut but do not appear to enter the tissues. Similarly, the persistent way of transmission involves the retention of viruses in the midgut tissues, further invading the salivary glands of the insects. Persistent virus transmissions are further sub-classified into two categories, viz. persistent-circulative and persistent-propagative. In the circulative way of transmission, viruses must escape the insect gut and spread to the neighboring organs to reach the salivary gland. This requires the physical interaction of viral proteins with the host (vector) factors. The virus propagates in the insect cells in persistent-propagative transmission.

Limited studies on virus-vector relationships make the management of virus diseases challenging. Moreover, most of the efficient insect vectors are tiny and develop quick resistance to conventional practices. A small population of vectors can spread the disease very efficiently and can have a devastating effect on crop cultivation and production. Most common approaches to managing the vectors of plant viral diseases are intended to eradicate or shift one or more of the primary participants in the transmission process (such as vector, virus inoculum, and host plant) or to prevent them from interacting with each other. However, the main hurdle at the farmers' level is to identify and characterize the virus disease because of the high similarities in disease symptom development (Jones 2014). Difficulties in the identification of the virus and insect vectors lead to the deceptive application of pesticides, which in turn aggravates the issue of pest resistance and equally costs the natural ecosystem. Plant virus transmission by insect vectors involves a combination of co-evolving biological factors in tripartite interactions (host, vector, and virus) in a given environment. Therefore, integrated pest management (IPM) approaches involving all the available components for effective management of insect vectors can reduce this grievous problem in crop production. The chapter summarizes the currently available tactics for the management of viruses and vectors.

5.2 DIFFERENT PLANT VIRUSES AND THEIR VECTORS

Plant viruses depend on vectors for their survival and dispersion. Out of 29 orders of class Insecta, 7 are known to act as vectors of several plant viruses, of which Hemipteroid accounts for the most. A number of plant hoppers, leaf hoppers, whiteflies, thrips, beetles, mites and nematodes are known to carry plant viruses (Table 5.1).

5.3 MANAGEMENT STRATEGIES FOR CONTROLLING INSECT VECTORS

Disease incidence due to viral pathogens and epidemiology of plant-vectored-pathogens relies upon the interaction between the plant, availability of an alternate host, pathogen diversity, viruliferous vectors, and a suitable environment. However, the interactions among the vector, pathogen and crop plants are not so simple. Thus, a clear understanding of the seasonal cycles of plants, pathogens, and vectors is very much needed. And the management of vector-borne plant viruses must involve integrated approaches.

5.3.1 Cultural Management

These are the age-old practices and most important tools for managing the viruliferous vectors of different plant pathogenic viruses. Several cultural practices have been shown to have a potential role in reducing the incidence of vectors and vector-borne diseases. This is a very feasible and easily adaptable practice. Farmers or growers can easily adjust their crop production system, e.g. the adjustment of planting time and seed sowing time, seed treatment, crop canopy management, trap

TABLE 5.1
Different major groups of plant viruses and their vectors

Virus groups	Genome	Mode of transmission	Persistence	Insect act as vector
Alfamovirus	ssRNA	Non-persistent	Few hours	Aphids
Badnavirus	dsDNA	Semi-persistent	Days	Mealybugsand leaf hopper
Begomovirus	ssDNA	Circulative	Weeks	Whiteflies
Carlavirus	ssRNA	Non-persistent	Few hours	Aphids
Caulimovirus	dsDNA	Non-persistent	Many hours	Aphids
Closterovirus	ssRNA	Semi-persistent	Many hours	Aphidsor mealybugs
Comovirus	ssRNA	Semi-persistent	Days	Beetles
Cucumovirus	ssRNA	Semi-persistent	Few hours	Aphids
Curtovirus	ssDNA	Circulative	Weeks	Leafhoppers
Emaravirus	ssRNA	Semi-persistent	Few hours	Eriophyid mite
Enamovirus	ssRNA	Circulative	Weeks	Aphids
Fabavirus	ssRNA	Non-persistent	Few hours	Aphids
Luteovirus	ssRNA	Circulative	Weeks	Aphids
Machlomovirus	dsDNA	Semi-persistent	Many days	Leafhoppers
Mastrevirus	ssDNA	Circulative	Weeks	Leafhoppers
Nepoviruses	ssRNA	Semi-persistent	Up to a year	Nematodes
Poleroviruses	ssRNA	Circulative	Weeks	Aphids
Potyvirus	ssRNA	Non-persistent	Few hours	Aphids
Sequivirus	ssRNA	Semi-persistent	Few hours	Aphids
Sobemovirus	ssRNA	Semi-persistent	Days	Beetles
Tospoviruses	ssRNA	Persistent Propagative	Many days to week	Thrips
Tymovirus	ssRNA	Semi-persistent	Days	Beetles
Waikavirus	ssRNA	Semi-persistent	Few days	Leafhoppers

Source: Modified from Raccah and Fereres (2009).

cropping, intercropping, keeping barrier crops, border crops or utilizing border netting technology, reflective mulches and the use of paints to repel the vectors population. Various protective row covers have encouraging impacts to reduce the incidence of vectors and their dispersal. Adjustment in planting dates might be a good option to avoid the incidence of specific arthropods. A recent study suggests that transmission of viruses by eriophyid mites and aphids can damage early-planted cereal crops.

Similarly, an aphid-transmitted barley yellow dwarf virus (BYDV) disease incidence can be avoided by simply delaying the planting of seeds in the Pacific Northwest region of North America. Additionally, intercropping of onion and garlic plants with tomato crops decreased the level of thrips infestation by 79–86 and 81–85 percent, respectively. Similarly, the disease incidence of yellow vein mosaic virus (YVMV) in okra and tomato leaf curl virus (ToLCV) infection in tomato will be greatly reduced by controlling the whitefly population and intercropping with soybean and coriander crops during the cultivation season, respectively. Barrier crops around the main crop are likely to be more effective in protecting plants from non-persistent types of transmission. For example, sunflower plants act as the barrier crop in tomato cultivation against potato virus Y (PVY). Similarly, maize and wheat were used as the barrier crop in potato cultivation, which trapped the aphid population and inhibited potato virus transmission (Schroder and Kruger 2014). Furthermore, sunflower, sesame, and pearl millet reduced the incidence of cucumber mosaic virus (CMV) in chilli, when used as a barrier crop. Other efficient ways are seed treatment through systemic insecticides like imidacloprid (4ST), thiamethoxam (30% FS), and imidacloprid (600 FS). Treatment of plants with these insecticides protects them from a wide range of insect vectors. Seedlings treated with these systemic insecticides also reduce the infestation of different types of sucking pest vectors and mite-mediated virus transmission. Canopy management steps are also used for the eradication of latent infection and rejuvenation of plants. A combination of cultural control and need-based insecticidal applications can mitigate the sucking pest populations in a judicious way. The growing of resistant and tolerant varieties is another effective way of managing vectors and vector-transmitted diseases. In field conditions, the viruliferous vector population can be drastically reduced with the use of an insect-resistant or tolerant cultivar by preventing the availability of virus inoculum during the farming season. To be on the safer side, farmers are always dependent on resistant/tolerant varieties as it is the cheapest and most reliable measure against pathogens. The efficacy of host plant resistance to the vector species is an important cultural operation and it depends on factors such as the host phenology, means of resistance, and the mode of transmission. Cultivar confers resistance that prevents feeding or repels the insects and can prevent the transmission of pathogens (Heck 2018). If the host-mediated resistance merely prevents or slows down the growth of the insect population, it cannot prevent primary transmission. However, it might have some effect on secondary transmission. It is very desirable to breed plant varieties that are resistant to vector species. Resistant/tolerant plant varieties are the most effective source of insect vector management, which ultimately helps in resisting plant viruses. Paratransgenesis is a novel approach, used to control the vector-borne disease. It is used for the genetic manipulation of host symbiotic microorganisms to achieve an array of objectives ranging from disease eradication to pest control. Other approaches include modifying crop density and plant nutrition through spacing management and balanced fertilizer use. However, plants possessing better growth with succulent foliage are preferred for feeding insect vectors and thus likely to suffer more from viral infections. The previous study suggests that when robust crops are fertilized with excessive nitrogen, they become more susceptible to tobacco leaf curl virus (TbLCV) infection. Furthermore, the reduction in the dose of nitrogen in the onion crops reduces the population of thrips species (Buckland et al. 2013).

5.3.2 Removal of Alternate Host

It is well known that insect vectors transmitting viruses are trying to replicate in an alternate host during the post-harvest to the next sowing or planting season. This allows them to survive in non-favorable conditions. Several vectors undergo diapauses on an alternate or on collateral host plants. The broader host range of the pathogen may determine whether the inoculum control is viable or not. Several weeds, ornamental and medicinal plant families such as Amaranthaceae, Apiaceae, Asclepiadaceae, Asteraceae, Brassicaceae, Caryophyllaceae, Chenopodiaceae (Goosefoot or Lamb's quarter), Cucurbitaceae, Euphorbiaceae, Fabaceae, Malvaceae, and Solanaceae serve as a potential alternate host for most of the common plant viruses and its vector too (Kumar and Chakraborty 2021). For example, tomato leaf curl New Delhi virus (ToLCNDV) transmitted by whiteflies, is one of the most devastating pathogens that cause severe disease in crops such as tomato, okra, capsicum, dahlia, duranta, coccinia, cotton, hollyhock, luffa, tulsi, brinjal, marigold, black gram, green gram and cowpea (Pant et al. 2018; Chakraborty and Kumar 2021). The population density of viniferous whiteflies (more than 100 insect/leaf) increases from April to December in the northern part of India. However, it is also reported that its population was much higher during June and August on Tulsi and Black gram plants in northern India. The population of a vector depends upon the availability of an alternate host plant, types of vector species, and environmental factors. Therefore, the management of alternate hosts is a foremost essential component for the control of viral diseases in the plant. The actual dimension of the host range can be reduced using barriers that prevent contact between vectors and hosts, and the unsynchronized seasonal timing between (1) available infected hosts in a viremic stage and feeding activity of vectors and (2) available uninfected species and infectious vectors in a given environment (Fermin 2018). Therefore, the removal and destruction of the alternate host in the vicinity of the target crop are strictly needed. To do so, the application of non-selective herbicide shall be advisable under field conditions. Characterization and identification of susceptible crops is an effective practice to reduce the risk of viral infection. Increasing the distance between the susceptible crop or variety from the desirable crop may be the simplest and most effective way of minimizing the number of viruliferous insects.

5.3.3 Mechanical and Physical Control

Mechanical and physical control focuses on managing the insect vector population by eliminating the infected plant. As most of the sucking pests are the active carrier of plant viruses and their different life stages (pupa, larva, and adults) they may be collected from the field and destroyed to reduce the population load of those particular pests. Installation of yellow sticky traps (20 cm x 15 cm), blue sticky traps (20 cm x 15 cm), predator-safe light traps @ 18 nos. per ha and 3 nos. per ha respectively may be utilized to trap the vectors transmitting plant viruses (Ghosal et al. 2021). If the population of the vector is 100 per each sticky trap then the crop may be sprayed at 10-day intervals commencing from the fourth week after seed germination. Banana bunchy top virus (BBTV) can be successfully controlled by rouging the diseased plants. Previously, the rouging of diseased plants of cassava has been practiced regularly to mitigate the cassava mosaic virus (CMV) disease severity (Thresh and Cooter 2005). Handpicking of infected shoots and leaves by aphids and jassids is an effective way to reduce the potato leaf roll virus (PLRV) and yellow mosaic virus (YMV) infection. Maintenance of good field sanitation is the easiest and most potent measure in managing the insect vector population for cash crops. Visual repellents like mulches, paints, and colored netting may be utilized. However, till now very little has been done with chemical repellents.

5.3.4 Biological Control

Several biotic organisms played a specific role in reducing the pest population. These natural enemies played an important role in vector management however, the efficacy of biological control has mixed opinions. The insect vector with high population density can be controlled by biological ways whereas, it is non-achievable for the low level of population density of insect vectors. Thus, a scatter response in the vector-prey situation might be evoked during pest management. Therefore, insect vector management using biological control alone might not be advisable in an open field system or for the perennial crops as it is known that only one viruliferous individual is competent to spread the disease in the field. The maintenance of the low population of insect vectors could be risky in that case. However, if it is integrated with other management tools it may bring quantitative results. Several predators and parasitoids are present in a particular existing field ecosystem and they could be conserved and utilized meticulously. The parasitoid *Aphelinus gossypii* plays a very vital role in reducing the population of aphids during the colder months of the year. A relatively higher degree of temperature control is generally exerted by these natural enemies if co-exist with other insects. Several insect vectors such as true bugs (family Lygaeidae), flower bugs (family Anthocoridae), and mites (class Arachnida) are the active predators of thrips species. In addition, the hairy (hirsute) varieties of cotton glabrous leaves are very much susceptible to jassid species. The susceptibility is associated with both density and length of the hairs on the abaxial surface of the leaves, especially on leaf veins where the jassids are used to feed and lay eggs. Biocontrol is not painstaking enough to have a considerable impact on the jassid species population, though several egg parasitoids have been documented previously. The parasitoid *Anagrus* sp. has been found to parasitize the jassids, however, the impact is not admirable. Release of *Chrysoperlaza strouviearaabica* or *Coccinella* sp. or *Scymnus* sp. or conservation of *Lynx*, *Neoscona* spider population, and some ant species like *Camponotus* sp. may be effective against different soft-bodied vectors. Aphelinid parasitoids

(*Encarsia* and *Eretmocerus*), lady beetles (order Coleoptera), and lacewings (order Neuroptera) were found to be the dominant natural enemies of *B. tabaci* and could be successfully exploited to reduce the whitefly population (Li et al. 2011). It is sometimes very difficult to achieve pest management through classical biological control methods; therefore, disruption of existing biological control measures might result in an increase in virus transmission via insect vectors.

5.3.5 Pathogen-free Seed/Planting Material and Quarantine Measures

Quarantine measures are preferred to inhibit the spread of plant viruses over a long distance by preventing novel or existing virus species from entering the new ecological area, where they are more prevalent. A vegetative production of infected plants or their parts such as bud wood scions, rootstocks, trees, shrubs, tubers, rhizomes, corms, and bulbs helps in the spreading of viral pathogens. For example, an aphid-transmitted citrus tristeza virus (CTV) was not found in countries, like the United States of America (USA), China, and South Africa, that grew citrus on a big scale for two reasons: resistant/tolerant cultivars and a lack of efficient vectors. However, quarantine measures are taken efficiently for CTV spread in the USA. In many countries, certification procedures for potatoes have been implemented to produce virus-free (potato leaf roll and potato virus Y) seeds. In India, Punjab is using the seed plot technique to produce virus-free potato seeds in anticipation of the advent of aphid vectors in the state. Virus indexing can be used to check for the presence of viruses in planting material. Many viruses are transmitted through seeds and planting materials. Therefore, to limit the initial buildup of infection in the field, approved seeds/planting material must be used. It is always advisable to use virus-free seeds or planting materials to ensure that the pathogen has no virus molecules in the source tissue to proliferate further. The vector would be unable to acquire the virus if the virus inoculum was removed properly. Previously, by producing lettuce mosaic virus-free seeds, the yield of the lettuce crop was shown to be significantly improved in the Salinas Valley area of California, USA.

In addition, various seed certification methods were used against plant viruses in several countries including India. In the USA, barley stripe mosaic virus (BSMV) has been attributed to major economic losses because millions of dollars were getting wasted due to severe barley strip disease in barley crops. In general, this is accomplished using a variety of methods, like thermotherapy (heat-induced), chemotherapy (chemical method), cryotherapy (cold-induced), and electrotherapy (using electric current). Furthermore, virus-free plants can be grown by using the micro-propagation of meristematic tissues of crops via the tissue culture technique. Planting resistant varieties against tungro virus disease is the most economical means of managing the disease. Uses of resistant varieties like IR 36, IR 50, ADT 37, Ponmani, Co 45, Co 48, Surekha, Vikramarya, Bharani, IR 36, and White ponni will help in managing the disease in the field. The most common method of controlling the tungro virus is by spraying insecticides. Spraying insecticides can suppress the populations of green leafhoppers which means it will reduce the speed of spreading the virus. Some known insecticides can be used to control green leafhoppers. However, once a rice plant is infected, it cannot be cured. Therefore, some preventive measures are more effective for the control of tungro as compared to direct disease control measures. Using insecticides to control leafhoppers is often not very effective because green leafhoppers continuously move to surrounding fields and spread tungro rapidly in a very short feeding time. Therefore, the most practical measures at present, include the growing of tungro leafhopper resistant varieties. This is the most economical means of managing the disease. There are tungro-resistant varieties available in the Philippines, Malaysia, Indonesia, India, and Bangladesh. One can contact their local agriculture office for up-to-date lists of varieties available. Farmers can also practice synchronous planting with surrounding farms. Delayed or late planting, relative to the average date in a given area, makes the field susceptible to tungro. Late-planted fields also pose a risk to early planting in the next season. It is advisable to adjust planting times to when green leafhoppers are not in season or abundant if known and plow the infected stubbles immediately after harvest to reduce the inoculum sources and destroy the eggs and breeding sites of the green leaf hopper.

Rice grassy stunt (RGS) is caused by a virus transmitted by the brown planthopper (BPH). The disease symptoms develop in 10–20 days. The BPH flies from rice grassy stunt-affected fields to newly planted rice fields in distant areas and disperses RGSV, as well as another BPH-borne virus, rice ragged stunt virus (RRSV). To control the RGSV the brown planthopper vectors need to be managed. This can be done either through the use of insecticides, brown plant hopper-resistant varieties or by synchronized crop establishment. Infected stubble needs to be plowed under after harvest to reduce the virus source as an inoculum. Close planting should be avoided. Farmers should provide 30 cm rogue spacing at every 2.5 to 3.0 m to reduce the pest incidence. There are varieties released by IRRI, which contain genes for BPH resistance, like IR26, IR64, IR36, IR56, and IR72, therefore plowing and harrowing the field to destroy stubbles should be undertaken straight after harvest in order to eradicate other hosts.

5.3.6 Chemical Control

Chemical management of vectors is sometimes considered the most effective control measure as the population of vectors gets apprehended directly. However, the chemical control of vectors will limit the spread of plant pathogens only when the vectors are being controlled. These vectors must be the colonizer of the crop and being treated, dispersal of insect vectors must come from within the treated area (secondary spread). The chemical control of insect vectors is sometimes very difficult because it simply prevents the population buildup of insects in the treated area, but it does not restrict the entry of immigrant vectors thus one viruliferous vector from the vicinity of a crop field might affect the disease incidence. So, an integrated approach to management is needed to achieve protection from viruliferous insect vectors. Chemical control of certain aphid-transmitted luteoviruses (barley yellow dwarf virus, potato leafroll virus) has been proven to be effective. However, sometimes a single missed spray or insecticide failure could result in an infection of the crops. In addition, some insecticidal spray induces insect resurgence by changing behavioral hormoligosis and inducing several biochemical changes in crop phenology. Therefore, the frequent applications of insecticide spray without the following recommendation led to the resurgence of the insect vectors population, and hence a judicious chemical application is very much important. Additionally, the application of a sub-lethal dose of insecticides should be avoided to reduce the chances of the development of resistance and resurgence (Ghosal and Chatterjee 2018).

Chemical insecticides still play the most potent role in virus management. Most of the vectors which transmit viruses belong to the sucking pest categories. So, the spraying of systemic insecticides like organophosphates (acephate 75 SP, dimethoate 30 EC), neonicotinoids (imidacloprid 17.8/200 S.L, and thiamethoxam 25 WG), pymetrozine 50 WG, flonicamid 50 WG, and spiromesifen 240 SC, can efficiently reduce the vector population. Interestingly, if the insecticide spraying tank is mixed with spiromesifen and imidacloprid (@ 60+30 g a.i./ha), or sulfoxaflor + buprofezin (@ 25 + 150 g a.i./ha), it might give a synergistic effect on sucking pest population control (Ghosal et al. 2018). Therefore, insecticides are usually required when the population of the vectors reaches its economic threshold level (ETL). The ETL of each vector species varies significantly and it also depends upon the types of the cultivar. For example, spraying of insecticide on cotton plants should be initiated if the jassid population reaches 2 to 3 nymphs/leaf, while for glabrous cultivars the threshold might be a little lowered, which is 1 nymph/leaf. In addition, horticultural oil can also reduce the style penetration of several sucking pests. The potential use of chemical repellents and anti-feedants is very useful in reducing the vector population (Fereres and Moreno 2009). Several biological insecticides and acaricides like neem oil (1%), fish oil resin soap (2.5%), and neem seed kernel extract (5%) will give effective control if it is used in an integrated approach. Microbial insecticides like *Beauveria bassiana,* and *Verticillium lecanii* show a better protection response against different sucking pest control (Ghosal 2018). The volume of insecticidal spray is another important factor in the management of insect vectors. In general, spraying fluid must be sprayed in the range of 250 l/ha to 500 l/ha and a power sprayer can be used for better coverage.

5.3.7 Molecular Approaches in Vector Management

In recent years, advancements in biotechnology have provided us with several new approaches, other than applying synthetic chemicals and the construction of first-generation genetically modified (GM) crops for integrated pest management. These new generations of insect-resistant crops provide more sustainable and long-lasting solutions. Previously, insect-resistant Bt crops were generated by overexpressing several crystalline protoxin (*cry*) genes in host plants. Now, these *cry* genes are used to accomplish transgene pyramiding to develop a better resistance against the insect vector (Table 5.1). Plants also evolved to defend insects by producing several metabolites. Some of them are now utilized for the formulation of insecticidal ingredients, such as nicotine (isolated from tobacco plant), capsaicin (extracted from chilli/peppers crops), pyrethrin (extracted from chrysanthemum leaves), and azadirachtin (isolated from the neem tree).

Additionally, the transgenic approach in several crops such as rice, maize, tomato and wheat was developed mostly against the aphid-transmitted insect vectors (Gutensohn et al. 2014). Other novel practices including RNA interference (RNAi), R-gene-based host-defense and CRISPR-based techniques are the most advanced form of molecular approaches for pest management. RNAi has been successfully used to understand the virus-vector relationships in several cases (Kanakala et al. 2019; Kaur et al. 2020; Priti et al. 2022; Thesnim et al. 2023). Topical spray of RNAi molecules is promising in managing vector populations as well as reducing the begomovirus transmission ability of whitefly (Chakraborty and Ghosh 2022)

5.3.8 Integrated Vector Management

A well-designed integrated solution for the management of insect vectors shall be carried out in a planned fashion. For any successful vector management programme, proper coordination and an integrative strategy are required. The identification of pathogens and their insect vectors, as well as knowledge of their biology, are essential for pest control. Furthermore, to determine the vector population and viral diversity, sampling methodologies, crop phenology and a monitoring system should be advised. The potential damage must be determined using an acceptable method and should be proportional to the total number of vectors population in the crop and the proportion of viruliferous vectors present in that environmental niche. Some novel models for disease prediction, crop phonology assessment and the management strategy should be developed. Recently, viral capsid protein has been used to deliver a toxin to insect vectors. Furthermore, some viral protein is also used to block the transmission and entry mechanisms of viruses into insect vectors (Whitfield et al. 2015). These novel mechanisms can be used for deciphering a new virus and vector management tactics.

Stillson et al. (2020) suggested that a decision support system tool for combined surveillance of the aster leafhopper, crop consultants, with faster disease diagnostics, and artificial intelligence (AI) based text-messaging system at the country-level will be very much needed for the IPM. Genetic alterations such as the strengthening of naturally occurring genes that encode for defensive response or the introduction of foreign genes into crop plants will provide protection from insect vectors. The IPM technique was proved to be effective in controlling the aphid population in zucchini crops and eventually led to the testing for papaya ring spot virus in Australia (Pionese et al. 1994). Similarly, an integrated approach for suppressing tospovirus infection in tomato crops by using combining moderately resistant/tolerant cultivars, cultural techniques, and chemical practices was recently used (Culbreath et al. 2003). Alternatively, white sticky traps, specialized floral volatile attractants, resistant varieties, synergism with natural enemies, and pesticide combinations were used to suppress the population of raspberry beetle (*Byturus tomentosus*) and aphids in the United Kingdom (Birch et al. 2004).

5.4 CONCLUSION

Management of plant viruses through vector management is not an easy task; adoption of a single-window management system cannot give them better control as it depends on multiple interactions and factors. A clear understanding of virus transmission is the most important aspect of adopting the management strategies. Therefore, the management of plant viruses through vector management is the only option available to farmers besides the use of advanced genetic tools. A large amount of plant-infecting viruses are transmitted by insect vectors, and integrated strategies should be adopted to manage the viruliferous vector population. To achieve this, a clear understanding of the identification and characterization of the plant virus (through visual symptom manifestation) and insect vectors is very important. Strategies for the management of vectors also rely upon the knowledge of their biology. Various sampling methods and monitoring techniques should be developed to estimate the abundance of insects and virus nature concerning the crops, which will lead to help in deciding on the support system generation (Stillson et al. 2020). However, for efficient and sustainable vector control, it is indispensable to consider the genetic diversity and characterization of virus populations and have specific, fast, and reliable diagnostic tools. Therefore, successful plant pathogenic viral disease management now relies on two approaches: one is immunization to get resistant plants and another is prophylactic measures to restrain the vectors spreading the various pathogen.

REFERENCES

Birch ANE, Gordon SC, Fenton B, Malloch G, Mitchell C, Jones AT, Griffiths DW, Brennan R, Graham J, Woodford JAT (2004) Developing a sustainable IPM system for high value Rubus crops (raspberry, blackberry) for Europe. *Acta Horticult* 649: 289–292.

Buckland K, Reeve JR, Alston D, Nischwitz C, Drost D (2013) Effect of nitrogen fertility and crop protection on onion growth and yield, thrips density, Iris yellow spot virus and soil properties. *Agric Ecosyst and Environ* 177: 63–74.

Chakraborty P, Ghosh A (2022) Topical spray of dsrna induces mortality and inhibits chilli leaf curl virus transmission by *Bemisia tabaci* Asia II 1. *Cells* 11: 833. doi: 10. 3390/cells11050833

Chakraborty S, Kumar M (2021) *Tomato leaf curl New Delhi virus (Geminiviridae)* (vol. 3. 4th edn). edited by Bamford DH, Zuckerman M. San Diego, CA: Academic Press, pp. 749–760.

Cooper JI, Jones RAC (2006) Wild plants and viruses: under-investigated ecosystems. *Adv Virus Res* 67: 1–47.

Culbreath AK, Todd JW, Brown SL (2003) Epidemiology and management of tomato spotted wilt in peanut. *Annu Rev Phytopathol* 41: 53–75.

Fereres A, Moreno A (2009) Behavioural aspects influencing plant virus transmission by homopteron insects. *Virus Res* 141: 158–168.

Fermin G (2018) Host range, host virus interactions, and virus transmission. In: Tennant P, Fermin G, Foster JE (eds) *Molecular biology, host interactions, and applications to biotechnology*. Academic Press, UK, pp. 101–134.

Garcia-Arenal F, Fraile A, Malpica JM (2001) Variability and genetic structure of plant virus populations. *Ann Rev Phytopathol* 39: 157–186.

Ghosal A (2018) Studies on the potency of some biopesticides against whitefly in cotton and tomato. *J Biopest* 11: 60–68.

Ghosal A, Chatterjee ML (2018) Insecticide induced resurgence study of whitefly in cotton and tomato. *Univ Sindh J Anim Sci* 2: 1–6.

Ghosal A, Chatterjee ML, Bhattacharyya A (2018) Field bio-efficacy of some new insecticides and tank mixtures against whitefly on cotton in new alluvial zone of West Bengal. *Pestic Res J* 30: 31–36.

Ghosal A, Das K, Kundu P (2021) Molecular characterization of whitefly *Bemesia tabaci* (Genn.) and development of IPM module against chili leaf curl complex. *Indian J Entomol* 84(3): 1–6.

Gutensohn M, Nguyen TTH, McMahon RD, Kaplan I, Pichersky E, Dudareva N (2014) Metabolic engineering of monoterpene biosynthesis in tomato fruits via introduction of the non-canonical substrate neryldiphosphate. *Metab Eng* 24: 107–116.

Heck M (2018) Insect transmission of plant pathogens: A systemsbiology perspective. *mSystems* 3: e00168–e00217.

Islam W (2017) Management of plant virus diseases; farmer's knowledge and our suggestions. *Hosts Viruses* 4: 28–33.

Jones RAC (2014) Plant virus ecology and epidemiology: historical perspectives, recent progress and future prospects. *Ann Appl Biol* 164: 320–347.

Kanakala S, Kontsedalov S, Lebedev G, Ghanim M (2019) Plant-mediated silencing of the whitefly *Bemisia tabaci* cyclophilin b and heat shock protein 70 impairs insect development and virus transmission. *Front Physiol* 10: 557. doi: 10.3389/fphys.2019.00557

Kaur R, Gupta M, Singh S, Joshi N, Sharma A (2020) Enhancing RNAi efficiency to decipher the functional response of potential genes in *Bemisia tabaci* Asia II-1 (Gennadius) through dsRNA feeding assays. *Front Physiol* 11: 123. doi:10.3389/fphys.2020.00123

Kumar M, Chakraborty S (2021) Viruses and satellites infecting Hollyhock (*Alcea rosea*) plants: an update. In: Raj SK, Gaur RK, Yin Z. (Eds.), *Virus diseases of ornamental plants*. Springer, Singapore, pp. 191–210. https://doi.org/10.1007/978-981-16-3919-7_11

Kumar M, Kumar VR, Chakraborty S (2020) Association of a begomovirus-satellite complex with yellow vein and leaf curl disease of hollyhock (*Alcea rosea*) in India. *Arch Virol* 165(9): 2099–2103.

Li B, Yang M, Shi R, Ye M (2019, July) Insecticidal activity of natural capsaicinoids against Several agricultural insects. *Nat Product Commun* : 1–7. doi: 10.1177/1934578X19862695

Li S, Xue X, Ahmed MZ, Ren SX, Zhou DY, Hui WJ, Cuthbertson AGS, Li QB (2011) Host plants and natural enemies of *Bemisia tabaci* (Hemiptera: Aleyrodidae) in China. *Insect Sci* 18: 101–120.

Malathi VG, Renukadevi P, Chakraborty S, Biswas KK, Roy A, Sivalingam PN, Venkataravanappa V, Mandal B (2017) Begomoviruses and their satellites occurring in india: distribution, diversity and pathogenesis. In: Mandal B, Rao G, Baranwal V, Jain R. (Eds.), A century of plant virology in India. Springer, Singapore. pp. 75–177.

Pant RP, Bhatnagar A, Lal M (2018) Role of alternate host plants in the transmission of apical leaf curl disease of potato caused by tomato leaf curl New Delhi virus–potato (ToLCNDV-pot.) in Northern India. *Indian J Agril Sci* 88: 1258–1262.

Pionese B, Lisle BI, Ramsey MD, Halfpapp KH, De Faveri S (1994) Control of aphid borne Papaya ring spot Potyvirus in zucchini marrow (*Cucurbita pepo*) with reflective mulch, mineral oil- insecticide sprays. *Int J Pest Manage* 40: 81–87.

Priti, Mukherjee SK, Ghosh A (2022) Silencing of *Thrips palmi* UHRF1BP1 and PFAS using antisense oligos induces mortality and reduces tospovirus titer in its vector. *Pathogens* 11(11): 1319. doi: 10.3390/PATHOGENS11111319

Raccah B, Fereres A (2009) *Plant virus transmission by insects*. Wiley Online Library. doi: 10.1002/9780470015902.a0000760.pub3

Schroder ML, Kruger K (2014) Preference of aphids (Hemiptera: Aphididae) for lucerne, maize, soybean and wheat andtheir potential as prospective border crops for potato virusY management in seed potatoes. *African Entomol* 22: 144–155.

Stillson PT, Bloom EH, Illán JG, Szendrei Z (2020) A novel plant pathogen management tool for vector management. *Pest Manag Sci* 76(11): 3729–3737. doi: 10.1002/ps.5922

Thesnim P, Jangra S, Kumar M, Ghosh A (2023) Effect of silencing *Bemisia tabaci* TLR3 and TOB1 on fitness and begomovirus transmission. *Front Plant Sci* 14: 1136262. doi: 10.3389/fpls.2023.1136262

Thresh M, Cooter RJ (2005) Strategies for controlling cassava mosaic virus disease in Africa. *Plant Path* 54: 587–614.

Varma A, Malathi VG (2003) Emerging geminivirus problems: a serious threat to crop production. *Ann Appl Biol* 142: 145–164.

Whitfield AE, Falk BW, Rotenberg D (2015) Insect vector-mediated transmission of plant viruses. *Virology* 479–480: 278–289.

6 Biotic Stress Management Caused by Fungi and Bacteria in Rice Crops

Vindeshwari Prasad, Manish Kumar Maurya, Shivam Singh, Saurabh Dixit, S.K.S. Rajpoot and Satyam Pathak

6.1 INTRODUCTION

Rice (*Oryza sativa* L.) belongs to graminae, a major staple food to half of the global population. Asia is known as the "Rice Bowl of the world" because more than 90% of global rice is grown and consumed there. There is a great demand for this crop in India because the average diet is based on rice. The gigantic production aim of rice to fulfill the needs of a rising global population has become immensely tough because of the ongoing loss of arable land to industrialization and human endowment (Laha et al. 2017). To achieve maximum production, there is massive use of nitrogenous fertilizer and pesticides resulting in loss of soil fertility, contamination of groundwater and the development of new strains of pests and pathogens that are resistant to pesticides or fungicides. Continuous change in climate results in minor diseases or pests becoming major ones (Gupta and Pathak 2016). To overcome these problems we need to adopt sustainable management tactics that are eco-friendly in nature. Several biotic and abiotic stresses constitute a key restriction in rice, reducing production by 20–100%, depending on disease severity (Shivappa et al. 2021). Major diseases are blast, bacterial leaf blight, sheath blight, brown leaf spot and a disease false smut which was previously minor in nature but presently has become major. This chapter discusses the information on economic importance, symptoms, disease cycle and sustainable management strategies of the important diseases of rice.

6.2 MAJOR DISEASES OF RICE

6.2.1 Blast

This fungal disease is one of the most devastating and widespread rice diseases globally and ranks among the top ten most destructive fungal pathogens (Dean et al. 2012). It severely affects low-lying rice in temperate and subtropical Asia and upland rice in tropical Asia. In India, it became significant after a severe epidemic in Thanjavur (The Rice Bowl of Tamil Nadu) delta in 1919. Average yield losses range from 10–30%, but under optimal conditions, it can devastate entire fields within 10 to 20 days, causing up to 100% yield loss (Asibi et al. 2019).

6.2.1.1 Symptoms

The fungus can infect and produce lesions on nearly all aerial parts of the plant including leaves, nodes and panicles (Yasuda et al. 2014). The symptoms of the disease appear in three phases.

 DOI: 10.1201/9781032631578-6

FIGURE 6.1 (a) Characteristic eye-shaped lesion produced on leaves. (b) Node blast (Laha et al. 2017) causal organism.

Leaf blast: Greyish or bluish dots of 1–3 mm diameter appear on the leaf blades. Later, the spots enlarge quickly and become elliptical or 'eye-shaped' with a grey or whitish center and brown or dark brown margin (Figure 6.1a) (Neupane and Bhushal 2021). In severe infection, seedlings and plants in the nursery may be completely blasted or killed. However, on the resistant cultivars, minute brown specks of pinhead size may be observed.

Node blast: The pathogen also infects the nodes that turn brown, greyish brown or black and the portion above the infection may be due to blockage of xylem and phloem which affects the water and nutrient supply followed by the death of all the plant parts (Figure 6.1b) (Neupane and Bhushal 2021).

Neck and panicle blast: During flowering, a greyish-brown lesion girdles the area near the panicle base, causing it to fall over if the infection is severe. The neck shrivels and is covered with grey mycelium. If infection occurs before the milk stage, the panicle may die prematurely, resulting in white, unfilled grains. Later infections lead to partial grain filling, chaffy spikelets, and poor milling quality. The pathogen also causes brown lesions on panicle branches and spikelet pedicels, leading to panicle blast.

The disease is caused by the filamentous fungus *Pyricularia oryzae* Cav. [synonym *P. grisea* Sacc, teleomorph *Magnaporthe oryzae* (Hebert) Barr] (Zhang et al. 2016). The mycelium of the fungus is hyaline to olivaceous and septate. Conidia are produced in clusters on long septate, olivaceous conidiophores. Conidia are pyriform to ellipsoid, attached at the broader base by a hilum. Conidia are hyaline to pale olive green, usually 3 celled. The perfect state of the leaf blast fungus is *Magnaporthe oryzae* producing perithecia. The ascospores are hyaline, fusiform, 4-celled and slightly curved.

6.2.1.2 Disease Cycle

The primary sources of infection for the pathogen include infected seeds, mycelia or conidia on diseased straw, and hosts like cereals or grasses. In regions with multiple crops per year, the pathogen continues its cycle on the rice crop. Symptoms can appear within 4–5 days of infection, with conidia produced 6–7 days after and spread by wind. Sporulation rates increase with higher humidity, while spore release and dispersal are enhanced by dew and wind speed. Secondary infection cycles can occur frequently, leading to severe disease. Favorable conditions include low night temperatures (below 24°C), day temperatures of 28–30°C, and high humidity (over 90%).

6.2.1.3 Management

- Use of tolerance/resistant cultivars is the best method of management of the disease.
- Cleaning of crop residues as it is a primary source of inoculums.
- Silicon fertilizer (Calcium silicate) can be applied to soil which is effective in reducing the severity of the disease (Sathe et al. 2021).
- Use of botanicals like neem oil and garlic extract for spray and bioagents *Pseudomonas fluorescens* @ 10 g/kg seed + spray 0.25% is also found effective.
- Spray of fungicide tricyclazole or azoxystrobin + difenoconazole or azoxystrobin + tebuconazole or picoxystrobin may be used judiciously for the control of blast, as per recommended dose (Balol et al. 2022).

6.2.2 Brown Spot

Brown spot is one of the devastating diseases of rice. Average grain yield loss ranging from 4 to 52% has been reported from different countries (Sunder et al. 2014). Losses of up to 90% were also observed in such cases where water supply is scarce, combined with nutritional imbalances, particularly nitrogen. This disease has historic importance as it caused heavy yield losses in Eastern India which is considered to be a major factor in the 'Bengal Famine' during 1942.

6.2.2.1 Symptoms

The disease appears on coleoptiles, leaf sheath and leaf blades as brown spots with a grey or whitish center, cylindrical or oval in shape similar to sesame seeds usually with a yellow halo on the leaves (Figure 6.2) (Aslam et al. 2022). Black or dark brown spots appear, which result in discoloration and shriveled grains on glumes (Chakrabarti 2001). Under favorable conditions, dark brown conidiophores and conidia develop on the spots giving a velvety appearance. Blackish lesions may also show on young roots. The pathogen may show blackish to dark brown colored lesions on the panicle at the joint of the flag leaf and stalk. These lesions usually extend downwards beneath the sheath resulting in severe wet rotting and the appearance of greyish mycelial growth.

FIGURE 6.2 Symptom of brown spot disease on leaves of rice.

6.2.2.2 Causal Organism

Drechslera oryzae, formerly *Helminthosporium oryzae* Breda de Haan and teleomorph *Cochliobolus miyabeanus* (Ito and Kunibayashi) produces intercellular and intracellular mycelium that forms a greyish to dark brown mat on infected tissues. In culture, the fungus appears grey to olive or black. Its sporophores are thick, erect, and dark olivaceous at the base, emerging in clusters through stomata. Conidia are 5–10 septate, slightly curved, widest in the middle, and mature to a fuliginous or brownish color. Mature conidia germinate with two polar germ tubes, while less mature spores may produce germ tubes from intermediate segments. Sporulation requires both light and dark periods, stimulated by near-ultraviolet light and inhibited by blue light. Sporulation can be poor under various environmental and nutritional conditions

6.2.2.3 Disease Cycle

The primary sources of inoculum of fungus are stubbles, straw and grains surviving in soil and infected plant parts. Many plant species including *Cynodondactylon*, *Digitarias anguinalis*, *Echinochloa colona*, etc. have been reported to be infected by the pathogen (Sunder et al. 2014). The primary infection is usually initiated by the infected seed as necrotic lesions on coleoptiles and the sheath of the first leaves. The subsequent lesions on leaves arise from secondary infection by airborne spores produced on primary lesions (Ou 1985). The disease is aggravated by prolonged leaf wetness, high soil pH, late planting, and nutritional imbalances.

6.2.2.4 Management

- Use of balanced fertilizer to avoid the incidence of disease.
- Use of resistant varieties like NDR-359, CR-1, CR-2.
- Seed treatments with *Trichoderma harzianum* and *T. viride* have reduced disease by 70% (Biswas et al. 2010).
- Spray of botanicals Neemazal @ 3 ml/l or Wanis @ 5 ml/l and *T. viride* @ 10 g/l is found effective.
- Seed treatment with *T. viride* + foliar spray of propiconazole was proved to be successful (Singh et al. 2017).
- The disease is effectively managed by a spray application of fungicides propiconazole or pyraclostrobin+ epoxyconazole @ 0.1% (Chauhan et al. 2021).
- To manage the secondary infection mancozeb + tricyclazole provide effective control.

6.2.3 Sheath Blight

Sheath blight, a devastating fungal disease affecting rice worldwide, was first reported by Miyake in Japan in 1910. In India, it was first noted in Gurdaspur (Punjab) in 1963 by Paracer and Chahal, and later in Uttar Pradesh by Kohli in 1966. Yield losses vary from 20% to 50%, potentially reaching 70% under severe conditions (Singh et al. 2016).

6.2.3.1 Symptoms

The first symptoms of sheath blight appear as ellipsoid or ovoid, slightly irregular lesions on the leaf sheath, ranging from 1 to 3 cm in length. Initially greenish-grey, these lesions turn greenish-white with brown margins as they develop (Figure 6.3a). Under favorable conditions, the infection spreads quickly to upper leaf sheaths and leaf blades, where sclerotia form. The lesion size, color, and sclerotia formation are influenced by environmental conditions. Symptoms also appear on leaf blades, starting as greenish-grey lesions that enlarge and turn greyish-white with brown margins (Figure 6.3b). In severe infections, white mycelial growth and small brown to black sclerotia cover the panicles. Infected grains become chaffy, partially filled, shriveled, lighter in weight, and discolored.

FIGURE 6.3 (a) Symptom of sheath blight on leaves. (b) Symptom on sheath of rice plant.

6.2.3.2 Causal Organism

The disease is caused by *Rhizoctonia solani* Kuhn (AG 1 IA) [Teleomorph: *Thanatephorus cucumeris* (Frank) Donk], Out of 14 anastomosis groups of *R. solani*, sheath blight fungus has been placed in AG-1 IA (Gonzalez-Vera et al. 2010). In India, Korea, the Philippines and the USA, *R. solani* isolates infecting rice belonged to AG-1 IA group while in Japan and Taiwan to AG-1 IA and AG-2-2. However, in Indonesia and China, they belonged to AG-1 IA, AG2-2, AG-4, and AG-1 IA, AG-1 IB, AG-1 IC, AG-4 and AG-Bb (binucleate isolates), respectively. Lore et al. (2015) observed that sheath blight can be caused by *R. solani* and *R. oryzae sativae* occurring singly or in combination in different parts of India, though *R. solani* is the dominant species. They also found that mixed inoculation of *Rhizoctonia* species has been found to aggravate the disease complex.

6.2.3.3 Disease Cycle

Sclerotia and fungal mycelium in plant debris are primary sources of infection. Sclerotia can survive long periods under unfavorable conditions, and their dispersal is facilitated by rainwater runoff and flood irrigation. Weeds like *Cynodon dactylon* and *Echinochloa crusgalli* in and around paddy fields also contribute to primary infection (Singh et al. 2012). The pathogen can be seed-borne, with seed infection rates reported between 10% to 39% (Dasgupta 1992). Sheath blight thrives in warm, humid conditions (28–30°C, 96–97% RH) and is exacerbated by the cultivation of high-yielding, nitrogen-responsive varieties, close planting, and heavy nitrogen fertilizer use, which increase humidity and plant-to-plant contact.

6.2.3.4 Management

- Optimum spacing should be maintained.
- Use of green manure / FYM to promote antagonistic microflora (Rodrigues et al. 2003).
- To avoid excess doses of nitrogenous fertilizers.
- Use of polyethylene glycol 400-based nano-formulations (0.5%) of these compounds demonstrated excellent *in vivo* antifungal activity under net-house conditions as compared to other chemicals (Siddhartha et al. 2020).
- Use of a combination of *T. harzianum* + *Pseudomonas fluorescens* + Herbal Kunapajala (Seed treatment + Soil drenching + foliar spray) was found effective in managing the disease (Kabdwal et al. 2023).

- Azoxystrobin (alone or in combination with propiconazole), trifloxystrobin plus propiconazole, pyraclostrobin and flutolanil fungicides @ 0.15% was found effective (Uppala and Zhou 2018; Behera et al. 2022).

6.2.4 False Smut

False smut is an emerging rice disease causing yield losses between 0.2% and 49% (Dodan and Singh 1996). First reported from Tirunelveli in Tamil Nadu in (Cooke 1878), it was once considered a symbol of a good harvest in southern India, known as the "Laxmi" (goddess of wealth and prosperity) disease. However, the extensive use of nitrogen fertilizers, hybrid cultivars, and climate change has turned it into a major problem (Ladhalaksmi et al. 2012, 2018; Wang et al. 2019; Maurya et al. 2021).

6.2.4.1 Symptoms

The pathogen produced typical symptoms only after flowering when it infected the florets and transformed individual healthy grains of the panicle into a whitish smut ball. Later changed to yellowish, yellowish-orange, green, olive green and finally to greenish black. Spore balls were small and restricted between the glume which slowly enlarged and became double the size of a healthy grain (Figure 6.4 a,b,c). The number of smut balls per panicle varied depending on the severity of the disease.

6.2.4.2 Causal Organism

The disease is caused by *Ustilaginoidea virens* (Cooke) Takahashi [Teleomorph: *Villosiclava virens* (Nakata) Tanaka and Tanaka]. *Ustilaginoidea virens*, belongs to the division Ascomycota, subdivision Pezizomycotina and class Sordariomycetes. The fungus grows slowly on many defined agar media with slow growth as 1 cm/week. The fungus usually forms sclerotia and also produces asexual spores, conidia. The sclerotia were compact and round, initially yellow-orange, later become olive green and then later stage formed conidia with jet black color. The size of conidia was 3.0-5.5 μm × 5.5-6.0 μm and the ultrastructure of conidia shows a projected spine in conidia with the 200–500 nm size. The pathogen grows at 27±2°C on the cultured media.

6.2.4.3 Disease Cycle and Epidemiology

Sclerotia serve as the primary inoculum for rice infection, releasing ascospores during crop anthesis that infect floral parts and produce chlamydospores responsible for secondary infections. These chlamydospores, both airborne and soil-borne, can reinfect rice panicles. In hilly regions, sclerotia initiate infection, while in plains, chlamydospores are primary. The pathogen also affects maize (Gohel et al. 2014) and various weeds like *Echinochloa crus-galli, Imperata cylindrical*, etc. (Wang et al. 2019). Disease development is promoted by low temperatures, especially at night, high humidity (>90%), and rainy conditions during the flowering period favor the disease (Raji et al. 2016).

6.2.4.4 Management

- Bunds around the field should be clean.
- The date of sowing was adjusted to avoid the coincidence of the booting stage with rainy days.
- Used a balanced dose of fertilizers.
- Grow early- and medium-duration varieties that escape the false smut disease.
- *B. subtilis* and *T. harzianum* have maximum potential to control false smut of rice (Baite et al. 2022).
- Spray of fungicides propiconazole @ 0.1% or trifloxystrobin + tebuconazole @ 0.04% at the booting stage was found highly effective against the disease (Ladhalakshmi et al. 2018).

(a)

(b)

(c)

FIGURE 6.4 (a) Symptom of false smut of rice with color change of smut balls – white. (b) Symptom of false smut of rice with color change of smut balls – yellow. (c) Symptom of false smut of rice with color change of smut balls – black.

6.2.5 Bakanae or Foolish Seedling Disease

Rice bakanae, also known as foolish seedling disease, is a major rice disease, especially at the seedling stage, prevalent globally but primarily in Asia (Rood 2004). The term "Bakanae" comes from Japanese, meaning foolish seedling. Significant yield losses due to bakanae outbreaks have been reported, particularly in western Uttar Pradesh, India (Jain et al. 2019), affecting aromatic and improved rice cultivars (Yadav et al. 2020). The disease, first reported in Japan in 1828 (Ito and Kimura 1931), can cause yield losses ranging from 20% to 95% under favorable conditions, with severe impacts observed in countries like Australia, India, Iran, etc (Singh and Sunder 2012).

6.2.5.1 Symptoms

The disease initially appears as elongated, slender, pale seedlings scattered in the field. Infected seedlings may become stunted and chlorotic, and show crown or foot rot symptoms, affecting both the nursery and main field stages. Symptoms include white or pink mycelial growth, adventitious roots from lower nodes, wider leaf angles, stem elongation, empty panicles, and grain discoloration (Figure 6.5) (Bashyal et al. 2022). Surviving plants bear only empty panicles, and in Japan, infected panicles often turn pink. The fungus can cause grain infection at maturity and pre-emergence seedling death, with symptoms influenced by the fungus strain, inoculum level, and soil conditions (Singh and Sunder 2012). Elongation symptoms are more prevalent in high moisture conditions, while rotting occurs in low soil moisture (Yadav et al. 2020). At the end of the season, lower sheaths may turn black with pycnidia and perithecia under favorable conditions.

6.2.5.2 Causal Organism

The disease is caused by the ascomycetous fungus *Fusarium fujikuroi* Nirenberg (teleomorph: *Gibberella fujikuroi* (Sawada) Wollenw). Perithecia of *G. fujikuroi* are dark blue and globose to conical, 200–400 × 150–400 μm in dimension. Asci are ellipsoid to clavate and contain 4–8 hyaline, elliptical, 1–3 septate ascospores measuring 15–18 × 6.7 μm. Microconidia are hyaline, oval to clavate, aseptate, single-celled, 5–12 × 3–3.5 μm and borne basipetally in chains of 15–32 spores on the host. Macroconidia are vaguely sickle-shaped to almost straight, with a foot-shaped

(a)

FIGURE 6.5 (a) Symptoms of bakanae disease of rice – elongation. (b) Symptoms of bakanae disease of rice – rotting at the maximum tillering stage. (c) Symptoms of bakanae disease of rice – empty panicles at the panicle emergence stage (Bashyal 2018).

(b)

(c)

FIGURE 6.5 (Continued)

basal cell, usually 3–5 septate, and measure 8.4–66 × 2.4–3.5 μm (Ou 1985). The fungus does not produce chlamydospores. Dark blue sclerotia are sometimes present.

6.2.5.3 Disease Cycle

The disease is seed-borne, spreading from field to field via infected seeds. Infection occurs through airborne conidia and ascospores from diseased plants, with the fungus being both internally and externally seed-borne. It grows intercellularly in rice stigma and anthers, reaching and covering the ovary within 48 hours. The fungus survives in soil for about four months as thick-walled hyphae or macroconidia and can remain viable in seeds and plant parts for 4–10 months (Pannu et al. 2012). It infects seedlings through roots or crowns, growing systemically and producing growth hormones like gibberellins and fusaric acid. The pathogen concentrates in vascular bundles and is favored by high temperatures (30–35°C), root injury during uprooting, and excessive nitrogen fertilizer. The sprouting period is highly susceptible to infection (Chan et al. 2004).

6.2.5.4 Management

- Always use clean seed or use salt water to separate light seed which may reduce seed-borne inoculums of the pathogen.
- Destruction and burning of crop residues which also help to limit the inoculums.
- Seeds soaked in 0.3% sodium hypochlorite solution for 12 hours were found effective (Shin et al. 2014).
- Disease is suppressed by the irradiation of infected rice seeds with atmospheric plasma (Ochi et al. 2017).
- Seed treatment with *Trichoderma* sp. @ 6 g/kg seed was found effective (Pal et al. 2019).
- Use of 0.2% Carbendazim as nursery drenching is effective (Anonymous 2018).
- Grow resistant varieties like Athadapunnu, Chandana, PAU 201, Pusa 1342, BPT 5204, Himju and Suphala (Fiyaz et al. 2014).
- Seedling treatment with Carbendazim 50% WP or Tebuconazole 50%+Trifloxystrobin 25% w/w (75 WG) was found effective against the disease (Bashyal et al. 2022).

6.2.6 Sheath Rot

Sheath rot is a serious rice disease affecting many regions, causing yield losses ranging from 3–90% (e.g., Taiwan 3–20%, Philippines 53%, Thailand up to nearly 100%, USA 11%, India 3–90%) (Manibhushan Rao 1996). The average loss is about 15%. First reported in India by Agnihothrudu in 1973, it has since been widely observed across the country.

6.2.6.1 Symptoms

The pathogen infects crops shortly after the flowering stage, causing rot on the uppermost leaf sheath enclosing the young panicles. These results in oblong lesions with brown margins and grey centers, which can enlarge, coalesce, and cover most of the sheath (Figure 6.6), hindering or aborting panicle emergence. Severe early infection may lead to "choking," where panicles remain compressed within the sheath, producing chaffy, discolored grains. This infection also causes stunted growth, premature plant death, and reduced seed viability and nutritional value (Gopalakrishnan et al. 2010). A diffuse reddish-brown discoloration may also be seen in the sheath. The plants may stunt and die prematurely.

FIGURE 6.6 (a) Rice sheath rot symptoms on sheath. (b) Rice sheath rot symptoms on panicle severe infection.

6.2.6.2 Causal Organism

Rice sheath rot is a disease complex that can be caused by various fungal and bacterial pathogens (Berhan et al. 2020). Two fungal pathogens, *viz., Sarocladium oryzae* (Sawada) Gams and Hawksworth and *Fusarium moniliforme* Sheldon, can cause sheath rot symptoms, *Sarocladium oryzae* being the most predominant. *Sarocladium oryzae* produces verticillate conidiophores with one to two appressed whorls of branches. Conidia are produced on phialides, cylindrical to slightly fusiform, often somewhat curved, hyaline and single-celled measuring 3.5–9×1–2.5 µm. The morphological characteristics of *F. moniliforme* isolates causing sheath rot are similar to those described under foot rot and bakanae. However, the strains of *Fusarium moniliforme* causing bakanae do not induce sheath rot.

6.2.6.3 Disease Cycle and Epidemiology

The pathogen *S. oryzae* survives as mycelium in infected seeds, crop residues (Bigirimana et al. 2015) and certain grasses and weeds like *Eleusine indica* and *Echinochloa crusgalli*. It infects rice at various stages, particularly after the booting stage, spreading intercellularly in susceptible cultivars. Disease severity increases in plants stressed by factors like viruses, pests, and nutrient deficiency (Pearce et al. 2001). The disease is more intense in late, densely planted fields (Pal and Mandal 2020) under conditions of moderate temperature (20–30°C), high humidity, cloudy days, and occasional rainfall.

6.2.6.4 Management

- Early sowing, i.e. sowing within the last week of June, is found to be the best sowing time to escape sheath rot disease (Pal and Mandal 2020).
- Gypsum-based formulation of biocontrol agents, *viz., Pseudomonas fluorescens* and *Bacillus subtilis* were found more effective against the disease (Manonmani et al. 2023).
- *Pleurotus djamor* was found to be an effective degrader of coirpith and reduced disease incidence (Theradimani et al. 2019).
- Three sprays of fungicide combination, *viz.,* trifloxystrobin 25%+tebuconazole 50 % (0.04 %) at 15-day intervals starting the first spray immediately after disease appearance was found to be most effective (Balgude et al. 2019).

FIGURE 6.7 Symptom of bacterial leaf blight of rice and bacterial oozing on the surface of the leaf of rice. (Laha et al. 2017.)

6.2.7 Bacterial Leaf Blight

Bacterial leaf blight was first identified in 1884 in southern Japan (Nino-Liu et al. 2006). Later, it was reported in other countries, including India, where it was first found in Maharashtra (Srinivasan et al. 1959). The disease spread rapidly with the cultivation of high-yielding, nitrogen-responsive varieties like TN-1, causing yield losses of 20% to 50% (Singh et al. 2011).

6.2.7.1 Symptoms

Leaf blight phase: water-soaked lesions appear at the leaf tips, spreading downward longitudinally. Initially pale green, these lesions turn yellow to straw-colored with wavy margins. Lesions may develop at the leaf edges, midrib, or elsewhere on the leaf blade. As the disease progresses, the entire leaf blade becomes white, eventually turning greyish or blackish due to saprophytic fungi growth. In humid conditions, yellowish, opaque bacterial ooze forms on young lesions, drying into small, yellowish beads (Figure 6.7).

Kresek or wilt phase: Early systemic infection in the nursery or from seed is the most destructive stage of this disease in tropical areas. The leaves are dropping completely, turning yellow or grey and finally, the tillers start to pull out. The affected hills may be completely destroyed, if the situation so requires.

6.2.7.2 Causal Organism

The disease is caused by the bacterium *Xanthomonas oryzae* pv. *oryzae* (Ishiyama) Swings et al. (Nino-Liu et al. 2006). The bacterium is rod-shaped 1.2 x 0.3–0.5 µm. They are single, occasionally in pairs, gram-negative, non-sporeforming and devoid of capsules. Their colonies are pale yellow, circular and smooth with an entire margin. They are convex and viscoid. Shekhar et al. (2020) reviewed the work on the epidemiology of BLB and its geographical distribution along with strategies for their management.

6.2.7.3 Disease Cycle

Weeds, rice stubble, and ratoons of infected plants are the primary inoculum. The pathogen survives briefly on infected seeds and in soil but is not a primary inoculum source. Secondary spread occurs through wounds, stomata, irrigation water, rain, and plant contact. Cloudy, rainy weather, floods, strong winds, excessive nitrogen fertilizer, and moderate temperatures (28–30°C) favor rapid disease development and spread.

6.2.7.4 Management

- Seed soaking for 8 h in agrimycin 0.025% and wettable ceresan 0.05% followed by hot water treatment at 52 to 54°C for 30 min.
- Seed treatment and foliar spray of Salicylic acid @ 100 mg/l can reduce the severity of disease by 71% (Thepbandit et al. 2023).
- Removal of infected plant debris, self-grown rice plants, ratoons and infected wild rice.
- Grow resistant varieties like Ajaya, TKM-6, IR-8, Improved Sambha Mahsuri and improved basmati, etc.
- Recommended doses of NPK (80:40:40) + higher doses of potash (24 kg/ha) + three sprays of fresh cow dung slurry @ 1 kg in 10 l water at 10-day intervals was found effective (Mandal et al. 2017).
- Three applications of 0.1 % Agricin at 10-day intervals starting from the jointing stage were found effective in controlling the disease (Sharma et al. 2022).
- A spray of *Paenibacillus polymyxa* @ 7.5 ml/l was found effective (Mumpuni and Rohmah 2021).
- Seed treatment and root dip of rice seedlings in *Pseudomonas fluorescens* with a spray of Datura leaf extract 15% was found quite effective.
- Seedling root dip with streptocycline (0.01%) and spraying of streptocycline (0.01%) along with copper oxychloride (0.2%) was found effective against the disease.

6.2.8 Bacterial Leaf Streak

Bacterial leaf streak (BLS) caused by *Xanthomonas oryzae* pv. *oryzicola* (Fang et al.) Swings et al. was first reported from the Philippines as bacterial leaf stripe. The disease is currently widespread and of economic importance in tropical Asian countries like India, Indonesia, Malaysia, Cambodia, Bangladesh and Thailand (Nino-Liu et al. 2006). Bacterial leaf streak is economically less important compared to bacterial blight of rice. It can seriously affect the quantity and quality of the rice (Chen et al. 2021).

6.2.8.1 Symptoms

Initial symptoms appear as fine, water-soaked to translucent interveinal streaks of various lengths anywhere on the leaves (Figure 6.8). The lesions enlarge, turn yellowish-orange to brown (depending on cultivar) and eventually coalesce lengthwise resulting in bigger streaks. With time, the leaves may be completely blighted. Numerous tiny yellow beads of bacterial exudate are commonly found on the lesions. In many cases, concurrent occurrence of both BB and BLS in the same field or on the same leaf has also been noted.

FIGURE 6.8 Symptoms of bacterial leaf streak showing translucent interveinal streaks. (Laha et al. 2017.)

6.2.8.2 Causal Organism

The morphological characters and taxonomic position of *Xanthomonas oryzae* pv. *oryzicola* are similar to that of *X. oryzae* pv. *oryzae*.

6.2.8.3 Disease Cycle and Epidemiology

The primary source of inoculum is infected seeds, where the pathogen can survive between seasons, though not in debris. Infected wild rice can also spread the disease. The bacterium enters leaves through stomata or wounds, colonizing intercellular spaces and eventually oozing out, spreading via contact, wind, and rain. Disease development is favored by drizzling, stormy weather, high humidity, moderate temperatures, and excessive nitrogen fertilizer use.

6.2.8.4 Management

- Use of disease-free seed.
- Early planting date (i.e. June) of rice can effectively control the disease (Zougrana et al. 2022).
- Use of varieties like Zenith, Tetep, H 4, S 67, Co 4, BJ 1 and DZ 60 has been reported to be resistant (Ou 1985).
- Seed treatment by soaking the seeds in 0.025% Streptocycline for 12 h or by hot water treatment at 50°C for 30 minutes can prove effective against this disease (Laha et al. 2017).

6.2.9 Bacterial Panicle Blight

Bacterial panicle blight is a severe and emerging rice disease first reported in Japan in 1956. It has since spread to many rice-growing regions of the world (Kumar et al. 2017), including South and Central America, and is widespread across Asia (Chompa et al. 2022). In severe cases, it can cause economic yield losses of up to 75%, posing a significant risk in tropical and subtropical countries (Ham et al. 2011).

6.2.9.1 Symptoms

Bacterial panicle blight can affect rice at any developmental stage, from seed to maturity. In early stages, it causes seed rot, stunted growth, and chlorosis in seedlings. During vegetative stages, it leads to sheath rot and grey lesions on leaves (Nandakumar et al. 2009). The most severe impact occurs when the bacteria infect reproductive tissue, causing spikelet discoloration and hindering grain development (Figure 6.9 a, b). In severe cases, entire panicles darken, remain erect, and fail to fill with grain, leading to significant yield losses as most panicles remain empty, affecting overall grain production (Fory et al. 2014).

(a)

(b)

FIGURE 6.9 (a) Symptoms of bacterial panicle blighting – panicle blighting and grain rot. (b) Symptoms of bacterial panicle blighting – flag leaf sheath rot. (Xianglong 2004.)

6.2.9.2 Causal Organism

This disease is caused by *Burkholderia glumae*, a rod-shaped, gram-negative, non-fluorescent, and polar flagellated bacterium. The bacterial colony appears greyish white or yellow, convex, glossy because of the pigment present, and 3.0 to 6.0 mm in diameter. The bacterium can grow at pH range of 5.5–8.5 (optimum 6.5 and 7.0). Its optimum growth temperature is around 30°C, but it can grow even at 41°C.

6.2.9.3 Disease Cycle and Epidemiology

It is primarily a seed-borne disease, though inoculum can survive in soil, leaf sheaths, and panicles. The pathogen infects seeds, invades through stomata or wounds, and proliferates during seed germination. It grows epiphytotically from the booting stage, spreading from plant to plant (Sayler et al. 2006). Warm night temperatures and high humidity increase BPB disease incidence (Cha et al. 2001), especially during the rice heading stage. The pathogen moves long distances via the vascular system, reducing grain weight (Jeong et al. 2003). Optimal growth occurs between 30–35°C, with a thermal death point at 70°C. The disease cycle remains partially understood

6.2.9.4 Management

- Early planting and appropriate use of nitrogen fertilizer.
- Rice seeds treated at 65°C of dry heat for 6 days can eradicate the BPB pathogens (Zeigler et al. 1989).
- An antagonistic *Pseudomonas* spp. strain when applied to seeds was effective in reducing the *B. glumae* populations in seeds (Inoue et al. 2001).
- Disease is suppressed by irradiation of infected rice seeds with atmospheric plasma (Ochi et al. 2017).
- Application of oxolinic acid at the booting stage to the heading stages reduced BPB infection by up to 88% (Zhou et al. 2011).

6.3 CONCLUSIONS

The rice crop suffers from several biotic and abiotic stresses that drastically reduce its production and productivity resulting in huge economic loss all over the world. Among biotic stress, the major factor is disease caused by fungi, bacteria, viruses and nematodes, etc. Among diseases, blast, bacterial leaf blight, brown leaf spot, sheath blight, bakanae disease, sheath rot, bacterial panicle blight and false smut are economically important. Recently, due to changes in the global temperature and climatic conditions minor diseases are becoming major (false smut of rice and rice root-knot nematode). Integration of all the management tactics (Integrated Disease Management) practices is the most feasible, effective and successful management of plant disease in rice. Though diseases like sheath blight, bakanae and false smut till now no resistant varieties have been developed. Chemical management is the only solution for this disease while a number of resistant varieties are available against blast and bacterial leaf blight. Therefore integration of integrated disease management practices like seed treatment, cultural practices, use of organic manures and application of balanced fertilizers, biological control and judicious use of chemical fungicides may applied for the protection of rice crops from these diseases.

REFERENCES

Agnihothrudu V (1973) *Acrocylindrium oryzae* Sawada sheath rot on paddy. *Kavaka* 1: 69–71.

Anonymous (2018) *IARI. Annual report 2017–18*. Indian Agricultural Research Institute, New Delhi, India.

Asibi AE, Chai Q, Coulter JA (2019) Rice blast: A disease with implications for global food security. *Agron* 9: 451. doi: 10.3390/agronomy9080451

Aslam, HMU, Khan NA, Hussain SI, Ali Y, Raheel M, Shahzad R, Jamil S, Yasin O, Ali S, Amrao L (2022) First report of brown leaf spot of rice (*Oryza sativa*) caused by *Bipolaris sorokiniana* in Pakistan. *Plant Dis* 106: 1750 doi: 10.1094/PDIS-05-21-1097-PDN

Baite MS, Prabhukarthikeyan SR, Raghu S (2022) Biological control of a fungus *Ustilaginoidea virens* causing false smut of rice. *BioControl* 67: 357–363.

Balgude YS, Kshirsagar CR, Gaikwad AP (2019) Evaluation on the efficacy of modern fungicides against blast and sheath rot of rice. *Int J Curr Microbiol App Sci* 8(3): 83–88.

Balol G, Talekar SC, Satish RG (2022) Efficacy of various fungicides on the management of rice blast disease caused by *Pyricularia oryzae* (Cav.) under field condition in Dharwad district of Karnataka. *Intl J Plant Sci* 17: 214–217.

Bashyal BM (2018) Etiology of an emerging disease: bakanae of rice. *Indian Phytopath* 71: 485–494.

Bashyal BM, Gupta AK, Parmar P, Yadav J, Choudhary R, Kumar R, Singh D, Aggrawal R (2022) Management of bakanae disease using fungicides and their effect on disease symptomatology. *Indian J Agri Sci* 92(9): 56–61.

Behera SK, Beura S, Rout MR (2022) Efficacy of fungicides for the management of sheath blight of rice. *Intl J Res Trends Innov* 7: 119–122.

Berhan M, Yalew D, Zeleke T (2020) Evaluation of fungicides efficacy against rice sheath rot disease (*Sarocladium oryzae*) in rain fed low land rice (*Oryzae sativa* L.) in fogera hub. *Int J Agri Biosci* 9(5): 221–225.

Bigirimana VDP, Hua GKH, Nyamangyoku OI, Hofte M (2015) Rice sheath rot: An emerging ubiquitous destructive disease complex. *Fron Pl Sci* 6: 1066. doi: 10.3389/fpls.2015.01066.

Biswas C, Srivastava SSL, Biswas SK (2010) Effect of biotic, abiotic and botanical inducers on crop growth and severity of brown spot in rice. *Indian Phytopath* 63: 187–191.

Cha KH, Lee YH, Ko SJ, Park SK, Park IJ (2001) Influence of weather condition at heading period on the development of rice bacterial grain rot caused by *Burkholderia glumae*. *Res Plant Dis* 7: 150–154.

Chakrabarti NK (2001) Epidemiology and disease management of brown spot of rice in India. In: Sreenivasaprasad S Johnson R (eds) *Major fungal diseases of rice recent advances*. Kluwer Academic Publisher, pp. 293–306.

Chan ZL, Ding KJ, Tan GJ, Zhu SJ, Chen Q, Su XY, Ma K, Wang AE (2004) Epidemic regularity of rice bakanae disease. *J Anhui Agri Univ* 31(2): 139–142.

Chauhan V, Kumar A, Singh R (2021) Evaluation of fungicides for management of brown leaf spot disease of rice. *Middle East J App Sci Tech* 4: 75–79.

Chen S, Zhang K, Zhao Y, Sun Y, Ban W, Chen Y, Zhuang H, Zhang X, Liu J, Yang T (2021) An approach for rice bacterial leaf streak disease segmentation and disease severity estimation. *Agric* 11: 420.

Chompa SS, Akter A, Sadeq ABM, Rahman ME, Rashid HO, Ibnat N, Hossain MB (2022) An overview of major bacterial diseases of rice and management strategies for their control in malaysia. *Global Scientific J* 10(4): 1074–1102.

Cooke MC (1878) Some extra–European fungi. *Grevillea* 7: 13–15.

Dasgupta MK (1992) Rice sheath blight: the challenge continues. In: Singh US, Mukhopadhayay AN, Kumar J, Chaube HS (eds) *Plant diseases of international importance: diseases of cereals and pulses*, vol I. Prentice Hall, Englewood Chiffs, pp. 130–157.

Dean R, Kan JALV, Pretorius ZA, Hammond-Kosack KE, Pietro AD, Spanu PD, Rudd JJ, Dickman M, Kahmann R, Ellis J, Foster GD (2012) The top ten fungal pathogens in molecular plant pathology. *Mol Pl Pathol* 13: 414–430.

Dodan DS, Singh R (1996) False smut of rice: Present status. *Agric Rev* 17: 227–240.

Fiyaz RA, Krishnan SG, Rajashekara H, Yadav AK, Bashyal BM, Bhowmick PK, Singh NK, Prabhu KV, Singh AK (2014) Development of high throughput screening protocol and identification of novel sources of resistance against bakanae disease in rice (*Oryza sativa* L.). *Indian J Gen* 74: 414–422.

Fory PA, Triplett L, Ballen C, Abello JF, Duitama J, Aricapa MG, Prado GA, Correa F, Hamilton J, Leach JE, Tohme J. Mosquera GM (2014) Comparative analysis of two emerging rice seed bacterial pathogens. *Phytopathology* 104(5): 436–444.

Gohel NM, Patil RK, Pandey RN, Prajapati HN (2014) A new report on occurrence of false head smut (*Ustilaginoidea virens* (Cooke) Takahashi) of maize (*Zea mays* L.) from Gujarat India. *The Bioscan* 9(3):1147–1149.

Gonzalez-Vera AD, Bernardes-de-Assis J, Zala M, McDonald BA, Correa-Victoria F, Graterol-Matute EJ, Ceresini PC (2010) Divergence between sympatric rice-and maize-infecting populations of *Rhizoctonia solani* AG-1 IA from Latin America. *Phytopathology* 100(2): 172–182.

Gopalakrishnan C, Kamalakannan A, Valluvaparidasan V (2010) Effect of seed borne *Sarocladium oryzae,* the incitant of rice sheath rot on rice seed quality. *J Plant Protec Res* 50: 98–102.

Gupta A, Pathak H (2016) *Climate change and agriculture in India*. Climate Change Programme, Department of Science & Technology, Ministry of Science & Technology Government of India, New Delhi, p. 19.

Ham JH, Melanson RA, Rush MC (2011) *Burkholderia glumae*: Next major pathogen of rice? *Mol Pl Pathol* 12(4): 329–339.

Inoue H, Nakaho K, Miyagawa H (2001) Efficacy of the antagonistic bacteria CAB-02 on prevention of bacterial grain rot and seedling blight of rice, applied together with seed disinfectants. *Phytopathology* 90: S37.

Ito S, Kimura J (1931) Studies on the Bakanae disease of the rice plant. *Rep Hakkaido Agric Exp Stn* 27: 1–95.

Jain SK, Khilari K, Dongre M, Pal S (2019) Occurrence of Bakanae disease of rice in western Utter Pradesh India. *Int J Curr Microbiol App Sci* 8: 207–212.

Jeong Y, Kim J, Kim S, Kang Y, Nagamatsu T, Hwang I (2003) Toxoflavin produced by Burkholderia*glumae* causing rice grain rot is responsible for inducing bacterial wilt in many field crops. *Plant Dis* 87: 890–895.

Kabdwal BC, Sharma R, Kumar A, Kumar S, Singh KP, Srivastava RM (2023) Efficacy of different combinations of microbial biocontrols agents against sheath blight of rice caused by *Rhizoctonia solani. Egypt J Biol Pest Control* 33: 29.

Kohli CK (1966) Pathogenicity and host range studies on paddy sheath blight pathogen *Rhizoctonia solani* Kuhn. *J Res PAU* 3: 7–40.

Kumar S, Meshram S, Sinha A (2017) Bacterial diseases in rice and their eco-friendly management. *Intl JAgril Sci Res* 7(2): 31–42.

Ladhalakshmi D, Laha GS, Singh R, Karthikeyan A, Mangrauthia SK, Sundaram RM, Thukkaiyannan P, Viraktamath BC (2012) Isolation and characterization of *Ustilaginoidea virens* and survey of false smut disease of rice in India. *Phytopathology* 40: 171–176.

Ladhalakshmi D, Madamsetty SP, Vellaichamy P, Donempudi K, Banda S, Singh R, Prasad V, Lore JS, Jain J, Mariappan S, Laha GS (2018) Geographic distribution of false smut disease of rice in India and efficacy of selected fungicides for its management. *Intl J Pest Manage* doi: 10.1080/09670874.2018.1494865

Laha GS, Singh R, Ladhalakshmi D, Sunder S, Prasad MS, Dagar CS, Babu VR (2017) Importance and management of rice diseases: A global prospective. In: Chauhan BS et al. (eds) *Rice production worldwide.* Springer International Publishing AG, Cham. doi: 10.1007/978-3-319-47516-5_13

Lore JS, Jain J, Hunjan MS, Gargos G, Mangat GS, Sandhu JS (2015) Virulence spectrum & genetic structure of *Rhizoctonia* isolates associated with rice sheath blight in the North region of India. *Eur J Pl Pathol* 143: 847–860.

Mandal D, Pal R, Mohanty AK (2017) Management of bacterial leaf blight of rice in an integrated way. *J Mycopath Res* 54: 539–541.

Manibhushan Rao K (1996) *Sheath rot disease of rice*. Daya Publishing House, Delhi, p. 71.

Manonmani K, Jayasekhar M, Muthusamy M (2023) Biocontrol efficacy of bacterial antagonists in the management of sheath rot disease of rice. *Indian J Agri Res* 42(3): 212–215.

Maurya MK, Singh SK, Yadav VK, Rahul SN, Vishwakarma SP (2021) Inhibitory effect of botanicals and bioagents against *Ustilaginoidea virens* (Cooke) Takah. causing false smut of rice. *Ann Phytomed* 10(2): 515–520.

Mumpuni RP, Rohmah AJ (2021) Application of biological control *Paenibacillus polymyxa* towards bacteria leaf blight disease in rice plant. *7th Int Conf Sustain Agric Environ* 637: 012040. doi: 10.1088/1755-1315/637/1/012040

Nandakumar R, Shahjahan AKM, Yuan XL, Dickstein ER, Groth DE, Clark CA, Cartwright RD, Rush MC (2009) *Burkholderia glumae* and *B. gladioli* cause bacterial panicle blight in rice in the southern United States. *Plant Dis* 93(9): 896–905.

Neupane N, Bhusal K (2021) A review of blast disease of rice in Nepal. *J Pl Pathol Microbiol* 11: 528. doi: 10.35248/2157-7471.20.12.528

Nino-Liu, DO, Ronald PC, Bogdanove AJ (2006) *Xanthomonas oryzae* pathovars: model pathogens of a model crop. *Mol Pl Pathol* 7: 303–324.

Ochi A, Konishi H, Ando S, Sato K, Yokoyama K, Tsushima S, Yoshida S, Morikawa T, Kaneko T, Takahashi H (2017) Management of bakanae and bacterial seedling blight diseases in nurseries by irradiating rice seeds with atmospheric plasma. *Plant Pathol* 66: 67–76.

Ou SH (1985) *Rice diseases*, 2nd edn. Commonwealth Mycological Institute, Kew, Surrey, p. 380.

Pal R, Mandal D (2020) Effect of different dates of sowing on the occurrence and severity of sheath rot disease in rice. *J Pharmacog Phytochem* SP6: 489–491.

Pal S, Khilari K, Jain SK, Singh J, Kumar A, Kumar A (2019) Management of bakanae disease of rice through combination of *Trichoderma* spp. and fungicides. *Int J Curr Microbiol App Sci* 8(11): 494–501.

Pannu PPS, Kaur J, Singh G, Kaur J (2012) Survival of *Fusarium moniliforme* causing foot rot of rice and its virulence on different genotypes of rice and basmati rice. (Abstr.). *Indian Phytopath* 65S: 149–209.

Paracer CS, Chahal DS (1963) Sheath blight of rice caused by *R. solani* Kuhn-A new record in India. *Curr Sci* 32: 328–329.

Pearce DA, Bridge PD, Hawksworth DL (2001) Species concept in *Sarocladium*, the causal agent in sheath rot in rice and bamboo blight. In: Sreenivasa Prasad S, Johnson R (eds) *Major fungal diseases of rice: recent advances*. Springer, Dordrecht, pp. 285–292.

Raji P, Sumiya KV, Dhanya S, Remya K, Narayanankutty MC (2016) Screening of rice varieties and *in vitro* evaluation of botanicals against false smut pathogen, *Ustilaginoidea virens*. *Int J Agri Sci Res* 6: 79–86.

Rodrigues FÁ, Vale FXR, Datnoff LE, Prabhu AS, Korndörfer GH (2003) Effect of rice growth stages and silicon on sheath blight development. *Phytopathol* 93: 256–261.

Rood MA (2004) Bakanae in-field yield loss. *Rice J* 15: 8–10.

Sathe AP, Kumar A, Mandlik R, Raturi G, Yadav H, Kumar N, Shivaraj SM, Jaswal R, Kapoor R, Gupta SK, Sharma TR, Sonah H (2021) Role of silicon in elevating resistance against sheath blight and blast diseases in rice (*Oryza sativa* L.). *Pl Physiol Biochem* 166: 128–139.

Sayler RJ, Cartwright RD, Yang Y (2006) Genetic characterization and real-time PCR detection of *Burkholderia glumae*, a newly emerging bacterial pathogen of rice in the United States. *Plant Dis* 90: 603–610.

Sharma P, Baidya S, Kandel S, Chaudhary S, Magar PB (2022) Management of bacterial leaf blight disease of rice in farmer's field condition at Bhaktapur districts of Nepal. *J Agric Nat Resources* 5: 105–112.

Shekhar S, Sinha D, Kumari A (2020) An overview of bacterial leaf blight disease of rice and different strategies for its management. *Int J Curr Microbiol Appl Sci* 9: 2250–2265.

Shin DB, Goh J, Lee BC, Kang IJ, Kang HW (2014) Use of sodium hypochlorite for the control of Bakanae disease in rice. *Res Plant Dis* 20(4): 259–263.

Shivappa R, Navadagi DB, Baite MS, Yadav MK, Rathinam PS, Umapathy K, Pati P, Rath PC (2021) Emerging minor diseases of rice in India: losses and management strategies. In: Huang M (ed) *Integrative advances in rice research*, Intech Open, London. doi: 10.5772/intechopen.99898

Siddhartha, Verma A, Bashyal BM, Gogoi R, Kumar R (2020) New nano fungicides for the management of sheath blight disease (*Rhizoctonia solani*) in rice. *Intl J Pest Man* 68(3), 217–226. https://doi.org/10.1080/09670874.2020.1818870.

Singh A, Singh VK, Singh SP, Pandian RTP, Ellur RK, Singh D, Bhowmick PK, Gopala Krishnan S, Nagarajan M, Vinod KK, Singh UD, Prabhu KV, Sharma TR, Mohapatra T, Singh AK (2012) Molecular breeding for the development of multiple disease resistance in Basmati rice. *AoB Plants* pls029. doi: 10.1093/aobpla/pls029

Singh L, Lal AA, Kumar P, Simon S, Kumar M (2017) Management of brown spot disease of rice by safer fungicides and some bioagents. *Plant Archiev* 17(2): 1020–1022.

Singh R, Sunder S (2012) Foot rot and bakanae of rice: an overview. *Rev Plant Pathol* 5: 565–604.

Singh R, Sunder S (1997) Foot rot and bakanae of rice: retrospect and prospect. *Int J Trop Plant Dis* 15: 153–176.

Singh R, Sunder S, Kumar P (2016) Sheath blight of rice: Current status and perspectives. *Indian Phytopath* 69: 340–351.

Singh UD, Gogoi R, Mondal KK (2011) Molecular approach for augmenting disease resistance in cereals: rice and maize. *Plant Pathol India: Vision* 2030: 21–30.

Srinivasan MC, Thirumalachar MJ, Patel MK (1959) Bacterial blight disease of rice. *Curr Sci* 28: 469–470.

Sunder S, Singh R, Agarwal R (2014) Brown spot of rice: an overview. *Indian Phytopath* 67: 201–215.

Thepbandit W, Papathoti NK, Daddam JR, Hoang NH, Le Thanh T, Saengchan C, Buensanteal K (2023) *In vitro* and *in vivo* studies of salicylic acid on systemic induced resistance against bacterial leaf blight disease and enhancement of crop yield. *J Integ Agric* 22: 170–184.

Theradimani M, Thangeshwari S, Arulsamy M, Parthasarathy S (2019) Composted *Coirpith* using edible fungi *(Pleurotus* spp.) for the management of rice sheath blight and sheath rot. *Res J Agri Sci* 10(3): 521–525.

Uppala S, Zhou XG (2018) Field efficacy of fungicides for management of sheath blight and narrow brown leaf spot of rice. *Crop Protec* 104: 72–77.

Wang W, Fan J, Jeyakumar JMJ (2019) Rice false smut: an increasing threat to grain yield and Quality'. In: Jia Y (ed) *Protecting rice grains in the post-genomic era*, Intech Open, London. doi: 10.5772/intechopen.84862

Xianglong Y (2004) Identification of bacterial pathogens causing panicle blight of rice in *Louisiana*. Louisiana State University and Agricultural and Mechanical College. LSU Master's Theses. pp. 4–5.

Yadav J, Bashyal BM, Sinha P, Aggrawal R (2020) Effect of different abiotic factors on *symptom* expression and severity of bakanae disease of rice (*Oryza sativa*). *Indian J Agri Sci* 90(2): 386–391.

Yasuda N, Mitsunaga T, Hayashi K, Koizumi S, Fujita Y (2014) Effects of pyramiding quantitative resistance genes Pi21, Pi34, and Pi35 on rice leaf blast disease. *Plant Dis* 99: 904–909.

Zeigler RS, Alvarez E (1989) Grain discoloration of rice caused by *Pseudomonas glumae* in Latin America. *Plant Dis* 73: 368.

Zhang N, Luo J, Rossman AY, Aoki T, Chuma I, Crous PW, Dean R, De Vries RP, Donofrio N, Hyde KD, Lebrun MH, Talbot NJ, Tharreau D, Tosa Y, Valent B, Wang Z, Xu JR (2016) Generic names in Magnaporthale. *IMA Fungus* 7: 155–159.

Zhou XG, Liu G, Kloepper JW, Reddy MS, Kumar KVK, Zhang S (2011) Management of bacterial panicle blight of rice with beneficial *Bacillus* strains. *Phytopathology* 101: S270.

Zougrana S, Koita K, Wonni I (2022) Sowing dates influence incidence of bacterial leaf streak and rice yield under irrigated production systems in Burkina Faso. *J Rice Res Dev* 5: 360–365.

7 Eco-friendly Approaches to Manage Biotic Stresses Caused by Fungal Pathogens in Wheat Crops

Om Prakash Gangwar, Subhash Chander Bhardwaj, Subodh Kumar, Pramod Prasad, Charu Lata and Sneha Adhikari

7.1 INTRODUCTION

Wheat is one of the most important cereal food crops in the world. India is the second largest producer of wheat in the world. In India, approximately 31 Mha area, is under wheat cultivation with a production of 107 million tons (Kumar et al. 2022). Wheat production has been threatened by several biotic stresses caused by pathogens such as fungi, bacteria, viruses and nematodes. These pathogens directly affect the growth and development of wheat plants by depriving them of nutrition resulting in reduced grain yield and production. Economically important biotic stresses on wheat are rusts (leaf/stripe/stem), powdery mildew, head blight, spot and leaf blotches, and wheat blast (Table 7.1). Of biotic stresses, pathogenic fungi pose a significant challenge to wheat production globally.

Historically, wheat rusts caused by fungi have been the biggest hindrance in efforts to sustain wheat production and are currently causing considerable yield losses in different regions of the world. Moreover, other threats like races of stem rust pathogens belonging to Ug99 lineage, virulence to *Yr9* in *Puccinia striiformis* f. sp. *tritici*, and wheat blast have emerged. Under conducive environmental conditions, both stem rust and stripe rust have the potential to cause 100% yield loss, whereas leaf rust results in up to 50% losses on susceptible cultivars (Bhardwaj et al. 2019). Crop losses due to powdery mildew generally range from 10% to 15% and in severe cases up to 50% grain loss may occur particularly when it affects the flag leaf during the heading and grain filling stages (Morgounov et al. 2012). Fusarium head blight (*Fusarium graminearum*) is of great concern to farmers due to its ability to decrease grain yield and health issues due to the production of baleful mycotoxins in infected grains. At a global level, 21.5% yield losses were estimated in wheat crops due to biotic stresses (Savary et al. 2019).

In modern agriculture, agrochemicals are being utilized haphazardly for managing biotic stresses caused by pathogens on wheat crops; however, these chemicals/pesticides adversely affect the health of both humans and animals due to the accumulation of chemical residues in plant tissues and grains. Additionally, pesticides have an adverse effect on soil microbiomes which are advantageous in various ways (Gangwar and Singh 2018). Thus, it is necessary to search for alternative methods for managing various pathogenic biotic stresses. In this chapter, eco-friendly approaches for managing the pathogenic biotic stresses on wheat crops through host resistance, cultural practices and biological methods, as well as innovative biotechnological tools will be discussed. Moreover, the

DOI: 10.1201/9781032631578-7

TABLE 7.1
Major wheat biotic stresses of global importance

S. No.	Biotic stress	Causal organism
1.	Yellow (stripe) rust	*Puccinia striiformis* Westend. f. sp. *tritici* Eriks. & Henn.
2.	Brown (leaf) rust	*Puccinia triticina* Eriks.
3.	Black (stem) rust	*P. graminis* Pers. f. sp. *tritici* Eriks. & Henn.
4.	Powdery mildew	*Blumeria graminis* (DC.) Speerf. sp. *tritici* (Em. Marchal)
5.	Scab/ Fusarium Head Blight (FHB)	*Fusarium graminearum* Schwabe
6.	Spot blotch	*Bipolaris sorokiniana* (Sacc.) Shoemaker
7.	Wheat blast	*Magnaporthe oryzae Triticum* Couch & Kohn
8.	*Septoria tritici* blotch (STB)	*Zymoseptoria tritici* (Desm.) Quaedvlieg & Crous.

remarkable achievements for mitigating the effect of economically important biotic stresses will be focused on.

7.2 DIAGNOSTIC FEATURES OF DIFFERENT MAJOR DISEASES OF WHEAT

7.2.1 Wheat Rusts

Rust diseases caused by the fungus *Puccinia* species are the main biotic constraints to sustaining and enhancing the wheat production. In conducive environmental conditions, both black and yellow rusts have the ability to cause 100% yield losses, whereas brown rust can incur yield loss of up to 50% (Bhardwaj et al. 2016). Rust fungi are macrocyclic and heteroecious in nature. However, in India, rust pathogens are reproducing asexually due to a lack of active alternate hosts (Gangwar et al. 2022). There are three main rust diseases in wheat, i.e. yellow (stripe), black (stem) and brown (leaf) rusts.

7.2.1.1 Yellow (stripe) Rust

Wheat yellow rust, caused by *Puccinia striiformis* Westend. f. sp. *tritici* Eriks. & Henn., infects leaves and leaf sheaths, and many times it can be observed on ears and even on awns. The initial symptoms are long, yellow to orange-colored narrow stripes on leaves, usually between veins in a horizontal manner. The symptoms also develop on leaf sheaths, glumes, and awns under favorable environmental conditions on susceptible wheat plants (Gangwar et al. 2017, Figure 7.1). The mature pustules break open and release yellow-orange masses of urediniospores. The infected plant tissues may become brown and dry as the plant matures. Cool-humid weather conditions with intermittent rains favor the occurrence of yellow rust. Heavy dew or intermittent rains accelerate the spread of the disease.

7.2.1.2 Black (stem) Rust

Black rust caused by *P. graminis* Pers. f. sp. *tritici* Eriks. & Henn., is characterized by reddish-brown, powdery, oblong pustules. The pustules have a characteristic torn margin (epidermis) that could be observed on any side of the leaf, on the stems and even on awns and glumes (Figure 7.2). At the end of the cropping season, the pustules turn black and produce black spores known as teliospores. These pustules occur mainly on the stems and leaf sheaths. Black rust infection leads to leaf defoliation, increase in transpiration water loss, loss of photosynthetic area, and poor grain filling. Warm-humid weather conditions with intermittent rains, hot days (25–30°C) and mild nights (15–20°C) are congenial for the appearance and spreading of black rust disease.

FIGURE 7.1 Yellow rust: (a) On leaf (b) On glumes and awns.

FIGURE 7.2 Black rust: (a) On stem and leaf (b) On glumes and awns.

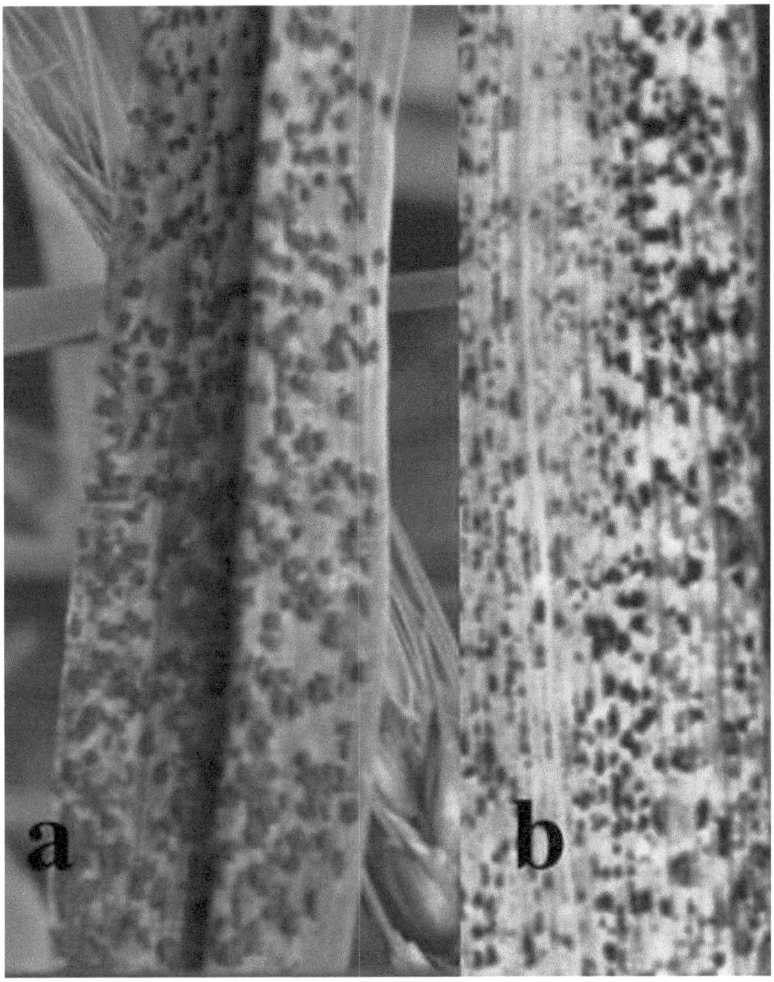

FIGURE 7.3 Brown rust: (a) On leaf (b) Telial sori on leaf.

7.2.1.3 Brown (leaf) Rust

Brown rust, caused by *P. triticina* Eriks., usually infects leaves; however, in severe disease conditions, it may appear on stems, leaf sheaths, glumes and awns. In the initial stages, pustules of brown rust look like small circular to oval orange-brown (rusty red) protuberances on the infected leaf surface (Figure 7.3). As the disease progresses, these pustules turn orange-colored, forming a ring-like sporulation that could be surrounded by a yellow halo. The pustules produce a large number of urediniospores that are easily dissipated by wind. As the disease progresses, the pustules turn to black telial sori, which results in a mixture of orange and black lesions on the same leaf. Warm-humid weather conditions with intermittent rains favor the occurrence of brown rust.

7.2.2 Powdery Mildew

Blumeria grarninis (DC.) Speer f. sp. *tritici* (Em. Marchal), causing wheat powdery mildew, remains dormant on infected plant debris as mycelium. The first obvious symptom of this disease appears as greyish white powdery or fuzzy growth of mycelia, and conidia on the upper surfaces of the leaf, sheath, stem, and floral parts (Figure 7.4). Glumes and awns are also infected when the

FIGURE 7.4 Powdery mildew: (a) On leaf (b) The fruiting bodies of the fungus (Cleistothecia).

disease is severe. Fungal growth is largely limited to outer plant surfaces and can be easily wiped off by rubbing a finger across the affected area. Beneath the mass of mycelial and conidial growth, host tissues turn chlorotic or necrotic and affected leaves may die in severe infections. The favorable temperature for powdery mildew development is 20–21°C with high relative humidity.

7.2.3 Fusarium Head Blight (FHB)

Fusarium graminearum Schwabe is the pathogen of FHB. Head blight symptoms appear on the outer glumes of infected florets which turn slightly darkened and oily. Ear heads or individual spikelets are bleached with the presence of superficial pink/orange mycelium/spores (Figure 7.5). The bleached heads contain scabby seeds (tombstones). The diseased spikelets may have a dark brown discoloration at the base and an orange fungal mass along the lower portion of the glume. Infected grains and the surface of the florets are covered by white, interwoven mycelia. Warm and humid weather during heading and thereafter favor the infection and disease development.

7.2.4 Spot Blotch (SB)

Spot blotch, caused by *Bipolaris sorokiniana* (Sacc.) Shoemaker is a seed and soil-borne disease of wheat. The symptoms of *Bipolaris* are common in seeds, leaves, stems and roots. Initially, the symptom appears as small brown spots, which give a blighted appearance when spots coalesce (Figure 7.6). During tillering the disease initiates from the lower leaves and then spreads to the whole plant with an increase in temperature. Under favorable conditions, the infection occurs all over the plant. Warm weather (high temperature, i.e. >25°C) and high relative humidity favor spot blotch infection.

FIGURE 7.5 Fusarium head blight.

FIGURE 7.6 Spot blotch: (a) At initial stage on leaf (b) At later stage on leaf.

7.2.5 *Septoria tritici* Blotch (STB)

Septoria tritici blotch is caused by *Zymoseptoria tritici* (Desm.) Quaedvlieg & Crous. Oval to elongated chlorotic spots (lesions) develop at initial infection sites (Figure 7.7). Lesions may have a yellow margin, but the intensity of yellowing varies among varieties. In old lesions, the centers of the spots become pale yellow or straw-colored, slightly necrotic along with numerous small black dots (pycnidia). The lesions of *Septoria nodorum* and *Septoria avenae* blotch are usually lens-shaped,

FIGURE 7.7 Septoria leaf blotch.

whereas *Septoria tritici* blotch tends to be linear and restricted laterally. Initial infections begin on the lower leaves and subsequently progress to the upper leaves and spikes under conducive environmental conditions. The dark, reproductive structures produced by the fungus are key diagnostic features and can often be seen without magnification. Cool temperatures (10°C to 15°C) and prolonged wet, cloudy weather favors these diseases.

7.2.6 Wheat Blast

Wheat blast, caused by the *Magnaporthe oryzae Triticum* Couch & Kohn, is mainly a disease of the wheat head and can cause up to 100% yield loss if favorable environmental conditions remain for an extended period. The characteristic symptoms of wheat blast range from small elliptical lesions to complete bleaching and empty spikes (Figure 7.8). The head infection results in complete or partial bleaching of the spike above the point of infection with either no grain or shrivelled grain with low test weight. The foliar lesions caused by *Magnaporthe oryzae Triticum* (*MoT*) have also been described; however, their role and significance in grain yield losses remain unknown. Warm and humid conditions (temperature about 25°C with at least 10 hours wetting period) favor wheat blast development.

7.3 ECO-FRIENDLY APPROACHES FOR MANAGING FUNGAL BIOTIC STRESSES

In modern high-input wheat farming, indiscriminate use of fertilizers and pesticides renders the crop vulnerable to various diseases and causes water/ air pollution, residual toxicity, resistance to pesticides in pathogens, mortality of beneficial insects and microbiomes, soil health deterioration, etc. To counter the baleful effects of chemical pesticides, some alternative eco-friendly approaches for managing biotic stresses (diseases) in wheat crops are discussed below-

FIGURE 7.8 Wheat blast.

7.3.1 Genetic Resistance

Genetic resistance is the basic tool to defend wheat crops from the biotic stresses particularly caused by rust and powdery mildew pathogens. It is the most effective, economic, environmentally safe and sustainable way to mitigate the devastating effect of biotic stresses. However, incorporating and maintaining adequate resistance against diseases is difficult. There are two types of genetic resistance – vertical (qualitative) resistance and horizontal (quantitative) resistance. Vertical resistance is controlled by a single or few genes (monogenic or oligogenic) and possesses specificity to a race/pathotype of a pathogen, whereas horizontal resistance is race-nonspecific type resistance, controlled by minor genes and detectable at adult plant stages. Interaction between a crop cultivar and a pathotype of a particular pathogen follows a gene-for-gene relationship. For an allele of a gene conditioning resistance (*R*-gene) in the plant, there is an allele of its corresponding gene (*Avr*-gene) in the pathogen regulating virulence. To date, more than 225 rust resistance genes have formally been catalogued from wheat and its wild relatives, most of them are all-stage resistance (ASR) genes. However, three pleiotropic loci *Yr18/Lr34/Sr57/Pm38/Ltn1*, *Yr29/Lr46/Sr58/Pm39/Ltn2* and *Yr46/Lr67/Sr55/Pm46/Ltn3* confer adult plant resistance to three rust diseases and powdery mildew of wheat (Ye et al. 2022). *Aegilops tauschii* (DD, 2n =14) is one of the wild species ancestors of the D genome of hexaploid wheat (*Triticum aestivum*, AABBDD). Rust resistance genes *Yr28, Lr21, Lr22a, Lr32, Lr39, Lr42, Sr33, Sr45 and Sr46* have been derived from *Aegilops tauschii* (Kou et al. 2023). In India, stripe rust resistance genes *Yr5, Yr10, Yr15, Yr16*, *Yr32* and *YrSp*, stem rust resistance genes *Sr26, Sr27, Sr31, Sr32, Sr35, Sr39, Sr40, Sr43, SrTt3* and *SrTmp* and leaf rust resistance genes *Lr24, Lr25, Lr29, Lr32, Lr39, Lr45,* and *Lr47* have been found effective against *Puccinia striiformis* f. sp. *tritici* (*Pst*), *P. graminis tritici* (*Pgt*) and *P. triticina* (*Pt*) populations, respectively (Prasad et al. 2022). Primitive cultivars and landraces are also rich sources of genetic resistance to various biotic stresses because they have been growing under extreme environmental conditions and low-input farming systems (Adhikari et al. 2022). Recently, two all-stages resistance genes *Lr80* and *Sr65* were identified from Indian local wheat landrace (*Hango-2*) which confer resistance to all the predominant and virulent pathotypes of *Puccinia triticina* and *Puccinia graminis* f. sp. *tritici,* respectively (Kumar et al. 2021; Norman et al. 2024). Wheat rusts could effectively be managed by avoiding large-scale planting of single cultivar/similar type of resistance and deployment of varieties possessing different combinations of rust resistance genes.

Besides rusts, *R*-genes have also been identified in wheat for powdery mildew, Karnal bunt, Fusarium head blight, and loose smut (Bakala et al. 2021). Genetic resistance to Fusarium head blight is mainly quantitative and governed by many moderate to minor effect genes. Moderate resistance to FHB was observed in cultivars Sumai 3 from China and Frontana from Brazil (Randhawa et al. 2019). *Septoria tritici* blotch (STB) can also be managed effectively using a single gene with major effects or multiple genes with minor effects. So far, 21 STB resistance genes with large effects have been characterized and mapped (Brown et al. 2015). Wheat blast is hard to manage using host resistance. Until now, no commercial wheat variety is resistant to wheat blast; however, some cultivars such as BH1146, BR18, IPR85, CD113, CNT8, BARI Gom33, BR8 and Milan revealed moderate levels of resistance. Eight blast resistance genes designated as *Rmg1-Rmg8* have been identified in wheat (Singh et al. 2021; Hussain 2022).

7.3.2 Biotechnological Tools

Molecular approaches will aid in devising better strategies for developing resistance in wheat crops. Different biotechnological tools such as marker-assisted selection (MAS), genomics, recombinant DNA technology TILLING (targeted induced local lesions in the genome), and VIGS (virus-induced gene silencing) are currently being used for enhancing resistance against biotic stresses in wheat crops in a short span of time. The non-transgenic method TILLING may become the most effective tool for developing rust resistant wheat cultivars (Hussain 2015). Additionally, RNAi-based host induced gene silencing (HIGS) is another strategy to create resistance in plants against pathogens. Recently, the clustered, regularly interspaced, short palindromic repeats (CRISPR)/ CRISPR associated protein 9 (Cas9) system facilitated the precise editing of genomes and apprehending the disease resistance mechanisms and developing resistant varieties (Savadi et al. 2017). DNA-based molecular markers have several advantages over the traditional phenotype trait selection. MAS is a powerful tool to facilitate gene pyramiding for developing resistant cultivars to multiple diseases in a short duration of time.

Yellow rust resistance gene (*Yr26*), powdery mildew resistance gene (ML_{91260}) and high-molecular-weight glutenin subunits Dx5 + Dy10 were pyramided into the dwarf mutant of an elite wheat cultivar Xiaoyan22 using MAS (Zheng et al. 2020). Molecular markers *KASP_17425, KASP_17148* and *barc124,* closely linked to leaf rust resistance gene *Lr80,* are robust and could be used for marker-assisted selection of this gene in wheat breeding programs (Kumar et al. 2021). Gene *TaNFXL1* represses *F. graminearum* resistance, it was demonstrated using both virus-induced gene silencing and CRISPR-mediated genome editing. This suggests that targeting the *TaNFXL1* gene could facilitate FHB-resistant wheat cultivars (Brauer et al. 2020). Homozygous triple mutant and double mutant lines were developed using the non-transgenic TILLING technology, and showed enhanced powdery mildew resistance in bread wheat without producing discernible pleiotropic phenotypes (Acevedo-Garcia et al. 2017). Biotic stress-related genes from different sources could be cloned and transferred into elite genotypes for developing transgenic wheat. Transgenic wheat plants were developed using RNAi technology and expressed strong resistance to FHB (Cheng et al. 2015). Expressing dsRNA complementary to mRNAs of *Puccinia triticina* MAP-kinase (PtMAPK1, 520 bp) or a cyclophilin (PtCYC1, 501 bp) showed effective silencing of the corresponding genes in the pathogen and significant reduction of the leaf rust pathogen growth and pathogenicity transgenic wheat (Panwar et al. 2018). Genes encoding a endochitinase (chit42) and a chitosanase (harcho) from *T. harzianum* were transferred into wheat which enhanced the resistance to powdery mildew (Rana et al. 2012). Black rust resistance gene *Sr33* introgressed from *Aegilo pstauschii* into wheat was cloned and transformed into a susceptible wheat cultivar. Transgenic

TABLE 7.2
List of biocontrol agents used to manage different wheat diseases

Disease	Biocontrol Agent	Reference
Powdery Mildew	*Bacillus* spp.	(Shahin et al. 2019) (Xiaoning et al. 2015) (Xie et al. 2021)
	Pseudomonas spp.	(Elad et al. 1996)
Fusarium Head Blight (FHB)	*Bacillus* spp.	(Palazzini et al. 2009; Zalila-Kolsi et al. 2016; Palazzini et al. 2022)
	Pseudomonas fluorescens	(Herrera et al. 2016)
	Trichoderma spp.	(Baroncelli et al. 2016; Winter et al. 2018)
	Trichoderma atroviride	(Cabrera et al. 2020)
Stripe rust	*Bacillus subtilis*	(Li et al. 2013; Kiani et al. 2021)
	Cladosporium cladosporioides	(Zhan et al. 2014)
Leaf rust	*Streptomyces viridosporus* and *Trichoderma harzianum* combinely	(El-Sharkawy et al. 2015)
	Combination of *Verticillium* sp. with *Paecilomyces* sp. and *Verticillium* sp. with *Beauveria* sp.	(Sheroze et al. 2003)
Stem rust	Arbuscular mycorrhizal fungi	(El-Sharkawy et al. 2018)
	Trichoderma spp.	(El-Sharkawy et al. 2018)
Septoria tritici blotch (STB)	*Bacillus* spp.	(Kildea et al. 2008; Allioui et al. 2021)
	Trichoderma harzianum	(Cordo et al. 2020)
Spot blotch	*Trichoderma* spp.	(Perello et al. 2009; Yadav et al. 2015; Hossain et al. 2016; Villa-Rodriguez et al. 2022)

variety carrying *Sr33* showed resistance to various races of black rust pathogen including Ug99 (Periyannan et al. 2013).

7.3.3 Biological Control

Biological control is an effective disease management strategy that utilizes beneficial microorganisms and their metabolites. Endophytic microbes (bacteria and fungi) live inside the plants and protect against pathogens by inducing systemic acquired resistance (SAR) and triggering metabolic pathways in the host. These endophytes restrict the growth of pathogens by producing various bioactive metabolites, antibiotics and hydrolytic enzymes. Microbial antagonists belonging to the genus *Bacillus, Pseudomonas* and *Trichoderma* are widely used biocontrol agents for managing wheat diseases (Table 7.2). *Trichoderma* strains are competent to mitigate the biotic stresses primarily by mycoparasitism or competition for nutrients and space, promoting plant growth and plant defensive mechanisms and production of antibiotics or enzymes. The severity of brown rust reduced significantly on two susceptible genotypes Morocco and Sids-1 of wheat under greenhouse and open field conditions using *Streptomyces viridosporus* and *T. harzianum* combined (El-Sharkawy et al. 2015). *Trichoderma gamsii* T6085 was effective in reducing the growth of Fusarium head blight pathogen and reduction of mycotoxin accumulation(Baroncelli et al. 2016). In field conditions, application of *T. harzianum* significantly reduced the spot blotch disease (*Bipolaris sorokiniana*) incidence and the leaf blotching caused by *Zymoseptoria tritici* (Perello et al. 2009; Yadav et al. 2015). Seed treatment with *T. harzianum* was as effective as fungicide treatment in controlling STB and yield responses (Cordo et al. 2020). *T. harzianum* TSM39 showed strong antagonistic activity, high growth rate, non-cytotoxicity and its application to wheat crop resulted in a significant reduction of spot blotch disease severity under highly conducive conditions for disease outbreak (Villa-Rodriguez et al. 2022).

Bacillus subtilis strain E1R-j is effective in managing stripe rust and powdery mildew diseases (Li et al. 2013; Gao et al. 2015). In field conditions, *Bacillus megaterium* strain MKB135 was effective in restraining STB development by up to 80% (Kildea et al. 2008). Combined treatment of *Bacillus velezensis* RC218 and chitosan (0.1%) resulted in a reduction of 54% FHB severity on bread wheat under greenhouse conditions, whereas in field trials, FHB disease severity was reduced by 39 and 36.7% during the two harvest seasons (Palazzini et al. 2022).

7.3.4 Cultural Practices

Several cultural practices could be adopted to lessen the severity of biotic stresses. Cultural control practices are usually preventative and act indirectly against pathogens. Crop rotation, field fallowing, application of organic amendments, seed-cleaning, field sanitation, alteration in planting dates and physical seed treatment are some of the cultural practices known to reduce the severity of diseases. Soil test-based application of balanced fertilizers, particularly nitrogen (N), reduces the feasibility of severe grain losses caused by rust and powdery mildew diseases. Organic amendments such as botanicals, microbial pesticides and organic minerals serve as soil conditioners, and stimulate induced systemic resistance and systemic acquired resistance against diseases (Akanmu et al. 2021). Eradication of green bridge (volunteer plants or wild accessory hosts) and plant debris harboring plant pathogens in its active or dormant stage will facilitate in reduction of the initial inoculum. Early-sown wheat crop escapes from rust diseases and hot water treatment is effective in managing loose smut. Bacterial streak (black chaff) could be managed effectively using certified, pathogen-free seeds, eradicating volunteer plants and grassy weeds, and optimizing irrigation.

7.4 CONCLUSION AND OUTLOOK

Modern breeding approaches based on whole genome sequences and high-density genetic maps including marker-assisted selection, genome-wide association studies, genomic selection, mutagenesis, transgenesis and genome editing will facilitate harnessing the resistance genes from older wheat varieties, landraces, and wild relatives. Although the performance of some strains of *Trichoderma* spp. has been proven in disease suppression under field conditions similar to that of agrochemical, there is a need to produce highly potent strains using molecular tools for sustainable management of biotic stresses. Genome sequencing, CRISPR/Cas9 and gene cloning information would facilitate the coping mechanism of mycoparasitism and develop the improved biocontrol strains for managing wheat diseases efficiently under extreme environmental conditions. The generation of transgenic wheat by genetic engineering and its adoption is one of the ways of overcoming many biotic stresses and breaking the yield barrier. A combination of genetic resistance, biological control and cultural practices along with innovative biotechnological tools will be the most operative and eco-friendly sustainable module for managing the biotic stresses in wheat.

REFERENCES

Acevedo-Garcia J, Spencer D, Thieron H, Reinstadler A, Hammond-Kosack K, Phillips AL, Panstruga R (2017) mlo-based powdery mildew resistance in hexaploid bread wheat generated by a non-transgenic TILLING approach. *Plant Biotechnol J* 15: 367–378.

Adhikari S, Kumari J, Jacob SR, Prasad P, Gangwar OP, Lata C, Thakur R, Singh AK, Bansal R, Kumar S, Bhardwaj SC, Kumar S (2022) Landraces-potential treasurefor sustainable wheat improvement. *Genet Resour Crop Evol* 69: 499–523.

Akanmu AO, Babalola OO, Venturi V, Ayilara MS, Adeleke BS, Amoo AE, Sobowale AA, Fadiji AE, Glick BR (2021) Plant disease management: leveraging on the plant-microbe-soil interface in the biorational use of organic amendments. *Front Plant Sci* 12:700507.

Allioui N, Driss F, Dhouib H, Jlail L, Tounsi S, Frikha-Gargouri O (2021) Two novel *Bacillus* strains (subtilis and simplex species) with promising potential for the biocontrol of *Zymoseptoria tritici*, the causal agent of *Septoria tritici* blotch of wheat. *BioMed Res Int* 6611657: 1–10. https://doi.org/10.1155/2021/6611657

Bakala HS, Mandahal KS, Sarao LK, Srivastava P (2021) Breeding wheat for biotic stress resistance: achievements, challenges and prospects. In: Ansari M (ed) *Current trends in wheat research*, Intech Open, London. https://doi.org/10.5772/intechopen.97359

Baroncelli R, Zapparata A, Piaggeschi G, Sarrocco S, Vannacci G (2016) Draft whole-genome sequence of *Trichoderma gamsii* T6085, a promising biocontrol agent of Fusarium head blight on wheat. *Genome Announc* 4(1): e01747–e01815. https://doi.org/10.1128/genomeA.01747-15

Bhardwaj SC, Singh GP, Gangwar OP, Prasad P, Kumar S (2019) Status of wheat rust research and progress in rust management-Indian context. *Agronomy* 9(12): 892.

Bhardwaj SC, Prasad P,Gangwar OP, Khan H, Kumar S (2016) Wheat rust research-thenand now. *Indian J Agric Sci* 86(10): 1231–1244.

Brauer EK, Balcerzak M, Rocheleau H, Leung W, Schernthaner J, Subramaniam R,Ouellet T (2020) Genome editing of a deoxynivalenol-induced transcription factorconfers resistance to *Fusarium graminearum* in wheat. *Mol Plant-Microbe Interact* 33(3): 553–560.

Brown JKM, Chartrain L, Lasserre-Zuber P, Saintenac C (2015) Genetics of resistance to *Zymoseptoria tritici* and applications to wheat breeding. *Fungal Genet Biol* 79: 33–41.

Cabrera M, Garmendia G, Rufo C, Pereyra S, Vero S (2020) *Trichoderma atroviride* as a biocontrol agent of Fusarium head blight by reducing the inoculum of the pathogen in wheat straw. *Terra Latinoamericana* 38(3): 629–651.

Cheng W, Song X, Li H, Cao L, Sun K, Qiu X, Xu Y,Yang P, Huang T, Zhang J, Qu B, Liao Y (2015) Host-induced genesilencing of an essential chitin synthase gene confers durable resistance to Fusarium head blight and seedling blight in wheat. *Plant Biotechnol J* 13: 1335–1345.

Cordo C, Altamirano R, Simon MR, Stocco M, Lampugnani G, Abramoff C, Kripelz N, Monaco C (2020) Biocontrol strategies to reduce the impact of *Septoria tritici* blotch in wheat. *Rev Fac Agron* 119 (2): 1–10. https://doi.org/10.24215/16699513e060

El-Sharkawy HHA, Rashad YM, Seham AI (2018) Biocontrol of stem rust disease of wheat using arbuscular mycorrhizal fungi and *Trichoderma* spp. *Physiol Mol Plant Path* 103: 84–91.

El-Sharkawy HHA, Tohamey S, Khalil AA (2015) Combined effects of *Streptomyces viridosporus* and *Trichoderma harzianum* on controlling wheat leaf rust caused by *Puccinia triticina*. *Plant Pathol J* 14(4): 182–188.

Elad Y, Malathrakis NE, Dik AJ (1996) Biological control of Botrytis-incited diseases and powdery mildews in greenhouse crops. *Crop Prot* 15(3): 229–240.

Gangwar OP, Bhardwaj SC, Kumar S, Prasad P, Khan H, Savadi S (2017) Overcoming stripe rust of wheat-A threat to food security. In: Singh DP (ed) *Management of wheatand barley diseases*. Apple Academic Press, USA, pp. 115–132.

Gangwar OP, Singh APK (2018) *Trichoderma* as an efficacious bioagent for combating biotic and abiotic stresses of wheat-a review. *Agric Rev* 39(1): 48–54.

Gangwar OP, Kumar S, Bhardwaj SC, Prasad P, Lata C, Adhikari S, Singh GP (2022) Elucidating the population structure and genetic diversity of Indian *Puccinia striiformis* f. sp. *tritici* pathotypes based on microsatellite markers. *Phytopathology* 112: 1444–1453.

Gao X, Gong Y, Huo Y, Han Q, Kang Z, Huang L (2015) Endophytic *Bacillus subtilis* strain E1R-Jis a promising biocontrol agent for wheat powdery mildew. *Biomed Res Int* 462645: 1–8.http://dx.doi.org/10.1155/2015/462645

Herrera SD, Grossi C, Zawoznik M, Groppa MD (2016) Wheat seeds harbour bacterial endophytes with potential as plant growth promoters and biocontrol agents of *Fusarium graminearum*. *Microbiol Res* 186: 37–43.

Hossain MM (2022) Wheat blast: A review from a genetic and genomic perspective. *Front Microbiol* 13: 983243. https://doi.org/10.3389/fmicb.2022.983243

Hossain MM, Hossain I, Khalequzzaman KM (2016) Biological control of leaf blight of wheat caused by *Bipolaris sorokiniana*. *Bull Inst Trop Agr, Kyushu Univ* 39: 43–51.

Hussain B (2015) Modernization in plant breeding approaches for improving biotic stress resistance in crop plants. *Turk J Agric For* 39: 515–530.

Kiani T, Mehboob F, Hyder MZ, Zainy Z, Xu L, Huang L, Farrakh S (2021) Control of stripe rust of wheat using indigenous endophytic bacteria at seedling and adult plant stage. *Sci Rep* 11(1): 14473.

Kildea S, Ransbotyn V, Khan MR, Fagan B, Leonard G, Mullins E, Doohan FM (2008) *Bacillus megaterium* shows potential for the biocontrol of *Septoria tritici* blotch of wheat. *Biol Control* 47(1): 37–45.

Kou H, Zhang Z, Yang Y, Wei C, Xu L, Zhang G (2023) Advances in the mining of disease resistance genes from *Aegilops tauschii* and the utilization in wheat. *Plants* 12: 880. https://doi.org/10.3390/plants12040880

Kumar S, Bhardwaj SC, Gangwar OP, Prasad P, Chakrabarty R, Kashyap PL, Khan H, Savadi S, Mahato BN, Kumar P (2022) Characterization of five new pathotypes of *Puccinia triticina* identified from Northeast India, Nepal, and Bangladesh. *Australas Plant Pathol* 51: 315–325. https://doi.org/10.1007/s13313-022-00857-w

Kumar S, Bhardwaj SC, Gangwar OP, Sharma A, Naeela Q, Kumaran VK et al (2021) *Lr80*: A new and widely effective source of leaf rust resistance of wheat for enhancing diversity of resistance among modern cultivars. *Theor Appl Genet* 134(3): 849–858.

Li H, Zhao J, Feng H, Huang L, Kang Z (2013) Biological control of wheat stripe rust by an endophytic *Bacillus subtilis* strain E1R-j in greenhouse and field trials. *Crop Prot* 43: 201–206.

Li H, Zhao J, Feng H, Huang L, Kang Z (2013) Biological control of wheat stripe rust by an endophytic *Bacillus subtilis* EIR-j in greenhouse and field trail. *Crop Prot* 43: 201–206.

Morgounov A, Tufan HA, Sharma R, Akin B, Bagci A, Braun HJ et al (2012) Global incidence of wheat rusts and powdery mildew during 1969-2010 and durability of resistance of winter wheat variety Bezostaya 1. *Eur J Plant Pathol* 132: 323–340.

Norman M, Chen C, Miah H, Patpour M, Sørensen C, Hovmøller M, Forrest K, Kumar S, Prasad P, Gangwar OP, Bhardwaj S, Bariana H, Periyannan S, Bansal U (2024) Sr65: A widely effective gene for stem rust resistance in wheat. *Theor Appl Genetics* 137: 1. https://doi.org/10.1007/s00122-023-04507-7

Panwar V, Jordan M, McCallum B, Bakkeren G (2018) Host-induced silencing of essential genes in *Puccinia triticina* through transgenic expression of RNAi sequences reduces severity of leaf rust infection in wheat. *Plant Biotechnol J* 16:1013–1023.

Palazzini J, Reynoso A, Yerkovich N, Zachetti V, Ramírez M, Chulze S (2022) Combination of *Bacillus velezensis* RC218 and chitosan to control Fusarium head blight on bread and durum wheat under greenhouse and field conditions. *Toxins* 14:499. https://doi.org/10.3390/toxins14070499

Palazzini JM, Ramirez ML, Alberione EJ, Torres AM, Chulze SN (2009) Osmotic stress adaptation, compatible solutes accumulation and biocontrol efficacy of two potential biocontrol agents on Fusarium head blight in wheat. *Biol Control* 51(3): 370–376.

Perello AE, Moreno MV, Monaco C, Simon MR, Cordo C (2009) Biological control of *Septoria tritici* blotch on wheat by *Trichoderma* spp. under field conditions in Argentina. *BioControl* 54: 113–122.

Periyannan S, Moore J, Ayliffe M, Bansal U et al (2013) The gene *Sr33*, an ortholog of barley *Mla* genes, encodes resistanceto wheat stem rust race Ug99. *Science* 341: 786–788.

Prasad P, Gangwar OP, Lata C, Adhikari S, Kumar S, Bhardwaj SC (2022) *Mehtaensis*: *six-monthly newsletter*. ICAR-Indian Institute of Wheat and Barley Research.

Rana IA, Loerz H, Schaefer W, Becker D (2012) Overexpression of chitinase and chi-tosanase genes from *Trichoderma harzianum* under constitutive and inducible promoters in order to increase disease resistance in wheat (*Triticum aestivum* L). *Mol Plant Breed* 3: 37–49.

Randhawa MS, Bhavani S, Singh PK, Huerta-Espino J, Singh RP (2019) Disease resistance in wheat: present status and future prospects. In: Wani SH (eds) *Disease resistance in crop plants*. Springer, Cham, pp. 61–81.

Savadi S, Prasad P, Bhardwaj SC, Gangwar OP, Khan H, Kumar S (2017) Management of rust diseases in wheat and barley: next generation tools. In: Singh DP (ed) *Management of wheat and barley diseases*, Apple Academic Press, USA, pp. 39–82.

Savary S, Willocquet L, Pethybridge SJ, Esker P, Mc Roberts N, Nelson A (2019) The global burden of pathogens and pests on major food crops. *Nat Ecol Evol* 3: 430–439.

Shahin AA, Ashmawy MA, Esmail SM, El-Moghazy SM (2019) Biocontrol of wheat powdery mildew disease under field conditions in Egypt. *Zagazig J Agric Res* 46(6B):225–2270.

Sheroze A, Rashid A, Shakir AS, Khan SM (2003) Effect of bio-control agents on leaf rust of wheat and influence of different temperature and humidity levels on their colony growth. *Int J Agri Biol* 5(1): 83–85.

Singh PK, Gahtyari NC, Roy C, Roy KK, He X, Tembo B, Xu K, Juliana P, Sonder K, Kabir MR, Chawade A (2021) Wheat blast: a disease spreading by intercontinental jumps and its management strategies. *Front Plant Sci* 12:710707. https://doi.org/10.3389/fpls.2021.710707

Villa-Rodriguez E, Lugo-Enríquez C, Ferguson S, Parra-Cota FI, Cira-Chávez LA, de los Santos-Villalobos S (2022) *Trichoderma harzianum* sensu lato TSM39: a wheat microbiome fungus that mitigates spot blotch disease of wheat (*Triticum turgidum* L. subsp. *durum*) caused by *Bipolaris sorokiniana*. *Biol Control* 175: 105055. https://doi.org/10.1016/j.biocontrol.2022.105055

Winter M, Samuels PL, Dong Y, Dill-Macky R (2018) Trichothecene production is detrimental to early root colonisation by *Fusarium culmorum* and *F. graminearum* in Fusarium crown and root rot of wheat. *Plant Pathol* 68: 185–195.

Xiaoning G, Yufei G, Yunxia H, Zhensheng K, Li H (2015) Endophytic *Bacillus subtilis* strain E1R-j is a promising biocontrol agent for wheat powdery mildew. *Bio Med Res Int* 462645. http://dx.doi.org/10.1155/2015/462645

Xie D, Cai X, Yang C, Xie L, Qin G, Zhang M, Huang Y, Gong G, Chang X, Chen H (2021) Studies on the control effect of *Bacillus subtilis* on wheat powdery mildew. *Pest Manag Sci* 77(10): 4375–4382.

Yadav B, Singh R, Kumar A (2015) Management of spot blotch of wheat using fungicides, bio-agents and botanicals. *Afr J Agric Res* 10(25): 2494–2500.

Ye B, Singh RP, Yuan C, Liu D, Randhawa MS, Huerta-Espino J, Bhavani S, Lagudah E, Lan C (2022) Three co-located resistance genes confer resistance to leaf rust and stripe rust in wheat variety Borlaug 100. *The Crop J* 10(2): 490–497.

Zalila-Kolsi I, Mohmoud AB, Ali H, Sellami S, Nasfi Z, Tounsi S, Jamoussi K (2016) Antagonist effects of *Bacillus* spp. strains *Fusarium graminearum* for protection of durum wheat (*Triticum turgidum* L sub sp. *durum*). *Microbiol Res* 192: 148–158.

Zhan G, Tian Y, Wang F, Chen X, Guo J (2014) A novel fungal hyperparasite of *Puccinia striiformis* f. sp. tritici, the causal agent of wheat stripe rust. *PLoS ONE* 9(11): e111484.

Zheng W, Li S, Liu Z, Zhou Q, Feng Y, Chaiet S (2020) Molecular marker assisted gene stacking for disease resistance and quality genes in the dwarf mutant of an elite common wheat cultivar Xiaoyan22. *BMC Genet* 21: 45. https://doi.org/10.1186/s12863-020-00854-2

8 Integrated Disease Management of Maize Crops

Rajender Jatoth, Lokesh Babu P., Omprakash S., Bharathi V., Priyadarshini G., Vinesh Banoth, Reddy C.R.K., and Dinesh Singh

8.1 INTRODUCTION

Maize is the third most important food crop in India, after rice and wheat, known as the queen of cereals due to its high genetic yield potential. Maize has a wider adaptability under varied agro-climatic conditions, is quick-growing, high-yielding, and provides palatable and nutritious forage, which can be fed to animals at any stage of growth without any risk. However, the average productivity and yield output of India's maize crop is lower than that of developed nations (Yadav et al. 2015). This is due to numerous biotic factors, such as insect pests, fungi, bacteria, phytoplasma, viruses, and nematodes, which affect maize production (CIMMYT 2004).

The key factors that render the maize production are diverse, with insects and closely related organisms such as mites being among the most significant. These pests can infest maize at any stage of crop development also in storage and attack any part of the plant, often causing severe damage in high valleys of the subtropics and tropics to the temperate and tropical lowland regions. The moth group (cutworms, armyworms, earworms, borers, and grain moths) is the most damaging to maize worldwide, followed by the beetles (rootworms, wireworms, grubs, grain borers, and weevils). However, in India, the four major pests of maize are spotted stem borer (*Chilo partellus* Swinhoe), pink stem borer (*Sesamia inferens* Walker), shoot fly (*Atherigona* spp.) and fall armyworm (Spodoptera frugiperda J. E. Smith). The group of insects that serve as carriers (vectors) for disease agents (viruses, phytoplasma, bacteria, and fungi) is also important. The sap-sucking bugs (leafhoppers and aphids) are the greatest problem among these vectors. More than 30 diseases have been reported from different regions of India, including turcicum leaf blight (*Exserohilum turcicum*), maydis leaf blight (*Drechslera maydis*), post-flowering stalk rots complex (PFSR: *Fusarium verticilloides*, *Macrophomina phaseolina*), banded leaf and sheath blight (*Rhizoctonia solani* f. sp. *sasakii*), sorghum downy mildew (*Peronosclerospora sorghi*), Rajasthan downy mildew (*Peronosclerospora heteropogoni*), bacterial stalk rot (*Erwinia chrysanthemi* pv. *zeae*) and brown stripe downy mildew (*Sclerophthora rayssiae* var. *zeae*) (Yadav et al. 2015).

8.2 DISEASES OF MAIZE CROPS

8.2.1 Turcicum Leaf Blight (Northern Corn Leaf Blight)

The disease was first reported in Italy by Passerini (1876) and it has since become a severe problem in many parts of the world, including the Northeastern United States, sub-Saharan Africa, Europe, Australia and in areas of North Korea, China and India (Wathaneeyawech et al. 2015). In India, it was first reported by Butler in 1907 on sorghum and later on both sorghum and maize from the Punjab by Mitra in 1923. The disease has been prevalent in all the major maize-growing regions of

DOI: 10.1201/9781032631578-8

India during both the rainy (*Kharif*) and winter (*Rabi*) seasons for the past two decades. However, Dharward in Karnataka, Kolahpur in Maharashtra, Karim Nagar in Andhra Pradesh, and Dholi in Bihar, Punjab, Himachal Pradesh and the Kashmir valley are among the hot spot locations. Due to the favorable conditions for disease, turcicum leaf blight disease has become a major limiting factor for maize production in the hills regions (Ahangar et al. 2016). The grain yield loss ranges from 24 to 91% depending on the severity of the disease (Nwanosike et al. 2015). The blighted leaves are not suitable for fodder due to their low nutritional value.

8.2.1.1 Symptoms

The maize plant is mainly affected by a fungus at a young stage. The disease initiates on lower leaves as small yellowish or greyish-green water-soaked round to oval spots and gradually moves upwards. The spots become larger elliptical or spindle-shaped and turn straw to greyish brown in the center with dark brown edges. The disease is characterized by long elliptical greyish-green or tan lesions on the leaves measuring 2.5 to 25 cm in length and up to 4 cm in width (De Rossi et al. 2015). The spots coalesce to form bigger spots, giving a blighted appearance. The surface is covered with olive-green velvety masses of conidia and conidiophores. Under high humidity, the whole leaf area becomes necrotic, and the plant appears dead (Figure 8.1). Lesions may extend to the husk.

8.2.1.2 Causal Organism

Turcicum leaf blight is a disease that affects maize and is caused by a fungus *Exserohilum turcicum* (Pass.) Leo and Suggs. This fungus is a hemibiotroph and has a sexual stage (teleomorph) known as *Setosphaeria turcica* (Luttrell) Leonard and Suggs. Although rarely, *Trichometasphaeria turcica* can occur in nature. The conidiophores of *Exserohilum turcicum* are grouped, geniculate, mid-dark brown, pale near the apex, and smooth. The conidia are distinctly curved, fusiform, and pale to golden brown with 5–11 pseudo septa.

8.2.1.3 Disease Cycle and Epidemiology

The fungus is transmitted through seeds, surviving in plant debris as chlamydospores and in the soil as saprophytes. It can also infect various plants such as Sudan grass, Johnson grass, sorghum, wheat, barley, oats, and sugarcane, and spores of the fungus are also found to associate with seeds of green gram, black gram, cowpea, and Teosinte. The fungus spreads through windborne conidia and can travel long distances (Ferguson and Carson 2004).

FIGURE 8.1 *Exserohilum turcicum* disease symptoms of the leaves of maize.

The optimum temperature for conidial germination is 18 to 27°C, with the presence of free water on the leaf. Turcicum leaf blight is more common in regions with high humidity (75–90%) and moderate temperatures (22–25°C) during the cropping season (Sibiya et al. 2013). Infection typically occurs early in the wet season. The increase in the disease incidence can be attributed to mono-cropping practice, high humidity, morning fogs and extended dew periods, minimum tillage and the use of uniform and highly susceptible varieties (Reddy et al. 2013).

8.2.1.4 Disease Management

To effectively manage the disease, an integrated approach is necessary. This includes crop rotation with non-host crops, treating seeds with Captan or Thiram at a rate of 4 g/kg, and seed treatment with mancozeb 75 WP @ 0.25% followed by two foliar sprays of propiconazole 25 EC @ 0.1% at 40 and 50 days after sowing. The fungicides prothioconazole + trifloxystrobin were found to be highly effective in controlling NCLB (Camera et al. 2019). However, the most efficient and cost-effective long-term remedy is to develop varieties of the crop that are resistant to *E. turcicum*. Qualitative resistance against NCLB can be conferred by single Ht genes (Ht1, Ht2, Ht3, and Htn), but this resistance can weaken under highly virulent races of *S. turcica*. Therefore, it is essential to pyramid multiple Ht genes to develop durable resistance against a variety of pathogen races.

8.2.2 Maydis Leaf Blight (Southern Corn Leaf Blight)

Maydis leaf blight is a disease that poses a significant threat to crops in warm and wet temperate and tropical regions. This disease has been known to cause yield losses of up to 70% (White 1999). In India, this disease was first reported from Maldah (West Bengal) by Munjal and Kapoor in 1960. Since then, it has become a serious problem in various states such as Jammu & Kashmir, Himachal Pradesh, Sikkim, Meghalaya, Punjab, Haryana, Rajasthan, Delhi, UP, Bihar, MP, Gujarat, Maharashtra, Andhra Pradesh, Karnataka and Tamil Nadu (Singh and Srivastava 2016).

8.2.2.1 Symptoms

Symptoms of maize leaf spot fungus vary depending on the causal race, and host germplasm. Race C is a specific race found only in China that affects cms-C cytoplasm. Race T is pathogenic to maize only if the Texas source of male sterility is present, while Race O is not. Lesions caused by Race O initially appear as small and diamond-shaped spots that elongate as they mature. However, their growth is restricted by leaf veins, and the final lesions are rectangular (2–6 × 3–22 mm) and tan in color. Race O only causes lesions on leaves. In contrast, Race T produces slightly larger oval lesions (6–12 × 6–27 mm) with dark brown borders. Race T causes lesions on all above-ground parts of the plant (including stems, sheaths and ears) and can also cause ear rots. Seedlings from seeds infected with Race T often wilt and die within 3 to 4 weeks. Under severe disease pressure, when infection occurs before silking, the lesions may coalesce and blight the entire leaf (Figure 8.2). In such cases, sugars may be diverted from the stalk for grain filling, thus predisposing the plant to lodging (Hooda et al. 2018). The center of maydis leaf spot lesion is straw-shaded to light brown with a dark brown margin. The lesions elongate over time, although their growth is limited by the leaf veins. The spore proliferation causes the lesions to turn dark grey, olive, or black. They initially appear on the lower leaves and increase in number as the plant grows, leading to complete foliage brightening. A heavily infected field presents a burnt appearance.

8.2.2.2 Causal Organism

Bipolaris maydis Nisikado & Miyake (Teleomorph: *Cochliobolus heterostrophus* Drechs.) incites southern leaf blight disease in maize. The fungal pathogen is a necrotrophic that actively kills the host tissues and thrives on the contents of dried and dead cells. The sexual stage of fungus produces ascospores in perithecium type ascocarp that are rare in nature or found in laboratory culture. The asci are cylindrical, hyaline, and ascospores are thin-walled 6–7 × 130–340 μm in size, dark, and have 5

FIGURE 8.2 Maydis disease symptoms on leaves of maize.

to 9 septa. The ascospores are olivaceous brown, spindle-shaped, and taper to round ends, with a size of 15–20 × 70–160 µm. Conidia are 5–11 septate and are characterized by bipolar germination. The pathogenic fungus has been identified in three physiological races: race 'O', race 'T' and race 'C. Race O pathotypes are widely distributed and can cause yield losses of up to 40%, while Race T pathotypes are only prevalent in varieties with Texas sources of male sterility (Byrnes et al. 1989).

8.2.2.3 Disease Cycle and Epidemiology

The conidial stage of the fungi is the primary source of inoculum during and at the beginning of the season. Fungus survives in the winter as mycelium in infected crop debris that remains on the soil surface between growing seasons. When the subsequent growing season begins, mycelia within the crop debris initiate conidia production in response to favorable temperatures and humidity. These spores are then disseminated through wind and rain splash to newly planted maize in the vicinity. Conidia germinate and infect plants through stomata and lead to characteristic lesions where more conidia are produced and cause secondary disease cycles. In favorable environmental conditions, the disease cycle can be completed within 72 hours.

Regions with warm temperatures between 20 to 32°C and water accumulation on leaves are particularly vulnerable to maydis leaf blight. However, the disease is less likely to occur during long, dry, sunny periods during the growing season.

8.2.2.4 Disease Management

The management of maydis disease in maize is similar to turcicum leaf blight as follows:

- Crop rotation with non-host crops reduces the primary inoculum of the pathogen.
- Removal of volunteer maize plants and plant debris from the field help to reduce incidence of the disease.

- Planting should be done with wider spacing to reduce relative humidity.
- Spraying of chemicals such as chlorothalonil @ 2.5 g and mancozeb @ 2.5 g/ l during the first lesion formation is recommended, followed by the second spray at 15-day intervals at the knee-high stage (Aggarwal et al. 2023).
- Resistance has been reported in inbred lines – CM 104, CM 105, LM 13, LM 14, CM 123, and LM 5.
- Three management modules as organic, chemical, and integrated disease management (IDM) were evaluated against maydis disease in maize at three hot spot locations like Ludhiana, Karnal, and Delhi. Among them, the chemical module, particularly foliar spray with Mancozeb 75WP@ 2.5 g/l water, was most effective in controlling the disease followed by the IDM module (Hooda et al. 2012).

8.2.3 Downy Mildews

Ten different species of oomycete fungi from the genera *Peronosclerospora*, *Scleropthora*, and *Sclerospora* are known to cause various types of downy mildews in maize (Rathore et al. 2002). In India, various agro-ecological regions have reported five types of downy mildew, including sorghum downy mildew, Philippine downy mildew, sugarcane downy mildew, brown stripe downy mildew, and Rajasthan downy mildew. Downy mildews caused by *Peronosclerospora* and *Sclerospora* species, are the most dangerous diseases for maize crops in the lowland tropics, especially in tropical Asian countries. The important five downy mildews are mentioned below:

8.2.3.1 Sugarcane Downy Mildew of Maize

In India, downy mildew disease on maize was first reported by Butler (1913). This disease was later recorded in different geographical states viz. Punjab, Bihar, and Delhi. Subramanian (1931) reported a wide occurrence of downy mildew disease in maize fields at Pusa, Bihar on sugarcane. Later, this disease was reported in major sugarcane-growing regions in the Tarai area of Uttar Pradesh.

8.2.3.1.1 Symptoms

The disease is characterized by the appearance of long and broad chlorotic stripes that run along the entire length of the leaf. Often, several stripes can be present on each leaf and extend to the entire length. In later stages of infection, the strips appear shorter and narrower. The color of the stripes changes from whitish to yellowish-white in the early stages and later it turns into a light- to dark-brown color. The leaves become puckered, and a downy growth is found on both surfaces of the leaf. This can often result in the shredding of the leaves in a linear pattern. The young maize leaves exhibit more prominent downey growth than the older leaves. Due to the shortening of the internodes, the plants exhibit a stunted and bushy appearance.

8.2.3.1.2 Causal Organism

The disease is caused by *Peronosclerospora sacchari* (Miyake) Shaw. This fungus appears as a white downy growth on both surfaces of the leaves, consisting of sporangiophores and sporangia. Sporangiophores are short and stout and branch out profusely into a series of pointed sterigmata that bear hyaline, oblong or ovoid sporangia (conidia). Sporangia germinate directly to infect the plants. In advanced stages, oospores are formed which are spherical, thick-walled and deep brown.

8.2.3.1.3 Disease Cycle and Epidemiology

The fungus survives in the soil, plant debris and graminaceous collateral hosts (*Sorghum bicolor*, *S. halapense*, etc). In Punjab, *Digitaria sanguinalis* serves as the primary source of infection.

Secondary spread is through airborne conidia. The oospores survive in the soil as well as in the infected plant debris. The development of disease is favored by high relative humidity (90%), water logging conditions, light drizzles, and a temperature of 20–25°C. The young plants are highly susceptible.

8.2.3.2 Brown Stripe Downy Mildew

Brown stripe downy mildew disease has gained importance as it causes yield loss ranging from 20–90% in many cultivated areas of India.

8.2.3.2.1 Symptoms

The disease only affects the leaves and is not systemic within the plant. During the early stages of infection, the infected leaves show narrow chlorotic or yellowish stripes, which are 3–7 mm wide. The lowermost leaves of the plant show early symptoms, and severely affected leaves may shed prematurely. However, affected leaves remain intact and do not shred after severe storms. As the disease progresses, the chlorotic stripes become reddish or purple. The disease occurs at an early stage of the crop, just before the reproductive stage. Seed development may be adversely affected, and the plant may die prematurely. In the early stages of the crop, the infected plant shows stunted growth symptoms.

8.2.3.2.2 Causal Organism

The disease is caused by *Sclerophthora rayssiae* var. *zeae* Payak and Renfro in maize.

8.2.3.3 Crazy Top Downy Mildew

The pathogen causes mainly systemic infections of its hosts. *Sclerophthora macrospora* causes downy mildew on a vast number of cereal crops including oats, rice, maize, and wheat as well as varieties of turf grass.

8.2.3.3.1 Symptoms

The most typical disease symptom is known as "crazy top". It results from the transformation of a partial or complete tassel into a mass of small narrow, twisted leafy structures. Ears or separate kernels can also degenerate into little leaves. Yellow semi-transparent stripes appear on "normal" leaves. The proliferation of axillary buds on the stalk of the tassel as well as the cobs is called crazy top.

8.2.3.3.2 Causal Organism

The disease is caused by *Sclerophthora macrospora* (Sacc.) Thirum, which produces oospores. The fungal pathogen's sporangia are formed in groups of two to six, arising in basipetal succession. These sporangia are hyaline, smooth-walled, ovate, obclavate, elliptic or cylindrical in shape. They possess a projecting, truncate, rounded, or tapering poroid apex, and are smooth-walled and papillate. The sporangia shed easily, but they have a persistent, straight or cuneate peduncle. They range in size from 18.5–26.0 μm × 29.0–66.5 μm and may be lens-shaped pores through which zoospores or cytoplasm may escape. The sporangia may encyst within or outside and 4–8 zoospores formed in it. Oospores are pleurotic, spherical or subspherical. They are hyaline and contain one prominent oil globule.

8.2.3.3.3 Disease Cycle and Epidemiology

The primary sources of inoculum for the pathogen *S. macrospora* are oospores, which overwinter in soil or plant debris or in infected seed. These oospores are sexual spores that are produced by the fusion of an antheridium and oogonium. In late spring to autumn, the oospore germinates and produces a pearly white and lemon-shaped sporangium on sporangiophores under damp conditions and disseminated by splashing water. When transported by water, the sporangium lands on a leaf surface and releases zoospores that infect the host plant by encysting near the stoma and penetrating the plant tissue with a germ tube. The leaf surface must remain damp for the sporangia to infect. The sporangia can liberate many zoospores in wet conditions. The pathogen's zoospores on the infected leaf provide a secondary source of infection, allowing the disease to be polycyclic. Oospores and mycelium also survive on dead leaves. In well-drained soils, oospores can persist for a few months. The surviving oospores produce more sporangia in late spring. Secondary spread of the disease occurs through sporangia produced by oospores, which are dispersed through wind and water splash. The sporangia contain 4–8 zoospores that initiate new infections.

8.2.3.4 Sorghum Downy Mildew

The downy mildew of sorghum was first identified and named *Scleropora graminicola* by Butler in 1907. It is prevalent in Gujarat, Maharashtra, A.P., Karnataka, and Tamil Nadu. Seedlings that are less than one month old are highly susceptible to this disease.

8.2.3.4.1 Symptoms

In systemic infection, the disease appears soon after the emergence of seedlings, and affected seedlings have narrow, pale yellow leaves that are covered with a fine downy growth consisting of the conidial stage of the fungus. Infected leaves have a downy growth, more commonly on the lower surface toward the basal part. Infected plants are chlorotic, stunted and sterile, with striped leaves due to formation of oospores. High populations of spores are produced on the leaf surface, which are short-lived and require extended periods of high-humidity infection. Overwintering spores produced between leaves' veins exist in the soil for long periods.

8.2.3.4.2 Causal Organism

Sorghum downy mildew caused by *Peronosclerospora sorghi* (Weston & Uppal) Shaw (Syn: *Sclerospora sorghi* Weston & Uppal) is a disease that infects both sorghum and maize crops. It is prevalent in tropical and subtropical regions of the world. The fungus is an obligate parasite that produces coenocytic mycelium. The sterigmata of *P. sorghi* is longer than that of *S. graminicola*, up to 16 µm. The major difference lies in the sporangia, which always germinate by germ tube and do not produce zoospores. The sporangium is commonly spherical instead of oval and lacks the apical papilla. The size of conidia ranges from 15–29 µm in diameter. The oospores are spherical and have three walls viz. exosporium, mesosporium and endosporium. The endosporium is smooth, yellow in color and of evenly thick. Presently, six pathotypes of the *P. sorghi* have been identified in the Americas, based on virulence phenotypes on 10 differential inbred lines of sorghum. Pathotypes 1–3 and 6 were found in Texas, pathotype 4 in Brazil, and pathotype 5 in Honduras (Prom et al. 2015). The line SC155 was initially thought to differentiate pathotype 6 from 3 (Isakeit and Jaster 2005), but this was not subsequently confirmed (Prom et al. 2015). Pathotype 6 isolates overcome the resistance of proprietary Pioneer brand germplasm resistant to pathotype 3 isolates. The maize cultivars of American were showed highly resistant to *P. sorghi*, but they are susceptible to other *Peronosclerospora* spp. found in Asia (Perumal et al. 2008).

8.2.3.4.3 Disease Cycle and Epidemiology

The primary source of infection of *P. sorghi* is oospores, which can survive in soil for subsequent growing seasons. The fungus mainly infects the underground parts of susceptible seedlings, leading to systemic infection. The airborne conidia produced on the surface of leaves can serve as a secondary source of infection, causing systemic infection in plants that are up to four weeks old. The pathogen can also be transmitted through seeds, but it is most easily spread from freshly harvested or immature seeds. However, infected plants are usually sterile and do not produce any viable seeds. Seeds that have been dried to below 13% are unlikely to carry viable fungal material.

8.2.3.5 Rajasthan Downy Mildew

The disease Rajasthan downy mildew was first time reported in 1968 in maize fields at the Regional Research Station in Vallabh Nagar, Udaipur, Rajasthan. This disease causes significant yield losses under conditions favoring infection. Infection can occur at the early stage (2–3 leaf) of crop growth. Once infected, the disease spreads quickly and can affect many plants. In some cases, systemically infected plants may produce rudimentary cobs, but these cannot be harvested for grain purposes and therefore result in the complete loss of yield (White 1999).

8.2.3.5.1 Symptoms

Typical symptoms of *P. heteropogoni* infection includes "*half diseased leaf*" which is characterized by the pale appearance of bases of the second and third diseased leaves of the seedling. On infected leaves, the yellow stripes formed at the base and extend up to the upper green portion. Severely infected plants give a yellowish appearance even from a distance. Most of the infected plants die at about knee-high stage. Under humid conditions, whitish fluffy growth formed due to abundant fructification of the fungus can be identified on the lower and upper leaf surfaces. No oospore formation by *P. heteropogoni* has been observed on maize. Systemically infected maize plants generally do not form a cob. The cobs were formed often, but these are small and poorly filled, and even if the tassel emerges, there is no pollen formation in the anthers.

8.2.3.5.2 Causal Organism

The disease is caused by *Peronosclespora heteropogoni* Siradhana et al.

8.2.3.5.3 Disease Cycle

The disease cycle of *P. heteropogoni* involves abundant conidia and oospores production on the leaves of its collateral host, *H. contortus* while producing only conidia on maize leaves.

8.2.3.5.4 Disease Management

- Destruction of plant debris, and deep summer plowing are major cultural practices to reduce the disease incidence of downy mildew.
- Avoid growing maize crops in low and wet fields and provide adequate soil drainage.
- The disease can be effectively managed by treating the seeds with Metalaxyl @ 3 g/kg seeds or Mefenoxam (Exconde 1975) or Metalaxyl (Apron 35 SD) @4g/kg seed, or foliar applications of Metalaxyl @ 0.2%.
- Spray the crop with Metalaxyl MZ (Ridomil MZ) @ 0.2% starting from the 20th day after sowing 3–4 times, Metalaxyl seed treatment (0.25– 0.6 ai /kg seed) has been reported to control the disease.
- Crop rotation with non-host crops like pulses should be followed to avoid build-up of infection.
- Collateral hosts and other grass weeds of the pathogen should be removed from the field and destroyed it properly to limit further spread of the disease.
- Grow resistant hybrids like DHM-1, DHM-103, DMR-5 and Ganaga II against Rajasthan downy mildew disease. In addition, 2 highly resistance lines (WLS-F54-2-1-2-B-2-B and (S92145-2EV-7-3-B*5/CML427)-F2-10-2-B) of maize can be used against both Rajasthan downy mildew and sorghum downy mildew (Rashid et al. 2013).
- Ten lines, *viz.,* CMH 108–154, JH-12114, IDX-2901, PAC–T99, BIO-265, JKMH-31314and KMH-3426 were highly resistant and 22 stocks were resistant to brown stripe downy mildew (Singh and Singh 2011, 2012). Resistant varieties against the disease are Prabhat, Kohinoor, ICI-703, PAC-9401, PMZ-2, SEEDTEC-2331. Infected plants must be rogued out at an early stage to reduce further spread of the disease.

8.2.4 Banded Leaf and Sheath Blight

This disease is globally prevalent and causes seed decay, damping off, stem canker, root rot, aerial blight, and seed or cob decay. The first recorded occurrence of this disease was on maize in Sri Lanka by Bertus in 1927. The banded leaf and sheath blight pathogen has been detected in the soil and was found in many maize-growing regions, such as Himachal Pradesh, Uttar Pradesh, Haryana, Punjab, Madhya Pradesh, and Rajasthan (Rani et al. 2013). The disease can lead to a substantially lower grain yield, with losses reaching up to 90%. Cobs heavily infected may experience a lack of grain production, potentially leading to the formation of resting spores, also known as sclerotia, inside the cobs (Singh et al. 2020).

8.2.4.1 Symptoms

Symptoms develop on all above-ground plant structures except for the tassel (Sharma et al. 2004). The disease is characterized by the presence of large, discolored patches alternating with uneven dark bands. Severe infection causes discoloration of the lower leaf sheath. Under extremely humid conditions, the infection spreads from the lower sheath to the higher leaf sheaths, leading to the decay of the leaf sheaths and the complete drying of the entire leaf. Basal sheath disease appears as straw-colored, water-soaked, irregular to roundish spots on the leaf surface. Abundant mycelium growth is observed in the affected region of the leaf sheath and between the stem and the leaf sheath (Sagar and Bhusal 2019). The lesions rapidly increase in size, resulting in discolored patches alternating with black bands. These changes become visible on the lower leaves after 7–8 days (Figure 8.3). The disease directly impacts premature mortality, stalk fracturing, leaf and leaf sheath damage, and the development of ear rot, all contributing to crop lodging (Lu et al. 2012; Yadav et al. 2022). The disease typically appears during the pre-flowering period. Symptoms also manifest on the stalk, with internodes fracturing at the site of infection. Ear rot symptoms, as depicted in Figure 8.3, are characterized by the appearance of cottony white mycelium on the plant's ear, small spherical black sclerotia, premature drying of ears, and caking of the ear sheath (Rajput and Harlapur 2014).

8.2.4.2 Causal Organism

Banded leaf and Sheath blight (BLSB) of maize is caused by *Rhizoctonia solani* f. sp. *sasakii* (Teleomorph: *Thanatephorus cucumeris*). The fungal colonies exhibit rapid growth and develop silky white colonies on PDA medium at a temperature of 28±1°C. Over time, these colonies gradually lose their shine and take on a dull appearance. The mycelium is initially colorless but turns brown as it matures. Microscopic analysis of the hyphae indicates that they are septate and multinucleate, and exhibit branching at right angles. Hyphae are divided into separate compartments by septa and are usually narrowed at the point where they branch. These compartments have doughnut-shaped pores that allow nuclei and mitochondria to move across cells. A large quantity of sclerotia is formed in the culture. Sclerotia are usually spherical with a diameter of 1–5 μm and have a dark brown to black color.

8.2.4.3 Disease Cycle and Epidemiology

The fungus persists in weeds and plant debris. The fungus survives in the form of sclerotia which are normally spherical with a diameter of 1–5 μm and have a dark brown to black color. These sclerotia, along with infected host debris, act as a source of primary inoculum for maize and other host crops. The pathogen has a broad spectrum of host specificity, infecting more than 32 families encompassing 188 species of plants. The following plant species were infected by this pathogen *Paspalum scrobiculatum*, *Pannisetum purpureem*, *Setaria italica*, *Panicum miliaceum*, *Coixlachryma jobi*, *Echnochola fromentacea*, *Pennisetum americanum*, *Zea maxicana*, *Zea mays*, *Oryza sativa*, *Saccharum officinarum*, *Sorghum bicolor*, *Arachis hypogea*, *Glycine max*, *Pisum*

FIGURE 8.3 Banded leaf and sheath blight disease symptoms on maize plants.

sativum, *Vigna radiata*, and *Lycopersicon esculentum*. The pathogen is transmitted through the soil and can also persist in a wide range of wild and cultivated plants (Sharma et al. 2002).

Optimal conditions for the development of BLSB disease in maize are temperatures between 25 and 30°C, with relative humidity of 90–100% (Singh and Shahi 2013). The disease progresses rapidly during the first two weeks of infection due to a high relative humidity (88–90%). However, the growth and spread of the disease are hindered when the relative humidity drops to 70%. Furthermore, rainfall exceeding 100 mm during the initial two weeks of infection promotes the early onset of infection and the progression of sickness (Ahuja and Payak 1981).

8.2.4.4 Disease Management

- Selection of a well-drained field, clean cultivation, destruction of crop debris, and crop rotation with non-host plants like pulses and stripping the second and third leaf sheaths from the ground level at the age of 35–40 days (Mehra et al. 2012) are important cultural practices that should be followed to reduce disease incidence.
- Planting on raised beds is an important aspect to avoid contact of water with seeds and promote faster growth of seedlings (Hooda et al. 2015).
- Seed dressing with fungicides (Bavistin 50WP @ 2.5 g/kg of seed, carboxin 35.5% + Thiram 37.5% @ 2.5 g/kg of seed, and Thiram 50 WP @ 2.5 g/kg of seed)

- Seed treatment with mancozeb 75 WP @ 0.25% followed by two foliar sprays with propiconazole 25 EC @ 0.1% at 40 and 50 days after sowing proved most effective.
- Several fungus species including *Trichoderma*, *Gliocladium*, and *Laetisaria*, bacteria such as *Pseudomonas fluorescens*, and nematodes like *Aphelenchus avenae* are found effective in controlling the disease. Seed treatment of the peat-based formulation @ 6 g/kg or as soil application @ 2.5 kg/ha or by spraying the liquid formulation twice @ 5 g/l of water (Sivakumar et al. 2000).
- Foliar application of *Bacillus subtilis* @ 0.5% has shown a reduction in disease incidence of BLSB. Validamycin antibiotic-based formulations have shown effective management of BLSB (Sharma et al. 2002).
- Foliar sprays of Propiconazole were most effective when applied at the initial stages, at the 30th or 40th day after planting and the second spray 10 days after the first. Foliar sprays of carbendazim showed ineffectiveness against BLSB (Saxena 2002).

8.2.5 Post-flowering Stalk Rot

Pammel (1914) initially reported this disease as a severe affliction affecting the roots and stalks, originating from the United States of America. A total of eight fungi and three bacteria were identified as the causes of stalk rots in maize and among them, Fusarium stalk rot (*Fusarium verticillioides*), charcoal rot (*Macrophomina phaseolina*), and late wilt (*Cephalosporium maydis*) are among the most common and damaging the crops in India (Khokhar et al. 2014).

8.2.5.1 Fusarium Stalk Rot

Fusarium stalk rot, caused by *Fusarium verticillioides* (Pammel 1914) and *F. moniliforme* is one type of stalk rot that affects maize. The disease was initially documented in Mount Abu, Rajasthan, India (Arya and Jain 1964), and is widespread in the majority of maize cultivation regions in the country where water scarcity is experienced during the flowering phase of the crop. Yield reductions range from 10 to 42%, and in certain areas, losses can reach up to 100%. These losses typically occur during the tasselling, silking, and grain formation stages. Decay spreads from the infected roots to the stem, leading to early drying, stem breakage, and ear dropping, ultimately reducing maize yields.

8.2.5.1.1 Symptoms

The occurrence of Fusarium stalk rot has been observed in plants in the age of 55–65 days, corresponding to tasselling and silking stages, followed immediately by grain development. This results in stalk breaking, rotting, lodging, and premature mortality of affected plants (Reis et al. 2004). Symptoms appear as softening and discoloration of the stalks near the first three internodes from the bottom. Upon splitting, the inner stalk exhibits a pale pink to tan discoloration, and the pith disintegrates while the vascular bundles remain intact. The plants show signs of desiccation, starting from the leaf edge and spreading towards the central vein, eventually covering the entire leaf surface. The eventual loss of all leaves leads to the desiccation of the entire plant before seed formation. These infections progressively intensify as the days pass after reaching physiological maturity. Stalk rot leads to decreased crop yields through two mechanisms: (i) damaged plants experience premature death, resulting in the production of ears that are light in weight and have poorly packed kernels, and (ii) plants with stalk rot are prone to lodging. Due to lack of water, as a result, the plants show wilting. This wilting is a precursor to stalk rot, which ultimately leads to the premature death of the plants.

8.2.5.1.2 Causal Organism

Two major species of *Fusarium i.e. Fusarium verticillioides* (Pammel 1914) and *F. moniliforme* are responsible for root and stalk rot of maize. Several other species of *Fusarium* such as *F. semitectum*,

F. avenaceum, *F. sulphurcum*, *F. acuminatum*, *F. roseum*, *F. merismoides*, *F. nivale*, and *F. solani* are reported to cause stalk rots. In India, so far, only *F. moniliforme* and *F. semitectum* are reported to be widespread in western Uttar Pradesh, Punjab, Rajasthan, and Assam (Borah et al. 2016). Macroconidia are hyaline, slightly curved, and tapered at both ends. They are 3 to 5 septate and measure 2–5 × 15–60 µm. Microconidia are single-celled and measure 2–3 × 5–12 µm. They are produced prolifically and are borne in chains.

8.2.5.1.3 Disease Cycle and Epidemiology

The fungi overwinter as mycelium, hard-wall developed chlamydospores, or sexual bodies in some species in corn residue, other dead plant residue, and in corn seed. This fungus is often found growing in healthy stalks and may cause rot only under certain conditions. Spores are spread by wind and splashing water; infection takes place through the roots, wounds in the stalk, or leaf scars. Disease development is favored by warm temperatures (26.7–37.8°C) and when a wet midseason follows early-season dry weather. The fungus attacks the stalk as mycelia by spores developed in a sexual or asexual way.

8.2.5.1.4 Disease Management

- Biocontrol agents *Bacillus siamensis* GL-02 (Zhang et al. 2022), *B. methylotrophicus* TA-1 (Cheng et al. 2019) and *Pseudomonas fluorescens* (Nayaka et al. 2009) have shown a good capability to inhibit Fusarium stalk rot and is a potential biocontrol agent for the control of maize stalk rot.
- Maize plants were artificially inoculated with *F. verticillioides*, and three potential *Trichoderma* strains (*T. harzianum* BThr29, *T. asperellum* BTas25, and *T. erinaceum* BTer43) were delivered as soil application (Trichoderma enriched FYM 10 tones ha−1), seed treatment (10 g/kg seeds), and drenching (10 g/liter water). In these treatments, *T. harzianum* BThr29 treated plants showed at par with fungicidal application and showed less than 50% discoloration in the inoculated internode of the maize crop (Jambhulkar et al. 2022).

8.2.5.2 Black Bundle Disease and Late Wilt

Late wilt is a vascular wilt disease affecting maize, caused by *Harpophora maydis*. The first recorded instance of late wilt disease was in Egypt in 1961–1962, gradually spreading to other countries. It is notably more prevalent and destructive in India including Pantnagar and Andhra Pradesh (Khokhar et al. 2014), with reported yield losses of up to 40% in susceptible cultivars.

8.2.5.2.1 Symptoms

Late wilt disease symptoms are characterized by the relatively rapid wilting of maize plants at the age of 70–80 days, occurring before tasselling and shortly before maturity (Degani and Cernica 2014). Initial symptoms appear approximately 60 days after sowing, manifesting as light green stripes on the leaves, which progressively become translucent. Entire leaves roll inward from the edges. Subsequently, upward drying-out of the plant occurs, accompanied by leaf yellowing, dehydration, and color alteration of the vascular bundles to a yellow-brown hue. Red-brown stripes appear on the lower internodes, with symptoms advancing to the fifth internode or beyond (Figure 8.4). With disease progression, the lower stem dries out, particularly at the internodes, displaying a shrunken and hollow appearance, along with dark yellow to brownish macerated pith and brownish-black vascular bundles. Late wilt disease, named so due to the delay in symptom appearance until flowering, is rarely associated with secondary invaders such as *H. acremonium*, *Sclerotium bataticola*, *Fusarium verticillioides*, and several bacterial rots, presenting as a "stalk rot complex." Vascular occlusion appears to be the principal cause of late wilt symptom development. Systemic infection affects the kernels, with severe cases resulting in the absence of cobs. If present, infected seeds are shrunken and undeveloped.

FIGURE 8.4 Symptoms of black bundle disease caused by *Harpophora maydis* in maize plants.

8.2.5.2.2 Causal Organism

Late wilt is caused by the soil-borne and seed-borne fungus *Harpophora maydis* (Samra, Sabet and Hingorani) Gams (Syn. *Cephalosporium maydis* Samra, Sabet and Hingorani and *Acremonium maydis*). The fungus forms unbranched, hyaline, septate hyphae, solitary, erect phialides formed directly on the hyphal tips, the hyphal ropes, or both. Phialides are parted from hyphae by a septum and taper towards their apices. Conidiophores of the fungus produce one-celled, hyaline conidia successively at the tip, usually embedded in a slimy drop, measuring 2–3 x 4–8 µm. Conidia typically appear in clusters, balls or rarely as fragile chains, bound by a gelatinous material. Rapid germination of conidia occurs through two polar germ tubes, occasionally forming one or three germ tubes. The fungus exhibits frequent anastomosis of germ tubes, and after approximately three weeks of growth on potato dextrose agar, conidia become scarce, with small botryoid sclerotia-like bodies appearing in old cultures.

8.2.5.2.3 Disease Cycle and Epidemiology

H. maydis survives as sclerotia on corn debris for 12–15 months and infects seedlings through the roots or mesocotyl. Disease spread primarily occurs through the infested soil, crop residue, or seed-borne inoculum. The secondary spread of the disease within fields is often associated with mechanical operations such as cultivation. In India, maximum disease occurs at a constant temperature of 24°C or when temperatures vary naturally between 20–32°C. Disease development is rapid at 20–32°C, with optimum development at 21–27°C. High temperatures and low soil moisture favor the disease progression.

8.2.5.2.4 Disease Management

- Moisture management and flood fallowing may be useful cultural controls for late wilt where they are economically practical.
- Crop rotation with non-host crops, sanitation of fields and avoid water stress at flowering are useful cultural practices to reduce the disease incidence significantly.
- Balanced fertilizer application can reduce disease severity, although it does not provide complete control.
- Seed treatment with carbendazim or @1g/kg of seed or thiram or captan @ 3.0 g/kg seed is also found effective in controlling the disease.

- Resistant varieties like DHM-103, DHM 105, Hi-Starch and Ganga Safed-2 may be grown to minimize the disease.
- Seed treatment with *T. viride, B. subtilis* and *P. fluorescens* @ 2.5 g each/kg of seed and soil application of 1.25 kg/h each fortified with 250 kg of FYM.

8.2.5.3 Charcoal Stalk Rot

Charcoal rot is a common stalk rot disease prevalent in warm, dry regions and causes yield loss of up to 70% in Africa. The disease thrives in drought years and arid regions where maize is commonly grown alongside other host crops, affecting over 500 plant species across more than 100 families, including soybean, sunflower, common bean, sorghum, maize, cotton, tobacco, and various vegetable crops (Ghosh et al. 2018). In India, an epidemic of charcoal rot disease was observed during the 1960 Kharif season in the Kashmir valley, followed by sightings during the rabi season of 1965–66 in Hyderabad (Andhra Pradesh, presently Telangana) and in 1966 in Pantnagar (Uttar Pradesh, presently Uttarakhand) during the Kharif season. The disease is particularly prevalent during the Rabi season when post-flowering temperatures rise significantly (35–45°C), leading to rapid development of charcoal rot.

8.2.5.3.1 Symptoms

After flowering, initial symptoms include abnormal drying of upper leaf tissue, stem lodging, and premature death. As plants near maturity, the lower stem internodes (typically limited to the first 5 nodes) exhibit a distinctive charcoal, grey-black discoloration, often giving the entire field a blackened appearance (Figure 8.5). Upon splitting the stem open, numerous minute black specks (micro-sclerotia) become visible on the shredded vascular bundles and the inner surface of the stem, leading to stalk breakage at the crown area and a charcoal decay, giving the stem's interior parts a charred appearance (Shekar et al. 2010). Brown, water-soaked lesions develop on the roots, later turning black. Additionally, the fungus may infect kernels, causing them to darken completely (Kumar and Shekar 2015).

8.2.5.3.2 Causal Organism

Charcoal stalk rot, caused by the fungus *Macrophomina phaseolina* (Tassi) Goid (Sclerotial stage: *Rhizoctonia bataticola*), exhibits characteristic features in its morphology and disease cycle. *M. phaseolina* is identifiable by its hyaline hyphae with thin walls, transitioning to light brown or dark brown hyphae with septa. Branches from the main hyphae typically form at right angles, with a constriction at the point of origin. The fungus produces microsclerotia, compact masses of

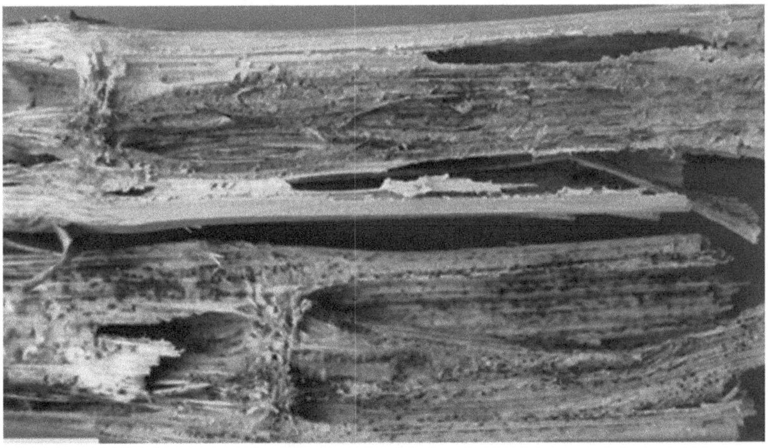

FIGURE 8.5 Symptoms of charcoal stalk rot disease in maize.

hardened fungal mycelium, which are initially light brown and become darker (brown to black) with age. Pycnidia, though rarely observed under natural conditions, are larger than microsclerotia, dark brown to black, rough, globose, or irregular, with a beaked and ostiolated appearance (Lakhran et al. 2018). Morphological heterogeneity may be attributed to fungal responses to environmental factors or host species variation (Tok 2019; Pandey et al. 2020).

8.2.5.3.3 Disease Cycle and Epidemiology

M. phaseolina survives as sclerotia, which are spherical and black in color, produced abundantly by the fungus. These sclerotia penetrate roots and lower stems during the growing season, and they overwinter as microsclerotia on lower stem residue. Microsclerotia can persist in dry soil for years but only a few weeks in moist soil. The primary mode of dispersal is through soil movement by tractor plow, farm machinery, and packing materials. Alternate hosts also serve as significant sources of inoculum for subsequent infections. Seed-borne infections are generally low, with limited evidence supporting transmission through seeds.

Maximum infection occurs in plants experiencing moisture stress during the post-flowering period. Stressors such as high plant populations, drought, nitrogen fertilizer applications, hail or insect damage, all contribute to disease development. The broad host range of *M. phaseolina* ensures a continuous source of inoculum for subsequent seasons. The disease is more prevalent in conditions of high soil temperature (30–42°C), low soil moisture, and low soil pH (5.4–6.0).

8.2.5.3.4 Disease Management

- Maize crops should be sown on ridges to avoid water logging and proper drainage maintained.
- Crop rotation with a poor host such as cotton or non-host plants such as wheat or barley for 1–2 years are found effective to manage the disease.
- Abiotic stressed plants are more susceptible; hence avoidance of stress factors like drought, and unbalanced nutrition may help to reduce damage.
- It is required to apply balanced fertilizers because adequate levels of available P and K reduce nutrient stress and encourage plant health.
- Effective water management especially as the crop reaches the flowering period is crucial to prevent stress and manage the disease.
- Early planting will aid in earlier canopy closure that will help reduce soil temperatures and therefore reduce the competitive ability of *M. phaseolina*.
- Minimizing exposure to stressors, such as drought and an imbalanced diet, may decrease the damage. It is advisable to prevent high plant populations as this might lead to elevated plant stress due to competition for water, especially during dry seasons.
- Removal of infected stalks will render the level of inoculum.

8.2.6 Bacterial Stalk Rot

It is a major disease in tropical and subtropical countries. In India, bacterial stalk rot disease was reported by Prasad in 1930. A severe outbreak of BSR occurred in the Mandi district of Himachal Pradesh in 1969 and its importance was realized (Kumar et al. 2017; Jatoth et al. 2022). Bacterial BSR of maize is a major hurdle in quality production and high yield in the crop causing 21–98% yield loss during epidemics. This disease is one of the four main diseases that affect maize in India. It has been reported in nature that the disease can occur in up to 80–85% of maize plants, and in artificial epiphytotics, it has been found to cause yield losses of 98.8% (Hooda et al. 2018).

8.2.6.1 Symptoms

The disease occurs in young as well as old plants. Water-soaked spots irregular in size and shape develop on any part of the plant from the soil line to the tassel. The presence of a mild, sweet

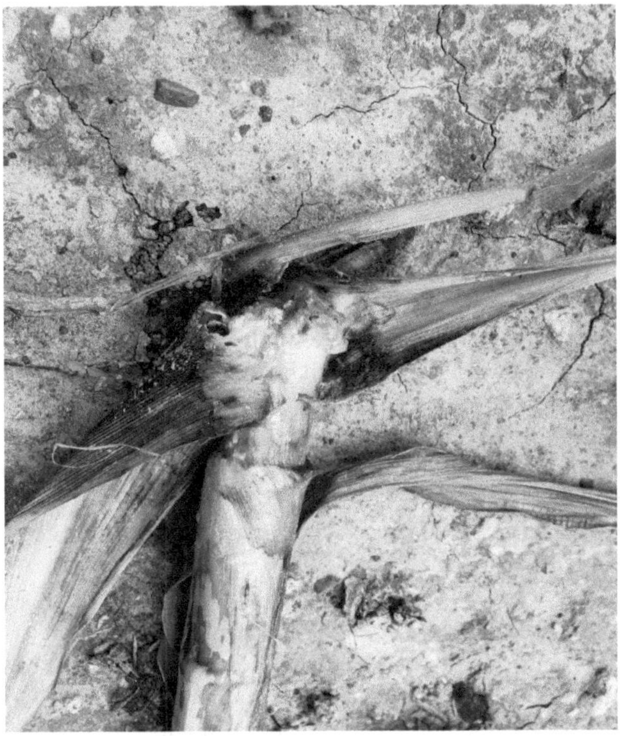

FIGURE 8.6 Symptoms of bacterial stalk rot disease in maize plant caused by *Dickeya zeae*.

fermenting odor is often associated with the rotting caused by bacterial stalk rot in maize plants. At times the rot involves rind only causing a dark soft decay, but mostly the rotted area is principally in the interior of the stalks. A mild sweet fermenting odor accompanies such rotting. With the advance of the stalk rot, leaves may start yellowing and drying. The infected tissues are at first soft, later turn into a dry mass of shredded, easily disjointed fibers and at this stage, the plant topples down (Figure 8.6). Spots often appear on leaf sheaths just above nodes and quickly develop yellow-green to reddish-brown discoloration. Infection may spread rapidly throughout the leaf sheath and cause sheath and leaf to wither under favourable conditions.

8.2.6.2 Causal Organism

Disease is caused by a gram-negative, soft rot bacterium *Dickeya zeae* belongs to Pectobacteriaceae family and was formerly named *Erwinia chrysanthemi* pv. *zeae* (Samson et al. 2005). Among eight currently established *Dickeya* species three species viz., *D. dadantii*, *D. paradisiaca* and *D. zeae* have been reported as maize pathogens. *D. zeae* is the major pathogen responsible for the maize stalk rot and rice foot rot diseases and it has ability to infect both monocotyledons and dicotyledons. *D. zeae* is a gram-negative, facultative anaerobic, non-fluorescent, motile, rod-shaped bacterium. There are 3–14, but more usually 8–11, peritrichous flagellae and cell size varying from 0.8–3.2 x 0.5–0.8 μm (Kumar et al. 2015). Bacterial colonies are round white-grey, shiny, non-mucoid with irregular margins (Prokic et al. 2020).

8.2.6.3 Disease Cycle and Epidemiology

Besides maize, *D. zeae* also causes soft rot in various monocots and dicots, including essential crops such as potato, rice and banana (Kumar et al. 2015; Prokic et al. 2020). *D. zeae* is soil-borne and may survive in crop residues in the soil, but the survival period varies from different environmental conditions and the bacteria can survive almost nine months in the soil containing naturally infected maize stalks. A longer survival period in sterilized soils and heavier soils was reported by (Kumar et al. 2017). Bacterium enter into the plant through its natural openings like the stomata, hydathodes, insect wounds in the leaf whorl, stalks, and roots and injuries made by winds or occurred during agronomical practices. It can also be transmitted from plant to plant and field to field through rain-water and its runoff. Certain insects like maize borer play a vital role in the initial infection and subsequent spread of the disease.

Bacterial stalk rot is prevalent in heavy soil under high temperature and humidity conditions. Disease incidences will be more in the month of August and September if warm temperature and frequent rains prevail in farmer's field. Frequent rainfall and overhead irrigation with sprinklers cause prolonged leaf wetness and accumulation of water in the whorl and this water heats, it can damage the plant tissues, providing opening through which infection can take place. Plants subjected to high temperatures or flooding may develop the symptoms first around the base of the plant. Irrigation water is believed to be the primary inoculum source. The disease is observed in maize when intermittent heavy rains are followed by high temperatures and humid conditions. A temperature of 35°C, 70% relative humidity and inoculum level of 2×10^8 CFU/ ml were more essential for disease development in maize plants of 15 to 30 days old and there was a positive correlation of disease with high dosage of nitrogen fertilizer (Kumar et al. 2017).

8.2.6.4 Disease Management

- A good drainage system should be maintained in the field and flooding avoided. Avoid high nitrogen applications and always split the application.
- High doses of phosphorus and potassium fertilizers should be given to decrease the incidence of the disease.
- Avoid irrigation during very hot periods of the day where water can collect in the plant whorl.
- Two inbred lines, CM101, CM-110, and two open-pollinated varieties, CM600, Basi, are found to be tolerant against *E. chrysanthemi* pv. *zeae*. In the Philippines, three composites, i.e. USM Var 16, USM Var 18, and USM Var 28, as well as two synthetics: USM Var 24 and USM Var 26, showed resistance against the disease. Additionally, one hybrid variety, USM Var 30 hybrid, was also introduced.
- Chlorination of irrigation water or soil drenching with bleaching powder (33% chlorine @ 10 kh/ ha) at the pre-flowering stage is recommended.
- 3–4 surface irrigations (one before maize planting) with chlorinated water encompasses chlorine @100ppm rendering the incidence of stalk rot in maize.
- Application of bleaching powder (25 kg/ha) at two stages, first at the flowering stage and the second 10 days after, was found significant for controlling the disease (Kumar et al. 2017).
- Streptocycline alone and in combination with Blitox-50 W (50% copper oxychloride) most effectively controlled stalk rot in a glasshouse and field experiments. However, stable bleaching powder is considered more suitable for soil drenching than Streptocycline because of its longer stability in soil.
- Spray with Streptocycline @ 2 g / lit and insecticide (Cypermethrin + Chloropyrifos @ 2.5 ml/ lit of water during knee height and subsequent spray after a 15-day interval reduces disease incidence (Subedi et al. 2016).
- *P. fluorescens* strains, MPF-5 strain showed effective results, with 8.07 log cfu/ml.
- Intercropping with legumes such as cowpea, soybean, red gram and green gram in a 2:1 ratio encourages the build-up of natural enemy population.
- Trap crop: Sorghum is preferred host for *C. partellus* and sowing 2–3 rows of trap crop on all sides of maize and uprooting it after 45 days can be minimized the infestation.
- Clipping of bottom leaves of maize on which majority of the eggs are laid.

- Crop rotation with non-host crops destroys the build-up of pests due to the non-availability of a host.
- Release of eggs parasitoid *T. chilonis* @ 8 cards per ha twice i.e. at 12 and 22 days after germination at one-week interval.
- Prophylactic spray of monocrotophos 36 SL @ 1.6 ml/l, at 10–15 DAG followed by carbofuran 3G @ 7.5 kg/ha in plant whorls in case of severity and cypermethrin @ 220–300 ml/dissolve in 250–300 lit of water/ acre.

8.3 CONCLUSIONS

The quality and yield of maize are greatly impacted by various diseases. Therefore, it is essential to perform disease diagnostics and identification to promptly intervene. Management of maize diseases by using eco-friendly approaches including cultural, resistant varieties against insect pests and diseases and biological control can improve the quality and economic efficiency of maize production. Breeding for resistance against existing and emerging diseases is a crucial component in achieving yield stability across different growing seasons.

REFERENCES

Aggarwal SK, Hooda KS, Kaur H et al. (2023) Comparative evaluation of management modules against Maydis leaf blight disease in maize (*Zea mays*). *Eur J Plant Pathol* https://doi.org/10.1007/s10658-023-02777-x.

Ahangar MA, Bhat ZA, Sheikh FA, Dar ZA, Lone AA, Hooda KS, Reyaz M (2016) Pathogenic variability in *Exserohilum turcicum* and identification of resistant sources to turcicum leaf blight of maize (*Zea mays* L.). *J Appl Natural Sci* 8(3): 1523–1529.

Ahuja SC, Payak MM (1981) Relationship of relative humidity and temperature level with the development of leaf and sheath blight of maize. *Z Pflakranth. Pflschutz* 88: 265–268.

Arya HC, Jain BL (1964) Fusarium seedling blight of maize in Rajasthan. *Indian Phytopath* 17: 51–57.

Bertus LS (1927) *A sclerotial disease of maize (Zea mays L.) due to Rhizoctonia solani Kuhn.* Yearbook, Department of Agriculture, Ceylon, pp 44–46.

Borah SN, Deka S, Sarma, HK (2016) First report of *Fusarium verticillioides* causing stalk rot of maize in Assam, India. *Plant Dis* 100(7): 1501.

Butler EJ (1913) The downy mildew of maize, *Sclerospora maydis* (Rae.) Butl. *Mem Dept Agric Bot Ser* 5: 275–280

Byrnes KJ, Pataky JK, White DG (1989) Relationships between yield of three maize hybrids and severity of southern leaf blight caused by race O of Bipolaris maydis. *Plant Dis*: 834–840.

Camera JN, Forcelini CA, Koefender J, Golle DP, Schoffe A (2019) Reaction of maize hybrids to Northern corn leaf blight and common rust, and chemical control of Northern corn leaf blight. *Plant Pathol* 86: 1–10.

Cheng XK, Ji XX, Ge YZ, Li JJ, Qi WZ, Qiao K (2019) Characterization of antagonistic *Bacillus methylotrophicus* isolated from rhizosphere and its biocontrol effects on maize stalk rot. *Phytopathology* 109: 571–581.

CIMMYT (2004) Maize diseases a guide for field identification. CIMMYT.

Degani O, Cernica G (2014) Diagnosis and control of *Harpophora maydis*, the cause of late wilt in maize. *Adv Microbiol* 4: 94–105.

De-Rossi RL, Reis EM, Brustolin R (2015) Conidial morphology and pathogenicity of *Exserohilum turcicum* isolates of corn from Argentina and Brazil. *Summa Phytopathol* 41(1): 58–63.

Ferguson LM, Carson ML (2004) Spatial diversity of *Setosphaeria turcica* sampled from the eastern United States. *Phytopathology* 94: 892–900.

Ghosh T, Biswas MK, Guin C, Roy P (2018) A review on characterization, therapeutic approaches and pathogenesis of *Macrophomina phaseolina*. *Plant Cell Biotechnol Mol Biol* 19: 72–84.

Hooda KS, Bagaria PK, Khokhar M, Kaur, H, Rakshit S (2018) *Mass screening techniques for resistance to maize diseases*. ICAR-Indian Institute of Maize Research, PAU Campus, Ludhiana- 141004, pp 93.

Hooda KS, Khokhar MK, Parmar H, Gogoi R, Joshi D, Sharma SS, Yadav OP (2015) Banded leaf and sheath blight of maize: Historical perspectives, current status and future directions. *Proc Nat Acad Sci India Sec B: Biol Sci* 87(4). doi: 10.1007/s40011-015-0688-5.

Hooda KS, Sekhar JC, Chikkappa GK, Kumar S, Gowda KTP, Sreeramsetty TA, Sharma SS, Kaur H, Gogoi R, Rangereddy R, Kumar P, Singh A, Devlesh RK, Chandrashekara C (2012) Identifying sources of multiple disease resistance in maize. *Maize J* 1(1): 82–84.

Isakeit T, Jaster J (2005) Texas has a new pathotype of *Peronosclerospora sorghi*, the cause of sorghum downy mildew. *Plant Dis* 89(5): 529. doi: 10.1094/PD-89-0529A.

Jambhulkar PP, Raja M, Singh B, Katoch S, Kumar S, Sharma P (2022) Potential native *Trichoderma* strains against *Fusarium verticillioides* causing post flowering stalk rot in winter maize. *Crop Prot* 152: 105838. https://doi.org/10.1016/j.cropro.2021.105838.

Jatoth R, Singh D, Geat N, Babu PL, Kesharwani AK (2022) Distribution of bacterial stalk rot disease of maize in India and identification of causal agent using biochemical and *fli*C gene-based marker and its sensitivity against chemicals and bacterial antagonist. *Indian Phytopath* 75(2): 517–525.

Khokhar MK, Sharma S, Gupta R (2014) Integrated management of post flowering stalk rot of maize caused by *Fusarium verticillioides*. *Indian Phytopath* 67: 228–233.

Kumar A, Hunjan MS, Kaur H, Rawal R, Kumar A, Singh PP (2017) A review on bacterial stalk rot disease of maize caused by *Dickeya zeae*. *J Appl Natural Sci* 9 (2): 1214–1225.

Kumar A, Hunjan MS, Kaur H, Singh PP (2015) Characterizing diversity of *Dickeya zeae* causing bacterial stalk rot of maize based on biochemical assays and antibiotic sensitivity. *Indian Phytopath* 68: 375–379.

Lakhran L, Ahir RR, Choudhary M, Choudhary S (2018) Isolation, purification, identification and pathogenicity of *Macrophomina phaseolina* (Tassi) Goid caused dry root rot of chickpea. *J Pharmacogn Phytochem* 7: 3314–3317.

Lu YL, Xu J, Yuan ZM, Hao, ZF, Xie CX, Li X, Shah T, Lan H, Zhang SH, Rong TZ, Xu YB (2012) Comparative LD mapping using single SNPs and haplotypes identifies QTL for plant height and biomass as secondary traits of drought tolerance in maize. *Mol Breeding* 30: 407–418.

Mehra R, Kamboj MC, Mehla JC, Madan L, Chand M (2012) Status of maize diseases and their management in Haryana. In: *Proceedings of national seminar on sustainable agriculture and food security: challenges in changing climate*. CCS HAU, Hisar, 27–28 Mar 2012. pp 217.

Murthy M, Shankara, Nagaraja SK (2014) Outbreak of maize leaf roller, *Cnaphalocrocis trapezalis* Guenee (*Marasmia trapezalis*) (Crambidae: Lepidoptera) on sorghum in Karnataka, South India. *Insect Environ* 19(4): 250–253.

Nayaka SC, Shankar AC, Reddy MS, Niranjana SR, Prakash HS, Shetty HS et al. (2009) Control of *Fusarium verticillioides*, cause of ear rot of maize, by *Pseudomonas fluorescens*. *Pest Manag Sci* 65: 769–775.

Nwanosike MRO, Mabagala RB, Kusolwa PM (2015) Disease intensity and distribution of *Exserohilum turcicum* incitant of northern leaf blight of maize in Tanzania. *Int J Pure Appl Biosci* 3(5): 1–13.

Pammel LH (1914) Serious root and stalk diseases of corn. *IOWA Agr* 15: 156–158.

Pandey AK, Burlakoti RR, Rathore A, Nair RM (2020) Morphological and molecular characterization of *Macrophomina phaseolina* isolated from three legume crops and evaluation of mungbean genotypes for resistance to dry root rot. *Crop Prot* 127: 104962. doi: 10.1016/j.cropro.2019.104962.

Passerini (1876) Lanebbia Delgranotur Co. Bol. Comiz, Agriculture. Parmense. 10: 3.

Perumal R, Nimmakayala P, Erattaimuthu SR, No E-G, Reddy UK, Prom LK, Odvody GN, Luster DG, Magill CW (2008) Simple sequence repeat markers useful for sorghum downy mildew (*Peronosclerospora sorghi*) and related species. *BMC Genet* 9(1):77–77.

Prokic A, Zlatkovic N, Kuzmanovic N, Ivanovic M, Gasic K, Pavlovic Z, Obradovic A (2020) Identification and characterization of *Dickeya zeae* strains associated with maize stalk soft-rot in northern Serbia. *Eur J Plant Pathol*. https://doi.org/10.1007/s10658-020-02019-4.

Prom LK, Perumal R, Montes-Garcia N, Isakeit T, Odvody GN, Rooney WL, Little CR, Magill C (2015) Evaluation of gambian and Malian sorghum germplasm against downy mildew pathogen, *Peronosclerospora sorghi*, in Mexico and the USA. *J Gen Plant Pathol* 81(1): 24–31. doi: 10.1007/s10327-014-0557-8.

Rajput LS, Harlapur S I (2014) Status of banded leaf and sheath blight of maize in North Karnataka. *Karnataka J Agric Sci* 27(1): 82–84.

Rani VD, Reddy PN, Devi GU (2013) Banded leaf and sheath blight of maize incited by *Rhizoctonia solani* f. sp. *sasakii* and its management- A review. *Int J Appl Biol Pharma Tech* 4: 52–56.

Rashid Z, Zaidi PH, Vinayan MT, Sharma SS, Srirama Setty TA (2013) Downy mildew resistance in maize (*Zea mays* L.) across *Peronosclerospora* species in lowland tropical Asia. *Crop Protec* 43: 183–191.

Rathore RS, Trivedi A, Mathur K (2002) *Rajasthan downy mildew of maize: The problem and management perspectives Proc 8th Asian Regional Maize Workshop*, Bangkok, Thailand: August 5–8, 2002.

Reddy TR, Reddy PN, Reddy RR (2013) Pathogenic variability of isolates of *Exserohilum turcicum*, incitant of leaf blight of maize. *Indian J Plant Protec* 41(1): 72– 75.

Reis EM (2004) *Manual de diagnose e controle de doenças do milho*. Lages: Graphel, p 141.

Sagar GC, Bhusal K, Rijal TR, Manandhar H (2019) Screening of maize genotypes/hybrids for identifying resistant sources against banded leaf and sheath blight (Blsb) in Rampur, Chitwan, Nepal. *Int J* 5(2): 260–264.

Samson R, Legendre JB, Christen R, Achouak W, Gardan L (2005) Transfer of *Pectobecterium chrysanthemi* (Burkholder et al. 1953) Brenner et al. 1973 and *Brenneria paradisiaca* to the genus *Dickeya* gen. nov. as *Dickeya chrysanthemi* comb. nov. and *Dickeya paradisiaca* comb nov. and delineation of four novel species, *Dickeya dadantii* sp. nov., *Dickeya dianthicola* sp. nov., *Dickeya dieffenbachiae* sp. nov. and *Dickey azeae* sp. nov. *Int J Syst Evol Microbiol* 55:1415–1427.

Saxena SC (2002) Bio-intensive integrated disease management of banded leaf and sheath blight of maize. In: *Proceed of 8th Asian regional maize workshop: new technologies for the New Millennium*, Bangkok, Thailand, pp 380–388.

Sharma RC, Vasal SK, Gonzalez F, Batsa BK, Singh NN (2002) Redressal of banded leaf and sheath blight of maize through breeding, chemical and biocontrol agents, In: *Proc 8th Asian Regional Maize Workshop: New Technologies for the New Millennium*, Bangkok, Thailand. pp 391–397.

Sharma RR, Gour HN, Rathore RS (2004) Etiology of banded leaf and sheath blight symptoms on maize. *J Mycol Plant Pathol* 34: 56–59.

Sibiya J, Tongoona P, Derera J (2013) Combining ability and GGE biplot analyses for resistance to northern leaf blight in tropical and subtropical elite maize inbred lines. *Euphytica* 191: 245–257.

Singh A, Shahi JP (2013) Banded leaf and sheath blight: an emerging disease of maize. *Maydica* 57: 215–219.

Singh A, Singh D (2011) Sources of resistance in maize (*Zea mays* L.) against Erwinia stalk rot and brown stripe downy mildew. *Plant Dis Res* 26 (2): 187.

Singh A, Singh D (2012) Screening of maize genotypes for resistances to BSR and BSDM. *Maize J* 1: 134–135.

Singh R, Srivastava RP (2016) Southern corn leaf blight-an important disease of maize: an extension fact sheet. *Indian Res J Extension Edu* 12(2): 324–327.

Singh S, Singh UB, Malviya D, Paul S, Sahu PK, Trivedi M et al (2020) Seed biopriming with microbial inoculant triggers local and systemic defense responses against *Rhizoctonia solani* causing banded leaf and sheath blight in maize (*Zea mays* L.). *Int J Environ Res Public Health* 17. doi: 10.3390/ijerph17041396

Singh SK, Patel MB, Thakker BN, Hooda KS, Alpesh Kumar Barad (2019) *Rhizoctonia solani* f. sp. *sasakii* Inciting banded leaf and sheath blight of maize and their management: An overview. *Int J Curr Microbiol Appl Sci* 8: 2858–2866.

Sivakumar G, Sharma, RC, Rai SN (2000) Biocontrol of banded leaf and sheath blight of maize by peat – based *Pseudomonas fluorescens* formulation. *Indian Phytopath* 53: 190–192.

Subedi S, Subedi H, Sarswati N (2016) Status of maize stalk rot complex in western belts of Nepal and its integrated management. *J Maize Res Dev* 2(1): 30–42.

Subramanian LS (1931). A note on downy mildew of sugarcane in India. *Agric Lives India* 1: 32–33.

Tok FM (2019) Relationship between morphologic, phenotypic and pathogenic characteristics in *Macrophomina phaselina* isolates from cucumber plants. *Int J Innov Approaches Agric Res* 3: 651–660. doi: 10.29329/ijiaar.2019.217.11.

Wani, NA, Saini MK, Sofi TA et al (2023) Taxonomic study of species of the genus *Psathyrella* (Psathyrellaceae, Agaricales) collected from the forests of the North Kashmir, India. *Indian Phytopath* 76: 383–399.

Wathaneeyawech S, Sirithunya P, Smitamana P (2015) Study of the host range of northern corn leaf blight disease and effect of *Exserohilum turcicum* toxin on sweet corn. *J Agric Technol* 11: 953–963.

White DG (1999) Downy mildews. In: *Compendium of Corn Diseases*, third ed. Amer. Phytopathol. Soc., USA, pp 25–32.

Yadav OP, Hossain F, Karjagi CG, Kumar B, Zaidi PH, Jat SL, Chawla JS, Kaul J, Hooda KS, Kumar P, Yadava P (2015). Genetic improvement of maize in India. *Agri Res* 4(4): 325–338.

Yadav P, Sharma SS, Yadav PD, Yadav SK (2022) Collection, isolation, purification and pathogenicity of banded leaf and sheath blight disease of maize Caused *Rhizoctonia solani* f. sp. *sasakii*. *J Plant Sci Res* 38: 133–138.

Zhang K, Wang L, Si H, Guo H, Liu J, Jia J, Su Q, Wang Y, Zang J, Xing J, Dong J (2022) Maize stalk rot caused by *Fusarium graminearum* alters soil microbial composition and is directly inhibited by *Bacillus siamensis* isolated from rhizosphere soil. *Front Microbiol* 13: 986401. doi: 10.3389/fmicb.2022.986401.

9 Important Plant Parasitic Nematodes (PPNs) of Major Cereals in India and Their Management

S.S. Vaish, Shiv Shankar Patel, Vandana Sahu, Ashwani Kumar Patel and Vinod Upadhyay

9.1 INTRODUCTION

Cereals are adaptable to different ecological habitats, are relatively easy to grow, and have higher yields and good nutritive values (Vasil 1999). These edible grains belong to Poaceae. Paddy (*Oryza sativa* L.), wheat (*Triticum aestivum* L.), maize (*Zea mays* L.), pearl millet (*Pennisetum glaucum* L.), sorghum (*Sorghum bicolor* L.), barley (*Hordeum vulgare* L.), and rye (*Secale cereal* L.), oats (*Avena sativa* L.) are considered the major cereal crops of India.

The total cultivated area of cereals in India is about 99.25 mha with production of 304.4 million metric tonnes (FAOSTAT 2023). Among these cereals, rice and wheat occupy the major positions in terms of production, acreage and source of nutrition, particularly, minerals and dietary fibers which makes them a staple food globally and contribute more than 50% of the world's cereal production. India is the world's second-largest producer of rice and wheat.

Rice occupies an area of 43.8 mha with a total production of 177.6 mt and productivity of 4,057 kg ha^{-1}, whereas wheat occupies a 29.3 mha area with 103.6 mt of production, and 3,533 $kgha^{-1}$ productivity (FAOSTAT 2021). The Indian products of these cereals have huge demand in the global market. Rice is a good source of calories and is rich in magnesium, phosphorus, manganese, selenium, iron, thiamin, niacin and folic acid, although has less fiber and fat content, is gluten-free and has several health benefits as it improves blood cholesterol levels and reduces the risk of related diseases (Fukagawa and Ziska 2019). Wheat is a good source of protein, vitamins (Vitamins B) and dietary fiber and has many health benefits, particularly, reducing the risk of cardiovascular disease, type 2 diabetes, and notably colorectal cancer (Peter et al. 2015).

There are several biotic and abiotic stresses that limit the production of cereals around the world. Among these biotic stresses, plant parasitic nematodes are also known as one of the major limiting factors affecting the cultivation of these crops directly. However, the infestation of soils with plant parasitic nematodes aggregates the problems of wilts and root rot caused by various fungal and bacterial plant pathogens by proving entry points (De Waele and Elsen 2007). Such circumstances result in disease complexes and cause synergistic effects in terms of damage that results in reduced yields and quality of crop products, high production costs and consequently income loss (Onkendi et al. 2014). Moreover, the infection of plants by plant parasitic nematodes predisposes them to foliar diseases (Pandey et al. 2016).

Over 4100 species of PPNs have been reported to belong within 197 genera, which accounts for 7% of the phylum Nematoda which are important constraints to global food security as they pose

DOI: 10.1201/9781032631578-9

a significant threat to crop production by causing substantial yield losses to a wide range of agricultural crops (Bridge et al. 2005; Decraemer and Hunt 2006). Root knot nematode (*Meloidogyne graminicola* L.), cereal cyst nematode (*Heterodera* spp.), wheat seed gall nematode (*Anguina tritici* L.), lance nematode (*Hoplolaimus galeatus* L.), root lesion nematode (*Pratylenchus* spp. L.), Tundu or yellow year rot (*Rathaybacter tritici* + *Anguina tritici*), rice root nematode (*Hirschmanniella* spp.), reniform nematodes (*Rotylenchulus* spp.), stem/bulb nematode (*Ditylenchus angustus* L.) and rice white tip/foliar nematode (*Aphelenchoides besseyi* L.) are some of the major plant parasitic nematodes (Jones et al. 2013). Moreover, the distribution of nematode species varies greatly, some are cosmopolitan, such as *Meloidogyne* spp., whilst others (*Heterodera* spp.) are restricted geographically or are highly host-specific.

Globally, the losses caused by plant parasitic nematodes have been estimated to the tune of $US 80 billion per year (Navia et al. 2017). Overall, plant parasitic nematodes cause 21.3% of crop losses amounting to Rs. 102,039.79 million (1.58 billion USD) annually. Out of the total estimated losses, 20% is due to infection by PPNs in rice. Cereal crops are infected by several genera of PPNs causing up to 10% yield losses annually (Dash et al. 2022). The economic loss caused by these destructive pathogens ranges from 10 to 100% depending on different agro-ecological conditions such as drought, heat stress, and cold stress (Seid et al. 2021). In India, the losses in 19 horticultural crops were assessed to the tune of Rs. 50,224.98 million, while for 11 field crops, it was projected at Rs. 51,814.81 million. Root knot nematode, *Meloidogyne graminicola*, is one the most economically important nematodes of rice that causes losses to the tune of Rs. 23,272.32 million (Kumar et al. 2020).

9.2 GENERAL CHARACTERISTICS OF PLANT PARASITIC NEMATODES

Plant parasitic nematodes are characterized by the presence of stylet, a hypodermic needle, by which they inject enzymes into plant cells, digest the cell contents and suck it as nutrients for their growth. PPN_S are triploblastic, bilaterally symmetrical, heterotroph, segmented, pseudocoelomates, ubiquitous, obligate and aquatic vermiform animals belonging to phylum *Nematoda* and order *Tylenchida*. The shape of the nematode varies for some species, for example females of *Meloidogyne, Heterodera, Rotylenchus* and *Nacobbus* (false root knot nematode). The intensity of symptoms was directly related to the inoculum levels. Soil fertility and texture have a significant influence on nematode abundance and diversity. Plant parasitic nematodes (e.g. *Meloidogyne, Rotylenchulus, Heterodera, Hoplolaimus*, etc.) generally prefer sandy loam or loam or sandy soils (Singh and Singh 2003) some (e.g. *Hirschmanniella* and *Pratylenchus*) are however, more prevalent in heavy soil. The abundance of nematodes and their diversity depends on soil fertility and texture. Root knot and cyst nematodes generally prefer sandy loam or sandy soil (Singh and Singh 2003).

9.3 SPREAD OF PLANT PARASITIC NEMATODES

Plant parasitic nematodes are known to spread by anything that moves and carries particles of soil, e.g., farm equipment, irrigation water, floods, drainage water, animals, birds and dust storms in local areas or from field to field. However, long-distance dispersal of PPN is caused by the movement of soil along with sapling plants as well as farm produce. Some foliar PPN may also spread by rain splashes or by overhead irrigation.

9.4 MAJOR PLANT PARASITIC NEMATODES

9.4.1 Root Knot Nematodes (*Meloidogyne graminicola* L.)

Root knot nematodes are present throughout the world, particularly in warm or hot climates and short or mild winters. There are more than 60 species of *Meloidogyne* around the world of which 12 are common in India. Out of these 12 species, *M. incognita*, *M. javanica*, *M. arenaria, M. hapla* and *M. graminicola* are more predominantly encountered. *M. graminicola* infects rice, barley and wheat (Singh and Singh, 2009) (Figures 9.1, 9.2 and 9.3). However, *M. incognita* infects several vegetables, pulses and coarse cereals (Singh et al. 2007; Vaish and Pandey 2012). All the species of RKNs produce a characteristic 'root gall' or 'knotted root symptom' which could be easily recognized by the naked eye resulting in poor growth, reduced tillering, chlorosis of young plants, and production of un-filled spikelets at maturity. On roots, the nematode incites the formation of hook, spiral, or horseshoe-shaped terminal galls (Hunt and Handoo 2009). Transformation of roots into galls leads to impairment of water and mineral uptake. As a result, seedlings in the nursery or plants in the field show yellowish leaves, sparse tillering, and stunted growth (Khan and Anwer 2011). The life cycle initiates from the egg that hatches to produce the second stage juvenile and following molting, the adult is formed (Figure 9.4). It completes its life cycle in 25 days at 27°C, but it may take longer at lower or higher temperatures. Each female lays nearly 500 eggs in a gelatinous matrix.

9.4.2 Cereal Cyst Nematodes (*Heterodera* spp.)

Twelve Heterodera species are considered of major economic significance in cereals, of which *Heterodera avenae*, *H. latipons* and *H. filipjevi* are the most important. The disease of wheat, barley and oats caused by cyst nematodes is also known as Molya disease, with a root system that is monkey pawlike and was first reported in Rajasthan in 1957.

Glistering white females can be observed adhering to roots, which fall off in the soil within 2 weeks before plants are ready for harvesting. Initially, small patches of 2–3 meters in diameter can be seen in infested fields. This nematode survives in soil by its cysts which serve as a source of primary inoculum. Cysts hatch to produce second-stage juveniles and migrate through soil to find a suitable host (Figure 9.5). Thereafter, they penetrate to the growing root tip, enter inside and establish within the host.

A temperature of 20–22°C is conducive for the maximum emergence of larvae. Well-aerated, light-textured soil with good moisture and drainage helps in the development of the disease. Vaish et al. (2011), during an extensive survey in the Ladakh region of India, found that cereal cyst nematode is a major problem in this area with a disease incidence between 5 to 20% at the 71 locations. Skampuk, Hunder, Deskit, Skurbuchan, Thickshey, Stakmo, Shey, Chuchot, Rambirpora, Khaltsi, Temisgam, Achinathang, and Hanuthang are major hotspots in this region.

9.4.3 Wheat Gall Nematode or Ear Cockle of Wheat (*Anguina tritici*)

Anguina tritici (Stein buch) Filipjevi was reported as the first plant parasitic nematode by Needham in 1973 from England. This nematode causes ear cockle disease in wheat. Now this nematode has become widely distributed and reported from many countries (Maqbool 1988). In India, it was first reported from Punjab in 1919 (Milne 1919). However, Singh et al. (2010) reported this nematode for the first time from a farmer's field in Jammu. This nematode is also associated with yellow ear rot *Clavibacter tritici* and causes Tundu disease in India. The ear cockle-infected wheat plant can be identified by the presence of slight enlargement at the basal portion of the stem, stunted growth of plants, twisting and curling of leaves and conversion of grain into hard dark brown to black cockles. These cockles serve as a source of survival through which nematodes can survive up to 28 years (Figure 9.6). This disease has also been reported from the Ladakh region at 20 locations with hotspots at Spituk, Phyang, Nemo and Basgo (Vaish et al. 2012a, 2012b).

9.4.4 Rice root nematode (*Hirschmanniella oryzae*)

Rice root nematode (*Hirschmanniella oryzae*) is a migratory endoparasite mainly distributed in tropical and subtropical regions of Asia (Bridge et al. 2005). This nematode has a preference to multiply

FIGURE 9.1 Illustrating symptoms of *Meloidogyne graminicola* that causes root knot disease of rice. a, rice field showing large patch of the diseased plants showing severe stunting and yellowing; b, rice hill showing presence of galls; c, collection of soil samples from the infested field; d and e, discussion with the farmers about the disease; f, g, & h, showing severe stunting, yellowing and wilting along with galls induced by M. graminicola; 1, j, & k, depicting formation of hook, spindle or club-shaped galls varied in size and color on the root system of rice plants.

FIGURE 9.2 Showing symptoms of *Meloidogyne graminicola* that causes root knot disease of barley. a, barley field showing several patches of the diseased plants showing stunting and yellowing and poor stand; b and c, severe stunting, yellowing and absence of heading of plants (Variety JB-206) in the infested plot (b) can be compared with the adjacent healthy plot (c); d. 15 d young plants naturally infected indicate the severe stunting presence of 40–65 galls per plant; e and g, roots of infected plant at tillering stage (27 d after sowing) are heavily laden with white to light brown round to pyriform galls 27 days after sowing; f, early infection at the time of germination result in stunting, yellowing. wilting, drying and death of infected plants.

FIGURE 9.3 Illustrating symptoms of *Meloidogyne graminicola* that causes root knot disease of wheat. a & b, wheat fields showing large patches of the diseased plants showing severe stunting and yellowing with poor stand; c & d, showing early infection of plants with stunting and yellowing along with formation of hook, spindle or club-shaped galls varied in size and color on the root system.

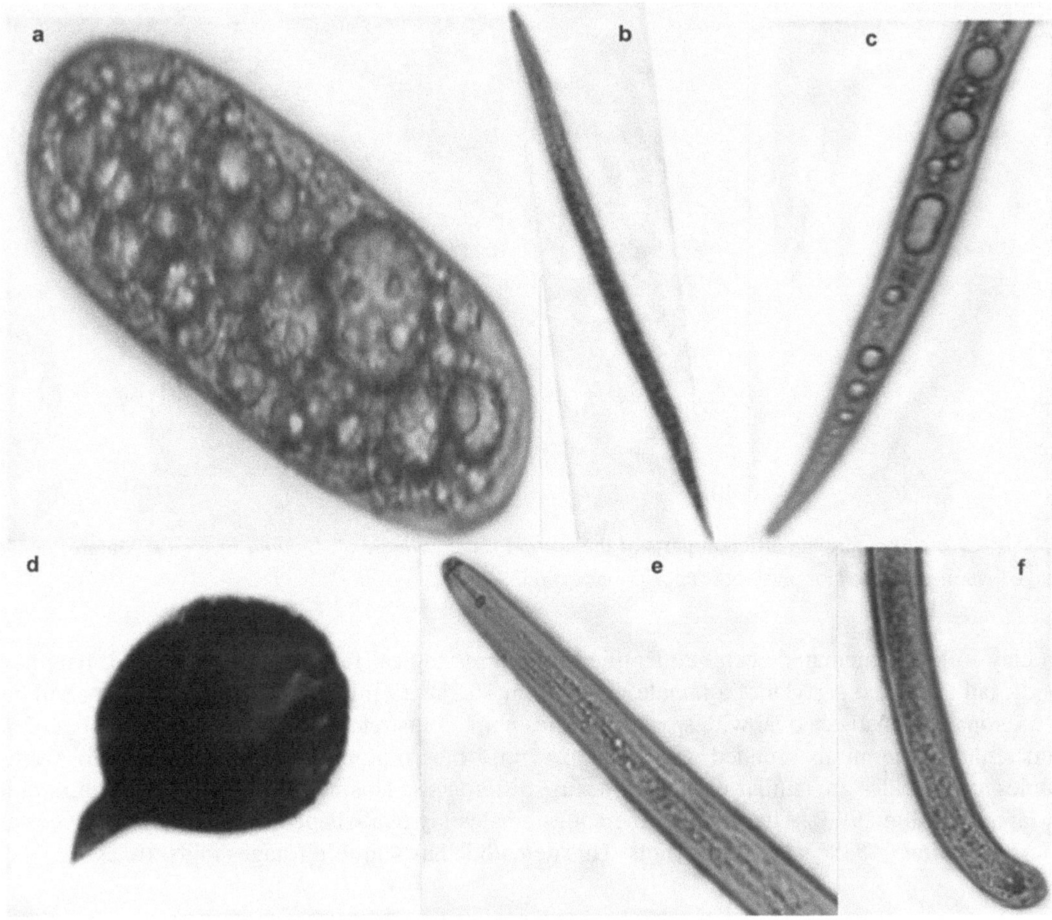

FIGURE 9.4 Illustrating the different stages of *Meloidogyne graminicola*. a, egg; b-c, second stage juvenile; d, adult female; e-f, adult male head and tail.

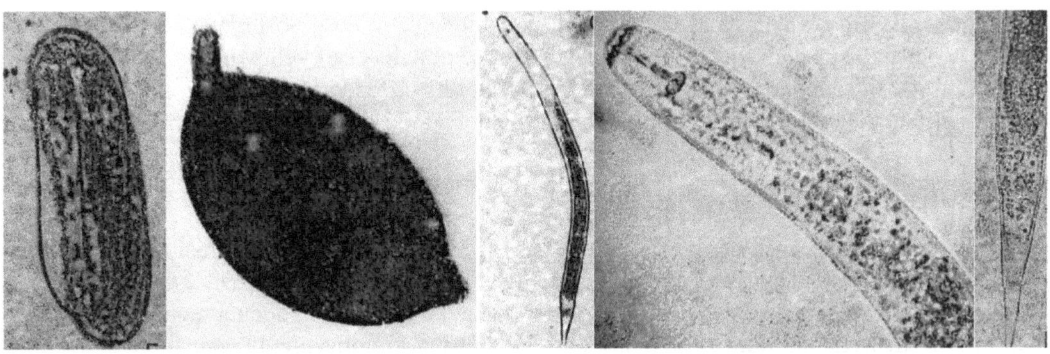

FIGURE 9.5 Presenting the different stages of cereal cyst nematode (*Heterodera avenae*). a, egg; b, cyst; c, second stage juvenile; d, anterior portion of second stage juvenile; e, posterior portion second stage juvenile.

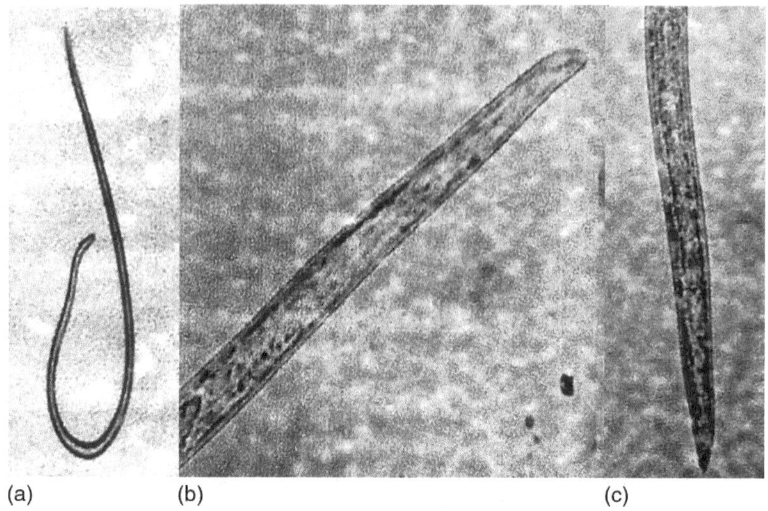

FIGURE 9.6 Showing the different parts of the second stage juvenile of *Anguina tritici*. (a) Mature second stage juvenile; (b) anterior portion; (c) posterior portion.

in clay soil. This nematode can be identified by morphological features of a stylet, tail, length of body, tail terminus: A stylet of a female varies from 16.2 to 19 microns with a rounded knob about 5 microns, spermatheca oval with sperms, ovaries paired outstretched. The tail is conoid about five times, the tail terminus rounded with a sharp ventral mucro along with a long anal body width. Males and females are similar except for sexual differences. This nematode is sexually dimorphic (Mai and Mullin 1996). After mating, the adult females lay oval-shaped eggs mainly in the cortex that hatch after 4–5 days inside the roots. This nematode has 4 molting stages in its life cycle.

9.4.5 Lance Nematodes (*Hoplolaimus* spp.)

Hoplolaimus species is characterized by the presence of a massive cephalic framework with a very strong stylet having a basal knob with anterior projections. The length of the body is moderate to large in size, varying from 1 to 2 mm. It infects plant roots growing in all types of soil and climate and completes its life cycle in 13 to 38 days. This nematode has a wide host range and may occur as a sedentary ectoparasite and is identified by the presence of patches of yellowing and drying similar to those caused by drought or nutrient deficiency. The ability of feeder roots and temperature play an important role in the buildup of the population of the nematode (Hailder and Nath 1992).

9.4.6 Lesion Nematodes (*Pratylenchus* spp.)

The genus *Pratylenchus* includes about 70 species, at least 8 of which are parasitic to wheat and barley (McDonald and Nicol 2005; Castillo and Vovlas 2007). *Pratylenchus crenatus*, *P. neglectus*, *P. penetrans*, and *P. thornei* are widely distributed in temperate, cereal-producing regions of the world (Castillo and Vovlas 2007; Mokrini et al. 2019). The typical symptoms involve impaired vigor, slight to severe stunting, reduced tillering, yellowing and premature death of lower leaves and reduced grain yield and quality with an increase in foliar temperature and delayed maturity (Thompson and Clewett 2021). Damage to foliage and roots is easily confused with other maladies such as nitrogen deficiency, low water availability, and root and crown diseases caused by fungi such as *Pythium*, *Rhizoctonia*, and *Fusarium* spp., (Smiley et al. 2009). This nematode completes its life cycle in 45 to 65 days, depending on the susceptible plant, favorable soil temperature and moisture (Figure 9.7).

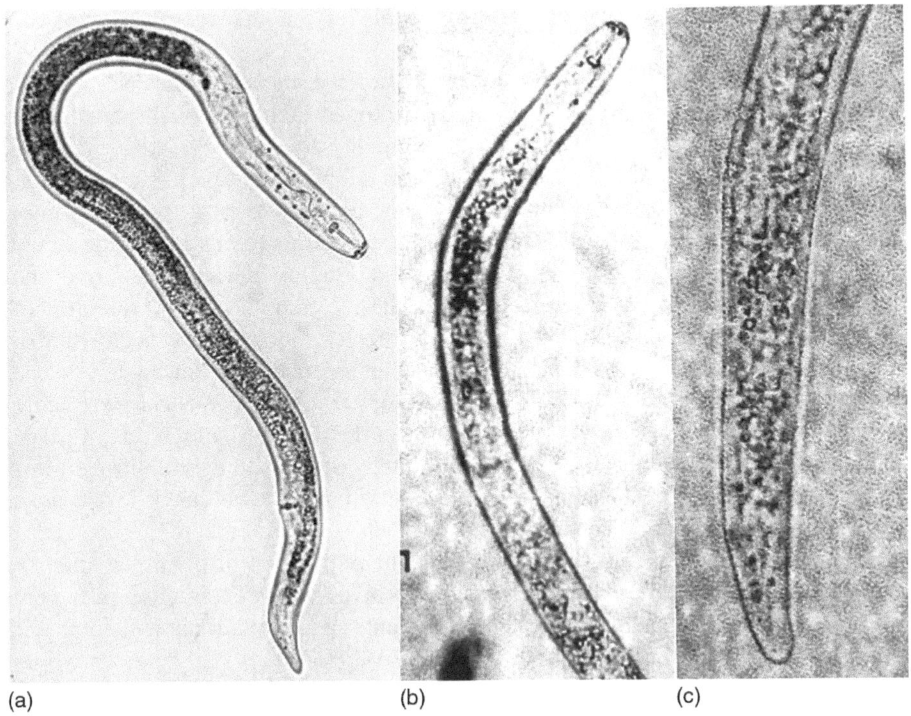

FIGURE 9.7 Showing the different parts of the adult female of *Pratylenchus*. (a) Mature female; (b) anterior portion; (c) posterior portion.

9.4.7 Rice White Tip/ Foliar Nematodes (*Aphelenchoides besseyi*)

Aphelenchoides besseyi is a widely distributed nematode and occurs in most rice-growing areas. The nematode has a large metacarpus (median bud) dorsally overlapping esophageal glands that open into the lumen of the esophagus in the metacarpus, just anterior to the pump chamber. This nematode has a slender stylet with small distinct knobs; the tail has a terminal spike or mucro. The male reproductive structure has no bursa and spicule. It is a seed-borne ectoparasite and survives as a dry seed, and the life cycle is completed within 8–12 days at 30°C. The nematode causes feeding at leaf tips that results in whitening of the top 3–5 cm of the leaf, leading to necrosis known as "White tip". The distortion of flag leaves that enclose the panicle can also be seen under severe conditions.

9.5 MANAGEMENT OF PLANT PARASITIC NEMATODES

Since the problem of plant parasitic nematodes has been found allied with the long-term use of inorganic fertilizers and pesticides causing disturbance to natural biological equilibrium, the application of non-chemical methods appears to be more important than chemical methods for eco-friendly and sustainable management. These methods include various cultural practices including the use of biological control agents. However, physical methods such as the use of a brine solution and physical separation of galls and cysts also provide effective approaches for the management of plant parasitic nematodes.

9.6 CULTURAL PRACTICES

Sanitation, crop rotation, fallowing, flooding, and summer plowing, along with organic substances, mulching, decoy trap crop, green manuring, plant spacing and the use of resistant varieties are common cultural practices that play an important role in the management of plant parasitic nematodes.

Cyst nematode can be managed by crop rotation of 3–5 years with sugar beet. The crop used in the crop rotation should be non-host against the pathogen including plant parasitic nematodes for reducing their population. Fallow periods in cropping sequences can also reduce nematode populations. Flooding or wet fallowing reduces the inoculum of plant parasitic nematodes as they are obligate aquatic pathogens. Similarly, dry fallowing and 2–3 deep summer plowing also help the management of plant parasitic nematodes. Juveniles are more sensitive than eggs. For the management of Molya disease, the infested field should be kept fallow during the summer months and plowed 2–3 times as the cysts are very susceptible to desiccation and cannot tolerate hot sun and hot summer winds. Sanitation by the uprooting of the infected root system of a crop and associated weed helps in reducing primary inoculum and has high significance in the management of soil-borne pathogens. Management of plant nutrition to protect crops from pathogens can be achieved by the avoidance of the nutrient strain, which may directly lower the resistance of a crop toward pathogens, and manipulation of nutrients to the relative disadvantage of the pathogen. Nitrogen in ammonia form is toxic to some plant parasitic nematodes. Particular forms as well as dose and proportion of NPK may affect the incidence of disease.

Decoy cropping has a significant role in reducing the population of plant parasitic nematodes. This practice involves the use of non-host crops, which involves the destruction of the crop before the pathogen can reproduce. The use of antagonistic plants against plant parasitic nematodes plays an important role in reducing their population. Wild *Brassica* (Mustard) is known to reduce the incidence of nematode disease by releasing toxic isothiocyanates along with root exudates. Similarly, marigold (*Tagetes* spp.) is known to reduce the population of root knot nematodes due to the presence of toxic thiophenes in their root exudates.

Industrial by-products, farmyard manure, vermicompost, oilcakes, and green manures are commonly used for amendment of soil for the management of plant parasitic nematodes. Oilcakes of neem, peanuts, mustard, cotton, and linseed are usually applied at the rate of 5q/ha for fifteen days before planting for the management of various plant parasitic nematodes. The amendment of soil with oilcakes enhances the population of predacious fungi which play an important role in nematode management. The use of organic manures also plays a decisive role in reducing the population of plant parasitic nematodes by enhancing the population of natural enemies (Singh and Singh 1992; Singh and Singh 2001; Singh 2006; Singh 2008).

The use of a trap crop is also an effective tactic for the management of plant parasitic nematodes. Trap cropping involves the use of a host crop of the pathogen with a view to attracting the pathogen or stimulating its development, but the crop is destroyed before the pathogen can reproduce or complete its life cycle, e.g., sowing of some wart-resistant potato cyst nematodes (*Globodera rostochiensis*) completes its life cycle. Applications of botanicals play a vital role in the management of plant pathogens including plant parasitic nematodes.

The use of tolerant/resistant varieties is one of the most practical approaches for the management of nematode diseases. It is the cheapest means of nematode management for the small-scale farmers. The development of resistant varieties using the recently identified dominant Mi gene imparting resistance to root knot nematode is one of the best options for the management of nematodes infecting different crops.

9.7 BIOLOGICAL CONTROL

Certain predacious fungi such as *Arthrobotrys* spp., *Dactyllaria* spp., *Dactylella* spp., *Monacrosporium* spp., *Catenaria anguillulae,* and *Trichothecium* spp. have been reported for their role in reducing the population of nematodes. These fungi capture nematodes by producing adhesive devices and trapping structures such as a network of adhesive branches, stalk adhesive knobs,

TABLE 9.1
Potential fungal and bacterial biological control agents used against plant parasitic nematodes of cereals

Host	Nematode species	Biological control agent	Reference
Rice	*Aphelenchoides besseyi*	*Purpureocillium lilacinum*	Tulek et al. 2018
	Meloidogyne graminicola	*Pseudomonas fluorescens* + *Trichoderma harzianum*	Narasimhamurthy et al. 2017
		Pseudomonas putida	Haque et al. 2018
		Aspergillus welwitschiae	Liu et al. 2019
Wheat	*Meloidogyne gaminicola* and *Anguina tritici*	*Dactylaria brochopaga* and *Catenaria anguillulae*	Singh et al. 2013
	Heterodera filipjevi	*Achromobacter xylosoxidans* and *Bacillus cereus*	Zhang et al. 2016
	Heterodera avenae	*Bacillus* sp.	Ahmed et al. 2018
Maize	*Pratylenchus zeae*	*Pasteuria thornei*	Confort and Inomoto et al. 2023
	Heterodera zeae	*Paecilomyces lilacinus* and *Pochonia chlamydosporia*	Baheti et al. 2017

non-constricting rings and constricting rings (Singh et al. 2001; Bandyopadhyay et al. 2001). The biocontrol potential of *Paecilomyces lilacinus* against *M. incognita* on okra has been found effective and economically better than nematicides.

Now mycorrhiza has shown its potential in the biological control of plant parasitic nematodes. A VAM-mediated defense response has been observed in direct-seeded rice against *M. graminicola* (Malviya et al. 2023). Recently, the use of *Pasteuria penetrans* has been given attention against plant parasitic nematodes (Jagadeeswaran et al. 2024). Seed treatment, soil application by drenching and bare root dip application with *Pseudomonas fluorescens*, *P. penetrans, Bacillus subtilis*, *B. polymyxa* effectively control plant parasitic nematodes (Ravichandra and Reddy 2008). In addition to their direct parasitic effects, *Bacillus* produces volatile substances that are toxic to nematodes and eventually cause their killing (Ali et al. 2023). Some other recent examples of biological control are presented in Table 9.1.

9.8 CHEMICAL CONTROL

Carbofuran 3G (Furadan), phorate 10G (thimet), fenamiphos (nemacur), fensulphothion (dasanit) and oxamyl have been recommended for the management of nematodes (Singh 2006). Seed treatment with nematicides has been used successfully to control root knot nematodes in various crops *A. tritici* infected seeds are placed in 20% salt solution in a suitable container and stirred vigorously for some time to reduce the seed inoculum. The seeds are then thoroughly washed in fresh water, dried, and then can be sown or stored. The implication of synthetic pheromones to disrupt nematode reproduction and chemicals that directly interfere with nematode chemoreceptors should be exploited for nematode population management.

9.9 REGULATORY MANAGEMENT

Numerous attempts have been made to prevent the introduction of nematodes into countries or provinces by means of quarantine. Quarantines are established by legislative action in parliament, etc., and usually give quarantine authorities power to make and enforce regulations to accomplish the purpose. Such regulations usually prohibit bringing infected seeds into protected areas where similar crops might become infected. This method is found very effective against seed and soil-borne diseases (ear cockle of wheat).

9.10 PHYSICAL METHOD

Soil solarization, a method of pasteurization can effectively suppress most species of nematode along with other microbes and weeds under field conditions (Singh 2008). Ear cockle nematode has been eradicated in some developed countries by modern seed sanitation techniques like gravity table seed processing or by sieving and freshwater flotation, however, it could easily be introduced in new cultures. Soil solarization with transparent polyethylene sheets has been attended as a means of raising the soil temperature to lethal levels to control soil pathogens (Sharma and Nene 1990). Soil solarization with double transparent sheets caused the maximum reduction of nematode population (84.11%) whereas double black, single transparent and single black sheets reduced nematode population 80.96, 75.51 and 70.46 respectively. The scientific principle involved in thermotherapy is that the pathogens present in soil and plant materials are inactivated or eliminated at temperatures non-lethal for the host tissues. Physical means of nematode control include heat treatment of soil, solar drying, steam sterilization, hot water treatment and soil solarization. Generally, sheets of 50–100 μm thickness are most suitable for raising the soil temperature. The use of transparent polyethylene yielded better results than black sheets, since transparent sheets transmit most of the incident radiation to soil. Additions of salt, sugar, charcoal, etc. create osmotic stress on nematodes and can be used for controlling nematodes. Ganguly et al. (1996) observed that mulching significantly increased rice seedling growth; the plant parasitic nematodes were reduced by 50–87% in the solarized soil.

9.11 APPLICATION OF MOLECULAR BIOLOGY AND BIOTECHNOLOGY

Chemical pesticides have been the mainstay of nematode management in cereal crops, but most of them are being phased out because of toxicity concerns except for the two new nematicide molecules fluopyram and fluensulfone. Genetic resistance to nematodes is still considered the most economical and effective way to manage nematode problems in crops. Modern methods of nematode management in cereal crops include the identification of a specific molecular target critical for nematode parasitism or life cycle and then disrupting the target by using modern molecular tools such as host-induced gene silencing, or using a chemical or approach to interfere with the nematode chemotaxis in soil thereby preventing nematode from parasitizing the plant roots (Dash et al. 2022). Integration of wheat-resistant varieties with appropriate agronomic practices is recognized as the safest and most practical, effective, and applicable management strategy. Nine resistance genes (Cre1–Cre9) to CCN are well-documented in the literature. CreR, CreV and CreZ genes are relatively recently characterized from wheat and confer resistance to CCNs. On the other hand, Rlnn1 is the only resistance gene characterized from wheat that is known to confer resistance towards RLNs. However, breeding for resistance to PPN_S has numerous challenges that originate from the narrow genetic diversity and difficulty in the process of transferring resistance gene(s) from the source to the target variety (Seid et al. 2021).

9.12 CONCLUSION

Nematodes parasitize the roots of our carefully nurtured crops, reduce their ability to produce yields, and make them weak and perhaps vulnerable to many pests. Crop resistance offers great importance in this directivity. In addition, an intensive search of plant germplasm collection for natural resistance to nematodes and their races is also required. An in-depth understanding is needed of the molecular basis of how and why plants are susceptible to nematodes. New progress is being made in studying changes in gene expression during the infection of plants in nematode host interactions where feeding sites are formed. There is a need to reduce these avoidable yield losses by developing new environment-friendly, economically acceptable and ecologically based management strategies as the current options become ineffective or unacceptable. The research on identifying the promising biocontrol agents, their mass production, application techniques and their behavior in the soil under varying agroclimatic conditions need to be intensified. Future nematode management must employ sustainable agricultural practices that take into account beneficial, detrimental and other nematode species in the rhizosphere and in soil.

REFERENCES

Ahmed S, Liu Q, Jian H (2018) Biocontrol potential of *Bacillus* isolates against cereal cyst nematode (*Heterodera avenae*). *Pak J Nematol* 36(2): 163–76.

Ali Q, Yu C, Wang Y, Sheng T et al (2023) High killing rate of nematode and promotion of rice growth by synthetic volatiles from *Bacillus* strains due to enhanced oxidative stress response. *Physiol Plant* 175(1): e13868. https://doi.org/10.1111/ppl.13868.

Baheti BL, Dodwadiya M, Bhati SS (2017) Eco-friendly management of maize cyst nematode, *Heterodera zeae* on sweet corn (*Zea mays* L.). *J Entomol Zool Stud* 5: 989–993.

Bandyopadhyay P, Kumar D, Singh VK, Singh KP (2001) Eco-friendly management of root-knot nematode of tomato by *Arthrobotrys oligospora* and *Dactylaria brochopaga*. *Indian J Nematol* 31:153–156.

Bridge J, Plowright RA, Peng D (2005) Nematode parasite of rice. In: *Plant Parasitic Nematodes in Subtropical and Tropical Agriculture*, Luc M, Sikora RA, Bridge J (eds). CAB International, Wallingford, UK, pp 87–130. https://doi.org/10.1079/9780851997278.0087

Castillo P, Vovlas N (2007) Pratylenchus, *Nematoda, Pratylenchidae: Diagnosis, Biology, Pathogenicity and Management. Nematology Monographs and Perspectives* 6. D. J. Hunt and R. N. Perry, series eds. Brill, Leiden,Boston.

Confort PM, Inomoto MM (2023) Effect of seed treatment with the biological agent *Pasteuria thornei* on the control of *Pratylenchus zeae* in Maize. *Eur J Plant Pathol* 1–10. https://doi.org/10.1007/s10658-023-02659-2.

Dash M, Ahuja A, Rao U, Somvanshi VS (2022) Modern approaches for the management of cereal cyst nematodes in wheat and barley. In: Kashyap PL, Gupta V, Gupta OP, Sendhil R, Gopalareddy K, Jasrotia P, Singh GP (eds). *New Horizons in Wheat and Barley Research*. Springer, Singapore, pp 321–336. https://doi.org/10.1007/978-981-16-4134-3_11

De Waele D, Elsen A (2007) Challenges in tropical nematology. *Annu Rev Phytopathol* 45: 457–485.

Decraemer W, Hunt DJ (2006) Structure and classification. In: *Plant Nematology*, Perry RN, Moens M (eds). CABI Publishing, Wallingford, pp 3–32.

FAOSTAT (2021, 2023) *Food and Agriculture Data*. Food and Agriculture Organization of the United Nations, Rome.

Fukagawa K, Ziska LH (2019) Rice: Importance for global nutrition. *J Nutr Sci Vitaminol* 65: S2–S3.

Ganguly AK, Pankaj, Sirohi A (1996) Effect of soil solarization of rice nursery beds to suppress plant parasitic nematodes. *Int Rice Res Notes* 21: 80–81.

Hailder MG, Nath RP (1992) Distribution of plant parasitic nematodes in maize fields and seasonal variations of *Hoplolaimus indicus*. *Curr Nematol* 2: 173–177.

Haque Z, Khan MR, Ahamad F (2018) Relative antagonistic potential of some rhizosphere biocontrol agents for the management of rice root-knot nematode, *Meloidogyne graminicola*. *Biol Control* 126: 109–116.

Hunt DJ, Handoo ZA (2009) Taxonomy, Identification and Principal Species. In: *Root-Knot Nematodes*, Perry RN, Moens M, Starr JL, (eds). CAB International, Wallingford, UK, pp 55–97.

Jagadeeswaran R, Singh B, Dubey J (2024) Isolation of *Pasteuria penetrans*, an obligate hyper-parasite, infecting root knot nematode, *Meloidogyne* spp. from the rhizosphere of pulses in India. *Egypt J Biol Pest Control* 34(9): 2–10.

Jones JT, Haegeman A, Danchin EGJ, Gaur HS, Helder J, Jones MGK, Kikuchi T, Manzanilla-Lopez R, Palomares-Rius JE, Wesemael WML, Perry RN (2013) Top 10 plant-parasitic nematodes in molecular plant pathology. *Mol Plant Pathol* 14: 946–961.

Khan MR, Anwer MA (2011) Occurrence of rice root-knot nematode andyield loss assessment in Aligarh and Hathras districts of Uttar Pradesh, India. *Indian J Nematol* 41: 34–40.

Kumar V, Khan MR, Walia1 RK (2020) Crop loss estimations due to plant-parasitic nematodes in major crops in India. *Natl AcadSci Lett* 43: 409–412.

Mai WF, Mullin PG (1996) *Plant-Parasitic Nematodes: A Pictorial Key to Genera*. Comstock Pub. Associates. A division of Cornell University Press, Ithaca and London, p 277.

Malviya D, Singh DP, Singh UB, Paul S et al (2023) Arbuscular mycorrhizal fungi-mediated activation of plant defense responses in direct seeded rice (*Oryza sativa* L.) against root-knot nematode. *Front Microbiol* 14: 933. doi: 10.3389/fmicb.2023.1104490.

Maqbool MA (1988) Present Status of Research on Plant Parasitic Nematodes in Cereals and Food and Forage Legumes in Pakistan. In: *Nematodes Parasitic to Cereals and Legumes in Temperate Semi-arid Regions*, Saxena MC, Sikora RA, Srivastava JP (eds). Aleppo: Workshop Proceedings, Larnaca, Cyprus, 1–5 March 1987. ICARDA, pp 173–180.

McDonald AH, Nicol JM (2005) Nematode parasites of cereals. In: *Plant Parasitic Nematodes in Subtropical and Tropical Agriculture*, Luc M, Sikora RA, Bridge J (eds). CAB International, Wallingford, U.K, pp 131–191.

Milne D (1919) Earcockle in wheat. *Bull Dep Agric Punjab* 1: 1–5.

Mokrini F, Viaene N, Waeyenberge L, Dababat AA, Moens M (2019) Root-lesion nematodes in cereal fields: Importance, distribution, identification, and management strategies. *J Plant Dis Prot* 126:1–11.

Narasimhamurthy HB, Ravindra H, Sehgal M, Ekabote SD, Ganapathi G (2017) Bio-management of rice root-knot nematode (*Meloidogyne graminicola*). *J Entomol Zool Stud* 5(4): 1433–1439.

Navia D, Delgado A, Viera W, Baez F, Trevor J (2017) Application of bio-products in Ecuadorian agriculture: Case banana. *Int J Biol Sci* 2(2): S7.Onkendi EM, Kariuki GM, Marais M, Moleleki LN (2014) The threat of root-knot nematodes (*Meloidogyne* spp.) in Africa: A review. *Plant Pathol* 63: 727–737.

Pandey SK, Vaish SS, SinghKP (2016) Root-knot disease (*Meloidogyne graminicola*) predisposes barley (*Hordeum vulgare* L.) to leaf spot blotch/blight (*Bipolaris sorokiniana*). *Curr Nematol* 27(1): 11–22.

Peter RS, Sandra JH (2015) The contribution of wheat to human diet and health. *Food Energy Secur* 4(3): 178–202.

Ravichandra NG, Reddy BMR (2008) Efficacy of *Pasteuria penetrans* in the management of *Meloidogyne incognita* infecting tomato. *Indian J Nematol* 38: 172–175.

Seid A, Imren M, Ali MA, Toumi F, Paulitz T, Dababat AA (2021) Genetic resistance of wheat towards plant parasitic nematodes: Status and future prospects. *Biotech Stud* 30(1): 43–62.

Sharma SB, Nene YL (1990) Effects of soil solarization on nematodes parasitic to chickpea and pigeon pea. *J Nematol* 22: 658–664.

Singh KP, Singh VK (1992) *Terminelia arjuna* leaf powder reduces population density of *Heterodera cajani*. *Int Pigeonpea Newsletter* 16: 17–18.

Singh SK, Singh KP (2009) A first record of root-knot nematode (*Meloidogyne graminicola*) on wheat in Uttar Pradesh, India. *J Mycol Plant Pathol* 39: 340–342.

Singh UB, Sahu A, Sahu N, Singh RK et al (2013) Nematophagous fungi: *Catenaria anguillulae* and *Dactylaria brochopaga* from seed galls as potential biocontrol agents of *Anguina tritici* and *Meloidogyne graminicola* in wheat (*Triticum aestivum* L.). *Biol Control* 67(3): 475–482.

Singh VK (2006) Management of root-knot nematode infesting tomato. *Indian J Nematol* 36: 126–127.

Singh VK (2008) Effect of soil solarization for management of plant parasitic nematodes. *Ann Plant Prote Sci* 16: 541–542.

Singh VK, Singh KP (2001) Effect of some medicinal plant leaves on the biology of *Heterodera cajani*. *Indian J Nematol* 31: 143–147.

Singh VK, Singh KP (2003) Penetration and development of *Heterodera cajani* on pigeon pea in relation to different types of soil. *Curr Nematol* 14(1, 2): 17–24.

Singh VK, Dwivedi, MC, Wali P (2010) Occurrence of wheat ear cockle nematode (seed gall nematode) *Anguina tritici* in Jammu. *Indian J Nematol* 41(1): 109–110.

Singh VK, Kalha CS, Kaul V (2007) New record of root- knot nematode *Meloidogyne graminicola* infecting rice in Jammu. *Indian J Nematol* 37(1): 94.

Smiley RW, Nicol JM (2009) Nematodes which challenge global wheat production. In: *Wheat Science and Trade*, Carver BF, Ames IA (eds). Wiley Blackwell, USA, pp 171–187.

Thompson JP, Clewett TG (2021) Impacts of root-lesion nematode (*Pratylenchus thornei*) on plant nutrition, biomass, grain yield and yield components of susceptible/intolerant wheat cultivars as determined by nematicide applications. *Agron* 11: 296. https://doi. org/10.3390/agronomy11020296.

Tülek A, Kepenekçi İI, Oksal E, Hazir S (2018) Comparative effects of entomopathogenic fungi and nematodes and bacterial supernatants against rice white tip nematode. *Egypt J Biol Pest Control* 28: 1–6.

Vaish SS, Ahmed SB, Prakash K (2011) First documentation on status of barley diseases from the high altitude cold arid Trans-Himalayan Ladakh region of India. *Crop Prot* 30: 1129–1137.

Vaish SS, Ahmed SB, Prakash K (2012b) Documentation of wheat diseases of the Trans-Himalayan Ladakh region of India. *J Wheat Res* 4(2): 29–40.

Vaish SS, Pandey SK (2012) First report of the root knot disease of barley caused by *Meloidogyne graminicola* from India. *Curr Nematol* 23(1, 2): 77–80.

Vaish SS, Pandey SK, Singh KP (2012a) Root knot disease caused by *Meloidogyne graminicola*: A limiting factor for growth and yield of barley (*Hordeum vulgare* L.). *Curr Nematol* 23(1, 2): 7–12.

Vasil IK (1999) *Molecular improvement of cereal crops* (Vol 5). Springer Science and Business Media, Germany, p 402.

Ying L, Zhong D, Peng D, Liu SM et al (2019) Evaluation of the biocontrol potential of *Aspergillus welwitschiae* against the root-knot nematode *Meloidogyne graminicola* in rice (*Oryza sativa* L.). *J Integr Agric* 18(11): 2561–2570.

Zhang J, Li Y, Yuan H, Sun B, Li H (2016) Biological control of the cereal cyst nematode (*Heterodera filipjevi*) by *Achromobacter xylosoxidans* isolate 09X01 and *Bacillus cereus* isolate 09B18. *Biol Control* 92: 1–6.

10 Innovative Plant Nematode Management Strategies in Pulse Crops

Rashid Pervez

10.1 INTRODUCTION

The largest producer of pulses in the world is nowadays India. The most significant pulse crops grown in the different parts of India include chickpeas, pigeon peas, lentils, mungbean, urdbean, field peas, rajmash, and lathyrus. In India, pulse crops were cultivated in an area of 2,245,667 ha, yielding 4,283 kg/ha and 961,743 tonnes of production (FAOSTAT, 2020). Pulses have unique importance in the diet, especially for vegetarians, as they are high-protein grains. They are a good source of energy, vital amino acids, and vitamins in people's daily diets, especially among the impoverished, in addition to being an excellent source of protein. Because they fix atmospheric nitrogen in the soil, pulses are crucial to sustainable agriculture. However, as agriculture has become more intensive and irrigation and other amenities have been introduced, the area and yield of pulses have fallen to the point that their per-capita availability has dropped from 69 g day-1 to roughly 36 g day-1 (Ali et al. 2005).

Among the many pests and diseases that are limiting factors in the production of pulse crops, plant nematodes diminish both the quality and quantity of these crops. They are regarded as a hidden enemy of crops because of their tiny size, hidden habitats, and lack of obvious symptoms on the aerial parts of the plants. The most prevalent plant nematodes root-knot nematode *Meloidogyne incognita* and *M. javanica*, cyst nematode *H. cajani*, lesion nematodes *Pratylenchus* spp. and reniform nematode *R. reniformis* infected pigeon pea, mungbean, urdbean, cowpea, chickpea, lentil, common bean, field pea, and a few other minor legumes are widespread throughout the nation's pulse-growing regions (Ali et al. 2006; Ali and Pervez 2006a; Ali and Pervez 2009a; Pervez and Ali 2010). Among the plant nematodes, root-knot nematodes *M. incognita* and *M. javanica* caused root-knot diseases to various pulse crops and severe plant damage, resulting in economic losses. This book chapter discusses root-knot disease and other nematode pests infesting various pulse crops, the symptoms and the innovative management strategies.

10.2 MAJOR NEMATODE PESTS

10.2.1 Root-knot Nematodes (*Meloidogyne* spp.)

The root-knot nematodes, which are stationary endoparasites, are the cause of the most common root gall disease. Affected plants are stunted, less vigorous, and yield less than healthy plants. *M. incognita* and *M. arenaria* are the primary species of root-knot nematodes that impact grain legumes. In India, yield losses in chickpeas owing to *M. incognita* and *M. javanica* range from 8 to 60 percent, while avoidable losses in pulse crops range from 20 to 35 percent (Gaur et al. 2001). According to Padgham et al. (2004), additional kinds of root-knot nematodes that are relevant in specific areas are

DOI: 10.1201/9781032631578-10

M. enterolobii in Central America and West Africa, *M. ethiopia* in Central-Eastern African nations, and *M. graminicola* in South East Asian systems that produce rice, wheat, and legumes. Other root-knot nematode species, like *M. chitwoodi* and *M. hapla*, infect pulse crops but have lower temperature needs. French beans are attacked by *M. chitwoodi* in the United States while chickpeas are attacked by *M. artiellia* Franklin in Syria and other Mediterranean nations.

Most species of root-knot nematodes have a large host range. Warm season legumes like cowpea, soybean, and pigeonpea are severely damaged by these nematodes (*M. incognita* and *M. javanica* in particular). On the other hand, cool-season pulses like chickpea, faba bean, lentil, grass pea, and field pea are generally spared severe nematode damage when these crops are sown when temperatures are less favorable to the nematodes. In regions where pulse crops are grown at the end of the rainy season or early in the fall following a sensitive summer crop, these worms frequently cause significant crop damage (Sharma et al. 2007).

Root-knot nematodes increase nodulation, whereas others have no discernible impact. *M. incognita* decreased soybean capacity to fix nitrogen (Coyne and Oyekanami 2007) and also decreased chickpea, pigeon pea, and green gram nodulation (Shafi et al. 2002). On the other hand, *M. incognita* was shown to accelerate the formation of nodules in cowpeas. For nematode-infected fodder pulses, inoculation with root-knot nematode markedly increased nodule weight and activity as well as the number of nodules on lablab roots (Ibewiro et al. 2000).

Sclerotium rolfsii, *Fusarium oxysporum*, *F. solani*, *Rhizoctonia solani*, and other fungal pathogens inflict severe root rots and wilts that are exacerbated by the root-knot nematodes. Root-knot nematodes influence how resistant pulse cultivars are to these fungi. *M. artiellia* infections in the Mediterranean region can overcome complete resistance to *Fusarium* wilt and significantly increase the severity of the disease in cultivars that are partially resistant to the fungus (Castillo et al. 2003).

10.2.2 Cyst Nematodes (*Heterodera* spp.)

Similar to root-knot nematodes, sedentary endoparasites are cyst nematodes. "Pearly root" disease is brought on by these nematodes. Cystic nematode females hold most or all of their eggs inside their bodies, which harden into protective cysts after they die. The adult female cyst nematodes have a lemon-like shape and protrude from the roots, resembling white pearls that eventually turn brown upon death. Similar to root-knot nematodes, cyst nematode infections have a similar effect on plant growth: afflicted plants are stunted, less vigorous, and yield less than healthy plants. Important nematode parasites of legumes include the pigeonpea cyst nematode (*H. cajani*), soybean cyst nematode (*H. glycines*), pea cyst nematode (*H. goettingiana*), and chickpea cyst nematode (*H. ciceri*).

In warmer climates, cyst nematode species typically have shorter life cycles less than a month and thus finish multiple generations during a crop season, in contrast to species found in cooler climates that take longer to finish their life cycles. Although the species of cyst nematodes that prey on legumes have a limited host range, they are more adaptive and can survive in the absence of a host. Cyst eggs can lie dormant for a number of years. Cyst nematodes typically experience a slower annual population decline than root-knot nematodes.

In India, *H. cajani* seriously harms cowpeas, pigeon peas, and other warm season pulses. Pigeon peas have a tolerance limit of roughly 2.6 eggs and juveniles/cm^3 of soil for nematodes. In India, *H. cajani* can cause yield losses to various pulses ranging from 16 to 30 percent (Gaur et al. 2001; Sharma et al. 2007). *H. cajani* increases the frequency of *Fusarium* wilt in addition to causing direct damage. In the western Indian state of Rajasthan, *H. swarupi* preys on chickpea. However, an estimate of the damage's extent has not yet been made (Ali and Sharma 2003).

10.2.3 Root-lesion Nematodes (*Pratylenchus* spp.)

Among the plant parasitic nematodes, root lesion nematodes are among the most common. Because they feed in the root cortex, these migratory endoparasite nematodes cause root lesion disease. The external manifestation of this damage is the development of small, striate lesions that range in color from dark orange to brown and measure a few millimeters. If the attack is severe, the entire root system might turn a dark brown color. Though these nematodes generally do not cause spectacular damage to legumes, their widespread distribution means that a significant amount of yield loss is anticipated globally.

Both the length and quantity of lateral, or feeder, roots are decreased because of the nematode infection. Nodulation can be hampered by *Pratylenchus* sp. in pulse crops. The symptoms of a root-knot and cyst nematode infection that appear above ground can be mistaken for or made worse by deficiencies in nutrients and/or water. It has been discovered that several *Pratylenchus* spp. are connected to pulse crops. Nonetheless, *P. thornei*, *P. penetrans*, and *P. neglectus* are the most prevalent and harmful. *Pratylenchus* spp. have evolved to thrive in a diverse range of soil types and temperatures.

In the Mediterranean region, *P. thornei*, *P. penetrans*, and *P. neglectus* are the most prevalent and harmful species on chickpeas, broad beans, and peas in India. *P. thornei*, *P. thornei* and *P. neglectus* play a significant role in chickpeas in Australia (Thompson et al. 2008). Vanstone et al. (2005) reported that field pea crops infected with high concentrations of *P. penetrans* suffered severe damage in Western Australia. *P. alleni* and *P. scribneri* are also highly pathogenic in French beans. In northern Australian cropping areas, a variety of summer legumes can be grown. However, their susceptibility to root lesion nematodes often limits the options available to incorporate them into rotations (Thompson et al. 2008). When it comes to *P. thornei*, pigeonpea is a poorer host than cowpea, mungbean, and navy bean. On the other hand, navy beans serve as moderate hosts for *P. neglectus*, while cowpea, mungbean, and soybean are poor hosts. Lablab is a poor host for both species of nematodes. The yield of succeeding cereal crops is significantly impacted by the rotation strategy chosen, which is determined by the species of root lesion nematode that is present (Owen et al. 2005; Thompson et al. 2008).

10.2.4 Reniform Nematode (*Rotylenchulus reniformis*)

A semi-endoparasite is the reniform nematode. In tropical and subtropical areas, it is widely distributed. It results in "dirty root" disease. This nematode is rather unusual in that the juveniles in the second stage emerge from the egg masses (each containing approximately 40–100 eggs), molt three times without eating, and eventually mature into the immature female infectious stage. The young female enters the roots, its head staying inside while the remainder of its body emerges from the outside. The portion of the body that emerges from the roots swells to resemble a kidney after feeding, hence the name "kidney." The eggs are laid in a mucilaginous matrix, to which soil particles adhere, giving the infected roots a filthy appearance.

Pulse crops grown in the warm season are the ones most impacted by *R. reniformis*. Despite the nematode's presence in the Mediterranean region, no reports of damage to legumes have been made. Rather, the American continent, the Indian subcontinent, and other Asian nations are severely impacted by the nematode species of French beans, cowpeas, and pigeonpeas. For cowpeas, the tolerance limits for this nematode are 0.12 nematodes/cm^3 soil, and for chickpeas, they are less than 1 egg/cm^3. There have been reports of yield losses in India for pigeonpea, chickpea, mungbean, and cowpea of 23%, 26%, and 11%, respectively (Gaur et al. 2001).

10.3 TYPICAL SYMPTOMS

The symptoms of plant nematode damage are sometimes confused with those of nutritional deficiencies, dryness, or salinity. A root and soil investigation is required to identify plant nematode damage.

Even though some nematodes are inextricably related to soil and roots, it is important to carefully examine the taxonomic identity, pathogenicity, and population density of various plant nematodes before evaluating whether or not they may cause crop loss.

Pulse root-knot disease is characterized by the formation of numerous small, easily visible knots on the root system, particularly on the root tips. The nematode species and genetic responsiveness influence knot sizes. *M. javanica* has fewer knots than *M. incognita*, however, the latter has numerous knots on the laterals and all along the roots (Ali and Pervez 2009a; Pervez and Ali 2011). Inconsistent growth in patches, poor stand, wilting, weak pod development, low yield, and fewer rhizobium nodules are typical symptoms of root-knot disease in pulses. Strongly impacted plants have reduced growth, fewer branches, and occasionally pale green leaves. In the field, roots are from plants with heavy galling wilt, and the seed they produce is typically harmed, under-developed, smaller in size, and lighter in weight. As the crop ages, the symptoms worsen, and low-fertility soil makes them even worse. Pulses grown in rabi season on fields affected with the root-knot disease could not germinate well and suffer from both root-knot and wilt disease.

Pigeonpea plants infected with the cyst nematode *H. cajani* exhibit a week-long delay in blooming and pod development. On the roots of these plants are pearly white cysts. The plants exhibit extensive necrosis, yellowing of the foliage, and early senescence. Stunting, yellowing, and a few blossoms are some of the visible symptoms of *H. ciceri* and *H. cajani* infestations in chickpea and pigeon pea crops, respectively. Necrotic sores develop on the roots when lesion nematodes *Pratylenchus* spp. invades them. Necrotic patches gather and spread throughout the root during a serious attack. *Pratylenchus* spp. rarely attacks chickpeas severely, and it does not appear above ground. Less notable than root-knot nematodes' damage is that induced by lesion nematodes. On the roots, the infection manifests as a dark brown lesion. On the surface, the signs and symptoms resemble those of malnutrition.

Infection with *R. reniformis*, resulted in dirty root disease in pigeon pea. This is because when the roots of infected plants were washed with water, the gelatinous matrix of soil was not entirely rinsed away. Such plants develop pale green or light-yellow leaves, and patches of stunted plants are visible in the field. In the early stages of the crop, plants of the chickpea, mungbean, urdbean, and lentil exhibit uneven growth, although this is challenging to identify any evident signs on the mature plant. Field pea and common bean species can exhibit early senescence even during the flowering period. Female nematodes can be seen on rootlets following a brief wash of the entire root system. This is the only noticeable sign of reniform nematode infection and is visible at the flowering and maturity phases.

10.4 ECONOMIC LOSSES

M. incognita and *M. javanica*, *R. reniformis*, *H. cajani*, and *P. thornei* are the most economically important nematode pests of pulse crops in India (Table 10.1). Other plant nematodes *Tylenchorhynchus* sp., *Helicotylenchus* sp., *Hoplolaimus* sp., and *Telotylenchus* sp. are also associated with pulse crops, and highly populated densities of these plant nematodes in the root or rhizosphere are usually correlated with patchy, stunted, chlorotic, and sickly developing crops (Ali and Pervez 2005; Ali et al. 2009; Pervez 2018). Regarding nematode species and population densities, numerous field tests on crop loss ranged from 5 to 60% (Ali and Pervez 2006b).

10.5 MANAGEMENT OF NEMATODE DISEASES

It has been shown that nematode population densities can be decreased and crop yield can be increased to varied degrees using physical, chemical, biological, and cultural management approaches. No strategy has been proven to be adequate or cost-effective for entirely regulating or sustaining nematode population densities below the economic tolerance level, except for very small regions or quantities of plant material. As a result, a careful blending of several strategies fitted to the needs of each place must be adopted.

TABLE 10.1
Major plant nematodes infesting pulse crops in India

Zone	Area	Pigeonpea	Chickpea	Other Pulses
1.	North East Plain Zone (Bihar, West Bengal, Eastern U.P., Orissa)	*Heterodera cajani*, *Meloidogyne* spp. and *Rotylenchulus reniformis*	*M. incognita*, *M javanica*, *R. reniformis* and *Tylenchorhynchus* spp.	*H. cajani* and *Meloidogyne* spp.
2.	North West Plain Zone (Haryana, Punjab, Rajasthan, Western U.P.)	*H. cajani*, *Meloidogyne* spp.	*M. incognita*, *M javanica*, *Tylenchorhynchus* spp., *Pratylenchus* spp. and *H. swarupi*	*H. cajani* and *Meloidogyne* spp.
3.	Central Zone (Gujarat, Maharashtra, M.P., Parts of Rajasthan)	*H. cajani*, *Meloidogyne* spp. and *R. reniformis*	*Meloidogyne* spp., *Pratylenchus* spp. and *H. swarupi*	*H. cajani* and *Meloidogyne* spp.
4.	South Zone (Andhra Pradesh, Tamil Nadu and Karnataka)	*H. cajani*, *Meloidogyne* spp. and *R. reniformis*	*Meloidogyne* spp. and *H. cajani*	*H. cajani* and *Meloidogyne* spp.

10.5.1 Soil Solarization

The density of eggs and juveniles of *H. cajani* by 93%, *H. retnsus* by 99%, *Pratylenchus* sp. by 98%, and *R. reniformis* by 100% were reduced by soil solarization. Two-season solarization had a similar annual impact on population densities as solarization that only occurred during one season. Solarization and the effects it caused on nematode populations didn't endure longer than a crop season. The pigeonpea crop parasites *Pratylenchus* spp. and *Tylenchorhynchus* spp. were significantly depleted throughout the summer months by solarization. Although summer fallow (February–June) had no influence on densities at 15–30 cm or 30–45 cm, it reduced juvenile *H. cajani* populations by 45% at 0–15 cm. However, *H. cajani*, *H. seinhorsti*, *H. retusus*, and *R. reniformis* survived in the soil inside buried plots without any hosts for 305 days.

10.5.2 Seed Treatment

Similarly, nematicidal seed treatments decreased the plant nematode population in the soil by 45.2 to 70.6%. *M. incognita* was successfully controlled, and chickpea yield was increased, by applying carbofuran 3G @ 2 kg a.i. ha-1 as seed and soil treatment. *M. incognita*-induced root galling was eliminated when pigeon pea seeds were treated with 2% carbofuran. Application of neem seed powder in soil @ 12.5 kg /ha + FYM @ 0.5 t/ha and seed treatment with carbosulfan @ 3% (w/w) controlled root-knot nematode and increased yield in urdbean. Seed treatment with 3% carbosulfan has been found most effective in the management of root-knot nematode in chickpea (Pervez and Ali 2011). Pigeonpea, mungbean, chickpea, field pea, and urdbean crops are demonstrated to have higher yields and lower nematode populations when their seeds are treated and when pesticides are applied for soil treatment (Table 10.2).

10.5.3 Botanicals

The seeds were soaked in aqueous extracts of *Cannabis sativa*, *Eclipta alba*, *Datura metal*, *Argemone mexicana*, and *Azadirachta indica* leaf extracts as well as oil seed cakes (neem, mustard, and karanj). Neem seed coat performed significantly better in terms of plant growth and reduced root-knot galling when Mojumder and Mojumder (2002) tested it as a soil and seed treatment against *M. incognita*. The reduction in root-knot galls was more when soil treatments were combined with seed

TABLE 10.2
Management of plant nematodes through seed treatment

Crop	Method	Dosage	Increase in yield (%)
Pigeonpea	Monocrotophos	0.1%	18.6
	Carbofuran	0.1%	38.9
	Latex of Calotropis	1%	20
Chickpea	Carbosulfan 25 ST	3%	35.10
	Carbofuran	3%	39.7
	Latex of Calotropis	1%	18
Mungbean	Monocrotophos	0.1%	36.34
Urdbean	Monocrotophos	0.1%	17.64–26.14
	Carbofuran	0.1%	17.64–26.14
Field pea	Carbofuran	0.1%	32.89

Source: Ali and Pervez (2006b).

treatments. All of the treatments considerably reduced the number of root-knot galls. *M. incognita*-infested chickpea can be managed by introducing 100 ml of aqueous neem seed kernel or neem cake extract into the soil. Ali and Pervez (2009b) reported that the latex of *Calotropis procera* is effective against *M. javanica* infesting chickpeas. Application of neem seed kernel and neem seed cake @ 20% w/w decreased major plant nematodes *M. incognita*, *R. reniformis*, *T. mashhoodi*, and *H. indicus* infesting mungbean and chickpea in the field.

10.5.4 Crop Rotation

Pulse crops can be grown in rotation with cereals and millets on soil infested with root-knot, reniform, or cyst nematodes. The rate of nematode population decline in the absence of the host determines the length of the rotation or avoidance of the host crop. However, far longer rotations may be necessary to achieve near-total control. Growing non-host legumes and other crops for two to four years or more has been suggested as a treatment for cyst nematodes that seriously damage most pulse crops. Non-host crops of pigeonpea cyst nematode such as linseed, pearl millet, maize, or wheat are required to grow for at least 2 years to reduce the cyst population in *H. cajani*-infested fields. Crop rotation has proven to be difficult to control since the root-knot nematode has such a wide host range.

10.5.5 Intercropping

When grown as intercrops, some nematode antagonistic crops, including *Tagetes* spp., sesame, mustard, and others, have been demonstrated to reduce root-knot and reniform nematode. It has been found that growing mustard alongside chickpeas lowers the invasion of root-knot nematodes in chickpeas and peas. Sorghum and pigeonpea intercropping proved very effective in reducing the population of *H. cajani* and increasing by approximately 40.4% yield (Upadhyay and Dwivedi 2001). Our most recent research, however, indicates that it might only be feasible in long-lasting pigeonpea. *M. javanica* populations were reduced when cowpea and maize were grown together rather than separately. Intercropping of mustard with chickpea has been found effective in the management of root-knot nematode. Urdbean + sorghum (3:1) followed by wheat, urdbean followed by wheat + mustard (8:1) reduced the population of root-knot nematode significantly (76–83%) (Pervez and Ali 2011).

10.5.6 Cropping System

Cropping systems could be a useful plant nematode management tactic because they affect nematode population control. Plant nematode populations rise sharply when vegetables are grown before pulse crops. Nematode populations are decreased when non-host crops are added to the cropping system. Cereals like wheat, sorghum, or pearl millet do not support root-knot nematodes. When areas are left fallow and weed-free, nematode numbers are drastically decreased. Traditional low-input cropping systems had three times fewer plant nematode levels than high-input cropping systems. It was discovered that sorghum/pigeon pea intercropping, or soybean/pigeon pea intercropping were suitable systems for the establishment of *H. cajani*. Plant nematode populations did not significantly rise during the rainy season in plots that were either fallowed or sown with pearl millet. A summer fallow period between February and June decreased soil concentrations of all plant nematodes by 42%. The populations of *H. cajani*, *Tylenchorhynchus* spp., and other plant nematodes were significantly higher in plots treated with inorganic fertilizers than in plots treated with farmyard trash. In irrigated areas, *R. reniformis* densities were higher.

Cereals or grasses have been successfully used in crop rotation to manage chickpea root-knot nematodes. Winter cereals, mustard, and sesame are not good hosts for *M. javanica* and *M. incognita*, therefore crop rotation every two to three years may help with disease control. The population of *M. javanica* was significantly decreased under the mustard-chickpea intercropping system in comparison to solo chickpeas. The intercropped sesame plants had the fewest nematode galls per plant.

10.5.7 Organic Amendments

It has been demonstrated that organic substances or manures can reduce plant nematode populations in soil. Neem (*Azadirachta indica*), karanj (*Pongamia glabra*), mahua (*Madhuca indica*), castor (*Ricinus communis*), and other non-edible deoiled seed cakes increase soil fertility and organic matter status while decreasing nematode population density due to their toxic or nematode-repelling components or byproducts of decomposition. The need for high doses (10–30 q ha^{-1}) is a drawback, but lesser doses used in conjunction with other techniques can be helpful. Among neem-based formulations, Nimbecidine (0.2%) is more effective in plant nematode management (Pervez and Ali 2011).

10.5.8 Biological Control

The use of biocontrol agents, in particular certain bacteria and fungi, to combat nematodes has received a lot of attention recently. Numerous microbes have demonstrated significant potential, including *P. penetrans*, *P. lilacinus*, *V. chaladosporium*, *T. harzianum*, *T. viridi*, *A. niger*, and others. Large-scale culture and regular release of biocontrol agents may therefore be quite expensive. But the simultaneous alteration of the soil environment enables proper colonization and establishment of native or alien microorganisms. Naturally, the process must include prior testing for human and animal health before suggesting such creatures

10.5.9 Chemical Control

Application of aldicarb, followed by phorate, demonstrated the strongest response in diminishing cyst population. As seed treatments, granules, or drench applications, fensulphothion, phorate, aldicarb, vydate, carbofuran, and DBCP all produced more developed plants than the untreated control. Phenomiphos, oxamyl, phosalone, and Danusban @ 1000 ppm/plant (soil, spray, and soil + spray) significantly reduced *H. cajani* population. Carbosulfan in combination with carbendazim (0.1%) is also effective in managing the plant nematodes.

TABLE 10.3
Resistance cultivars against major plant nematodes infesting pulse crops

Crop	Nematode	Resistance cultivars
Pigeonpea	*M. javanica*	MA-7, TT-10, H-84-5, TDRG-4, IPA-8F, BRG-11-1, BRG 10-2, RVKT 260, ICWP 016, ICWP 062, ICWP 086, ICWP 097
	H. cajani	MAL-9, ICWP 062, ICWP 069
Chickpea	*M. incognita* and *M. javanica*	GL 88341, BG 360 and RSG 524
Mungbean	*M. javanica*	LMR/13-26, IC 349701, IC 331454, TCR-86, IC 251442
Urdbean	*M. javanica*	PLU 648, MU-44, NUL-205
Lentil	*M. javanica*	L-4076, DPL-15, L-4710, L-4076, IPL-332, DPL-15, LL 1397
Field pea	*M. javanica*	IPFD 11-10, IPFD99-13, NDP 1200, RFP 2009-4, KPF 1028, RFP 72, HFP 8909, HFP 715, Pant P 244, KPF 10, Pant P-243

Source: Pervez and Ali (2011).

10.5.10 Host Plant Resistance

Nematology and plant breeding research led to the development of nematode-resistant pulse crop varieties. Researchers keep in mind that a novel variety must be resistant to plant nematodes since plant nematodes can destroy the resistance if it is only resistant to fungus by changing the host's physiology. As a result, breeders need to keep both objectives in mind before releasing a new variety. By decreasing the pace of reproduction, delaying the parasite's maturity, increasing the number of males generated, and preventing the nematodes from finishing their development, resistant plants act as a barrier for nematodes (Ali and Ahmad 2000). A list of pulse-resistant cultivars of specific plant nematodes that have been evaluated in various places has been provided (Table 10.3).

10.6 FUTURE THRUSTS

A regional assessment of enhanced cropping techniques for managing root-knot, lesion, and cyst nematodes in pulse crops. Developing integrated control techniques for the wilt and root-rot disease complex, which comprises both nematode and fungal pathogens, in the pigeon pea, chickpea, and lentil crops. Determining genotypes with high levels of resistance to cyst and the root-knot nematode. Determine nematode hotspots in significant pulse-growing regions using surveys, and make nematode control knowledge available to farmers. For polyspecific nematode environments, an integrated nematode management strategy is being developed to investigate and monitor the spread of significant nematode pests.

REFERENCES

Ali SS, Ahmad R (2000) Screening of chickpea germplasm against nematode. *International Chickpea and Pigeonpea Newsletter* 7: 8–9.

Ali SS, Ahmad R, Hussain MA, Pervez R (2005) *Pest management of pulses through entomopathogenic nematodes*. Indian Institute of Pulses Research, Kanpur, p 59.

Ali SS, Pervez R (2005) Attraction behaviour of *Tylenchorhynchus mashhoodi* towards chickpea roots. *Indian Journal of Pulses Research* 18(2): 43–49.

Ali SS, Pervez R (2006a) Distribution of plant parasitic nematodes associated with pulse crops from Hamirpur district, Uttar Pradesh. *Indian Journal of Pulses Research* 19(2): 100–101.

Ali SS, Pervez R (2006b) Ecofriendly management of phytonematodes in pulses. In: Rajvanshi I, Sharma GL (Eds.) *Ecofriendly management of phytonematodes*, Oxford Book Company, Jaipur, pp 1–26.

Ali SS, Pervez R (2009a) Biotic stress in pulse crops. In: Sharma GL (Ed.) *Phytonematodes management of field crops*, Oxford Book Company, Jaipur, pp 83–96.

Ali SS, Pervez R (2009b). Effect of plant extract on mortality and activity of the second-stage juveniles of *Meloidogyne javanica*. *Trends in Biosciences* 2(2): 42–44.

Ali SS, Pervez R, Shaheen A (2006) Community analysis of plant Parasitic and Predatory nematodes associated with pulse crops from Hamirpur districts, Uttar Pradesh. *Indian Journal of Nematology* 36(1): 1–6.

Ali SS, Shaheen A, Pervez R (2009) A new species of *Hoplolaimus* (*Basiriolaimus*) (Hoplolaiminae: Tylenchida) from pigeonpea ecosystem of Bundelkhand region. *Trends in Bioscience* 2(1): 36 – 38.

Ali SS, Sharma SB (2003) Nematode survey of chickpea production areas in Rajasthan, India. *Nematologia Mediterranea* 31: 147–149.

Castillo P, Navas-Cortes JA,Gomar-Tinoco D, di Vito M, Jiminez-Diaz RM (2003) Interactions between *Meloidogyne artiella*, the cereal and legume root-knot nematode, and *Fusarium oxysporum* f. sp. *ciceris* race 5 in chickpea. *Phytopathology* 93: 1513–1523.

Coyne DL, Oyekanami EO (2007) Symbiotic nitrogen fixation of two soybean genotypes as affected by root-knot nematode and microsymbionts. *Journal Biological Science* 7: 1221–1226.

FAOSTAT (2020) *Food and Agriculture Organization of the United Nations*, 2020, 02–28. www.fao.org.

Gaur HS, Singh RV, Kumar S, Kumar V, Singh JV (2001) Search for nematode resistance in crops. In A.I.C.R.P.: plant parasitic nematodes with integrated approach for their control, I.A.R.I., New Delhi. P. 84.

Ibewiro B, Sanginga N, Vanlauwe B, Merckx R (2000) Influence of phytoparasitic nematodes on symbiotic N_2 fixation in tropical herbaceous legume cover crops. *Biology and Fertility of Soils* 31: 254–260.

Mojumder V, Mojumder V (2002) Effect of neem seedlings on major phytoparasitic nematodes associated with chickpea and on crop yield. *Pesticide Research Journal* 14: 150–152.

Owen KJ, Clewett TG, Thompson JP (2005) Long-term choice of rotation crops maximises wheat yields in *Pratylenchusthornei*-infested fields. *Proceedings of the 15th Biennial Conference of the Australasian Plant Pathology Society*, Geelong, Victoria, September 2005. p. 104.

Padgham JL, Duxbury JM, Mazid AM, Abawi GS, Hussain M (2004) Yield loss caused by *Meloidogyne graminicola* on lowland rainfed rice in Bangladesh. *Journal of Nematology* 36: 42–48.

Pervez R (2018) Chemoattraction of *Tylenchorhynchus mashhoodi* towards pigeonpea (*Cajanus cajan* (L.) Millspaugh) and various factors influencing on the attraction. *Annals of Plant Protection Sciences* 26(1): 207–212.

Pervez R, Ali SS (2010) Biodiversity and community analysis of nematodes associated with chickpea from Jhansi district, Uttar Pradesh, India. *Trends in Bioscience* 3(1): 71–73.

Pervez R, Ali SS (2011) Phytonematodes, a thrust in pulse crops. In: Sharma GL (Ed.) *Phytonematodes in Crops: their identification, treatment and management*, New India Publishing Agency, New Delhi, pp 25–54.

Shafi A, Mahmood I, Siddiqui ZA (2002) Integrated management of root-knot nematode *Meloidogyne incognita* on chickpea. *Thai Journal of Agricultural Science* 35: 273–280.

Sharma SB, Greco N, Gaur HS, Abawi GS, Ali SS (2007) Monitoring, forecasting and management of nematodes in food legumes. In: Kharakwal MC (Ed.) *Food legumes for nutritional security and sustainable agriculture* Proceedings of the Fourth International Food Legumes Research Conference, October 18–22, 2005, New Delhi, India

Sharma SB, Smith DH, McDonald D (1992) Nematode constraints of chickpea and pigeonpea production in the semi-arid tropics. *Plant Disease* 76: 868–874.

Thompson JP, Owen KJ, Stirling GR, Bell MJ (2008) Root-lesion nematodes (*Pratylenchus thornei* and *P. neglectus*): a review of recent progress in managing a significant pest of grain crops in northern Australia. *Australasian Plant Pathology* 37: 235–242.

Upadhyay KD, Dwivedi K (2001) Population dynamic of *Heterodera cajani* associated with pigeonpea in various cropping systems of central Uttar Pradesh. *Indian Journal of Nematology* 2: 120–121.

Vanstone VA, Kelly SJ, Hunter HF, Gilchrist MC (2005) *Pratylenchus penetrans* in Western Australian field crops. *Proceedings of the 15th Biennial Conference of the Australasian Plant Pathology Society*, Geelong, Victoria, September 2005. P. 102.

11 Biotic Stress Management Strategies in Pulse Crops

Raj K. Mishra, Sonika Pandey and Abhishek Tiwari

11.1 INTRODUCTION

Pulse crops have a significant impact on preserving soil fertility and providing protein to the substantial vegetarian community within the nation. India holds the distinction of being the foremost producer and consumer of pulses on a global scale. Consequently, pulses offer immense potential in addressing the pressing issues of protein deficiency and malnutrition, particularly prevalent among the underprivileged segments of society. The greater importance of pulses in offering protein-rich food is evident in India, as the majority of the population relies on pulse protein and avoids eating meat due to the religious taboos associated with it. The main emphasis on research and development to manage pulse diseases is on developing host resistance and chemical control wherever appropriate, and to a certain extent, these means of disease management are often practiced in isolation from each other. In recent times, a dramatic change in scientific thoughts and practice in the management of legume diseases has been seen, because of which an increased focus has been placed on identifying, evaluating, and integrating region-specific components of integrated disease management. Here, efforts were undertaken to address the major biotic stresses, with a major focus on the diseases of chickpea, lentil, field pea, pigeonpea, mungbean (green gram), and urd bean (black gram) and their management strategies in integrated ways. These pulse crops, grown worldwide, are prone to attack by a substantial amount of biotic and abiotic stresses, which subsequently result in substantial economic losses (Table 11.1).

11.2 YIELD LOSSES DUE TO DISEASES

The pulse crops of both the cool (*rabi*) season, *viz.*, chickpea, lentil, rajmash and field pea, and the warm (*kharif*) season, *viz.*, pigeonpea, mungbean, urdbean and cowpea, are prone to larger numbers of fungi, bacteria, phytoplasmas and viruses which subsequently results in economic losses. The dry root rot, wilt, phytophthora stem blight, powdery mildew, rust, stem rot and yellow vein mosaic disease are recognized as key diseases of utmost significance in majority of the pulses (Table 11.1). Over 50% of the crop's yield is lost in pulses due to biotic stresses, mainly insect and disease incidence. Fusarium wilt in chickpea results in reduced seed yield and seed weight. The fusarium wilts in chickpea, pigeonpea and lentil cause over 20–25 % yield losses all over the country. Similarly, dry root rot of chickpea causes 10 to 100 % yield loss annually. Stem rot of pulses is found to cause almost 100% losses annually around the world. The fusarium wilt disease can occur at all the stages of pigeonpea and 30 to 100% yield losses in susceptible cultivars have been reported (Dhar et al. 2005). Chickpea is found to be most vulnerable to BGM infection during flowering and podding stages and infection at these stages may lead to a complete failure of the crop. Yield losses of up to

DOI: 10.1201/9781032631578-11

TABLE 11.1
The major diseases of pulses and their causal agents

S.N.	Crops	Diseases	Causative organisms	References
1.	Pigeonpea	Phytophthora stem blight	*Phytophthora drechsleri* f. sp. *cajani*	Iqbal et al. 2019
		Wilt	*Fusarium oxysporum* f. sp. *udum* (Butler) Snyd. and Hans.	Iqbal et al. 2019
		Leaf spot	*Cercospora* spp.	Iqbal et al. 2019
		Powdery mildew	*Erysiphe polygoni* DC	Iqbal et al. 2019
2.	Chickpea	Wilt	*F. oxysporum* f. sp. *ciceri* Matuo and Sato	Jimenez-Díaz 2015; Iqbal et al. 2019
		Dry root rot	*Rhizoctonia bataticola*	Iqbal et al. 2019
		Stem rot	*Sclerotinia sclerotiorum*	Iqbal et al. 2019
		Rust	*Uromyces ciceris-arietinii* (Grogn.) Jacz. & Boy.	Iqbal et al. 2019
		Wet root rot	*Rhizoctinia solani*	Iqbal et al. 2019
		Botrytis grey mold	*Botrytis cinerea*	Iqbal et al. 2019
3.	Lentil	Wilt	*F. oxysporum* f. sp. *lentis*	Iqbal et al. 2019
		Root rot	*Macrophomina phaseolina*	Iqbal et al. 2019; Pérez et. al. 2003
		Rust	*Uromyces fabae*	Iqbal et al. 2019
		Ascochyta blight	*Ascochyta fabae* f. sp. *lentis*	Iqbal et al. 2019
4.	Pea	Rust	*U. fabae*	Iqbal et al 2019
		Powdery mildew	*E. polygoni*	Iqbal et al. 2019
		Root rot	*R. solani*	Iqbal et al. 2019
5.	Mungbean and urdbean	Root rot	*M. phaseolina*	Iqbal et al. 2019; Broughton 2003
		Leaf spot	*Cercospora* spp.	Iqbal et al. 2019

50 % were reported under endemic situations and if favorable conditions of BGM infections prevail during vegetative and reproductive growth of plants, the losses may reach up to 100%.

11.3 MAJOR DISEASES

11.3.1 Wilt

Wilt is a major problem all over the country. Wilt is caused by *Fusarium* spp. and has been a major biotic stress of chickpea, pigeonpea and lentil crops leading to over 20–25% yield losses. The infested soil serves as a primary source of primary inoculum for the emergence of fusarium wilt epidemics in pulses.

11.3.1.1 Symptoms

The pathogen invades the vascular bundles of plant roots, leading to a decrease or cessation of water absorption by the foliage. Seedlings infected with fusarium wilt will wilt and fall over, maintaining a faded green hue. Mature plants will exhibit classic wilt signs such as drooping petioles, rachis, and leaflets (Figure 11.1 A, B & C). The wilting plants do not exhibit any external rotting in their roots. However, upon vertical splitting, a dark brown discoloration of the internal xylem becomes apparent (Dubey 2002). The pods of wilted plants appear normal, but the seeds are typically smaller, wrinkled, and discolored. It is worth noting that even a visually normal-looking seed harvested from wilted plants may still carry the wilt pathogen.

FIGURE 11.1 Wilt disease symptoms on pulses. A: Pigeonpea, B: Chickpea, C: Lentil.

11.3.1.2 Host Range

The pathogenic fungus exhibits a broad spectrum of hosts, infecting various types of pulses. It is a significant pathogen responsible for inducing wilt disease in numerous plant species globally. By producing chlamydospores, the fungus is able to persist for extended periods, even in the absence of a host plant.

11.3.2 Phytophthora Stem Blight

Phytophthora stem blight (PSB) is a recurring and economically significant disease of pigeonpea (*Cajanus cajan*), caused by *Phytophthora drechsleri* f. sp. *cajani*. This disease becomes particularly problematic when there are excessive rains within a short period of time and hot and humid weather persists throughout the crop season. PSB is the third most important disease of pigeonpea, following fusarium wilt (*Fusarium udum* Butler) and pigeonpea sterility mosaic disease (SMD). Studies by Pande et al. (2006) and Sharma et al. (2006) have highlighted the severity of PSB in this region.

11.3.2.1 Symptoms

Visible foliage blight symptoms are observed on seedlings that are at least one month old. Initially, blight symptoms manifest as water-soaked lesions on the primary and triplicate leaves, which turn necrotic within a week under conditions of high humidity (≥ 80%) and temperatures ranging from 20 to 30°C. The circular to irregularly shaped leaflet lesions can grow as large as 1 cm in diameter. Stem symptoms present as brown to dark brown lesions that are clearly distinguishable from the healthy green parts on the main stem, branches, and petioles (Figure 11.2 A & B).

11.3.2.2 Host Range

The pathogen *Phytophthora drechsleri* Tucker var. *cajani* possesses a broad spectrum of hosts, *viz.*, lucerne, safflower, and skeleton weed, *Atylosia scarabaeoides* and *A. platycarpa* (wild pigeon pea).

11.3.3 Dry Root Rot

Rhizoctonia bataticola, is responsible for causing dry root rot (DRR) in chickpeas. The Pycnidial stage of this pathogen leads to disease in green gram and black gram as well. This destructive

FIGURE 11.2 Phytophthora stem blight infected plants. (A) Pigeonpea field and (B) plants.

disease significantly impacts chickpea productivity, resulting in an annual yield loss of 10 to 20%. The presence of *R. bataticola* is observed in various regions including the Indian subcontinent, Australia, Ethiopia, Iran, and other countries (Nene et al. 1991).

11.3.3.1 Symptoms

Typically, when we remove a plant from the ground, the lower parts of its tap root tend to stay in the soil. The deceased roots become rigid and fragile, displaying signs of bark peeling. On the exposed roots or at the base of the stem, one can observe small dark sclerotial bodies. Moreover, if we vertically divide the colored area, we may also notice mycelium or tiny sclerotia within the central core.

11.3.3.2 Host Range

The pathogen can be transmitted through seeds and soil. It has a wide host range particularly belonging to the *Leguminosae* family. High day temperatures (~30°C) with moisture stress conditions at the podding and flowering stage rapidly aggravate disease incidence (Gurha et al. 2003; Singh and Sharma 2002).

11.3.4 Powdery Mildew

11.3.4.1 Symptoms

The initial sign of infection manifests on the leaves before spreading to other green parts of the plant. The attack is identified by the development of white powdery patches on both sides of the leaf, as well as on tendrils, pods, and stems (Figure 11.3). These patches begin as small, discolored spots, from which a powdery substance spreads in all directions. As the disease progresses, extensive areas of the plant's above-ground parts may become covered in these white powdery patches.

11.3.4.2 Host Range

Powdery mildew is a most important disease of pulses including, field pea, French bean, cowpea, mungbean and urdbean.

FIGURE 11.3 Powdery mildew disease symptoms on field pea.

FIGURE 11.4 Rust infected (a) field pea (b) lentil (c) chickpea and (d) French bean plants.

11.3.5 Rust

Rust is caused by *Uromyces* spp. It is a self-sufficient fungus that undergoes its entire life cycle on the same host. The infection typically takes place during the flowering or early podding stage, appearing as aecia, which can later transform into secondary aecia or uredia. The fungus survives in summer as teliospores. High humidity and cloudy weather with temperatures of 20–22°C favor disease development. The plants give a dark brown or blackish appearance visible as patches in the field (Mishra et al. 2010) (Figure 11.4).

Integrated management of rust includes control of volunteer plants over the summer and removal of infected plant debris. It is advisable to use clean seeds without rust contaminations, and to treat the seed with a suitable fungicide such as diclobutrazole. For prevention purposes sprays of mancozeb (0.25%), Propiconazole (0.1%) and Hexaconazole (0.1%) at the early disease development stage have been recommended (Mishra and Pandey 2009). Host plant resistance is the most effective method for managing rust.

11.3.6 Ascochyta Blight

Ascochyta rabiei Labrousse [Teleomorph: *Didymella rabiei* (Kovachevski) Arx.] is the causative agent of ascochyta blight. This particular pathogen plays a significant role as a foliar pathogen in chickpea cultivation. This disease is a seed and is stubble-borne producing both airborne ascospores and water-splashed conidia during the cropping season (Armstrong et al. 2001). The occurrence of this disease has been documented in major chickpea-growing regions worldwide.

11.3.6.1 Symptoms

Infection on petioles, leaflets and young branches of chickpea includes early symptoms such as epinasty and loss of turgor, followed by water soaking. This pathogen secretes a toxin which maintains the symptom. The chickpea seeds with light infection by *A. rabiei*, appear usually discolored with very low seed weight.

11.3.6.2 Host Range

Chickpea is commonly infected by *A. rabiei*, making it the crop most vulnerable to ascochyta blight. Lentil, cowpea, pea, and common bean have also been reported to be affected by this disease.

11.3.7 Botrytis Grey Mold

Botrytis grey mold (BGM) is caused by the causal organism *Botrytis cinerea* Pers. ex. Fr. It is a significant disease that affects chickpea crops. This disease is commonly found in regions with cool, cloudy, and humid weather conditions.

11.3.7.1 Symptoms

Flower petals can easily become infected with Botrytis. Blossom blight frequently results in fruit rotting. The infected fruit will appear water-soaked and mushy. A dense grey mold will quickly cover them. Bulbs and stems, along with other plant parts, can also be impacted.

11.3.7.2 Host Range

This disease can lead to significant economic losses for both field and greenhouse crops. *Botrytis cinerea*, the causal agent, can infect mature or senescent tissues, plants before harvest, or seedlings. A wide range of hosts are susceptible to this pathogen, e.g. chickpeas, lettuce, broccoli, beans, and small fruit crops e.g. grape, strawberry, raspberry, blackberry) are particularly vulnerable to grey mold.

11.3.8 Stem Rot/White Rot

Sclerotinia sclerotiorum (Lib.) de Bary is a widespread fungal pathogen that infects a wide range of broadleaf crops and weeds, including beans, mustard, potatoes, and sunflowers. It primarily enters fields through contaminated seed lots and survives as a mass of fungal hyphae covered by a resilient black rind called sclerotia. The hyphae can also directly penetrate plant tissues, particularly when there is plant-soil contact. The main survival structure of this disease is the sclerotia, which can endure for several years and germinate to produce apothecia.

11.3.8.1 Symptoms

The first signs of the disease manifest as small, irregular black lesions on the outer leaf sheath close to the water level (Figure 11.5). As the disease advances, the lesions continue to spread. Inside the infected plants, there are numerous small white and black sclerotia as well as mycelium. In cases of severe infection, the tillers may die.

11.3.8.2 Host Range

This fungus is a soil-borne fungus that attacks plants in all stages of growth including young seedlings, mature plants and their harvested products distributed worldwide (Agrios 2005). It is a major disease of several leguminous crops such as field pea, chickpea, pigeonpea, lentil, soybean, faba bean and alfalafa (Mishra et al. 2015). This fungus may cause almost 100% losses to the crop and as a result, millions of dollars are lost annually around the world owing to the occurrence of this disease.

11.3.9 Collar Rot

The disease is caused by *Sclerotium rolfsii* and the pathogen survives in soil in the form of sclerotia. Disease can be noticed during at the seedling stage (up to 6 weeks after sowing). Affected seedlings turn yellow and young seedlings may collapse. When seedlings are uprooted at the collar region show rotting. The rotten portion is covered with whitish mycelial strands. A white mycelial coating can be seen on the tap root of completely dried seedlings and presence of mustard seed like sclerotia

FIGURE 11.5 Stem rot disease symptoms on different plants of pulses.

can be seen at the site of infection infected seedlings usually appear in small, scattered patches in the field.

11.3.10 Sterility Mosaic Virus

Pigeonpea sterility mosaic virus (PPSMV) is a type of emaravirus that causes sterility mosaic disease in pigeonpea plants. This disease, also known as the 'green plague', is characterized by infected plants remaining in a vegetative state without producing flowers. It has been reported in India and a few other South-East Asian countries. The symptoms of SMD include stunted and bushy plants, smaller leaves with chlorotic rings or mosaic patterns, and a decrease or complete halt in flower production. The causal agent of the disease is PPSMV. It is transmitted by an eriophyid mite called *Aceria cajani* in a semi-persistent manner. PPSMV can infect various genotypes of cultivated and wild pigeonpea relatives. The epidemiology of SMD involves the virus, mite vector, cultivar, and environmental conditions.

11.3.11 Yellow Vein Mosaic

The mungbean yellow mosaic virus (MYMV), a member of the Gemini group of viruses, is responsible for causing this disease. It is transmitted by the whitefly (*Bemisia tabaci*). The YMV has a wide host range and is not limited to the Fabaceae family. It has been reported to infect nearly 200 species across 14 different families, including chickpeas, pigeonpea, faba beans, field peas, lentils, mungbean, and urdbean. The tender leaves initially display yellow mosaic spots, which gradually intensify as the disease progresses, eventually leading to complete yellowing. This yellowing hampers flowering and pod development. Infections that occur early on often result in plant death. As the symptoms advance, the plants become stunted, and the symptoms typically appear in patches and along the edges of paddocks (Figure 11.6).

11.3.12 Management Strategies for Diseases

11.3.12.1 Cultural Practices

It is advised to utilize nutritious seeds, practice proper cultivation techniques, and implement a three-year crop rotation system for effective disease management. Disease-free seeds can be obtained through surface irrigation in semiarid areas with high temperatures and low humidity, which hinder fungal infection. Alternatively, a pedigreed seed program can be employed, where seed plots are isolated and thoroughly inspected to ensure disease-free seeds. To achieve maximum yield and minimize disease occurrence, it is crucial to sow various pulses/legume crops at the appropriate times during the rabi, kharif, and spring seasons.

11.3.12.2 Chemical Method

The control of wilt disease through chemical means is not practical or cost-effective due to the fact that the pathogen is primarily found in the soil and its chlamydospores can survive for up to 6 years. Therefore, the most effective and economical method for managing wilt is through the use of host resistance. Various systemic and non-systemic fungicides have been utilized by researchers to reduce seed infections of different diseases. Additionally, foliar sprays of specific fungicides have been recommended for managing wilt. Some fungicides, such as Benlate and Bavistin, have been found to persist in the plant for up to 15–20 days after seed treatment. While certain EBI fungicides have shown effectiveness against the pathogen under laboratory conditions, they have been found to be phytotoxic to *P. vulgaris* seedlings when used as seed treatment. However, the combination of carbendazim with thiram as seed treatment followed by sprays of mancozeb at specific intervals effectively managed the disease. Another combination of seed treatment with tebuconazole and

FIGURE 11.6 Yellow vein mosaic diseases symptoms on leaves of (A) Mungbean (B) Urdbean and (C) Pigeonpea.

foliar sprays of tricyclazole reduced disease severity on leaves and pods, with the highest seed yield observed in the case of seed treatment and foliar sprays of carbendazim.

11.3.12.3 Biological Approaches

Considerable efforts were made to manage different diseases of pulses by incorporating bioagents colonized natural substrates into the soil. Bacterial endophytes like *Bacillus* and *Paenibacillus* showed their antifungal activity against major soil-borne phytopathogens like *Rhizoctonia bataticola*, *Macrophomina phaseolina*, *Fusarium udum*, and *Sclerotium rolfsii* infecting pulse crops (Sentilkumar et al. 2009). Among the antagonistic fungi *Trichoderma* species are the most studied bioagents in agricultural ecosystem including pulse ecosystem (Mishra et al. 2018a, 2020, 2023a, b). They are successfully used to control wilts, root rots, collar rot and stem rot diseases incited by different *Fusarium* spp., *Rhizoctonia solani*, *R. bataticola*, *Sclerotium rolfsii*, *Sclerotinia sclerotiorum*, *Ascochyta*, *Cercospora*, *Alternaria* spp., *Phytophthora* spp. and *Pythium* spp., in different pulses and other field crops. Fungi reported to suppress the fungal diseases in pulses are *Aspergillus candidus*, *A. niger*, *Pacilomyces lilacinus*, *Glomus mosseae*, *Verticillium chlamydosporum*, *Absidia cylindrospora*, *Coniothyrium minitans*, *Talaromyces flavus* and *Trichothecium roseum*. Chaudhary et al. (2004) compiled that many bio-control agents have effectively managed different phytopathogens of pulses (Table 11.2).

11.3.12.4 Resistant Varieties

The importance of disease-resistant varieties of pulses has been well realized and therefore, research projects on the development of resistant varieties of pulses are accorded the highest priority by the pulses researchers. Several varieties of pulse crops (chickpea, pigeonpea, lentil, fieldpea, mungbean,

TABLE 11.2
Bio-control agents effective against major pathogens of pulse crops

Crop	Pathogens	Effective biocontrol agents
Chickpea	*F. oxysporum* f. sp. *ciceri* *Sclerotium rolfsii* *Rhizoctonia bataticola* *Rhizoctonia solani* *Aschochyta rabei*	*Trichoderma harzianum*, *T. viride*, *T. koningii*, *T. asperellum*, *T. afroharzianum*, *T. longibrachiatum*, *T. virens*, *Gliocladium virens*, *Pseudomonas* sp. *Bacillus* sp.
Pigeonpea	*Fusarium udum* *Phytophthora drechsleri* f. sp. *cajani*	*T. harzianum*, *T. viride*, *G. virens*, *T. asperellum*, *T. afroharzianum*, *T. longibrachiatum*, *Bacillus subtilis*, *T. koningii*, *T. hamatum*, *Glomus mosseae*, *Verticillium chlamydosporum*, *Glomus mosseae*, *P. fluorescens*
Lentil	*Fusarium oxysporum* f.sp. *lentis* *Macrophomina phaseolina* *Macrophomina phaseolina* *Ascochyta fabae* f. sp. *lentis*	*T. viride*, *T. harzianum*, *G. virens*, *P. fluorescens* *Rhizobium leguminoserum* *T. asperellum*, *T. afroharzianum*
Field pea	*Uromyces fabae* *Erysiphe polygoni* DC *Rhizoctonia solani*	*T. harzianum*, *T. viride*, *T. virens*, *T. hamatum* *R. leguminoserum*
Mungbean/Urd bean	*Macrophomina phaseolina* *Cercospora* spp.	*T. viride*, *T. harzianum*, *T. koningii*, *T. longibrachiatum*, *P. fluorescens*, *B. subtilis*, *A. niger* and *Pencillium citrinum*, *T. asperellum*, *T. afroharzianum*

(Sources: Chaudhary et al. 2004; Mishra et al. 2018a, b, 2020, 2023a, b.)

urdbean, rajmash, etc.) developed by different ICAR institutes and SAUs are resistant to biotic stresses. The use of these disease-resistant varieties is accepted as the most cost-effective and best eco-friendly method of avoiding losses caused by the diseases.

11.3.12.5 Integrated disease management

Integrated disease management (IDM) strategies are the most adopted and effective integrative approaches worldwide. Crop rotation, cultural practices, and summer plowing are important components of IDM, which may suppress the population of pathogenic microbes in the soil, ultimately reducing the intensity of diseases. Sowing dates, timely sowing, chemical fertilizer dose application, and intercrops with non-hosts are other effective components of integrated disease management. Raised-bed planting of pigeonpea is another effective integrated approach for the effective management of phytophthora stem blight disease. Biological control agents based microbial formulations and consortia/bio-stimulants are another important component of IDM that is widely accepted and used worldwide for plant disease management as well as plant growth promotion. Application of some new molecules of safer fungicides with the integration of cultural and biological components can be effective IDM strategies for pulse crop protection as well as better soil health and higher productivity.

11.4 CONCLUSIONS

Pulse crops are primarily cultivated in rainfed environments, resulting in relatively low productivity levels. This is primarily attributed to significant losses caused by various diseases throughout the crop's growth stages. To combat these diseases and achieve higher yields for maximum profitability, farmers and extension workers typically rely on chemical pesticides. However, it is widely recognized that the use of agricultural chemicals not only contributes to environmental pollution but also poses significant health risks. Their improper use often leads to toxicity in humans, plants, animals, and wildlife. Therefore, renewed knowledge is needed to assess the current severity of these problems and apply bio-intensive management (BM) strategies to protect crops in a cost-effective and eco-friendly manner. These bio-intensive (BM) management packages of food legumes have been successfully refined and validated in partnership with stakeholders and end users. Despite the development of various bio-intensive modules to tackle economically important diseases of food legumes, a considerable knowledge gap persists between scientists and farmers, particularly in developing countries. This is due to dynamic agro-eco systems and socio-economic conditions and the dynamic nature of host-pathogen interactions/co-evolution that necessitates refinement of management strategies. Therefore, urgent need is there for consistent refinement, validation, transfer and adoption of bio-intensive modules. Future activities will focus on the expansion of bio-intensive technology by increasing the awareness of farmers, and also on the importance of quality of crop residues of food legumes used as an essential component of the prevailing mixed crop livestock system. In addition, novel/refined BM strategies need to be in place for tackling emerging pathogens such as phytophthora blight in pigeonpea and new pathotypes of ascochyta blight of chickpea. Experience over the last few decades has shown that plant viruses cannot be controlled by preventive measures only. The complicated ecology of many of the viruses which affect legume crops, and for this matter crops in general, calls for a strategy that is long term, sustainable, economically acceptable to the resource-poor farmers, and friendly to the environment. The promising approach is to harness all available measures and deploy them in an integrated fashion. Effective virus-disease control has always been and will continue to be through integrated management. Future bio-intensive management strategies will also account for the effects of climate variability on pathogen dynamics that are contributing to the emergence and resurgence of new strains/pathogens. This in turn necessitates appropriate refinement for disease control. HPR, fungicides, natural plant products, bio-fungicides, botanicals and agronomic practices will remain the potentially viable options for BM.

REFERENCES

Agrios GN (2005) *Plant Pathology* (Fifth edition), Academic Press, London.

Armstrong CL, Chongo G, Gossen BD, Duczek LJ (2001) Mating type distribution and incidence of the teleomorph of *Ascochyta rabiei* (*Didymella rabiei*) in Canada. *Can J Plant Pathol* 23: 110–113.

Broughton J, Hernandez G, Blair M, Beebe S, Gepts P, Vanderleyden J (2003) Beans (*Phaseolus* spp.): Model food legumes. *Plant Soil* 252: 55–128.

Chaudhary RG, Sharma N, Prajapati RK (2004) Biological control of soil borne diseases: An updates in pulse crops. In *Ecofriendly management of plant diseases*, Shahid A and Narain U (eds), 178–200. Daya Publishing House.

Dhar V, Reddy MV, Chaudhary RG (2005) Major diseases of pigeonpea and their management. In Advances in pigeonpea research, Ali M, Kumar S (eds), 229–261. Kanpur: Indian Institute of Pulses Research.

Dubey SC (2002) Bioagent based integrated management of collar rot of French bean. *Indian Phytopath* 55: 230–231.

Gurha SN, Singh G, Sharma YR (2003) Diseases of chickpea and their management. In: Chickpea research in India, Ali M, Kumar S Singh NB (eds). 195–227. Lucknow, India: Army Printing Press.

Iqbal M, Dubey M, Broberg A, Viketoft M, Jensen DF Karlsson M (2019) Deletion of the nonribosomal peptide synthetase gene nps1 in the fungus Clonostachys rosea attenuates antagonism and biocontrol of plant pathogenic Fusarium and nematodes. *Phytopathology* 109(10): 1698–1709.

Jimenez-Díaz RM, Castillo P, del Mar Jimenez-Gasco M, Blanca B. Landa b, Navas-Cortes, JA (2015) Fusarium wilt of chickpeas: Biology, ecology and management. *Crop Prot* 73: 16–27.

Mishra RK, Bohra A, Kamaal N, Kumar K, Sujayanand GK, Saabale PR, Naik SJ, Sarma BK, Kumar D, Mishra M, Srivastava DK, Singh NP (2018b) Utilization of biopesticides as sustainable solutions for management of pests in legume crops: Achievements and prospects. *Egyptian J Bioll Pest Control.* doi: 10.1186/s41938-017-0004-1.

Mishra RK, Mishra KK, Pandey KK, Singh UP (2010) Biochemical changes in pea infected by *Uromyces fabae*. *Indian Phytopath* 65(2): 125–128.

Mishra RK, Monika M, Naimuddin A, Krishna K (2018a) *Trichoderma asperellum*: A potential biocontrol agents against wilt of pigeonpea caused by *Fusarium udum* Butler. *J Food Legumes* 31: 50–53.

Mishra RK, Naimuddin A, Saabale PR (2015) First report of stem rot in pigeonpea from India. *Pulses News Letter* 7: 1–7.

Mishra RK, Pandey PK (2009) Efficacy of fungicides against pea rust caused by *Uromyces fabae*. *Ann Plant Protec Sci* 17(2): 54–55.

Mishra RK, Pandey S, Hazra KK, Mishra M, Naik S, Bohra A, Parihar AK, Rathore US, Naimuddin, Kumar K, Singh B, Singh NP (2023a) Biocontrol efficacy and induced defense mechanisms of indigenous *Trichoderma* strains against *Fusarium* wilt [*F. udum* (Butler)] in pigeonpea. *Physiol Mol Plant Pathol* 1127: 102122. 10.1016/j.pmpp.2023.102122.

Mishra RK, Pandey S, Mishra M, Rathore US, Naimuddin, Mohammad A, Krishna K Singh B (2020) Molecular identification, enzyme profiling and biocontrol potential of *Trichoderma* isolates against wilt disease of major pulse crops. *J Food Legumes* 3: 50–54.

Mishra RK, Pandey S, Rathore US, Mishra M, Kumar K, Kumar S, Manjunatha L (2023b) Characterization of plant growth-promoting, antifungal, and enzymatic properties of beneficial bacterial strains associated with pulses rhizosphere from Bundelkhand region of India. *Brazilian J Microbiol*. https://doi.org/10.1007/s42770-023-01051.

Nene YL, Reddy MV, Haware MP, Ghanekar AM, Amin KS (1991) Field diagnosis of chickpea diseases and their control. In: *Information Bulletin* no. 28. ed. by Int. Crops Res. Inst. for the Semi Arid Tropics, Patancheru, India.

Pande S, Rao JN and Sharma M (2006) Establishment of the chickpea wilt pathogen *Fusarium oxysporum* f. sp. ciceris in the soil through seed transmission. *Plant Pathol J* 23: 3–6.

Pérez-Hernandez F, Ayala-Burgos AJ, Belmar-Casso R (2003) Performance of growing lambs supplemented with Mucuna pruriens. *Trop Subtrop Agroecosyst* 1: 119–122.

Senthilkumar M, Swarnalakshmi K, Govindasamy V, Lee YK, Annapurna K (2009) Biocontrol potential of soybean bacterial endophytes against charcoal rot fungus, *Rhizoctonia bataticola*. *Curr Microbiol* 58: 288–293.

Sharma M, Pande S, Pathak M, Rao JN, Kumar PA, Reddy DM, Benagi VI, Mahalinga DM, Zhote KK, Karanjkar PN, Eksinghe BS (2006) Prevalence of Phytophthora blight of pigeonpea in the Deccan Plateau of India. *Plant Pathol J* 22: 309–313.

Singh, G. and Sharma, Y.R. (2002) Fungal diseases of pulses. In: Gupta, V.K. and Paul, Y.S. (eds) Diseases of field crops. Indus Publishing, New Delhi, India, pp. 155–192.

12 Integrated Disease Management Strategies in Soybean Crops

Sanjeev Kumar, Shubham Mishra and Jayant Bhatt

12.1 INTRODUCTION

One of the most widely cultivated and utilized oilseeds are soybeans. Human and animal foods, industrial products, ingredients, and precursor materials are just a few examples of the uses. Separating the oil from the protein and fiber in soybeans is a crucial step in using them. The comparatively high-quality oil and protein that make up about 20 and 40 percent of soybean correspondingly are appreciated globally (Clemente and Cahoon 2009). However, current initiatives to reduce carbon emissions and rely less on fossil fuels have led to the usage of soybean oils as a biodiesel product. The oils are still used in cooking and baking (Mofijur et al. 2013). The value of soybeans comes primarily from the meal, which is consumed in 97 percent of animal and poultry feed. Soybean is used extensively in the industrial sector, serving as crucial ingredients in paints and adhesives (Raghuvanshi and Bisht 2010). Farmers are worried about reducing soybean production losses due to the high market volatility and significant soybean acreage. Unfortunately, a variety of diseases and pests can seriously reduce the output of soybean. According to CPN 2020, (*https://loss.cropprotectionnetwork.org/*) 14.6 million tonnes (536 million bushels) of soybeans were lost in 2018 alone owing to soybean diseases. There are four types of diseases that harm soybeans: fungi, bacteria, nematodes, and oomycetes. This book chapter addresses the key annual problems, substantial intermittent dangers, and important new challenges to soybean output. Particular diseases are routinely reasons for concern in both the northern and southern producing regions, despite the fact that potential threats to soybean output vary by growing region (Allen et al. 2017). These include seedling diseases, charcoal rot, and the soybean cyst nematode. Phytophthora stem and root rot, sudden death syndrome (SDS), and Sclerotinia stem rot, on the other hand, are intermittent hazards that can be significantly influenced by environmental circumstances. These diseases affect soybeans economically. Other diseases that pose a threat to soybean production in addition to these have recently become more prevalent issues in both northern and southern areas. Root-knot and reniform nematodes, frogeye leaf spot, and Diaporthe infections are among the illnesses that are becoming more prevalent. Research on the biology, epidemiology, and management of various diseases has been prompted by the economic implications of these conditions. The soybean cyst nematode is arguably the most ubiquitous and persistent danger to the nation's soybean agriculture. Similar connections between soybean and *Phytophthora sojae* have shown a close gene-for-gene relationship (Flor 1955), which has sparked a lot of study and publications on this illness (Dorrance 2018). Despite the fact that the yield danger is still quite substantial, research on other important diseases has gotten less attention.

Some of the most common disease management approaches in row-crop agriculture include genetic resistance, crop rotation, seed treatments with fungicides and nematicides, mid-season fungicide

DOI: 10.1201/9781032631578-12

sprays, and biocontrol. Other cultural approaches, like as tillage, row spacing, plant population, and irrigation, can also be optimized. For effective disease reduction, each condition necessitates a particular, and frequently unique, combination of therapeutic measures. Some management options for the same disease may work better in some regions than others because environmental conditions can vary from region to region. An extensive study indicated that for every 1°C increase in the world mean temperature; there can be expected to be a 3.1 percent loss in soybean yield, underscoring the critical role that the environment plays in soybean growth and development (Zhao et al. 2017). This estimate, however, does not account for yield losses brought on by diseases, which also have an impact on the environment. Plant disease cannot exist unless all three components of the triangle for plant disease – a virulent pathogen, a susceptible host, and a conducive environment are present (McNew 1960).

We discuss one of the biggest risks to current and future agriculture in each of the sections below: the change in global climate patterns and the rise in the frequency of extreme weather events. Convective systems, or very massive thunderstorms, are predicted to occur three times more frequently and provide an additional 30 to 80 percent more precipitation by the end of the century, according to forecasting models (Prein et al. 2017). Heavy rains, severe flooding, and the filing of insurance claims for "prevented plant" on over 8 million hectares (20 million acres) due to extreme weather events in 2019 resulted in a financial burden of almost $4 billion (USDA 2019). Weather changes are anticipated to raise the risk of diseases in the future decades; however, it is difficult to forecast the effects these types of changes will have on plant diseases because each pathogen has a distinct disease cycle (Garrett et al. 2016: Juroszek et al. 2020). The environmental part of the disease triangle may be altering the frequency and severity of disease outbreaks in some areas due to concerns over the changing climate in the twenty-first century. Farmers need to know if they are susceptible to diseases brought on by pathogens found in their area and if their habitat is receptive to the introduction of other pathogens that could spread disease. Precipitation in the Upper Midwest region of the United States has increased by 37% from 1958 to 2012. If current climate trends hold, diseases like soybean rust and anthracnose, which have a significant impact on soybean production in other nations, may also pose greater threats in the United States (IPCC 2014). We intend to accomplish the following with this chapter: 1) a description of the impact and disease cycle for common soybean diseases, 2) a thorough review of the current state of the science regarding management strategies associated with every soybean disease, and 3) insights into each pathogen's epidemiology and the likelihood of outbreaks and spread of major pathogens to new regions as a result of changing climate trends.

12.2 DISEASE CHALLENGES TO SOYBEAN PRODUCTION – FREQUENTLY RECURRING

A soybean crop is always susceptible to bacteria, fungi, viruses or nematodes causing disease development. Depending on the disease, the climate, the time of infection, and the soybean product, the potential yield loss caused by an infection may be slight or severe. Making crucial management choices to maintain yield potential requires being aware of a field's disease history and being up to date on any reported disease outbreaks during the growing season.

12.2.1 Soybean Cyst Nematode – *Heterodera Glycines* Ichinohe (Tylenchida: Heteroderidae)

In the entire world, nematodes provide the greatest danger to soybean yield (Allen et al. 2017). From 2006 to 2014, soybean cyst nematode was the most economically significant soybean pest in both the northern and southern states of the United States, causing an estimated average annual yield loss of more than 3.4 million tonnes (125 million bushels) (Allen et al. 2017). Between 2000 and

2020, with an average soybean market price of \$330 per tonne (\$9 per bushel) (USDA NASS 2020, supplemental file 1), yield loss was predicted to have cost the economy more than \$1.1 billion. The obligatory plant-parasitic nematode known as the soybean cyst lives in a variety of soil types and geographical areas. On a susceptible host, it can complete its life cycle in 3–4 weeks, producing numerous generations and a sizable rise in nematode population density in a single season (Niblack et al. 2006). However, in many instances, aboveground symptoms are completely absent, even when soybean cyst nematode is causing large yield losses (Niblack et al. 2006). Foliar symptoms linked with soybean cyst nematode damage are frequently misdiagnosed as nutritional deficits (Wang et al. 2003). The soybean cyst nematode overwinters as eggs inside a cyst (a dead female), which enables them to endure unfavorable circumstances in the soil for at least ten years (Niblack et al. 2006) The eggs hatch when the surrounding environment is favourable, and second-stage juveniles (J2) emerge from the eggs in search of soybean roots. Warmer soil temperatures (24–32°C, or 75–90°F) and enough soil moisture are ideal for hatching, and the presence of root exudates from a suitable host, such as soybeans, accelerates the rate of egg hatching (Riga et al. 2001). The J2 stage of the nematode, which is the only infectious stage, travels through the soil in water films that envelop soil particles until it comes into contact with a host root. Using a specialized mouthpiece known as a stylet, it enters the root beyond the root tip, moves through the cortical tissue into the plant vascular system, and forms a feeding location known as a syncytium (Niblack et al. 2006). Before a final molt transforms the J2 into an adult male or female soybean cyst nematode, it goes through two further juvenile stages (J3 and J4) at the syncytium. The heads of adult females still protrude from the roots near the feeding cell as they grow larger and rupture through the root epidermis at the feeding location where they remain stationary. In order to mate with one or more females, adult males exit the root and go back to the ground. Females are capable of producing several hundred eggs, most of which stay inside the adult female body. However, some females may deposit an additional 50 to 100 eggs in an egg mass, an exterior gelatinous matrix (Niblack et al. 2006). The cyst forms after the female dies, shielding the eggs from outside influences and extending their lifespan. Since soybean cyst nematode cysts and eggs can survive for 10 years or longer without a suitable host (Niblack et al. 2006), and even the juveniles can survive for up to 7 months in flooded soil, up to 29 months in dry soil, and up to 90 months in soil at water capacity, it is difficult to predict how extreme weather events may affect the nematode (Slack et al. 1972). However, because low temperatures (15°C or 59°F) prevent egg hatching and soybean cyst nematode reproduction, rates of soybean cyst nematode reproduction and egg hatching are projected to rise as global temperatures rise (IPCC 2014). Extremely high temperatures (over 36°C or 97°F) can also prevent soybean cyst nematode eggs from hatching and adults from reproducing. Although pathogens are also more likely to travel poleward with changing global temperatures, predicted temperature rises could result in increasing soybean acreage in northern regions where soybeans have not historically been farmed (Bebber et al. 2013). Therefore, if temperatures do rise further, soybean cyst nematode, which already poses a problem in many parts of world, could become more of an issue, lowering soybean yields worldwide.

12.2.1.1 Genetic Resistance

The most widely used and successful method for controlling soybean cyst nematodes is genetic resistance. Commercial soybean lines have a large amount of genetic resistance that comes from the plant introduction line PI88788 (McCarville et al. 2017). Although other soybean breeding lines have several soybean cyst nematode resistance genes, PI88788 is the most commonly detected one (Mitchum et al. 2013). Although the resistance strategies and causal genes are unknown, all resistant lines exhibit quantitative genetic resistance. Resistance to Heterodera glycines 1 (Rhg1) is a quantitative trait locus (QTL) on chromosome 18 responsible for resistance in PI88788 (Concibido et al. 2004). It is interesting to note that the Rhg1 locus in susceptible varieties has fewer copies of these three unrelated genes than the Rhg1 locus in PI88788 (Cook et al. 2012). There are now more QTLs for soybean cyst nematode resistance, some of which overlap with QTLs for resistance to other significant soybean diseases (Swaminathan et al. 2018). Repeated planting of cultivars with PI88788 resistance led to significant selection pressure and the emergence of soybean cyst nematode populations that can override Rhg1 resistance, making the identification of additional QTLs linked to strong soybean cyst nematode resistance an important area of research. A greater understanding of the mechanisms behind soybean cyst nematode virulence has resulted from recent attempts to analyse the genomes of several populations of the parasite (Masonbrink et al. 2019a). An open-access soybean cyst nematode genome browser is the result of cooperative efforts to combat the parasite (scnbase.org, Masonbrink et al. 2019b). These coordinated efforts enable earlier genetic pathways underlying soybean resistance to harmful pests and diseases to be discovered. Different biotypes or races of the parasite have arisen as a result of the soybean cyst nematode's ability to overcome genetic resistance.. The HG typing system was created to classify these biotypes according to how well they can replicate on soybeans that are known to have sources of resistance (Niblack et al. 2002). With the aid of this HG typing method, farmers can identify the types of soybean cyst nematode populations they are dealing with and choose which genetically resistant cultivars will be most useful for controlling the nematode in their crops. This technique has been discovered to offer a significant yield gain when selecting cultivars, especially for early-maturing soybean cultivars (Rincker et al. 2017).

12.2.1.2 Cultural Control

Reduced population densities of the soybean cyst nematode and increased soybean yields can result from long-term crop rotations that include non-host crops (Poromarto and Nelson 2010). Weeds such as wheat, corn, and alfalfa are not hosts. Other typical legume crops like dry beans (*Phaseolus vulgaris*), cowpeas (*Vigna unguiculata*), and clover (*Trifolium* spp.) are susceptible to soybean cyst nematode colonization (Bond et al. 2019). Several common winter annual weeds, such as chickweed (*Stellaria media*), bittercress (*Cardamine parviflora*), purple deadnettle (*Lamium purpureum*), henbit (*L. amplexicaule*), pennycress (*Thlapsi arvense*), and shepherd's-purse (*Capsella bursapastoris*), can act as hosts for soybean cyst nematode (Creech et al. 2008). Unfortunately, short-duration rotation sequences, like the well-known 2-year corn-soybean rotation, are frequently insufficient for controlling soybean cyst nematode populations. This is because the eggs in soybean cyst nematode cysts can survive in the soil for several years without a host. The use of cover crops as "trap crops" or potential rotational crops has been frequently advocated. A trap crop would ideally encourage the hatching of soybean cyst nematode eggs and draw J2 to roots, but it would prevent soybean cyst nematode from completing its life cycle, keeping it inside the root where it would be unable to procreate. The consistency of these effects in the field is frequently lacking (Harbach 2019), and needs to be further assessed with additional field studies. Some crimson clover (*Trifolium incarnatum*) varieties may attract soybean cyst nematode J2 and some other cover crops may inhibit soybean cyst nematode reproduction in the greenhouse. Compared to conventional tillage methods including disking, chisel-plowing, and moldboard plowing, no-till practices can result in lower soybean cyst nematode levels at harvest, according to some findings (Grabau et al. 2017). However, one study was unable to determine any effects of tillage or row spacing on the population density of soybean cyst nematodes (Chen et al. 2001b). The presence of wheat residue in the field has been shown to reduce soybean cyst nematode populations during harvest. Tillage with a disc, molbord plow, or chisel plow has been shown to influence the capacity of numerous fungal species to parasitize the soybean cyst nematode, potentially improving soybean cyst nematode management. However, tillage practices alone are insufficient for controlling soybean cyst nematode in soybean-producing systems.

12.2.1.3 Biological Control

While some microorganisms are direct parasites of soybean cyst nematode, others compete with it for connections with soybean roots (Bernard et al. 1996). Commercial seed treatments with specific strains of these bacteria have been developed to help suppress the soybean cyst nematode and other nematodes. Products include Votivo (*Bacillus firmus* strain 1–1582, BASF), Clariva pn (*Pasteuria nishizawae* strain Pn1, Syngenta Crop Protection), Aveo EZ Nematicide (*Bacillus amyloliquefaciens* strain PTA-4838, Valent BioSciences, Libertyville, IL), and BIOST Nematicide 100 (heat-killed *Burkholderia* spp. strain A3). Notwithstanding being a parasite of the soybean cyst nematode J2 (Chen 2007), the fungus *Hirsutella rhossiliensis* hasn't been widely used or commercialized as a biocontrol agent. According to a recent study by Strom et al. (2020), both crop rotation and nematode density can have an impact on the fungi that live on soybean and corn roots. These fungi produce substances that are harmful to soybean cyst nematodes.

12.2.1.4 Chemical Control

Nematodes and other soilborne pathogens are difficult to chemically control because the timing and distribution of the chemistry into the soil are complex, and chemistries can negatively affect beneficial soil bacteria as well (Moorman 1989). New chemicals created by agricultural businesses, however, have more focused modes of action that are more effective against their intended diseases and often use lower doses of the active ingredient. Additionally, a lot of chemicals are now created as seed treatments, which means that they are dispersed into the soil with the seed. This helps to reduce the amount of active ingredients required, direct the chemistry where it is needed, and safeguard seedlings as they are germinating. Abamectin (Avicta, Syngenta Crop Protection, Greensboro, NC), fluopyram (ILEVO, BASF, Ludwigshafen, Germany), pydiflumetofen (Saltro, Syngenta Crop Protection), and tioxazafen are some of the common chemicals used to manage soybean cyst nematode at the moment (NemaStrike, Bayer CropScience, Kansas City, MO). The natural substance abamectin, which is effective against the soybean cyst nematode, is generated by the bacteria *Streptomyces avermitilis*. It interferes with neuron communication in nematodes and other arthropods (Vitti et al. 2014). As a succinate dehydrogenase inhibitor (SDHI), fluopyram has an impact on cellular respiration. Fluopyram has demonstrated effectiveness in field experiments against numerous fungal infections in addition to the soybean cyst nematode, reducing the quantity of soybean cyst nematodes in roots and reducing soybean cyst nematode reproduction in several soybean cultivars (Bissonnette et al. 2020; Roth et al. 2020). Fluopyram decreases soybean cyst nematode movement and prevents root penetration, according to additional laboratory tests (Beeman and Tylka 2018; Beeman et al. 2019). Pydiflumetofen and tioxazafen are more modern chemicals that have just recently been approved for use as nematicide seed treatments in soybean, although reports evaluating their effectiveness in comparison to other products have not yet been widely disseminated.

12.2.2 Charcoal Rot – *Macrophomina phaseolina* (Tassi) Goid (Botryosphaeriales: Botryosphaeriaceae)

At least 500 plant species, including key crops like corn and grain sorghum, are pathogenic to *Macrophomina phaseolina* (Kunwar et al. 1986). The disease has been observed worldwide, although historically charcoal rot was predominantly a southern United States problem (Cummings and Bergstrom 2013). In both the northern and southern U.S. areas, charcoal rot has consistently rated among the top-five soybean illnesses that limit productivity in recent years (Allen et al. 2017). In 2012, charcoal rot was the second-most common disease affecting soybean yields in North America, resulting in an estimated yield loss of 2.0 million tonnes (73 million bushels) (Allen et al. 2017). Small, black microsclerotia that germinate between 20°C and 40°C that's how *M. phaseolina* persists in soil. The roots of seedlings and young plants can become infected by the developing mycelia. After the initial infection, the fungus slowly spreads via the roots and shoots until the soybean plant enters the reproductive stage, at which point *M. phaseolina* growth rises significantly. Eventually, fungal mycelia and microsclerotia choke vascular tissue, causing wilting and early plant death (Khan 2007). The freshly created microsclerotia are dispersed back into the field with the leftover soybeans during harvest, where they will act as inoculum the next year. Although a study conducted in Brazil showed that microsclerotia could become nonviable in just 5 months (Reis et al. 2014), it is generally accepted that microsclerotia can survive in the soil for up to 15 years (Gupta et al. 2012) depending on a variety of environmental and soil conditions. Additionally, *Macrophomina phaseolina* can spread through infected seeds, which are sometimes difficult to detect since the seed coat might conceal the fungus (Mengistu et al. 2015).

12.2.2.1 Genetic Resistance

The introduction and characterization of cultivars resistant to *M. phaseolina* has received a lot of attention because there are now no effective cultural methods or chemical treatments for controlling charcoal rot. Around 850 soybean genotypes have been tested for resistance to charcoal rot over the past 20 years, and 23 were found to have moderate resistance (Pawlowski et al. 2015). Since charcoal rot has historically been a disease where later maturity groups are produced, the majority of the resistant genotypes are in maturity group III or later. There have been reports of *M. phaseolina* in northern soybean-growing regions where it has started to flourish, presumably as a result of hospitable circumstances brought on by a warming climate (Bebber et al. 2013). Northern regions yield earlier-maturing soybeans, which are more likely to be vulnerable to charcoal rot. This may help to explain why there have been more outbreaks of charcoal rot in northern states. Resistance has been found in maturity groups I to III according to recent study (Scott and Singh 2018). Furthermore, a recent study that identified possible QTLs in maturity group IV soybeans linked to resistance to *M. phaseolina* serves as the foundation for upcoming targeted breeding initiatives in soybean (Da Silva et al. 2019).

12.2.2.2 Crop Rotation

There are limited opportunities for rotation with non-host crops because *M. phaseolina* has a large number of hosts (Livingston 1945). Contrary to continuous soybean production systems, rotations with cotton can reduce the amount of inoculum in the soil (Francl et al. 1988). Genetically varied *M. phaseolina* populations can develop in a field as a result of crop rotations with corn, although these diversified isolates are frequently still virulent on soybean (Almeida et al. 2008).

12.2.2.3 Cultural Control

Irrigation has demonstrated potential for controlling charcoal rot (Kendig et al. 2000). Drought stress is linked to high disease severity and yield losses; therefore, irrigation could lessen the severity of charcoal rot this year. Asymptomatic soybeans may still experience root colonization, which could result in more microsclerotia in the soil and higher disease pressure, or increased inoculums, in subsequent years (Mengistu et al. 2011b). Throughout numerous management studies, the application of tillage has had erratic results. According to certain studies, boosting the pace at which soil microsclerotia decompose may help tillage prevent illness. According to additional research, microsclerotia germination is reduced when they are buried more deeply than 15 cm (5.9 inches) through plowing (Baird et al. 2003). Other research has shown that no-till systems are superior to tillage in terms of disease pressure or inoculums load (Mengistu et al. 2009). However, since drought stress encourages a higher occurrence of disease, no-till farming methods do allow for better soil moisture retention, which may help to prevent disease.

12.2.2.4 Biological Control

At the moment, there is no chemical control technique that consistently prevents charcoal rot (Reznikov et al. 2016). Biocontrol, though, has showed promise. In one investigation, *Rhizobacterium* (strain: FPT721)-treated seed decreased the prevalence of disease (Choudhary 2011). Two endophytic bacteria, *Paenibacillus* (strain:HKA-15) and *Bacillus* (strain:HKA-121), have also demonstrated a robust antagonistic relationship with *M. phaseolina*, indicating the possibility for usage as a biocontrol agent (Senthilkumar et al. 2009). There are currently no commercially accessible biocontrol therapies for charcoal rot, despite the fact that it may be managed.

12.2.3 Diseases of Seedlings

Every year, soybean producers in the growing region struggle with seedling infections. The most prevalent pathogens in this group of diseases, which includes oomycetes and fungi, include *Rhizoctonia solani*, *Fusarium* species, and *Pythium* species (Rizvi and Yang 1996). These diseases are collectively known as "damping-off" because diminished crops stand by either decaying seeds or by killing the growing seedling shortly after emergence. They are frequently more common in cool and wet years. Whenever farmers are forced to replant entire fields due to severe damping-off, crop profitability can be drastically reduced.

12.2.3.1 *Rhizoctonia solani* Kuhn (Cantharellales: Ceratobasidiaceae)

Reddish lesions on the hypocotyl of seedlings close to the soil line are signs of this disease. Despite the fact that this illness can cause seedling death, some seedlings may survive and grow to become plants with stunted growth. This fungus survives the winter in agricultural debris and soil as sclerotia and mycelia, sprouting in the spring to infect more plants (Tsror 2010). Due to the high genetic diversity of the fungus *Rhizoctonia solani*, certain plant hosts may be more susceptible to its virulence. The *R. solani* species complex is unique in that it can form anastomosis groups when it undergoes hyphal fusion with other strains (AG).

12.2.3.2 *Fusarium* spp. (Hypocreales: Nectriaceae)

Fusarium spp. is responsible for Fusarium root rot. This disease's seedling symptoms include diminished nodule production and reddish to brown browning of the taproot and lateral roots. Stunting, yellowing, and wilting are examples of the aboveground signs that the disease may be present in mature plants. Despite the fact that several *Fusarium* spp. are frequently found in agricultural settings, only a few of them are harmful to crops (Ellis et al. 2013a). These, however, can infect a wide variety of host crops. *F. solani*, *F. oxysporum*, *F. graminearum*, and *F. acuminatum* are a few of the pathogenic species that affect soybean seedlings most frequently (Diaz Arias et al. 2013a). Each species' individual isolates might vary substantially in terms of aggression (Daz Arias et al. 2013b). When plants have experienced extreme stress from issues like herbicide damage, compacted soil, or nematode pressure, *Fusarium* spp. are frequently found infecting them. They prefer cool, well-drained soil (Han et al. 2017). Through the tip of the root or an open cut, *F. oxysporum* enters the plant and finds its way to the vascular tissue where it will ascend the plant. Outside of vascular tissue, other *Fusarium* species migrate within the stem. The fungus may remain for several years in crop residues and soil as chlamydospores or mycelia (Nelson 2015).

12.2.3.3 *Pythium* spp. (Peronosporales: Pythiaceae) Pringsheim

The oomycetes family of fungus-like organisms includes *Pythium* spp. *Pythium* spp. can lead to seed rot, seedling blight, root rot, and damping-off in soybean. In cool, moist soils, this disease tends to flourish (Rothrock et al. 2015). Infected seedlings display water-soaked lesions that first emerge on the cotyledons or hypocotyl, as well as decaying roots that make it simple to take the seedlings from the ground. *Pythium* species can persist without a host by forming oospores in the soil or in crop waste. These oospores create fruiting bodies (sporangia) when the soil is wet, and these sporangia release motile zoospores that travel through water films and infect plants. More oospores are released once the infection and colonization of soybean root tissue has taken place, and the cycle continues. Across various crops, a diverse range of *Pythium* spp. exhibit varying levels of aggression. For instance, while some are more aggressive with soybeans than corn, others are more aggressive with corn (Rojas et al. 2017a, b). Different *Pythium* species virulence is influenced by temperature (Matthiesen et al. 2016; Rojas et al. 2017a). As global temperatures rise, the pathogenic Pythium species that cause damping-off in soybeans may shift.

12.2.3.4 Genetic Resistance

There are currently no soybean cultivars with high genetic resistance to *Fusarium* spp. However, the soybean line Archer has been shown to be resistant to *Pythium* species, and this resistance has been defined using the initial resistance gene known as Rpa1 (resistance to *P. aphanidermatum*-1) (Rosso et al. 2008). Additional QTLs for both qualitative and quantitative resistance have been found after the discovery of Rpa1 (Rosso et al. 2008; Lin et al. 2018). In addition to resistance, several soybean cyst nematode-resistant cultivars showed a decrease in the prevalence of *Pythium* spp.-caused disease, which may point to the presence of some cross-resistance mechanisms in soybean (Murillo-Williams and Pedersen 2008). Rhizoctonia root rot induced by *R. solani* was reduced in severity thanks to the transgenic insertion of antifungal genes into potato, which raises the prospect of transgenic methods in soybean to avoid losses (Fernandez Bidondo et al. 2019).

12.2.3.5 Crop Rotation

Crop rotation hasn't shown to be very effective at controlling seedling disease. Numerous different hosts are susceptible to these diseases, and some have been isolated from both weed and agronomic crop species (El Bakali et al. 2000). A corn-soybean crop rotation is unlikely to eradicate these pathogenic species from a field because all three genera of seedling diseases can infect both soybeans and corn (Murillo-Williams and Pedersen 2008). Crop rotation may affect the soil microbiome or enhance the population of microorganisms that are antagonistic to plant diseases, which is an intriguing but mostly investigated idea that may lead to future management measures (Rojas et al. 2019).

12.2.3.6 Cultural Control

The development of seedling disease brought on by *Pythium* spp. and *R. solani* depends on high soil moisture, and the current climate trends in much of the soybean production area include rising temperatures and soil wetness, especially during the spring planting season (Delgado-Baquerizo et al. 2020). By limiting the amount of moisture that remains in the soil after construction of extensive tile drainage systems, *Pythium* species and *R. solani* infections may be less likely to occur. This approach may perhaps increase the risk of seedling disease because *Fusarium* spp. likes well-drained soils (Han et al. 2017). Due to numerous advantages for soil conservation, reduced or no-till approaches have become more and more common for soybean cultivation (Uri 2000). Increased soil moisture retention is a result of less tillage. Soybean germination and early growth would then have access to this moisture, but it could also foster a microenvironment that is more favorable to seedling illnesses. Additionally, minimal or no-till soil conservation techniques permit agricultural residue to be kept on the soil's surface, which may help these species survive (Leslie et al. 1990).

12.2.3.7 Chemical Control

Fungicide seed treatments work best when planted early when soil temperatures are low, in moist circumstances, in areas with less tillage, or in fields with a history of seedling illnesses. The public has access to a summary of frequently applied fungicide or oomicide seed treatments and their overall effectiveness on diseases that affect soybean seedlings and roots (CPN 2019). *R. solani* caused stand losses, while certain seed treatments reduced those losses but did not eliminate completely yield losses (Dorrance et al. 2003). The majority of *R. solani* isolates from soybeans were positive for SDHI fungicide sensitivity in the lab; however, they may or may not also test positive for DMI fungicide sensitivity (Ajayi-Oyetunde et al. 2017). In the greenhouse, Rhizoctonia root rot appears to be well managed by seed treatments with SDHI and DMI fungicides (Ajayi-Oyetunde et al. 2017). Fludioxinil, an active component, outperformed other popular fungicide seed treatments like azoxystrobin, trifloxystrobin, and captan in reducing mycelial growth in *Fusarium* spp. (Broders et al. 2007a). But there have been numerous cases of fludioxinil resistance in isolates of *Fusarium* spp. Because there are now more oomicide products available in a variety of chemical classes, treating seeds with oomicides to reduce *Pythium* in soybeans has grown increasingly widespread (Broders et al. 2007a, b; Matthiesen et al. 2016; Scott et al. 2020). Unfortunately, fungicide and oomicide classes differ in that some pathogen populations are inherently resistant to some fungicides, while other populations may acquire or develop resistance to these fungicides (Peng et al. 2019). The Fungicide Resistance Action Committee (FRAC) studies the development of pathogen fungicide resistance over time based on the mechanisms of action of the various chemical classes and makes recommendations for the most efficient application (Brent and Hollomon 2007).

12.2.3.8 Biological Control

Actinomycete bacteria and binucleate *Rhizoctonia* spp. have both shown promising against *R. solani*. Another opponent, *Verticillium biguttatum*, parasitizes *R. solani* and prevents the growth of sclerotia, which lowers the inoculum in subsequent years (Van den Boogert and Deacon 1994). For both *R. solani* and *F. oxysporum*, *Trichoderma harzianum* colonization reduces disease symptoms (Thangavelu et al. 2004) for *F. oxysporum* f. sp. *adzuki* and *Pythium arrhenomanes*, *T. viride* reduces disease severity (John et al. 2010). When applied as a seed treatment on soybeans, *Pseudomonas putida* strain N1R reduced the occurrence of damping-off brought on by *P. ultimum* at temperatures over 12°C (Paulitz et al. 1992). Although there has been substantial research in this field, commercial versions of these biological agents have sadly not been successfully sold.

12.2.4 Phytophthora Root and Stem Rot – *Phytophthora sojae* Kaufm. & Gerd. (Peronosporales: Peronosporaceae)

Phytophthora sojae has long been acknowledged as a serious hazard to the production of soybeans. From 2006 to 2014, this disease caused an estimated annual loss of over one million tonnes (37 million bushels) of soybeans (Allen et al. 2017). Like *Pythium* spp., *Phytophthora* spp. are oomycetes, also known as "water-molds" (Tyler 2007). In comparison to its other life phases, *P. sojae* overwinters as oospores that are relatively large and robust (Dorrance 2018). The oospores develop into sporangia, which are stuffed with zoospores, during periods of high moisture, such as periods of intense precipitation, irrigation, or flooding. These zoospores are mobile, smaller than oospores, and have two flagella for swimming (Dorrance 2018). The flagella of a zoospore detaches when it comes into interface with a soybean root, allowing the zoospore to infect the root directly (Tyler 2007). *P. sojae* can infect any portion of a soybean plant at any age, despite the fact that it is soil borne and mostly affects the roots (Tyler 2007). Root and stem rot occur when *P. sojae* infects soybean. A young seedling's early wilting and mortality are frequently caused by damping-off or severe colonization of the root and stem (Tyler 2007; Dorrance 2018). When *P. sojae* infects mature plants, vulnerable cultivars experience extensive root rotting and water-soaked sores on the stem (Schmitthenner 2000). *P. sojae* is substantially more aggressive at 20°C than 13°C, demonstrating that temperature has a considerable impact on the pathogenicity of oomycete infections in general (Matthiesen et al. 2016). *P. sojae* abundance and disease pressure are predicted to rise if temperatures continue to rise and there is substantial flooding in soybean crop areas.

12.2.4.1 Genetic Resistance

P. sojae has a very small range of hosts, and soybean is its only significant commercial host (Tyler 2007). Based on certain genes present and expressed by different races of *P. sojae*, resistance to *P. sojae* (Rps) genes in soybean are responsible for resistance to these *P. sojae* infections (Birch et al. 2006). 33 Rps genes have been found in various soybean cultivars to date (Zhong et al. 2019).

TABLE 12.1
Sources and resistance genes for *Phytophthora sojae*

R genes	Source of R genes	Commercial lines have been deployed.	References
Rps6	Altona	(+)	Athow and Laviolette 1982
Rps11	PI594527	(-)	Ping et al. 2016
RpsUN1	PI567139B	(-)	Li et al. 2016
RpsUN2	PI567139B	(-)	Li et al. 2016
RpsYu25	Yudou 25	(-)	Sun et al. 2011
RpsYD29	Yudou 29	(-)	Zhang et al. 2013b
RpsYD25	Yudou 25	(-)	Fan et al. 2009
RpsJS	Nannong 10–1	(-)	Sun et al. 2014
RpsWY	Wayao	(-)	Cheng et al. 2017
RpsHC18	Huachun 18	(-)	Zhong et al. 2018
RpsX	Xiu94-11	(-)	Zhong et al. 2019
RpsGZ	Guizao1	(-)	Jiang et al. 2020

(-) No R genes have been deployed in commercial lines.; (+) R genes have been deployed in commercial lines.

Commercial cultivars typically have six of these genes, which offer effective disease management for Phytophthora root and stem rot (Dorrance 2018). These Rps genes help to produce a very strong qualitative resistance to particular *P. sojae* races. Even though they are not traditional Rps genes, soybeans may possibly have other genes that support partial resistance. Through processes such as the creation of fewer lesions, smaller lesions, or enabling decreased oospores production, these genes offer partial resistance (Dorrance 2018) (Table 12.1).

12.2.4.2 Crop Rotation

Wheat and corn, which are not hosts for *P. sojae*, are frequently cycled with soybeans. In contrast to when soybean was cultivated in monoculture, rotation with corn or wheat led to a more diversified population of *P. sojae* that was made up of less virulent races (Li-ming et al. 2016). However, long-term crop rotations are more likely to minimize *P. sojae* disease than 2-year corn-soybean or wheat-soybean rotations due to the high persistence and survivability of oospores in the soil.

12.2.4.3 Cultural Control

Phytophthora sojae pathogen requires free water to complete its life cycle effectively (Tyler 2007). The water allows motile zoospores to swim towards soybean roots efficiently. Therefore, heavy irrigation is not recommended in fields with a history of Phytophthora stem and root rot. Increased soil drainage using tillage or field tile can significantly reduce disease caused by *P. sojae* (Dorrance et al. 2009).

12.2.4.4 Chemical Control

Many chemicals have proven effective in controlling *P. sojae* (CPN 2019). Metalaxyl, or its pure active isomer, mefenoxam, is one typical fungicide with activity against oomycetes (oomicide) that is used in commercial seed treatments (Dorrance 2018). Mefenoxam is poisonous to oomycetes by an unknown mechanism, however evidence points to the fact that it inhibits RNA polymerase and stops oomycetes from expressing vital genes for survival (Parra and Ristaino 2001). Metalaxyl or mefenoxam seed treatments have demonstrated effectiveness against *P. sojae*, especially in locations that have recently seen considerable rainfall or irrigation (Dorrance et al. 2009). Other chemicals, including as ethaboxam and oxathiapiprolin, are approved for use as seed treatments and have shown effectiveness in the lab against specific oomycete species. (Radmer et al. 2017). According to research by Noel et al. (2019), some oomycete species exhibit innate resistance to ethaboxam. This resistance was genetically created in the lab, giving *P. sojae* ethaboxam resistance (Peng et al. 2019). Oxythiopiprolin resistance has been imparted in oomycetes by related genetic procedures (Miao et al. 2018). Either ethaboxam or oxathiapiprolin have not yet been found to be inherently resistant to *Phytophthora* isolates in the field, therefore using these chemicals as seed treatments may be effective in preventing Phytophthora root and stem rot. To sustain the efficacy of these newer chemistries, vigilant stewardship is required because regular application of these chemicals may select for natural changes that confer resistance to them in the field.

12.2.4.5 Biological Control

There is ongoing research towards biocontrol of *Phytophthora* (Liu et al. 2019), but no commercial formulations or seed treatments are available for general usage in soybean fields at this time. Many researchers are examining micro biomes and trying to find new bacteria that have helpful properties for Phytophthora stem and root rot and other crop diseases (Boscaiu et al. 2019).

12.2.5 Sclerotinia Stem Rot – *Sclerotinia sclerotiorum* (Lib.) de Bary (Helotiales: Sclerotiniaceae)

It can infect approximately 400 plant species. Sclerotinia stem rot, also known as white mold, is caused by the pathogen in soybean. The disease can cause farmers to lose money by reducing the amount of seeds, seed weight, and seed quality (Danielson et al. 2004) and classified in the top ten most yield-limiting diseases of soybean from 2010 to 2013, a major outbreak in 2014 resulted in a top-five ranking in that year (Allen et al. 2017). The rise in severity was mostly attributable to favorable meteorological conditions for severe Sclerotinia stem rot (Fall et al. 2018a). *Sclerotinia sclerotiorum* prefers cool, damp conditions and is abundant in the Upper Midwest, where such circumstances are common. Higher humidity is caused by rising average temperatures and more frequent rainfall events, which may increase the frequency of Sclerotinia stem rot in the northern soybean production belt (Bebber et al. 2013). Incidence and severity in some previously problematic areas may also decrease when temperatures rise, limiting pathogen development. Beginning around the R5 growth stage, Sclerotinia stem rot emerges on soybean as white, fluffy mycelia on the main stem and lateral branches. The phrase "white mold" comes from this white development. The pathogen, on the other hand, begins its life cycle in the soil as melanized, seed-like survival structures known as sclerotia. To germinate, these sclerotia require a 1- to 2-week period of soil moisture at a water potential of roughly 100 kPa, as well as an extended chilly period of 8–21°C (46–70°F) (Shahoveisi and del Ro Mendoza 2020). The fruiting structures, known as apothecia, resemble tiny, tan mushrooms with a cup-like crown where ascospores form. When these circumstances are met, apothecia start to emerge from sclerotia in the top 2–3 cm (0.8–1.2 inches) of the soil. Ascospores are aggressively released into the crop canopy when completely formed apothecia are exposed to a modest reduction in moisture tension, which often occurs after the morning dew dries. When the temperature is between 15 and 25°C (59–77°F) and the leaves have been wet for 2–4 hours, ascospores, the primary source of inoculum, settle on senescing plant parts and germinate. The fungus usually spreads by senescing blossoms, although it can also spread through wounds, natural openings, and contact with surrounding plants.

12.2.5.1 Genetic Resistance

Sclerotinia stem rot genetic resistance, discovered by germplasm screening and QTL mapping, serves as a disease control tool (McCaghey et al. 2019). However, due to the quantitative character of the resistance, producing resistant cultivars has been challenging (Vuong et al. 2008). Although the genetic basis for S. sclerotiorum resistance is unknown, specific control of genes in the phenylpropanoid pathway appears to promote resistance (Ranjan et al. 2019). Efforts have also been made to identify genes related with soybean characteristics that may assist plants escape infection, such as less lodging, increased plant height, and altered plant maturity.

12.2.5.2 Crop Rotation

Crop rotation is unlikely to be extremely useful in the management of Sclerotinia stem rot (Mueller et al. 2002b). Sclerotia can survive in soil for up to 8 years, and *S. sclerotiorum* has a diverse host range. Crop rotation is unlikely to diminish sclerotial density in a field enough to regulate disease incidence, while one study found that corn and wheat could reduce the number of apothecia in a field (Gracia-Garza et al. 2002).

12.2.5.3 Cultural Control

Higher plant densities of more than 432,000 plants per hectare (175,000 plants per acre) have been shown to worsen the severity of Sclerotinia stem rot because they reduce airflow and increase moisture retention in the plant canopy, which favors the germination of apothecia and the spread of disease (Fall et al. 2018b). In comparison to row spacing between 25 and 38 cm (10–15 inch), wide row spacings of 76 cm (30 inch) reduced the illness severity index. Because the canopy closes more quickly in narrow row spacing, apothecial formation is encouraged and the likelihood of Sclerotinia stem rot growth is increased (Fall et al. 2018b). The use of irrigation has been shown to increase Sclerotinia stem rot in soybean, likely by providing the moisture that sclerotia need to germinate. According to a recent study, rye that has been rolled and crimped and planted with soybeans has a lower incidence of Sclerotinia stem rot because nonfunctional apothecia are produced (Pethybridge et al. 2019). It is unknown, however, how many additional cover crops may have a comparable impact or even how many might result in a rise in disease levels given the large host range of *S. sclerotiorum*.

12.2.5.4 Chemical Control

Quinone outside inhibitor (QoI), 2, 6-dinitro-anilines, DMI, SDHI, and methyl benzimidazole carbamates (MBC) are some of the registered fungicide classes for treating Sclerotinia stem rot (Willbur et al. 2019a, b). To prevent S. sclerotiorum from infecting senescing flowers, fungicides should be used between the R1 and R3 growth stages (Mueller et al. 2002a). Boscalid or picoxystrobin fungicide regimens produced the greatest disease reduction and the highest yield (Willbur et al. 2019b). When a disease severity index of 20–25 percent is established, yield losses start to happen, with significant losses starting at 68 percent (Willbur et al. 2019b). Intriguingly, the incidence of Sclerotinia stem rot has decreased in areas where the illness is prevalent as a result of the application of the herbicide lactofen for post-emergence broadleaf weed control. Application of lactofen decreases Sclerotinia stem rot incidence by delaying the closure of plant canopy and stimulating soybean production of glyceollin, a phytoalexin that has known antifungal activity (Kim et al. 2010).

12.2.5.5 Biological Control

S. sclerotiorum fungal pathogen *Coniothyrium minitans* infects sclerotia in the soil, which reduces the inoculum load or pathogen abundance (Huang and Hoes 1976). According to Smoliska and Kowalska (2018), Contans WG, a commercial formulation of *Coniothyrium minitans* from Bayer Crop Science, destroyed up to 95% of sclerotia while reducing disease severity by 68 percent. In field tests in Iowa, *Sporidesmium sclerotivorum*, a second potential biological control agent, reduced Sclerotinia stem rot. *Streptomyces lydicus* and *T. harzianum* are two other biological agents that demonstrate potential for lowering Sclerotinia stem rot incidence. These two agents were able to reduce disease severity by 43 and 38 percent and decompose sclerotia in the soil by 90 and 70 percent, respectively (Zeng et al. 2012).

12.2.5.6 Predictive Modeling

The most cost-effective approach is to only use fungicides when there is a high danger of contracting a disease. Farmers can choose the best time to apply fungicides by using epidemiological modeling of Sclerotinia stem rot growth (Willbur et al. 2019b). Many Sclerotinia spp. predictive models have been developed for a wide range of cropping systems, including canola, lettuce, and peanut (Clarkson et al. 2014). Based on research suggesting that the location of apothecia in a field is closely related to local disease incidence, the prediction of apothecia presence has been used to forecast disease potential for Sclerotinia stem rot control in soybean. The most recent Sclerotinia stem rot model was tested in a number of cities and states, and it is still being refined (Willbur et al. 2018). These models are critical in helping farmers choose fungicides.

12.2.6 Frogeye Leaf Spot

The fungus persists in form of mycelium in contaminated seed and crop debris. The seed's discoloration is linked to infection, and more discoloration is linked to poorer germination (Wise and Newman 2015). The ensuing seedling, if the infected seed is able to germinate, will frequently be stunted and have lesions developing on the cotyledons. On infected cotyledons, spores may form and act as inoculums to infect nearby plants' leaves. One to two weeks after the initial infection, lesions will start to form. Conidia can be generated within 48 hours of lesions being present in humid settings. Conidia are then spread by wind or raindrops to infect nearby plants and leaves. New infections are often particularly likely to affect young leaves. On mature leaves, infections produce fewer, less severe lesions. Additionally, lesions can form on stems and pods, which could result in seeds that are ill. This disease is polycyclic and can infect huge areas of fields over the course of a season in favorable conditions including high temperatures, high humidity, and numerous rainfall or irrigation episodes.

12.2.6.1 Genetic Resistance

Up until the late 1950s, when *C. sojina* race 2 first appeared, the adoption of resistant cultivars was quite successful in controlling frogeye leaf spot disease. Races 3 and 4 then debuted in North Carolina in the 1960s, and race 5 debuted in 1978, all after this. The discovery of 11 distinct races of C. sojina suggests greater genetic variety than previously anticipated (Mian et al. 2008). The number of races that are currently known to exist means that host resistance genes may not be efficient in preventing disease in all races. For better disease control, it is crucial to introduce various resistance genes into cultivars. Rcs1, Rcs2, and Rcs3 are the three resistance genes that now give resistance to commercially significant races of *C. sojina*, albeit each gene responds differently to each race. Race 1 is given resistance by Rcs1, Race 2 is given resistance by Rcs2, and all known races are given resistance by Rcs3. The Rcs3 gene was still effective in giving resistance to present isolates of *C. sojina*, according to Zhang and Bradley's (2017) findings, despite the fact that these genes were found more than 30 years ago.

12.2.6.2 Cultural Control

Since *C. sojina* has no known alternate hosts, crop rotations with other crops like corn, alfalfa, or small grains may help reduce the pathogen's ability to survive. Long crop rotations over several years may be required if the illness has been severe in some fields to avoid the possibility of it reemerging (Bradley et al. 2016). Crop residue may break down more quickly or completely with tillage, which would decrease the chance of *C. sojina* surviving (Bradley et al. 2016).

12.2.6.3 Chemical Control

When used between the R1 and R5 growth phases, foliar fungicides have demonstrated effectiveness in preventing and treating frogeye leaf spot and preserving production. For successful fungicide control of frogeye leaf spot, active components from fungicides other than the QoIs must be used. In traditionally tilled fields but not in no-till operations, foliar fungicide applications decreased the severity of frogeye leaf spot disease and boosted yields, indicating that cultural practices may interact with foliar fungicide applications (Mengistu et al. 2014).

12.2.6.4 Biological Control

It has been proposed that the *Bacillus amyloliquefaciens* strains BNM340 and BNM122 can control *C. sojina* biologically. The severity of the disease was less when administered to the seedlings' leaves than when it wasn't (Simonetti et al. 2012).

12.2.7 Bacterial Blight

One of the first foliar diseases that affect soybeans, bacterial blight is brought on by the pathogen *Pseudomonas savastanoi* pv. *glycinea*. Bacterial blight rarely results in significant yield loss. Young leaves are most vulnerable, therefore symptoms typically start in the top canopy. A yellow halo surrounds a cluster of tiny, angular, reddish-brown lesions. Large, asymmetrically formed dead patches are frequently produced as the disease advances and lesions frequently grow together. Older lesions frequently lose their centers, giving the impression that the leaves are torn. Rain and wind disseminate the bacteria, which overwinter in agricultural debris and seed. In favorable conditions bacteria can persist on leaf surfaces throughout the season and infect plants. Stomates (natural leaf apertures that allow air exchange) and wounds brought on by wind, hail, etc. are the sources of infection. As a result, outbreaks frequently follow rainstorms with strong winds. Warmer temperatures will prevent or inhibit the spread of disease, while cooler temperatures will encourage bacterial blight.

12.2.7.1 Integrated Disease Management

Using clean seeds is one of the most effective control measures for the bacterial blight of soybeans, which can be spread by infected seeds. Clean seeds can be purchased commercially at any of the world's major seed stores. Purchasing seeds from commercial suppliers has the benefit of ensuring the absence of pathogenic activity in the seeds. The effectiveness of crop rotation in reducing outbreaks of bacterial blight in soybean fields has also been demonstrated. Crop rotation reduces the quantity of both living and dead tissue that the pathogen can infect, which in turn reduces the overall incidence of the disease in a particular land region. Soybeans should be alternated with non-sensitive crops. Another strategy frequently employed by farmers to prevent disease outbreaks is avoiding susceptible soybean varieties. Young plants can also be successfully chemically controlled by being sprayed with a mixture of streptocycline and copper oxychloride, albeit this is not a frequent practice due to the increased expense of treatment. But in recent years, researchers have found that neem, ginger, garlic, and onion leaf extracts can also dramatically lessen the effects of bacterial blight in soybeans.

12.2.8 Bacterial Pustule

Xanthomonas citri pv. *glycines* Constantin et al. 2016 is the cause of underlying bacterial pustules. In the middle to end of the season, when it is warmer, this disease strikes. On the outer leaves of the mid- to upper canopy, lesions can be seen. Lesions begin as tiny, raised, pale green dots that later enlarge into broad, irregularly shaped diseased patches. Each lesion is surrounded by a greenish-yellow halo. In some lesions, typically on the lower leaf surface, a pustule may develop in the middle. Like bacterial blight, bacterial pustule won't make leaves wilt. Similar to bacterial blight, bacterial pustule survives the winter in crop residue and is spread to plants by splashing water from

TABLE 12.2
Comparison of bacterial disease characteristics of soybean

Characteristics	Bacterial Blight	Bacterial Pustule
Overwinters	Crop residue	Crop residue
Initial Symptoms	Yellow-brown spots with water-soaked underside of leaf lesion	Tiny pale green spots on leaves
Later Symptoms	Irregularly shaped lesions with Circular spots with elevated centers water-soaked tissue and yellow halo	

the ground or wind-driven rain. Additionally, the disease can spread while the leaves are damp during cultivation. The bacterium will get into the plant through wounds and naturally occurring holes. This disease is encouraged to develop by warm, rainy weather. Warm temperatures do not inhibit the development of bacterial pustules, unlike bacterial blight. Its ideal development temperature is between 86 and 92 degrees Fahrenheit.

12.2.8.1 Integrated Disease Management

In fields with a history of bacterial disease issues, cultivars that are not very vulnerable to the disease should be taken into consideration. Crop rotation is an effective method of avoiding inoculum from previously diseased plants. Tillage reduces the amount of inoculum available to infect plants in the spring by integrating crop residues. Cultivate only when the leaves are dry to prevent disease transmission. To be successful, copper fungicides must be used early in the disease cycle to suppress bacterial blight on soybeans (Table 12.2).

12.2.9 Soybean Mosaic Virus

A virus belonging to the plant viral genus Potyvirus is the soybean mosaic virus (SMV) (family Potyviridae). It mostly affects members of the Fabaceae plant family, while it has also been observed to affect other commercially significant crops. The illness known as soybean mosaic, which affects all regions of the world where soybeans are produced, is brought on by SMV. One of the most significant sources of proteins and edible oil is soy (*Glycine max*), and pathogenic infections are to blame for annual production losses. The severity of the symptoms varies depending on the host genotype, virus strain, age of the plant at the time of infection, and environment. Symptoms are typically more noticeable on young, fast-growing leaves. The tissue where the viral infection first appears and is localized is in the leaves. Plants that appear to be symptom-free to leaves that are badly mottled and malformed can all exhibit macroscopic signs. The majority of the infected cultivars develop mild stunting and display fewer, occasionally undersized and flattened pods devoid of hairs and seeds. Trifoliate leaves exhibit noticeable mottling and mosaic symptoms, with light and dark green regions that may subsequently develop into elevated or blistered patches along the major veins. Chlorosis, particularly in the spaces between the dark green regions, has also been noted as an SMV infection symptom. In some cultivars, necrotic local lesions that may subsequently converge into veinal necrosis, followed by yellowing and leaf abscission, can cause the leaves to appear curled or wavy. Due to the systemic spread of the viral infection, some strains can cause severe stunting, systemic necrosis, leaf yellowing, petiole and stem necrosis, terminal necrosis, and defoliation, which results in the death of the plant. Aphids are the primary means through which SMV is transmitted. The non-persistent transmission of SMV has been demonstrated in 32 aphid species from 15 distinct genera. This means that the virus is transferred through the aphid's stylet rather than being incorporated into the aphid's tissue. In terms of effective transmission, *Acyrthosiphon pisum*, *Aphis fabae*, *A. glycines*, *Myzus persicae*, and *Rhopalosiphum maidis* are the most significant species. The soybean aphid (*A. glycines*), which was recently introduced to North America, has raised serious concerns due to its high transmission efficiency. It has not been demonstrated, therefore, that the presence of the aphid and other SMV-transmitting, migratory, non-colonizing aphids has considerably enhanced the incidence of SMV in the area.

12.2.9.1 Integrated Disease Management

Preventive measures are the primary management techniques for preventing yield decline and serious plant damage. To prevent excessive vector populations when plants are still young, it is essential to purchase certified virus-free seeds and arrange your planting season carefully. The most significant method of SMV transmission is through infected seeds, which also act as the main source of infection for aphid dissemination later on. Aphid populations are higher when planting is delayed, which could enhance the risk of virus transmission to early seedlings. When compared to infection later in the life cycle, infection in the early growth stages has the greatest influence on yield loss and seed quality. It is possible to check seed lots for virus presence using molecular and serological methods. Utilizing resistant cultivars should be the foundation of the most efficient disease control. For SMV resistance, at least three naturally occurring separate loci (Rsv1, Rsv3, and Rsv4) have been found and mapped. In order to provide resistance against the SMV strains G1–G7, these three loci were chosen. The loci Rsc4, Rsc5, Rsc7, Rsc8, Rsc15, and Rsc20 have been designated in China to provide resistance against SMV-SC strains. Due to the considerable diversity of SMV, using a single resistance gene could be risky; therefore pyramiding the available sources of resistance is advised to control the disease. The levels of infection should be greatly reduced if aphid vectors can be controlled. However, no effective aphid control strategies have yet been created. As the aphids present at the time of spraying are killed, the field is quickly recolonized by winged aphids, and viral transmission resumes, insecticides are not thought to be helpful in limiting the transmission of SMV by aphids. While destroyed when they come into touch with insecticide residues on leaf surfaces, aphids are still capable of transmitting viruses while alive. Growers should avoid using insecticides below the soybean threshold because doing so will simply suppress the disease's vector and maybe exacerbate the viral issue.

12.3 SUMMARY AND CONCLUSION OF IDM IN SOYBEAN

For the successful decrease of a disease, each condition necessitates a particular, and frequently unique, combination of management measures as mentioned in Table 12.3.

The pathogens that cause diseases like frogeye leaf spot and charcoal rot may be able to survive winters in more northern soybean-growing regions if temperatures rise as projected. Similar to this, warmer soil temperatures might also offer many plant-parasitic nematodes a more hospitable environment. Phytophthora root and stem rot may become more common throughout the growing zone as precipitation levels rise. This rise in temperature may also alter the typical spectrum of several of these diseases or even lessen their prevalence or severity. For instance, in its current geographic distribution, rising temperatures may result in decreased apothecial germination and soybean infection by the Sclerotinia stem rot pathogen, *S. sclerotiorum*. The association between the presence of infections and warming circumstances supports prediction models' claims that plant pathogens are becoming more common worldwide (Delgado-Baquerizo et al. 2020). Agronomic pests like weeds, insects, diseases, and nematodes would certainly exert more pressure as a result of the overall rising temperature and rainfall. Certain soybean viruses may become more prevalent and economically significant as a result of increased insect vector pressure. Additionally, it is unknown how these weather-related incidents may affect other elements like genetic resistance or pharmacological efficacy.

TABLE 12.3
An overview of integrated management strategies for disease management that have been effective

	Cultural Practices				R Genes, Chemicals And Biological Control Practices			
Name of Pathogen / Disease	Crop rotation	Irrigation	Row spacing/ plant population	Tillage	Genetic resistance	Seed treatments	Mid-season fungicide spray	Biocontrol
Soybean Cyst Nematode	E	NE	NE	E	E	E	NE	E
Frog eye leaf spot	E	NE	NE	E	E	NE	E	NE
Charcoal rot	NE	E	NE	NE	NE	NE	NE	NE
Phytophthora root and stem rot	E	E	NE	E	E	E	NE	E
Sclerotinia stem rot	E	E	E	E	E	NE	E	E
Seedling diseases	E	NE	NE	E	E	E	NE	E

NE – Not effective strategies; E – effective/ moderate strategies.

REFERENCES

Ajayi-Oyetunde OO, Butts-Wilmsmeyer CJ, Bradley CA (2017) Sensitivity of *Rhizoctonia solani* to succinate dehydrogenase inhibitor and demethylation inhibitor fungicides. *Plant Dis* 101: 487–495.

Allen TW, Bradley CA, Sisson AJ, Byamukama E, Chilvers MI, Coker CM, Collins AA, Damicone JP, Dorrance AE, Dufault NS, et al (2017) Soybean yield loss estimates due to diseases in the United States and Ontario, Canada, from 2010 to 2014. *Plant Health Prog* 18(1):19–27.

Almeida AMR, Sosa-Gomez DR, Binneck E, Marin SRR, Zucchi MI, Abdelnoor RV, Souto ER (2008) Effect of crop rotation on specialization and genetic diversity of *Macrophomina phaseolina*. *Trop Plant Pathol* 33: 257–264.

Athow KL, Laviolette FA (1982) Rps6, a major gene for resistance to *Phytophthora megapserma* f. sp. *glycinea* in soybean. *Phytopathology* 72: 1564–1567.

Baird, RE, Watson CE, Scruggs M (2003) Relative longevity of *Macrophomina phaseolina* and associated mycobiota on residual soybean roots in soil. *Plant Dis* 87: 563–566.

Bebber DP, Ramotowski MAT, Gurr SJ (2013) Crop pests and pathogens move polewards in a warming world. *Nat Clim Change* 3: 985–988.

Beeman AQ, Tylka GL (2018) Assessing the effects of ILeVO and VOTiVO seed treatments on reproduction, hatching, motility, and root penetration of the soybean cyst nematode, *Heterodera glycines*. *Plant Dis* 102: 107–113.

Beeman AQ, Njus ZL, Pandey S, Tylka GL (2019) The effects of ILeVO and VOTiVO on root penetration and behavior of the soybean cyst nematode *Heterodera glycines*. *Plant Dis* 103: 392–397.

Bernard EC, Self LH, Tyler DD (1996) Fungal parasitism of soybean cyst nematode, *Heterodera glycines* (Nemata: Heteroderidae), in differing cropping-tillage regimes. *Appl Soil Ecol* 5: 57–70.

Birch RI, Haas JH, Crawford LG, Vaartaja O (2006) Soybean Phytophthora rot in southwestern Ontario. *Can Plant Dis Surv* 57: 68–70.

Bissonnette KM, Marett CC, Mullaney MP, Gebhart GD, Kyveryga PM, Mueller TA, Tylka GL (2020) Effects of ILeVO seed treatment on *Heterodera glycines* reproduction and soybean yield in small-plot and strip-trial experiments in Iowa. *Plant Dis* 104(11): 2914–2920.

Bond J, Bird G, Warner F, Chen S, Malvick D, Dorrance A, Harrison K, Esker P, MacGuidwin A, Faghihi J (2019) *Soybean cyst nematode management guide*. Fifth edition. https://soybean research info.com/pdf_docs/SCNGuide_5thEd.pdf.

Boscaiu M, Estruch M, Fita A, Plazas M, Prohens J, RodríguezBurruezo A, Verdeguer M, Vicente O (2019) Creating products and services in plant biotechnology. In: Matei F, Zirra D (eds.), *Introduction to biotech entrepreneurship: from idea to business*. Springer, Cham, Switzerland. pp. 251.

Bradley C, Chilvers M, Freije A, Giesler L, Mueller D, Sikora E, Sisson A, Smith D, Tenuta A, Wise K (2016) *Frogeye leaf spot*. Crop Protection Network: Soybean Disease Management. doi.org/10.31274/cpn-20190620-013

Brent KJ, Hollomon DW (2007) Fungicide resistance*: t*he assessment of risk. FRAC Monograph No.2 second, (revised) edition. Aimprint, UK. www.frac.info/docs/default-source/publications/monographs/monograph-2.pdf?sfvrsn=749d419a_8

Broders KD, Lipps PE, Paul PA, Dorrance AE (2007b) Characterization of *Pythium* spp. associated with corn and soybean seed and seedling disease in Ohio. *Plant Dis* 91: 727–735.

Broders KD, Lipps PE, Paul PA, Dorrance AE (2007a) Evaluation of *Fusarium graminearum* associated with corn and soybean seedling diseases in Ohio. *Plant Dis* 91: 1155–1160.

Constantin EC, Cleenwerck I, Maes M, Baeyen S, Van Malderghem C, De Vos P, Cottyn B (2016) Genetic characterization of strains named as *Xanthomonas axonopodis* pv. *dieffenbachiae* leads to a taxonomic revision of the *X. axonopodis* species complex. *Plant Pathol* 65: 792–806.

Chen S (2007) Suppression of *Heterodera glycines* in soils from fields with long-term soybean monoculture. *Biocontrol Sci Technol* 17: 125–134.

Chen S, Stienstra WC, Lueschen WE, Hoverstad TR (2001b) Response of *Heterodera glycines* and soybean cultivar to tillage and row spacing. *Plant Dis* 85: 311–316.

Cheng Y, Ma Q, Ren H, Xia Q, Song E, Tan Z, Li S, Zhang G, Nian H (2017) Fine mapping of a Phytophthora-resistance gene RpsWY in soybean (*Glycine max* L.) by high-throughput genome-wide sequencing. *Theor Appl Genet* 130: 1041–1051.

Choudhary DK (2011) Plant growth-promotion (PGP) activities and molecular characterization of rhizobacterial strains isolated from soybean (*Glycine max* L. Merr) plants against charcoal rot pathogen, *Macrophomina phaseolina*. *Biotechnol Lett* 33: 2287–2295.

Clarkson JP, Fawcett L, Anthony SG, Young C (2014) A model for *Sclerotinia sclerotiorum* infection and disease development in lettuce, based on the effects of temperature, relative humidity, and ascospore density. *PLoS One* 9: e94049.

Clemente TE, Cahoon EB (2009) Soybean oil: Genetic approaches for modification of functionality and total content. *Plant Phys* 151: 1030–1040.

Concibido VC, Diers BW, Arelli PR (2004) A decade of QTL mapping for cyst nematode resistance in soybean. *Crop Sci* 44: 1121–1131.

Cook DE, Lee TG, Guo X, Melito S, Wang K, Bayless AM, Wang J, Hughes TJ, Willis DK, Clemente TE, et al (2012) Copy number variation of multiple genes at Rhg1 mediates nematode resistance in soybean. *Science* 338: 1206–1209.

CPN (2019) Fungicide Efficacy for Control of Soybean Seedling Diseases. *SW, Washington, D.C.* vol 2: 20250-9410.

Creech JE, Westphal A, Ferris VR, Faghihi J, Vyn TJ, Santini JB, Johnson WG (2008) Influence of winter annual weed management and crop rotation on soybean cyst nematode (*Heterodera glycines*) and winter annual weeds. *Weed Sci* 56: 103–111.

Cummings JA, Bergstrom GC (2013) First report of charcoal rot caused by *Macrophomina phaseolina* in soybean in New York. *Plant Dis* 97: 1506–1506.

Da Silva MP, Klepadlo M, Gbur EE, Pereira A, Mason RE, Rupe JC, Bluhm BH, Wood L, Mozzoni LA, Chen P (2019) QTL mapping of charcoal rot resistance in PI 567562A soybean accession. *Crop Sci* 59: 474–479.

Danielson GA, Nelson BD, Helms TC (2004) Effect of Sclerotinia stem rot on yield of soybean inoculated at different growth stages. *Plant Dis* 88: 297–300.

Delgado-Baquerizo M, Guerra CA, Cano-Díaz C, Egidi E, Wang JT, Eisenhauer N, Singh BK, Maestre FT (2020) The proportion of soil-borne pathogens increases with warming at the global scale. *Nat Clim Chang* 10: 550–554.

Díaz Arias MM, Leandro LF, Mukvold GP (2013b) Aggressiveness of *Fusarium* species and impact of root infection. *Phyotpathology* 103: 822–832.

Díaz Arias MM, Mukvold GP, Ellis ML, Leandro LFS (2013a) Distribution and frequency of *Fusarium* species associated with soybean roots in Iowa. *Plant Dis* 97: 1557–1562.

Dorrance AE (2018) Management of *Phytophthora sojae* of soybean: a review and future perspectives. *Can J Plant Pathol* 40: 210–219.

Dorrance AE, Kleinhenz MD, McClure SA, Tuttle NT (2003) Temperature, moisture, and seed treatment effects on *Rhizoctonia solani* root rot of soybean. *Plant Dis* 87: 533–538.

Dorrance AE, Robertson AE, Cianzo S, Giesler LJ, Grau CR, Draper MA, Tenuta AU, Anderson TR (2009) Integrated management strategies for *Phytophthora sojae* combining host resistance and seed treatments. *Plant Dis* 93: 875–882.

El Bakali MA, Martin MP, García FF, Moret BA, Nadal PM (2000) First report of *Rhizoctonia solani* AG-3 on potato in Catalonia (NE Spain). *Plant Dis* 84: 806.

Ellis ML, Díaz Arias MM, Cruz Jimenez DR, Munkvold GP, Leandro LF (2013a) First report of *Fusarium commune* causing damping off, seed rot, and seedling root rot on soybean (*Glycine max*) in the United States. *Plant Dis* 97: 284.

Fall ML, Willbur JF, Smith DL, Byrne AM, Chilvers MI (2018b) Spatiotemporal distribution pattern of *Sclerotinia sclerotiorum* apothecia is modulated by canopy closure and soil temperature in an irrigated soybean field. *Plant Dis* 102: 1794–1802.

Fan A-Y, Wang X-M, Fang X-P, Wu X-F, Zhu Z-D (2009) Molecular identification of Phytophthora resistance gene in soybean cultivar Yudou 25. *ACTA Agron Sin* 35: 1844–1850.

Fernandez Bidondo L, Almasia N, Bazzini A, Colombo R, Hopp E, Vazquez-Rovere C, Godeas A (2019) The overexpression of antifungal genes enhances resistance to *Rhizoctonia solani* in transgenic potato plants without affecting arbuscular mycorrhizal symbiosis. *Crop Prot* 124: 104837.

Flor HH (1955) Host parasite interaction in flax rust-its genetics and other implications. *Phytopathology* 45: 680–685.

Francl AJ (1998) Soil organic carbon sequestration and agricultural greenhouse gas emissions in the southeastern USA. *Soil Tillage Res*. 83:120–147.

Garrett KA, Nita M, De Wolf ED, Esker PD, Gomez-Montano L, Sparks AH (2016) Plant pathogens as indicators of climate change. In: Letcher TM (ed.), *Climate Change,* 2nd ed. Elsevier B.V., Amsterdam, Netherlands, pp. 325–338.

Grabau ZJ, Vetsch JA, Chen S (2017) Effects of fertilizer, nematicide, and tillage on plant-parasitic nematodes and yield in corn and soybean. *Agronomy J* 109: 1651–1662.

Gracia-Garza JA, Neumann S, Vyn TJ, Boland GJ (2002) Influence of crop rotation and tillage on production of apothecia by *Sclerotinia sclerotiorum. Can J Plant Pathol* 24: 137–143.

Gupta GK, Sharma SK, Ramteke R (2012) Biology, epidemiology and management of the pathogenic fungus *Macrophomina phaseolina* (Tassi) Goid with special reference to charcoal rot of soybean (*Glycine max* (L.) Merrill). *Phytopathology* 160: 167–180.

Han G, Kandel YR, Leandro LFS, Helmers MJ, Schoot LR, Mueller DS (2017) Influence of drainage on soybean seedling health. *J Water Soil Conserv* 72: 266–271.

Harbach CJ (2019) Investigating the interactions between cover crops and the soybean cyst nematode through lab, greenhouse, and field studies. Ph.D. dissertation, Iowa State University, Ames, IA.

Huang HC, Hoes JA (1976) Penetration and infection of *Sclerotinia sclerotiorum* by *Coniothyrium minitans*. *Can J Bot* 54: 406–410.

(IPCC) Intergovernmental Panel on Climate Change (2014) *Climate Change 2014: Synthesis Report. Contribution of Working Groups I, II and III to the Fifth Assessment Report of the Intergovernmental Panel on Climate Change* [Core Writing Team, R.K. Pachauri and L.A. Meyer (eds.)]. IPCC, Geneva, Switzerland, pp. 151.

Jiang B, Cheng Y, Cai Z, Li M, Jiang Z, Ma R, Yuan Y, Xia Q, Nian H (2020) Fine mapping of a Phytophthora-resistance locus RpsGZ in soybean using genotyping-by-sequencing. *BMC Genomics* 21: 280.

John RP, Tyagi RD, Prevost D, Brar SK, Pouleur S, Surampalli RY (2010) Mycoparasitic *Trichoderma viride* as a biocontrol agent against *Fusarium oxysporum* f. sp. *adzuki* and *Pythium arrhenomanes* and as a growth promoter of soybean. *Crop Prot* 29: 1452–1459.

Juroszek P, Racca P, Link S, Farhumand J, Kleinhenz B (2020) Overview on the review articles published during the past 30 years relating to the potential climate change effects on plant pathogens and crop disease risks. *Plant Pathol* 69: 179–193.

Kendig SR, Rupe JC, Scott HD (2000) Effect of irrigation and soil water stress on densities of *Macrophomina phaseolina* in soil and roots of two soybean cultivars. *Plant Dis* 84: 895–900.

Khan SN (2007) *Macrophomina phaseolina* as a causal agent for charcoal rot of sunflower. *Mycopath* 5: 111–118.

Kim HJ, Suh H-J, Lee CH, Kim JH, Kang SC, Park S, Kim J-S (2010) Antifungal activity of glyceollins isolated from soybean elicited with *Aspergillus sojae*. *J Agr Food Chem* 58: 9483–9487.

Kunwar IK, Singh T, Machado CC, Sinclair JB (1986) Histopathology of soybean seed and seedling infection by *Macrophomina phaseolina*. *Phytopathology* 76: 532–535.

Leslie JF, Pearson CAS, Nelson PE, Toussoun TA (1990) *Fusarium* species from corn, sorghum, and soybean fields in the central and eastern United States. *Phytopathology* 80: 343–350.

Li L, Lin F, Wang W, Ping J, Fitzgerald JC, Zhao M, Li S, Sun L, Cai C, Ma J (2016) Fine mapping and candidate gene analysis of two loci conferring resistance to *Phytophthora sojae* in soybean. *Theor Appl Genet* 129: 2379–2386.

Li-ming Z, Shuang L, Zhe S, Jing H, Qiu-ming C, Bing S, Jun-jie D, Wei-ting L, Jing-zhi W (2016) Impact of crop rotation on pathotype and genetic structure of *Phytophthora sojae* in fields. *J Northeast Agr Uni* 23: 1–12.

Lin F, Wani SH, Collins P, Wen Z, Gu C, Chilvers MI, Wang D (2018) Mapping quantitative trait loci for tolerance to *Pythium irregulare* in soybean. *G3* 8: 3155–3161.

Liu D, Li K, Hu J, Wang W, Liu X, Gao Z (2019) Biocontrol and action mechanism of *Bacillus amyloliquefaciens* and *Bacillus subtilis* in soybean Phytophthora blight. *Int J Mol Sci* 20: 2908.

Livingston JE (1945) Charcoal rot of corn and sorghum. *Nebr Agric Exp Stn Bull* 136. Lincoln, NE.

Masonbrink R, Maier TR, Muppirala U, Seetharam AS, Lord E, Juvale PS, Schmutz J, Johnson NT, Korkin D, Mitchum MG (2019a) The genome of the soybean cyst nematode (*Heterodera glycines*) reveals complex patterns of duplications involved in the evolution of parasitism genes. *BMC Genomics* 20: 119.

Masonbrink R, Maier TR, Seetharam AS, Juvale PS, Baber L, Baum TJ, Severin AJ (2019b) SCNBase: a genomics portal for the soybean cyst nematode (*Heterodera glycines*). *Database* 2019: 1–9.

Matthiesen RL, Ahmand AA, Robertson AE (2016) Temperature affects aggressiveness and fungicide sensitivity of four *Pythium* spp. that cause soybean and corn damping off in Iowa. *Plant Dis* 100: 583–591.

McCaghey M, Willbur J, Smith DL, Kabbage M (2019) The complexity of the *Sclerotinia sclerotiorum* pathosystem in soybean: virulence factors, resistance mechanisms, and their exploitation to control sclerotinia stem rot. *Trop Plant Pathol* 44: 12–22.

McCarville MT, Marett CC, Mullaney MP, Gebhart GD, Tylka GL (2017) Increase in soybean cyst nematode virulence and reproduction on resistant soybean varieties in Iowa from 2001 to 2015 and the effects on soybean yields. *Plant Health Prog* 18: 146–155.

McNew GL (1960) The nature, origin, and evolution of parasitism. In Plant Pathology: An Advanced Treatise, ed. JG Horsfall, AE Dimond, pp. 19–69. New York: Academic Press.

Mengistu A, Kelly HM, Bellaloui N, Arelli PR, Reddy KN, Wrather AJ (2014) Tillage, fungicide, and cultivar effects on frogeye leaf spot severity and yield in soybean. *Plant Dis* 98: 1476–1484.

Mengistu A, Reddy KN, Zablotowicz RM, Wrather AJ (2009) Propagule densities of *Macrophomina phaseolina* in soybean tissue and soil as affected by tillage, cover crop, and herbicide. *Plant Health Prog* 10. doi:10.1094/PHP-2009-0130-01-RS.

Mengistu A, Smith JR, Ray JD, Bellaloui N (2011b) Seasonal progress of charcoal rot and its impact on soybean productivity. *Plant Dis* 95: 1159–1166.

Mengistu A, Wrather A, Rupe JC (2015) Charcoal rot. In: Hartman GL, Rupe JC, Sikora EJ, Domier LL, Steffey JA (eds.), *Compendium of soybean diseases*, 5th ed. APS Press, St. Paul, MN. pp. 67–69.

Mian MAR, Missaoui AM, Walker DR, Phillips DV, Boerma HR (2008) Frogeye leaf spot of soybean: a review and proposed race designations for isolates of *Cercospora sojina. Hara Crop Sci* 48: 14–24.

Miao J, Chi Y, Lin D, Tyler BM, Liu X (2018) Mutations in ORP1 conferring oxathiapiprolin resistance confirmed by genome editing using CRISPR/Cas9 in *Phytophthora capsici* and *P.* sojae. *Phytopathology* 108: 1412–1419.

Mitchum MG, Hussey RS, Baum TJ, Wang X, Elling AA, Wubben M, Davis EL (2013) Nematode effector proteins: An emerging paradigm of parasitism. *New Phytol* 199: 879–894.

Mofijur KAV, Balasuriya BMCM, Arachchige USPR (2013) Comparison of performance of various homogeneous alkali catalysts in transesterification of waste cooking oil. *Asian J. Chem. 34*: 3157–3161.

Monkiedje A, Spiteller M (2002) Effects of the phenylamide fungicides, mefenoxam and metalaxyl, on the microbiological properties of a sandy loam and a sandy clay soil. *Biol Fert Soils* 35: 393–398.

Moorman TB (1989) A review of pesticide effects on microorganisms and microbial processes related to soil fertility. *J Prod Agric* 2: 14–23.

Mueller DS, Dorrance AE, Derksen RC, Ozkan E, Kurle JE, Grau CR, Gaska JM, Hartman GL, Bradley CA, Pedersen WL (2002a) Efficacy of fungicides on *Sclerotinia sclerotiorum* and their potential for control of Sclerotinia stem rot on soybean. *Plant Dis* 86: 26–31.

Mueller DS, Hartman GL, Pedersen WL (2002b) Effect of crop rotation and tillage system on Sclerotinia stem rot on soybean. *Can J Plant Pathol* 24: 450–456.

Murillo-Williams A, Pedersen P (2008) Early incidence of soybean seedling pathogens in Iowa. *Agron J* 100: 1481–1487.

Nelson BD (2015) Fusarium blight or wilt, root rot, and pod and collar rot. In: Hartman GL, Rupe JC, Sikora EJ, Domier LL, Steffey JA (eds.), *Compendium of soybean diseases and pests*, 5th ed. APS Press, St. Paul, MN. pp. 69–73.

Niblack TL, Arelli PR, Noel GR, Opperman CH, Orf JH, Schmitt DP, Shannon JG, Tylka GL (2002) A revised classification scheme for genetically diverse populations of *Heterodera glycines*. *J Nematol* 34: 279–288.

Niblack TL, Lambert KN, Tylka GL (2006) A model plant pathogen from the kingdom Animalia: *Heterodera glycines*, the soybean cyst nematode. *Annu Rev Phytopathol* 44: 283–303.

Noel ZA, Sang H, Roth MG, Chilvers MI (2019) Convergent evolution of C239S mutation in Pythium spp. β-tubulin coincides with inherent insensitivity to ethaboxam and implications for other *Peronosporalean oomycetes*. *Phytopathology* 109: 2087–2095.

Parra G, Ristaino JB (2001) Resistance to mefenoxam and metalaxyl among field isolates of *Phytophthora capsici* causing Phytophthora blight of bell pepper. *Plant Dis* 85: 1069–1075.

Paulitz TC, Anas O, Fernando DG (1992) Biological control of Pythium damping-off by seed-treatment with *Pseudomonas putida*: relationship with ethanol production by pea and soybean seeds. *Biocontrol Sci Techn* 2: 193–201.

Pawlowski ML, Hill CB, Hartman GL (2015) Resistance to charcoal rot identified in ancestral soybean germplasm. *Crop Sci* 55: 1230–1235.

Peng Q, Wang Z, Fang Y, Wang W, Cheng X, Liu X (2019) Point mutations in the β-tubulin of *Phytophthora sojae* confer resistance to ethaboxam. *Phytopathology* 109: 2096–2106.

Pethybridge SJ, Brown BJ, Kikkert JR, Ryan MR (2019) Rolled– crimped cereal rye residue suppresses white mold in no-till soybean and dry bean. *Renew Agr Food Syst* 34: 1–9.

Ping J, Fitzgerald JC, Zhang C, Lin F, Bai Y, Wang D, Aggarwal R, Rehman M, Crasta O, Ma J (2016) Identification and molecular mapping of Rps11, a novel gene conferring resistance to *Phytophthora sojae* in soybean. *Theor Appl Genet* 129: 445–451.

Poromarto SH, Nelson BD (2010) Evaluation of northern-grown crops as hosts of soybean cyst nematode. *Plant Health Prog* 11. doi:10.1094/ PHP-2010-0315-02-RS.

Prein AF, Liu C, Ikeda K, Trier SB, Rasmussen RM, Holland GJ, Clark MP (2017) Increased rainfall volume from future convective storms in the US. *Nat Clim Change* 7: 880–884.

Radmer L, Anderson G, Malvick DM, Kurle JE, Rendahl A, Mallik A (2017) *Pythium, Phytophthora*, and *Phytopythium* spp. associated with soybean in Minnesota, their relative aggressiveness on soybean and corn, and their sensitivity to seed treatment fungicides. *Plant Dis* 101: 62–72.

Raghuvanshi RS, Bisht K (2010) Uses of soybean: products and preparation. In: Singh G (eds.), *The soybean: botany, production, and uses*. CABI International, Wallingford, United Kingdom. pp. 404–426.

Ranjan A, Westrick NM, Jain S, Piotrowski JS, Ranjan M, Kessens R, Stiegman L, Grau CR, Conley SP, Smith DL, Kabbage M (2019) Resistance against *S. sclerotiorum* in soybean involves reprogramming of the phenylpropanoid pathway and up-regulation of antifungal activity targeting ergosterol biosynthesis. *Plant Biotechnol J* 17: 1567–1581.

Reis EM, Boaretto C, Danelli ALD (2014) *Macrophomina phaseolina*: density and longevity of microsclerotia in soybean root tissues and free on the soil, and competitive saprophytic ability. *Summa Phytopathol* 40: 128–133.

Reznikov S, Vellicce GR, González V, de Lisi V, Castagnaro AP, Ploper LD (2016) Evaluation of chemical and biological seed treatments to control charcoal rot of soybean. *J Gen Plant Pathol* 82: 273–280.

Riga E, Welacky T, Potter J, Anderson T, Topp E, Tenuta A (2001) The impact of plant residues on the soybean cyst nematode, *Heterodera glycines*. *Can J Plant Pathol* 23: 168–173.

Rincker K, Cary T, Diers BW (2017) Impact of soybean cyst nematode resistance on soybean yield. *Crop Sci* 57: 1373–1382.

Rizvi SSA, Yang XB (1996) Fungi associated with soybean seedling disease in Iowa. *Plant Dis* 80: 57–60.

Rojas JA, Jacobs JL, Napieralski S, Karaj B, Bradley CA, Chase T, Esker PD, Giesler LJ, Jardine DJ, Malvick DK (2017a) Oomycete species associated with soybean seedlings in North America— Part I: identification and pathogenicity characterization. *Phytopathology* 107: 280–292.

Rojas JA, Jacobs JL, Napieralski S, Karaj B, Bradley CA, Chase T, Esker PD, Giesler LJ, Jardine DJ, Malvick DK (2017b) Oomycete species associated with soybean seedlings in North America— Part II: diversity and ecology in relation to environmental and edaphic factors. *Phytopathology* 107: 239–304.

Rojas JA, Witte A, Jacobs JL, Noel ZA, Chilvers MI (2019) Diversity and characterization of oomycetes associated with corn seedlings in Michigan. *Phytobiomes* 3: 224–234.

Rosso ML, Rupe JC, Chen P, Mozzoni LA (2008) Inheritance and genetic mapping of resistance to pythium damping-off caused by *Pythium aphanidermatum* in 'Archer' soybean. *Crop Sci* 48: 2215–2222.

Roth MG, Jacobs JL, Napieralski S, Byrne AM, Stoffer-Hopkins A, Warner F, Chilvers MI (2020) Fluopyram suppresses population densities of *Heterodera glycines* in field and greenhouse studies in Michigan. *Plant Dis*. 104: 1305–1311.

Rothrock CS, Avanzato MV, Rupe JC (2015) Pythium seed rot, damping-off, and root rot. In: Hartman GL, Rupe JC, Sikora EJ, Domier LL, Davis EL, Steffey KL (eds.), *Compendium of soybean diseases*, 5th ed. APS Press, St. Paul, MN. pp. 76–79

Schmitthenner AF (2000) Phytophthora rot of soybean. *Plant Health Prog* 1. doi:10.1094/PHP-2000-0601-01-HM.

Scott B, Singh AK (2018) Breeding charcoal resistance into soybeans. *Farm Progress Rep* 2017: 9–10.

Scott K, Eyre M, McDuffee D, Dorrance AE (2020) The efficacy of ethaboxam as a soybean seed treatment toward *Phytophthora, Phytopythium*, and *Pythium* in Ohio. *Plant Dis* 104: 1421–1432.

Senthilkumar M, Swarnalakshmi K, Govindasamy V, Lee YK, Annapurna K (2009) Biocontrol potential of soybean bacterial endophytes against charcoal rot fungus, *Rhizoctonia bataticola. Curr Microbiol* 58: 288–293.

Shahoveisi F, del Río Mendoza E (2020) Effect of wetness duration and incubation temperature on development of ascosporic infections of *Sclerotinia sclerotiorum. Plant Dis* 104: 1817–1823.

Simonetti E, Carmona M, Scandiani M, García A, Luque A, Correa O, Balestrasse K (2012) Evaluation of indigenous bacterial strains for biocontrol of the frogeye leaf spot of soya bean caused by *Cercospora sojina. Lett Appl Microbiol* 55: 170–173.

Slack DA, Riggs RD, Hamblen ML (1972) The effect of temperature and moisture on the survival of *Heterodera glycines* in the absence of a host. *J Nematol* 4: 263–266.

Smolińska U, Kowalska B (2018) Biological control of the soil-borne fungal pathogen *Sclerotinia sclerotiorum. J Plant Pathol* 100: 1–12.

Strom N, Hu W, Haarith D, Chen S, Bushley KE (2020) Corn and soybean host root endophytic fungi with toxicity towards the soybean cyst nematode. *Phytopathology* 110: 603–614.

Sun J, Li L, Zhao J, Huang J, Yan Q, Xing H, Guo N (2014) Genetic analysis and fine mapping of RpsJS, a novel resistance gene to *Phytophthora sojae* in soybean [*Glycine max* (L.) Merr.]. *Theor Appl Genet* 127: 913–919.

Sun S, Wu XL, Zhao JM, Wang YC, Tang QH, Yu DY, Gai JY, Xing H (2011) Characterization and mapping of RpsYu25, a novel resistance gene to *Phytophthora sojae. Plant Breed* 130: 139–143.

Swaminathan S, Abeysekara NS, Knight JM, Liu M, Dong J, Hudson ME, Bhattacharyya MK, Cianzio SR (2018) Mapping of new quantitative trait loci for sudden death syndrome and soybean cyst nematode resistance in two soybean populations. *Theor Appl Genet* 131: 1047–1062.

Thangavelu R, Palaniswami A, Velazhahan R (2004) Mass production of *Trichoderma harzianum* for managing fusarium wilt of banana. *Agr Ecosyst Environ* 103: 259–263.

Tsror L (2010) Biology, epidemiology and management of *Rhizoctonia solani* on potato. *J Phytopathol* 158: 649–658.

Tyler BM (2007) *Phytophthora sojae*: Root rot pathogen of soybean and model oomycete. *Mol Plant Path* 8: 1–8.

Uri ND (2000) Perceptions on the use of no-till farming in production agriculture in the United States: An analysis of survey results. *Agric Ecosyst Environ* 77: 263–266.

(USDA) U.S. Department of Agriculture Farm Service Agency (2019) *Acreage data as of December 1, 2019*. USDA, Beltsville, MD.

(USDA) U.S. Department of Agriculture National Agricultural Statistics Service (NASS) (2020) *Soybeans – Price Received, Measured in $ / Bu*. USDA, Beltsville, MD.

Van den Boogert PHJF, Deacon JW (1994) Biotrophic mycoparasitism by *Verticillium biguttatum* on *Rhizoctonia solani*. *Eur J Plant Pathol* 100: 137–156.

Vitti AJ, Rezende Neto U da R, de Araújo FG, Santos LC, Barbosa KAG, da Rocha MR (2014) Effect of soybean seed treatment with abamectin and thiabendazole on *Heterodera glycines*. *Nematropica* 44: 74–80.

Vuong TD, Diers BW, Hartman GL (2008) Identification of QTL for resistance to Sclerotinia stem rot in soybean plant introduction 194639. *Crop Sci* 48: 2209.

Wang J, Niblack TL, Tremain JA, Wiebold WJ, Tylka GL, Marett CC, Noel GR, Myers O, Schmidt ME (2003) Soybean cyst nematode reduces soybean yield without causing obvious above ground symptoms. *Plant Dis* 87: 623–628.

Willbur JF, Mamadou L, Christopher Bloomingdale F, Byrne AM, Chapman SA, Isard SA, Magarey RD, McCaghey MM, Mueller BD, Russo JM, Schlegel J, Chilvers MI, Mueller DS, Kabbage M, Smith DL (2018) Weather-based models for assessing the risk of *Sclerotinia sclerotiorum* apothecial presence in soybean (*Glycine max*) fields. *Plant Dis* 102:1, 73–84.

Willbur JF, McCaghey MM, Kabbage M, Smith DL (2019a) An overview of the *Sclerotinia sclerotiorum* pathosystem in soybean: Impact, fungal biology, and current management strategies. *Trop Plant Pathol* 44: 3–11.

Willbur JF, Mitchell PD, Fall ML, Byrne AM, Chapman SA, Floyd CM, Bradley CA, Ames KA, Chilvers MI, Kleczewski NM (2019b) Meta-analytic and economic approaches for evaluation of pesticide impact on Sclerotinia stem rot control and soybean yield in the North Central United States. *Phytopathology* 109: 1157–1170.

Wise KA, Newman ME (2015) Frogeye leaf spot, pp. 43–45. In: Hartman GL, Rupe JC, Sikora EJ, Domier LL, Steffey JA (eds.), *Compendium of soybean diseases and pests*, 5th ed. APS Press, St. Paul, MN. p. 256.

Zeng W, Kirk W, Hao J (2012) Field management of Sclerotinia stem rot of soybean using biological control agents. *Biol Control* 60: 141–147.

Zhang G, Bradley CA (2017) Comparison of quinone outside inhibitor fungicide-resistant and -sensitive isolates of *Cercospora sojina*. *Crop Prot* 94: 59–63.

Zhang J, Xia C, Wang X, Duan C, Sun S, Wu X, Zhu Z (2013b) Genetic characterization and fine mapping of the novel Phytophthora resistance gene in a Chinese soybean cultivar. *Theor Appl Genet* 126: 1555–1561.

Zhao C, Liu B, Piao S, Wang X, Lobell DB, Huang Y, Huang M, Yao Y, Bassu S, Ciais P (2017) Temperature increase reduces global yields of major crops in four independent estimates. *Proc Natl Acad Sci USA* 114: 9326–9331.

Zhong C, Li Y, Sun S, Duan C, Zhu Z (2019) Genetic mapping and molecular characterization of a broad-spectrum *Phytophthora sojae* resistance gene in chinese soybean. *Int J Mol Sci* 20: 1809.

Zhong C, Sun S, Li Y, Duan C, Zhu Z (2018) Next-generation sequencing to identify candidate genes and develop diagnostic markers for a novel Phytophthora resistance gene, RpsHC18, in soybean. *Theor Appl Genet* 131: 525–538.

13 Plant Nematode Diseases and Their Innovative Management Strategies in Major Spice Crops

Rashid Pervez

13.1 INTRODUCTION

India is known as the "capital of spices" and has historically played a significant role in the production, use, and export of spices. These high-value but low-volume commodities are heavily relied upon by the global food industry to produce a wide range of goods with distinctive variants to meet consumer needs. Additionally, spice crops provide superior natural flavors and colors to inexpensive synthetic ones, which is significant to consumers who are concerned about their health. Spice demand is anticipated to increase at a rate of approximately 3.19 percent per year, which is a little faster than the rate of population growth in the world. The majority of India's states that produce spices are Kerala, Andhra Pradesh, Gujarat, Maharashtra, West Bengal, Karnataka, Tamil Nadu, Orissa, Madhya Pradesh, and Rajasthan, while all other states and union territories also produce spices.

India has a long history of producing, using, and exporting spices. In India, the area devoted to growing spices was 7,37,876 ha, with a yield of 20,326 kg/ha and productivity of 14,99,826 tonnes (FAOSTAT 2020). There were no significant changes in production or area under spices the following year. According to projections, India will produce 10536 thousand tonnes from 4487 thousand hectares in 2020–21 (DSAD 2022).

The economic significance of spice crops is substantial. They serve as a critical source of income for millions of individuals, particularly in tropical and subtropical regions. India exports spices to bolster its agricultural and overall economic growth. This economic sector has been on a steady incline due to the growing demand for diverse and exotic flavors in the food and beverage industry, as well as an increasing interest in traditional medicine and herbal remedies.

Spices such as black pepper, cardamom and turmeric have long been treasured for their ability to enhance the taste of dishes, and they have often been celebrated for their medicinal properties. The use of spices is deeply rooted in the traditions and cultures of many regions, infusing distinct and delightful flavors into local cuisines. Moreover, the cultivation and trade of these spices have significantly influenced historical events and global geopolitics.

The production of spices in India has been decreasing for several reasons. Among them plant nematodes causing diseases play a vital role in significant yield losses, particularly in major Indian spices like black pepper, cardamom, and turmeric (Pervez 2018a). This chapter focuses on the diseases affecting black pepper, cardamom, and turmeric crops and their innovative management strategies.

DOI: 10.1201/9781032631578-13

13.2 BLACK PEPPER (*PIPER NIGRUM* L.)

One of the biggest producers, consumers, and exporters of black pepper worldwide is India. Black pepper is mostly grown in Kerala, Karnataka, and Tamil Nadu, with Maharashtra, the North Eastern provinces, and the Andaman and Nicobar Islands having a smaller part. The berries of the plant are used as a spice. The crop was grown on approximately 2,59,18 hectares and produced 65,000 tonnes in 2019–20 (SPICEBOARD 2021). This crop of spices is quite lucrative and is regularly sold to important Western nations. Several plant nematodes affect the black pepper crops (Pervez and Eapen 2015, 2016; Pervez et al. 2016), and rice root nematode *Hirschmanniella* sp. has also been reported in black pepper rhizospheres from Kerala (Pervez et al. 2017). Black pepper is harmed by disease caused by plant nematodes. These diseases can cause severe crop loss if ignored if prompt and appropriate plant protection measures are not taken, or if no timely and effective plant care is provided.

13.2.1 Slow Decline Disease

Black pepper's slow decline disease is brought on by *R. similis*, *M. incognita*, or *Phytophthora capsici*, either singularly or in combination, damaging feeder roots. Slow decline describes a gradual fall in the vine's vitality and productivity that results in death over a few years. By developing small, elongated lesions on the young, sensitive roots, the nematodes seriously damage the roots. The primary signs include pale yellow, whitish, discolored leaves, the typical orange to purple color lesion on young roots, extensive root system rot, main roots devoid of fine feeder roots that rot quickly, extensive necrosis of longer lateral roots, and yellow patches that later turn into barren standards that have lost their vines or standards supporting dead vines. These signs are more obvious when the soil is less damp. In general, foliar yellowing and defoliation were lowest in July and highest from April to May. Symptoms are well pronounced when soil moisture is reduced and the plant dies within 3–5 years (Bhai et al. 2017).

13.2.1.1 Management

Severely affected vines should be pulled from the plantation and terminated when a high plant nematode population is found, as it is impossible to recover them. The base of the plant should be treated with carbofuran* 3 G @ 50 g or phorate* 10 G @ 15 g before planting. Nematode-free rooted cuttings grown in fumigated or steam-sterilized nursery mixture should be utilized for field planting. Phorate* 10 G @ 30 g or carbofuran* 3 G @ 100 g/vine should be applied throughout May/June (when the southwest monsoon begins) and September/October (Thomas and Rajeev 2015a). Apply aqueous leaf extract of *Strychnos nuxvomica* and its constituents, brucine and strychnine to control the *R. similis* (Pervez et al. 2012). Drenching of carbosulfan (Marshal) 25 EC (0.1%) @ 50 ml/polybag containing 1.5 kg. potting mixture is recommended for the management of plant nematodes in black pepper nurseries (Pervez and Eapen 2021) (*non-banned zones).

13.2.2 Root-knot Disease

The root system of the black pepper plant develops large galls, or "knots," as a result of root-knot nematodes (RKN) feeding on plant roots. The condition is known as a root-knot disease because it is brought on by *Meloidogyne* spp. Black pepper fields are where Kerala's two RKN species, *M. incognita*, and *M. javanica*, are most frequently found. Juveniles of the second stage of the RKN are often found in soils. They are vermiform and infectious phases. The root system of a plant infected with RKN is heavily galled, has egg masses with females stuck deep inside the roots, and has few galls in many cultivars compared to cultivars that have smoother, larger galls. Even though the majority of RKN damage takes place underground, many indicators are visible above ground.

Lower agricultural yields were seen in the fields where severely affected plant leaves displayed yellowing, chlorosis, and stunting. In April and May as well as in December and January, RKN populations in black pepper roots reach their maximum levels. Low soil temperature, sufficient soil moisture, and the availability of young, delicate roots throughout September and October all contribute to the population's growth. On the other hand, the nematode is present all year long. RKN populations were impacted by several variables, including temperature and rainfall (Devasahayam et al. 2012a).

13.2.2.1 Management

In regions where root-knot disease is a serious problem, cuttings of the resistant cultivar "Pournami" can be planted. *Pochonia chlamydosporia* or *Trichoderma harzianum*, for example, can be administered twice a year @ 50 grams per vine as biocontrol agents (during April–May and September–October). The substrate must have a fungal load of 10^8 cfu/g (Eapen et al. 2005). Nematicides should be applied by gently raking the soil in the basin of the vine without damaging the root system, then spreading the nematicide uniformly throughout the basin and quickly covering it with soil. It should be ensured that the soil is sufficiently moist at the time of nematicide treatment. As soon as the disease first appears, control measures should be put in place. *S. nuxvomica* leaf and its constituents were effective against *M. incognita* (Leela et al. 2012).

13.3 CARDAMOM (*ELETTARIA CARDAMOMUM* MATON)

The most significant spice crop is cardamom, also referred to as the "queen of spices," and it is extensively grown in India's Western Ghats. The Zingiberaceae family of the natural order Scitaminae includes the dried capsule of commercial cardamom. The crop was grown on roughly 69,190 hectares in 2020–21 and produced 22,520 tonnes (SPICEBOARD 2021). This crop of spices is quite lucrative and is regularly sold to important Western nations. Cardamom is harmed by numerous viral, nematode-infesting, and fungal diseases. Several causes have contributed to the decline in cardamom production in India during the past few years. Nematode-infested illnesses are one of these primary restrictions, which significantly reduce yields (Devasahayam et al. 2012b). The following is a description of the plant nematode that causes root-knot diseases that are significant economically.

13.3.1 Root-knot Disease

The root-knot nematode *M. incognita* is a serious problem and causes root-knot disease in cardamom. In every farm and nursery in the South Indian region that grows cardamom, they are present. In India, nurseries and plants producing cardamom have been reported to be overrun with the root-knot nematodes *M. incognita* and *M. javanica* (Koshy et al. 1976; Ali 1986; Raut and Pande 1986). Root-knot disease in a plantation causes stunting, impaired tillering, yellowing, early drying of leaf tips and margins, narrowing of leaf blades, delayed blooming, immature fruit drop, and decreased production (Eapen 1992, 1994, 1995). Galling of roots is not a noticeable symptom in cardamom mature plants, in contrast to several other plant species. However, the infected roots display an excessive branching resembling a witch's broom.

Due to infection of the radicle and plumule by second-stage juveniles of the root-knot nematode, more than half of the seeds that germinate in primary nurseries do not emerge. Affected seedlings exhibit minor leaf yellowing and dryness at the two-leaf stage as well as substantial root galling. When they are moved to a secondary nursery, the unopened leaves begin to curl. These leaves often develop following the opening of the pseudostem. Up to 40% of these seedlings in the secondary nursery fail to establish. Plants infected with the disease in secondary nurseries are yellowed and stunted, have poor tillering, dry leaf tips, and have significant root galling (Eapen 1995). Young seedlings experience RKN attacks more frequently than adult plants do, and seedling galling is more obvious (Eapen 1992).

13.3.1.1 Management

Adopting integrated nematode disease management techniques can aid in disease control. The following methods are recommended by various scientists for treating this condition (Rajeev and Thomas 2015; Thomas et al. 2022a).

13.3.1.2 Nursery

Use the biocide methyl bromide or any granular insecticides (Carbofuran*/Phorate*) for 3–7 days to disinfect nursery beds covered in polythene under the supervision of government specialists or experts recognized by the Plant Protection Advisor to the Government of India. In areas where carbofuran and phosphate are not prohibited, carbosulfan (2 mL/liter) may be used in their place (Sharon et al. 2020) (* non-banned areas).

13.3.1.3 Plantation

Make sure to plant nematode-free seedlings. Mulching is recommended, especially in exposed areas. Applying organic manures, such as neem cake, twice a year at a rate of 250–1000 g depending on the size of the clump helps reduce nematode infestation (Pervez 2018b). Depending on the size of the plant, use granular insecticides such as carbofuran*/phorate* @ 15–50 g twice a year in May/June and September (Ali and Koshy 1982; Ali 1985). Applying nematicides before the monsoon and then neem cake during the middle of the monsoon is the safest method for decreasing residue in capsules (* non-banned areas).

Aldicarb* reduced nematode population by 90% when applied at a very high dose of 15 kg a.i./ha. However, in a different study, phorate* application @ 2.5–5.0 g a.i./plant decreased the plant nematode population and boosted yield by more than 40% (Eapen 1995). It is advised to manage root-knot infestation by applying phorate 10G* at 2.5 g a.i./clump during May/June and November/December, or aldicarb or carbofuran* at 5–10 kg a.i/ha (Sharon et al. 2020) (*non-banned area).

13.4 TURMERIC (*CURCUMA LONGA* L.)

Turmeric also referred to as "the golden spice," is an important ingredient in many different industries around the world, including food, medicine, cosmetics, dyeing, and coloring. The largest producer and exporter of turmeric worldwide is India. The most significant states in India for turmeric farming are Andhra Pradesh, Orissa, Karnataka, West Bengal, Tamil Nadu, Gujarat, Meghalaya, Assam, and Maharashtra, with Andhra Pradesh accounting for 38.0 percent of the area and 58.5 percent of the production (Thomas and Rajeev 2015b). The crop was grown on around 2,94,542 hectares and produced 11,01,920 tonnes in 2020–21 (SPICEBOARD 2021). Due to several issues, India's production of turmeric has been dropping for some time. Nematode-infested diseases are one of the most severe limitations because they cause significant productivity losses. The diseases that have a substantial economic impact and the pathogens that cause them are covered here.

13.4.1 Root-knot Disease

Root-knot disease in turmeric is brought on by the root-knot nematode *M. incognita*, which causes significant crop losses. Heavy infestations leave plants with reduced growth, poor tillering, and chlorotic leaves with marginal necrosis. A poor crop stand is produced at harvest because the affected turmeric plants mature and die more quickly than healthy turmeric plants. Rhizomes that have been infected are both a source of infection and a means of disease dissemination. Root-knot disease-infected plants showed galling and rotting of roots and underground rhizomes. Second-stage juveniles of *M. incognita* enter the rhizome through the sheaths of the leaves near the tip of the stem (Chattopadhyay et al. 2018).

13.4.1.1 Management

In areas where nematode problems are widely known, root-knot disease in turmeric caused by the root-knot nematode *M. incognita* is controlled by select turmeric cultivation sites free from plant nematodes. Deep summer plowing in turmeric fields during April–May to minimize the plant nematodes. Soil solarization using transparent polythene sheets (100 gauge) during the summer season for 40–50 days can also reduce the RKN population in the soil. Use only healthy and plant nematode-free rhizomes as planting material. Application of Glyricidia compost or neem seed cake @ 1 t/ha or FYM @ 25–30 t/ha will increase organic content and reduce nematode population. Biocontrol agent *Pochonia chlamydosporia* was applied to the beds at the time of sowing @ 20 g/bed (10^6 cfu/g) for the management of RKN (Eapen et al. 2005). Select IISR Pragati for turmeric cultivation where the RKN problem is very pronounced. Application of Carbofuran* 3G @1kg a.i/ha at 3 and 5 months of sowing can effectively control plant nematodes. Intercropping with repellent plants like marigolds, glyricidia, asparagus, dahlia, etc., @ 5:1 (5 rows of turmeric followed by one row of repellent plants) helps in reducing plant nematode population. Crop rotation with marigold, chrysanthemum, sesbania, *Crotalaria* spp., gaillardia, cluster bean, and *Desmodium* spp. can be followed to manage RKN (Thomas and Rajeev 2015b; Thomas et al. 2022b) (*non-banned area).

13.4.2 Brown Rot Disease

The turmeric brown rot disease was found in the germplasms of *Curcuma aromatica* and *C. longa*. The infected plant has been shown to contain the lesion nematode *Pratylenchus* sp., which produces brown rot in turmeric caused by *Fusarium* sp. In the early stages of infection, rhizomes have a dark color, and necrotic lesions are discrete, isolated, and extend 2–5 mm into the rhizome. The symptoms of the infected rhizomes were unusual browning and rotting of mature rhizomes. In the advanced stage, the rhizome loses weight, and wrinkles, displays dry rot, and has less turgid symptoms, and the lesions merge to form larger necrotic patches that gradually spread throughout the rhizome. Rhizomes' fingers were more severely impacted than the mother rhizomes.

13.4.2.1 Management

Turmeric crops infested with brown rot disease caused by *Fusarium* sp. and the lesion nematode *Pratylenchus* sp. can be managed by planting properly, selected disease, and nematode-free planting material. To maintain track of disease outbreaks, inspect the turmeric field frequently. This disease can be managed with phytosanitation. In places with a lot of plant nematode issues, only nematode-free rhizomes were used for planting. By raising the soil's organic content, nematode multiplication is prevented. Application of @ 3 kg ai/ha of carbofuran*, the gall index drops (* non-banned regions).

13.5 CONCLUSION

Massive crop losses are occurring as a result of diseases plagued by plant nematodes that are wreaking havoc on spice farms. An investigation of the production barriers in the region of India where spice crops are grown the most revealed that diseases are a significant problem for the crop. Due to soil-borne illnesses, effective control is difficult. Farmers are forced to use chemical management, which is frequently ineffectual once symptoms show in the field and leave considerable chemical residues, due to a lack of resistant kinds.

13.6 FUTURE THRUST

The control of *R. similis*, *M. incognita*, and *Pratylenchus* sp. in spice crops is evaluated by enhanced cropping techniques based on location. Creating integrated management plans for the slow decline, root-knot, and brown rot disease complex, which comprises both nematode and fungal pathogens, for black pepper, cardamom, and turmeric. Identifying genetic variants with high levels of resistance to the root-knot and burrowing nematodes. Maximizing output with minimal effort and effective plant nematode control. Determine nematode hotspots in important spice-growing regions using surveys, and make worm control knowledge available to growers.

REFERENCES

Ali SS (1985) Preliminary observations on the effect of some systemic nematicides and neem oil cakes in cardamom field infested with root knot nematodes. In: Sethuraj MR (ed.), *Proceedings of PLACROSYM-VI 1984*. Indian Society for Plantation Crops, Kasaragod, Kerala, India, pp 215–223.

Ali SS (1986) Occurrence of root knot nematodes in cardamom plantations of Karnataka. *Indian Journal of Nematology* 16(2): 269–270.

Ali SS, Koshy PK (1982) A note on use of methyl bromide for control of root-knot nematodes in cardamom nurseries. *Indian Journal of Nematology* 12(1): 147–150.

Bhai RS, Eapen SJ, Kumar A, Aravind R, Pervez R, Varghese EM, Krishna PB, Sreeja K (2017) Mitigating *Phytophthora* foot rot and slow decline diseases of black pepper through the deployment of bacterial antagonists. *Journal of Spices and Aromatic Crops* 26(2): 69–82.

Chattopadhyay A, Purohit J, Patel DS, Tetarwal ML (2018) Management of diseases of spice crops through microbes. In: Singh D, Chakraborty BN, Pandey RN, Sharma P (eds.), *Biological control of crop diseases: Recent advances and perspectives*. Today & Tomorrow's Printers and Publishers, New Delhi, India, pp 249–270.

DASD, Directorate of Arecanut and Spices Development Calicut (2022) www.dasd.gov.in/adminimage/Spices_area_and_production1.pdf,.

Devasahayam S, Eapen SJ, Jacob TK, Pervez R. (2012a) Insect and nematode pests. In: Singh HP, Parthasarthy VA, Srinivasan V, Saji KV(eds.), *Piperaceous crops production and utilization*. Westville Publishing House, New Delhi, pp 144 –161.

Devasahayam S, Eapen SJ, Jacob TK, Pervez R (2012b) Pests. In: Singh HP, Parthasarthy VA, Kandianan K, Krishnamurthy KS (eds.), *Zingiberaceous crops (Present and future Cardamom, Ginger, Turmeric and others)*. Westville Publishing House, New Delhi, pp 332–347.

Eapen SJ (1992) Influence of plant age on root knot nematode development in cardamom. *Nematologia Mediterranea* 20(2): 193–195.

Eapen SJ (1994) Pathogenicity of *Meloidogyne incognita* on small cardamom, *Elettaria cardamomum* Maton. *Indian Journal of Nematology* 24(1): 31–37

Eapen SJ (1995) A note on the incidence of *Rotylenchulus reniformis* in cardamom plants. *Indian Journal of Nematology* 25(2): 213

Eapen SJ, Beena B, Ramana KV (2005) Tropical soil microflora of spice based cropping systems as potential antagonists of root-knot nematodes. *Journal of Invertebrate Pathology* 88(3): 218–225.

FAOSTAT (2020). *Food and Agriculture Organization of the United Nations*, 2020, 02-28. www.fao.org.

Koshy PK, Radhakrishnan Nair CP, Sosamma VK, Sundararaju P (1976) On the incidence of root knot nematode in cardamom nurseries. *Indian Journal of Nematology* 6(2): 174–175.

Leela NK, Pervez R, Rammanna V, Rosana OB, Eapen SJ (2012) Nematicidal activity of *Strychnos nuxvomica* leaf and its constituents against the root-knot nematode, *Meloidogyne incognita*. *Nematol Mediterranea* 40(2): 116–123.

Pervez R (2017) Plant Parasitic nematodes: Hidden enemy of black pepper (*Piper nigrum* L.). *Indian Journal of Arecanut, Spices and Medicinal Plants* 19(4): 24–28.

Pervez R (2018a) Current status of the plant parasitic nematodes and their management infesting major spice crops. *Trends in Horticulture* 1(2): 1–5.

Pervez R (2018b) Plant health management in major Spice crops. In: A Sharangi (ed.), *Indian spices, The Legacy, production and processing of India's Treasured Export*. Springer International Publishing AG, USA, pp 205–248.

Pervez R, Eapen SJ (2015) Distribution of plant parasitic nematodes associated with black pepper rhizosphere from Idukki district (Kerala), India. *Annals of Plant Protection Sciences* 23(1): 192–193.

Pervez R, Eapen SJ (2016) Community structure of plant parasitic nematodes associated with black pepper in Idukki district (Kerala), India. *Current Nematology* 27(1): 73–76.

Pervez R, Eapen SJ (2021) Management of burrowing nematode, *Radopholus similis* (Cobb, 1893) Thorne, 1949 in black pepper nurseries. *Indian Journal of Nematology* 51(2): 175–180.

Pervez R, Eapen SJ, Jacob TK, Hamza S, Srinivasan V (2016) Diversity and community analysis of nematodes associated with *Piper nigrum* from Idukki District (Kerala), India. *Indian Journal of Nematology* 46(1): 7–73.

Pervez R, Leela NK, Eapen SJ (2012) Nematicidal activity of *Strychnos nuxvomica* leaf and its constituents against the *Radopholus similis* infesting black pepper (*Piper nigrum* L.). *Current Nematology* 22(1,2): 49–53.

Pervez R, Niraul IM, Eapen SJ (2017) *Hirschmanniella* sp. (Nematoda: Pratylenchidae) associated with black pepper (*Piper nigrum* L.): A first report. *Annals of Plant Protection Sciences* 24(2): 385–391.

Rajeev P, Thomas L (2015) *Cardamom*, ICAR- Indian Institute of Spices Research, Kozhikode (Kerala), pp 26.

Raut SP, Pande VS (1986) A first report of *Meloidogyne incognita* causing root knot disease of cardamom in Konkan. *Indian Journal of Mycology and Plant Pathology* 16(2): 167–168.

Sharon A, Biju CN, Krishnamurthy KS, Radha E, Nissar VA, Aarthi S and Eapen SJ (2020) *ICAR- AICRPS Technologies, ICAR-All India Coordinated Research Project on Spices*, ICAR-IISR, Kozhikode, Kerala, India, pp 39.

SPICEBOARD (2021) Spice Board of India, Ministry of commerece and industries, Govt. of India. www.indianspices.com/sites/default/files/majorspicewise2021.pdf.

Thomas L, Rajeev P (2015a) *Black pepper*, ICAR- Indian Institute of Spices Research, Kozhikode (Kerala), pp 27.

Thomas L, Rajeev P (2015b) *Turmeric*, ICAR- Indian Institute of Spices Research, Kozhikode (Kerala), pp 15.

Thomas L, Rajeev P, Dinesh R (2022a) *Cardamom,* ICAR- Indian Institute of Spices Research, Kozhikode, Kerala, pp 32.

Thomas L, Rajeev P, Dinesh R (2022b) *Turmeric,* ICAR- Indian Institute of Spices Research, Kozhikode, Kerala, pp 32.

14 Major Insect Pests Infesting Seed Spice Crops and Their Innovative Management

Narottam Kumar Meena

14.1 INTRODUCTION

Intensification and diversification are the pressing need of today for the sustenance of the seed spice crops sector in the country. India is blessed with 63 spices crops, among them, 20 spices are considered as seed spices, growing in 40 percent of the total area occupied by spices cultivation in the country. Cumin (*Cuminum cyminum* L.), fennel (*Foeniculum vulgare* Mill.), coriander (*Coriandrum sativum* L.), and fenugreek (*Trigonella foenum-graecum* L.) are the four major, and ajwain (*Trachyspermum ammi* L.), dill (*Anethum graveolens* L.), nigella (*Nigella sativa* L.), celery (*Apium graveolens* L.), anise (*Pimpenella anisum* L), and caraway (*Carum carvi* L.) are categorized as seed spices of minor importance. India is one of the largest producers and consumer nations in the world, producing 19.98 lakh tonnes of seed spice from the 20.88 lakh hectare area in 2019–20 (Anonymous 2021a). Gujarat, Rajasthan, Andhra Pradesh, Madhya Pradesh, and Uttar Pradesh are the leading states of the country in seed spice production. India alone exported 355.92 thousand tonnes of seed spices amounting to rupees 4404.74 crores during 2019–20 (Anonymous 2021b). The acreage and production of a few seed spices, particularly cumin and coriander, is increasing subsequently year after year. Seed spices are used as spices in the preparation of vegetables and many spicy dishes like pizza, burgers, chat-bhaji, pav-bhaji, beverages, etc., for the enrichment of their taste, flavor and aroma. It also has an enormous source of nutritional (Peter 2004; Meena et al. 2017) and medicinal (Tatsuta and Iishi 1993; Malhotra and Vashishtha 2008; De et al. 2009; Pathak et al. 2011; Rajeshwari and Andallu 2011) properties, used in allopathic along with Ayurved for remedies of various ailments.

Seed spice crops receive numerous insect pest attacks including aphid species *Hyadaphis coriandri* Das, *Aphis gossypii* Glover, *Aphis craccivora* Koch and *Myzus persicae* Sulzer; jassids *Empoasca* sp.; thrips *Thrips tabaci, Scirtothrips dorsalis, Frankliniella schultzei* Tryb., and *Caliothrips indicus* Bag.; seed wasp *Systole albipennis* Walker; whitefly *Bemisia tabaci* Genn.; stink bug *Nezara viridula* Linn., painted bug *Bagrada hilaris* Burm.; seed bug *Nysius* sp.; leaf minor *Liriomyza trifolii*; cut worm *Agrotis* spp., fruit/leaf-eating caterpillars *Heliothis armigera* Hub; *Spodoptera litura* Fab., grasshopper, and termite starting from seed germination to crop harvesting, cause significant yield and quality loss (Singh and Baswana 1984; Araujo 1986; Jain and Yadav 1988; Meena et al. 2017; Meena et al. 2018; Meena and Meena, 2020; Meena et al. 2020). Amongst them, aphids, thrips, jassids, leaf minors, seed wasps and seed bugs are rated as severe loss-causing insects in various seed spice crops, described as follows.

DOI: 10.1201/9781032631578-14

14.2 ECONOMIC IMPORTANT INSECT PESTS

14.2.1 Aphid (Hemiptera: Aphididae)

The aphid is also known as *Chainpa* or *Manhu* in north India. *Hydaphis coriandri* (Das) infestation was reported on coriander, fennel and dill crop in India (Jain and Yadava, 1986). *Myzus persicae* (Sulzer), *Hyadaphis coriandri* (Das) and *Aphis gossypii* Glover species cause severe crop damage in the absence of timely control measures and create a momentous loss in cumin yield and seed quality (Meena et al. 2018; Kant et al. 2010). Pest occurrence starts on these seed spice crops in the first fortnight of December and lasts till March. Apart from seed spices, it also feeds on various crop plants like parsley (*Petroselinum crispum* (Mill.), horsemint (*Mentha longifolia* (Linnaeus) Huds.), soybean (*Glycine max* (Linnaeus) Merr.), spiny pigweed (*Amaranthus spinosus* Linnaeus), and blond psyllium (*Plantago ovata* Forssk.) (Blackman and Eastop 1984). Adult aphids are oval or pear-shaped, yellow-green or black in color, and dusted with a greyish wax. They have a pair of short, dusky, slightly swollen, cornicles on the posterior portion of the abdomen.

14.2.2 Damage

They form dense and sometimes often damaging colonies on leaves, stem shoots, umbels, and even developing seeds. The young ones and adults suck the cell sap, usually from the tender portion of stems, leaves and inflorescence. Early infestation of insects creates deformation in plant foliage, turns leaves yellow and ultimately reduces their vigor. When the infestation occurs at flowering times and control measures are neglected then the seed formation in flowers is adversely affected and the yield is reduced considerably.

14.2.3 Management

The aphids can be controlled by using timely sowing (between the first fortnight of October and the first fortnight of November) of seed spice crops, to escape from heavy invasion of aphids. Apply nitrogenous fertilizers at the recommended dose. Field sanitation is the most important tool of cultural practice to minimize the aphid incidence on the main crop. Use yellow sticky trap @ 5–6 traps/acre. Grow 3–4 lines of dill (*Anethum sowa* Roxb) surrounding the coriander field as a trap crop to minimize the aphid incidence. Enhance the activities of aphid predators like *Coccinella septempunctata* L., *Menochilus sexmaculata* (Febricious), *Ischiodan scutellaris* (Febricious), *Syrphus serarius* (Wiedmann) and *Chrysoperla zastrowi sillemi* (Esben-Petersen) and parasitoid like *Aphidius* sp. in seed spices ecosystem to minimize the aphid population. Botanical products are also effective in minimization of aphid population, i.e. apply any one of the following botanical products: eco-neem 3000 ppm @ 3.5 ml/lit or NSKE 5% solution @ 5ml/lit., or *Artemisia* leaf extract @ 10 ml/lit or garlic extract 5% provide good protection against aphids and repeat the spray of the same product at 7 to 10 day-intervals if required. There are no chemical insecticides to use in seed spices; however, one of the insecticides should be applied only when aphid infestation is very high. Foliar application of imidacloprid 17.8 SL @ 0.003% or thiamethoxam 25 WG @ 0.005% or dimethoate 30EC @ 0.03% or acetamiprid 20SP @ 0.2 g/lit., gives efficient management of aphids.

14.2.4 Thrips (Thysanoptera: Thripidae)

A number of thrips species cause damage to different crops of seed spices i.e. *Thrips tabaci* Lind., *Frankliniella schultzei* (Trybom)] *Aeolothrips collaris, Scirtothrips oligochaetus*, *Haplothrips* spp., and *Diarthothrips nimbus*. They are also known as thunder flies, because they sometimes take to the air in huge numbers, especially in thundery weather. Thrips infestation starts in the early plant growth stage of many crops of seed spices (15–20 days after seed germination) in the first week of

November and lasts 50–60 days after germination. Sometimes, thrips infestations are seen on completely grown crops also. *Frankliniella schultzei* has a very wide distribution and is mainly found in tropical and subtropical areas all over the world (Vierbergen and Mantel 1991). *Thrips tabaci* is world-wide and also found throughout India as a pest of spices and many vegetables like onion, cabbage, beans, etc. Thrips are tiny insects. Thrips in adult stages are slender-shaped, yellowish-brown colored, just 1–2 mm long. Adults have two pairs of narrow, fringed wings, although some insects are wingless.

14.2.5 Damage

Thrips infestation initiates almost 10–15 days after seed germination in most of the seed spices and preferably feeds on the newly emerged leaves. Thrips feed on succulent plant parts with a punch-and-suck behavior in both crawler and adult stages, resulting in chlorophyll loss, and due to this, white to silver patches and streaks appear on the damaged portions of the plants. Under severe infestation conditions, leaves take on a silvery cast, and ultimately, plant growth is entirely checked, and plants appear deformed. Insects initiate their breeding during the winter months from November to March or sometimes till April (under favorable weather conditions) and complete several generations in a year.

14.2.6 Management

Thrips are a serious problem in all seed spice crops throughout the country and cause significant yield loss; hence, efficient management can be done by using the following pest management practices. As much as possible, ensure clean cultivation by removal of weeds, regular hoeing and providing surface irrigation of infested fields to check thrips population. Apply the recommended dose of nitrogenous fertilizers, excess nitrogenous fertilizers invite heavy infestation of thrips. Use sticky traps: male thrips are most attracted to yellow sticky traps (15 traps/hectare) while female thrips were more attracted to pink sticky traps (Yaku et al. 2007). Natural toxicants from plants and biopesticides have also been used for their control because of the adverse effects of chemical pesticides. Apply any one of the following botanical products, i.e. eco-neem 3000 ppm @ 3.5 ml/lit or NSKE 5% solution @ 5ml/lit. or ker plant extract @10 ml/lit., for the control of thrips. Chemical insecticides should apply when the thrips population reaches above a threshold level. Spray any one of the given insecticides: imidacloprid 17.8 SL @ 0.003% or dimethoate 30EC @ 0.03% or acetamiprid 20 SP @ 0.2 g/lit. Farmers should always try to enhance the potential natural enemies (*Coccinella septempunctata* L., *Menochilus sexmaculata* (F.), *Ischiodan scutellaris* (F.), *Chrysoperla zastrowi sillemi* (Esben-Petersen) of thrips in seed spices ecosystem.

14.2.7 Jassids (Hemiptera: Cicadellidae)

Empoasca spp. was reported to feed on coriander, fenugreek, fennel and many more seed spice crops during the early crop developmental stage (Meena et al. 2017). *Empoasca kerri* Pruthi, a well-known species of jassid was reported to infest coriander under open field conditions (Ghadage et al. 2010). However, a total of forty jassid species have been reported from Western Maharashtra, India on various crops (Kamble et al. 2016). It is a small-sized, wedge-shaped, light-greenish-yellow-colored insect and walks diagonally.

14.2.8 Damage

Jassid infestation starts on coriander crops 15–20 days after the sprouting of the seed under favorable environmental conditions. Its infestation initiates under the lower surface of leaves and later spreads on tender leaves and shoots. Both the stages of insects (nymphs and adults) suck the cell sap from succulent plant parts with the help of piercing and sucking type of mouth parts and affect the growth and yield of seed spice crops.

14.2.9 Management

The pest can be controlled by using botanical pesticides azadirachtin 0.03EC @ 3.5 ml/liter of water of neem oil @ 5ml/liter of water or achook 3000 ppm @ 3 ml/liter of water. If needed, apply any one or the following insecticides like dimethoate 30EC@ 2 ml/liter of quinalphos 25EC @ 1.5 ml/liter or thiamethoxam 25 WG@ 0.12g/liter of water when the jassids population reaches the economic threshold level.

14.2.10 Seed Wasp, (Hymenoptera: Eurytomidae)

Seed wasp (*Systole albipennis* Walker) is a specific pest of coriander, fennel, cumin and ajwain and many other spice crops that has been reported from the Indian subcontinent, Asia, Africa and Europe. However, the serious losses of coriander fruit caused by seed wasp *S. albipennis* have been reported in Ethiopia. In Rajasthan's semi-arid regions, its infestation occurs in the field at seed forming stage and is sustained till crop harvesting. The larvae damage the seed and survive inside the fruits, hence further infestation of this insect reaches storage where it acts as storage pest. It's a minute hymenopteran wasp, about 1–1.5 mm long, having transparent wings on a black-colored body.

14.2.11 Damage

The newly emerged larvae feed on developing fruits and destroy the seed embryo or endosperm which results in loss of seed viability. The yield loss incurred was approximately 30 percent in coriander. However, under severe infestation conditions, the yield loss of coriander reached up to 50 percent. The adult females lay eggs on the immature seeds and upon egg hatching, they develop larvae and feed continuously inside the seed. Larvae complete their whole instars inside the seed and turn into pupae. Adults emerge by making holes in the seed cover. The insects complete their entire life span within 25 days from egg to adult.

14.2.12 Management

This insect can be managed efficiently by using the following pest management practices. Store the seeds of susceptible seed spices under airtight storage conditions and apply fumigants to kill the immature stages of this pest. Test the seeds' viability for germination and use the recommended seed rate for sowing. Sow the crop with a suitable crop geometry (between row and plant distance) as per the package of practices recommended for seed spices. Manipulate the date of sowing and microenvironment within the crop by maintaining proper plant-to-plant and row-to-row distance to enhance the plant growth and effectiveness of bio-control agents of *Systole albipennis* in the crop ecosystem. Spray crops with NSKE 5% @ 5ml/lit or neem oil 2% or karanj oil 2% to provide effective management of seed wasp. Apply any one from the following insecticides, i.e. imidacloprid 17.8 SL 0.003% or thiamethoxam 25 WG @ 0.005% during vegetative growth stages and fipronil 5 SC @ 1.5–2.0 ml/lit., at the flowering stage to control the seed wasp.

14.2.13 Leaf/Fruit-eating Caterpillar, *Helicoverpa armigera* (Hub.)

This pest has a minor status in coriander and a few other seed spice crops due to its irregular appearance, but sometimes causes significant yield loss. Its infestation occurs on crops at the time of flowering or the seed formation stage in winter months (January–February) under semi-arid conditions in Rajasthan. It is a cosmopolitan pest, widely distributed throughout the world on various crops like ornamental, field crops, vegetables and weed plants. Adult moths are yellowish brown in color and their body is stoutly built.

14.2.14 Damage

There is a dark speck and area near the outer margin of each forewing and greyish wavy lines and black spots on the upper side and a round spot on the underside of the forewings. Full-grown caterpillars are 2.5–3 cm long having two yellowish green parallel lines on the sides of the body. Only the caterpillars cause damage by grazing inflorescence and developing grains.

It is a polyphagous species (Robinson et al. 2010). The most important crop hosts are coriander, cotton, chickpea, pigeon pea, tomato, cowpea and sorghum. Other host plants include peas, field beans, lucerne, okra, soybeans, groundnut, etc. Other than the described food crops, the insect also feeds on some weed plants like makoi (*Solanum nigrum*), chaulai (*Amaranthus viridis*), bindweed (*Convolvulus arvensis*), *Datura stramonium* and some forest tree plants.

14.2.15 Management

The leaf/fruit-eating caterpillar can be managed efficiently by using the following pest management practices. Cultural practices: Summer deep plowing of the field is an effective management practice to expose the immature stages of the insect and reduce the further incidence on the crop. Hand-collecting caterpillars from the crop can destroy them to check the infestation. Removal and destruction of alternate host plants, growing of 3–4 rows of tomato or marigolds surrounding the coriander field as a trap crop to attract the insect and protect the coriander crop from pest infestation. Use pheromone traps with 'Heli lure' @ 10–15 traps/hectare to monitor the pest activity. A light trap is also effective for attracting moths and then killing them with kerosinized water. Install T-purchase in the seed spices field to allow the birds to hunt the caterpillars. If required, spray the crop with any one of the following insecticides, i.e. quinalphos 25 EC @ 2.5 ml/lit. or chlorantraniliprole 18.5% SC @0.3 ml/lit and repeat the same schedule at 10-day intervals if needed to control the caterpillars. Release of *Trichogramma chilonis* @ 50000 eggs/ha., during flowering through Tricho-cards is also effective in the management of pests. Encourage the activity of natural enemies in coriander and other crop fields.

REFERENCES

Anonymous (2021a) Anonymous. Spices: area, production and productivity in India. Directorate of Arecanut and Spices Development (GOI), Calicut, Kerala, 2021. www.dasd.gov.in

Anonymous (2021b) Spices Board, Ministry of Commerce and Industry, Govt. of India, Kochi, Kerala. www.indianspices.com/sites/default/files/majorspicewise2021.

Araujo CZDde (1986) Occurrence of *Myzus persicae* (Sulzer) (Homoptera: Aphididae) on a crop of coriander *Coriandrum sativum* (Umbelliferae) [in Port.]. *Anais da Sociedade Entomologica do Brasil* 15: 173–174.

Blackman RL, Eastop VF (1984) *Aphids on the World's Crops*, J. Wiley and Sons, Chichester, p 466.

De ML, De FV, Fratianni F, Nazzaro F (2009) Chemistry, antioxidant, antibacterial and antifungal activities of volatile oils and their components. *Natural Product Communications* 4: 1741–1750.

Ghadage SM, Vyas HJ, Joshi MD (2010) Loss estimation in coriander due to sucking pests and screening of coriander genotypes against jassids. *Agricultural Science Digest* 30: 299–301.

Jain PC, Yadava CPS (1986) Effect of date of sowing on the incidence of insect pests of coriander. *Indian Journal of Agricultural Sciences* 56(1): 56–59.

Jain PC, Yadav CPS (1988) Relative susceptibility of coriander varieties to coriander aphid (Hyadaphis coriandri Das). *Indian Journal Entomology* 50: 535–536.

Kamble CS, Sathe TV (2016) Diversity, seasonal abundance, distribution and damage of jassids to crops from Western Maharashtra. *Biolife* 4: 357–363.

Kant K, Sharma YK, Meena SS, Mehta RS, Meena SR (2010) Saving seed spices from insect enemies. *Indian Horticulture* 55: 25–27.

Malhotra SK, Vashishtha BB (2008) *Package of Practices for Production of Seed Spices*, Book Publisher Director, National Research Centre on Seed Spices, Ajmer, pp 3–98.

Meena NK, Meena RS (2020) Pests problem in coriander (*Coriandrum sativum* L.) cultivation and their management for quality production-a review. *Journal of Plant Health Issues* 1(3): 69–74.

Meena NK, Lal G, Meena RD, Choudhary MK (2020) Pest status on fennel (*Foeniculum vulgare* Mill) under organic production system in semi-arid region of Rajasthan, India. *Journal of Entomology and Zoology Studies* 8(6): 181–184.

Meena NK, Lal G, Kant K, Meena RS, Meena SR (2018) Pest scenario of cumin (*Cuminum cyminum* L.) and population dynamics in semi-arid region of Rajasthan. *International Journal of Seed Spices* 8(1): 80–83.

Meena NK, Lal G, Meena RS, Harisha CB, Meena SR (2017) Pest scenario of coriander (*Coriandrum sativum* L.) and population dynamics in semi-arid region of Rajasthan. *International Journal of Tropical Agriculture* 35: 779–783.

Pathak NL, Kasture SB, Bhatt NM, Rathod JD (2011) Phytopharmacological properties of *Coriander sativum* as a potential medicinal tree: An overview. *Journal of Applied Pharmaceutical Sciences* 1: 20–25.

Peter KV (2004) *Hand Book of Herbs and Spices*, Vol. 2, Woodhead Publishing Ltd, Cambridge, pp 158–174.

Rajeshwari U, Andallu B (2011) Medicinal benefits of coriander (*Coriandrum sativum* L.). *Spatula* 1: 51–58.

Robinson GS, Ackery PR, Kitching IJ, Beccaloni GW, Hernandez LM (2010) "Hosts – a database of the world's lepidopteran host plants". London: Natural History Museum, p 234.

Singh GK, Baswana S (1984) Screening of coriander germplasm against chalcid fly (*Systole albipennis*), tests of agrochemicals and cultivars. *Annals of Applied Biology Supplement* 104(5): 114–115.

Tatsuta M, Iishi H (1993) Effect of treatment with liu-jun-zi-tang (TJ-43) on gastric emptying and gastrointestinal symptoms in dyspeptic patients. *Alimentary Pharmacology & Therapeutics* 7: 459–462.

Vierbergen G, Mantel WP (1991) Contribution to the knowledge of *Frankliniella schultzei* (Thysanoptera: Thripidae). *Entomologische Berichten (Amsterdam)* 51: 7–12.

Yaku A, Walter GH, Najar-Rodriguez AJ (2007) Thrips see red – flower colour and the host relationships of a polyphagous anthophilic thrips. *Ecological Entomology* 32: 527–535.

15 Biotic Stress Management Strategies in Sugarcane Crop Using Eco-friendly Approaches

Latha Paramanandham, Rajaswaminathan Vairavan and Suresh J.

15.1 INTRODUCTION

Sugarcane, an important commercial crop in the world is being cultivated in many countries. Brazil stands first in the production of sugarcane, followed by India and China. Sugarcane is the major sugar-contributing crop, with about 70% of the demand for sugar around the world being met through crushing sugarcane. In India, the production of sugarcane is around 355.52 million tonnes of cane per year, with a production area of 5.15 million hectares. Uttar Pradesh is the leading state in sugarcane production where sugarcane is being cultivated on 22.47 lakh hectares with productivity of about 59.6 tonnes per hectare (Economic Survey 2021–2022). Changes in agricultural farming systems and adverse changes in environmental conditions, along with the exploitation of chemicals and pests, pathogens and weeds have led to a severe loss in the production of sugarcane. Among the various factors which lower its production and productivity is the prevalence and abundance of diseases. Sugarcane is a vegetatively propagated crop that is exposed to complex diseases during the course of its germination, growth and maturity, which reduces both quality as well as quantity. There are around 120 sugarcane diseases, providing biotic stress to the crop by infection (Rott 2000; Viswanathan 2018). Diseases can inflict loss of sugarcane production ranging from 10–15%. Sugarcane is infected by various causal organisms such as fungi, bacteria, viruses, and phytoplasma. A list of diseases infecting sugarcane crops collected from literature is given in Table 15.1.

DOI: 10.1201/9781032631578-15

TABLE 15.1
List of sugarcane diseases

S.No.	Disease	Causal Organism	Major Symptoms	Reference
1	Red rot	*Colletotrichum falcatum*	Dull red internodes with rotting appearance. Internal tissues damaged and rotten (Figure 15.1).	Duttamajumder 2008
2	Smut	*Ustilago scitaminea*	Smut sori present in leaf. Whip (Long sori) formed from apex of the plant (Figure 15.2).	Bhuiyan et al. 2021
3	Wilt	*Fusarium sacchari*	Complete drying of plants. Pinkish red/brownish red discoloration in internal tissues (Figure 15.3).	Viswanathan et al. 2012
4	Pineapple disease/ Sett rot	*Ceratocystis paradoxa*	Reddening of tissue. Blackening of setts internal tissue and hollowing in severe stage (Figure 15.4).	-
5	Eye spot	*Helminthosporium sacchari*	Water-soaked spots on leaves. Spot has reddish brown centre with yellow halo (Figure 15.5).	Huang and Li 2016
6	Pokkahboeng disease	*Gibberella moniliformis*	Large, yellowish white spot in leaf. Leaves become deformed and twisted (Figure 15.6).	Huang and Li 2016
7	Brown stripe	*Cochliobolus stenospilus*	Water-soaked spots formed parallel to the veins in leaf. The spot turns brownish red with yellow halo (Figure 15.7).	Huang and Li 2016
8	Yellow spot	*Cercospora koepkei*	Irregular yellow and red spots appear on the leaf. At later stage, the whole leaf turns reddish yellow, and the leaf turns necrotic from tip to base (Figure 15.8).	Huang et al. 2018
9	Brown rust	*Puccinia melanocephala*	Brown flecks appear on the leaf, later turn to black in color. The flecks combine and forms a large lesion (Figure 15.9).	Huang et al. 2018
10	Orange rust	*Puccinia kuehnii*	Orange-brown flecks on both surface of the leaves, with yellow halo (Figure 15.10).	Huang et al. 2018
11	Brown spot	*Cercospora longipes*	Oval/linear shape spot surrounded by an yellow halo (Spot ring) (Figure 15.11).	Huang et al. 2018
12	Sooty mold	*Capnodiumn* sp. and *Fumago* sp.	Black moldy film covering the leaf and stalks partially/fully (Figure 15.12).	Huang et al. 2018
13	Stalk rot	*Phaeocytostroma sacchari*	Rind discolored to yellowish green, which later turned to reddish brown. Nodal region changes to red color (Figure 15.13).	Viswanathan et al. 2003
14	Ring spot	*Leptosphaeria sacchari*	Oblong spot with yellow halo on the leaf blade (May also occur on leaf sheath). Rarely reddish-brown patches occur along with the spot (Figure 15.14).	Huang et al. 2018
15	Downy mildew	*Peronosclerospora sacchari*	Presence of yellow or withered portion in leaves (Figure 15.15).	Sumael et al. 2020
16	Sheath rot	*Cytospora sacchari*	Reddish brown sheath which later turns necrotic. Drying and drooping of leaves (Figure 15.16).	Saumtally et al. 2000
17	Red stripe	*Acidovarax avenae*	Water-soaked, long, chlorotic streaks appear on the leaf. The stripe occurs mostly near the midrib (Figure 15.17).	Yonzone et al. 2018
18	Leaf scorch	*Stagonospora sacchari*	Reddish brown spots on leaf which later elongate and forms a spindle shape, with yellow halo (Figure 15.18).	Huang et al. 2018
19	Leaf scald	*Xanthomonas albilineans*	Narrow white/yellow line appears on the leaf (Like pencil line) (Figure 15.19).	Ricaud et al. 2012
20	Ratoon stunting	*Leifsonia xyli* sub sp. *xyli*	Stunting happens in the number of stalks reduced in ratoon. Stalks become thinner and poor growth happens (Figure 15.20).	Huang et al. 2018
21	Mosaic	*Sugarcane mosaic virus* (SCMV) and *Sorghum mosaic virus (SrMV)*	Longitudinal stipes parallel to the veins in the leaf (Figures 15.21 & 22).	Huang et al. 2018

TABLE 15.1 (Continued)
List of sugarcane diseases

S.No.	Disease	Causal Organism	Major Symptoms	Reference
22	Streak mosaic	*Sugarcane streak mosaic virus* (SCSMV)	Yellow and green patches in leaf forming mosaic pattern (Figure 15.23).	Huang et al. 2018
23	Yellow leaf syndrome	*Sugarcane yellow leaf virus* (SCYLV)	Occurs on older leaves, showing change in midrib color from green to yellow. The midrib color changes to whitish green/pink and later turns to yellow (Figure 15.24).	Huang et al. 2018
24	Leaf freckle/ Bacilliform disease	*Sugarcane bacilliform virus* (SCBV)	Spots and mottles appear on the leaves. Shrunken leaves and chlorotic spots might appear (Figure 15.25).	-
25	Streak	*Maize streak virus*	Chlorotic lesions were elongated and appears in the leaf border (Figure 15.26).	Van Antwerpen et al. 2008
26	White leaf	*Sugarcane white leaf phytoplasma* (16Sr XI group)	Chlorosis occurs in the leaves, proliferating tillers and stunting occurs (Figure 15.27).	Rao et al. 2012
27	Green grassy shoot	*Ca.* Phytoplasma	Thin, slender, adventitious tillers appear which cluster together giving a grassy appearance (Figure 15.28).	Rao et al. 2012
28	Leaf yellows	*Ca.* Phytoplasma	Yellowing of leaf. Midrib turns bright yellow (Figure 15.29).	Gaur et al. 2012

15.2 IMAGES OF SUGARCANE DISEASES

FIGURE 15.1 Red rot symptom in the internal tissue of the cane. (Huang et al. 2018)

FIGURE 15.2 Whip caused by the sugarcane smut disease. (Bhuiyan et al. 2021)

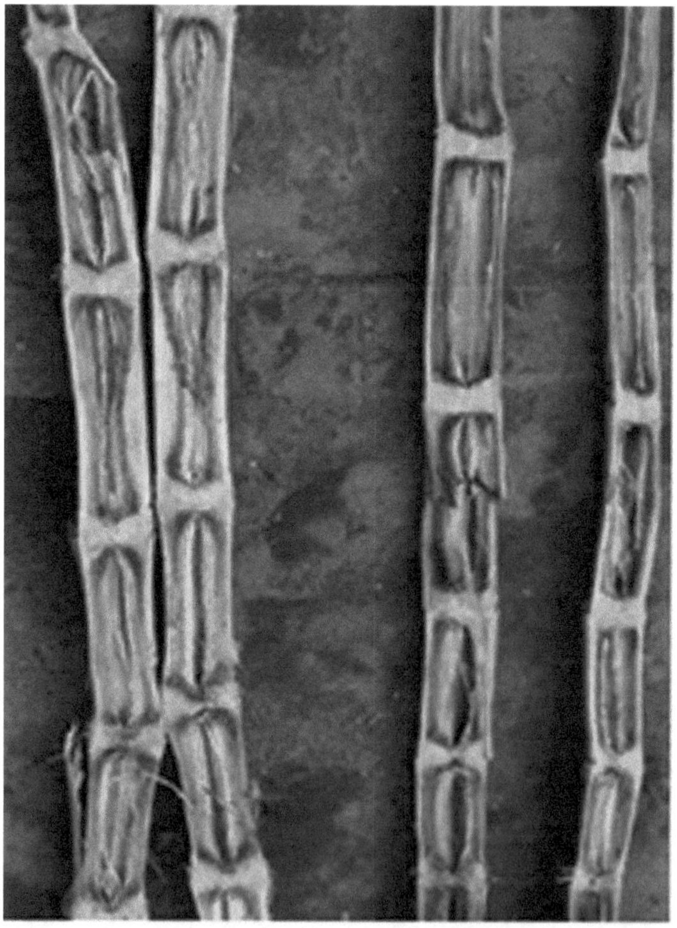

FIGURE 15.3 Destruction of the vascular bundle due to sugarcane *Fusarium* wilt. (Viswanathan et al. 2012)

FIGURE 15.4 Internal tissue discoloration due to pineapple disease. (Huang et al. 2018)

FIGURE 15.5 Eye spot on leaf. (Huang et al. 2018)

FIGURE 15.6 Infected leaf – Pokkah Boeng disease. (Huang et al. 2018)

FIGURE 15.7 Brown color stipes in leaves due to Brown Stripe disease. (Huang et al. 2018)

FIGURE 15.8 Yellow spot disease. (Huang et al. 2018)

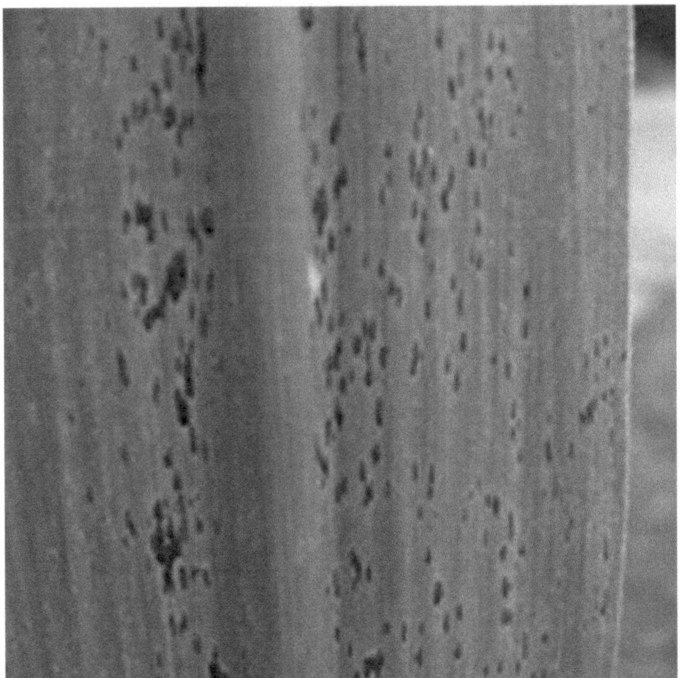

FIGURE 15.9 Brown rust-infected leaf. (Huang et al. 2018)

FIGURE 15.10 Orange rust-infected plants. (Huang et al. 2018)

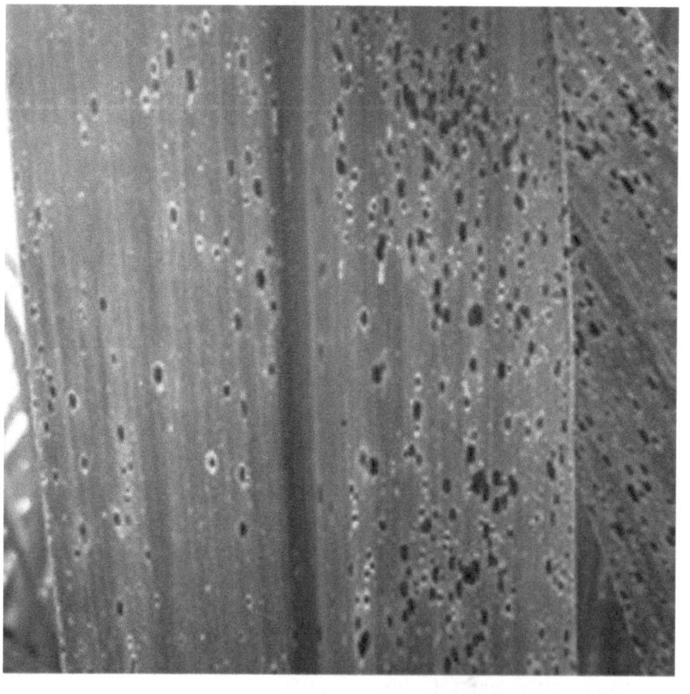

FIGURE 15.11 Brown spot on leaf. (Huang et al. 2018)

FIGURE 15.12 Sooty mold-covered leaf surface. (Huang et al. 2018)

FIGURE 15.13 Stalk rot disease. (Viswanathan et al. 2003)

FIGURE 15.14 Ring spot disease. (Huang et al. 2018)

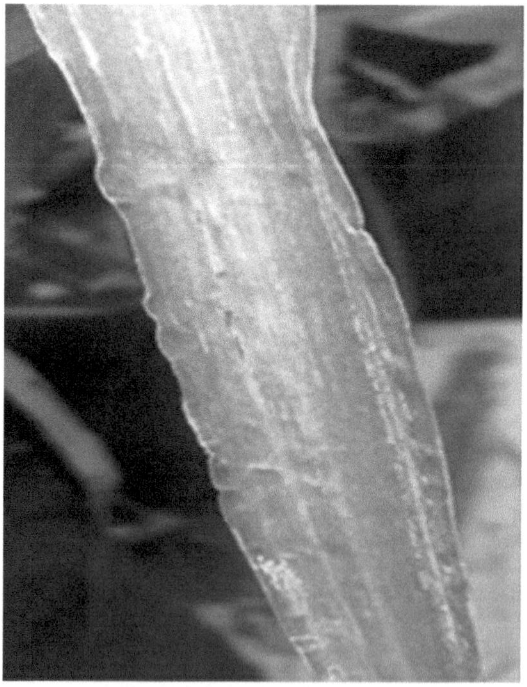

FIGURE 15.15 Sugarcane downy mildew in leaf. (Sumael et al. 2020)

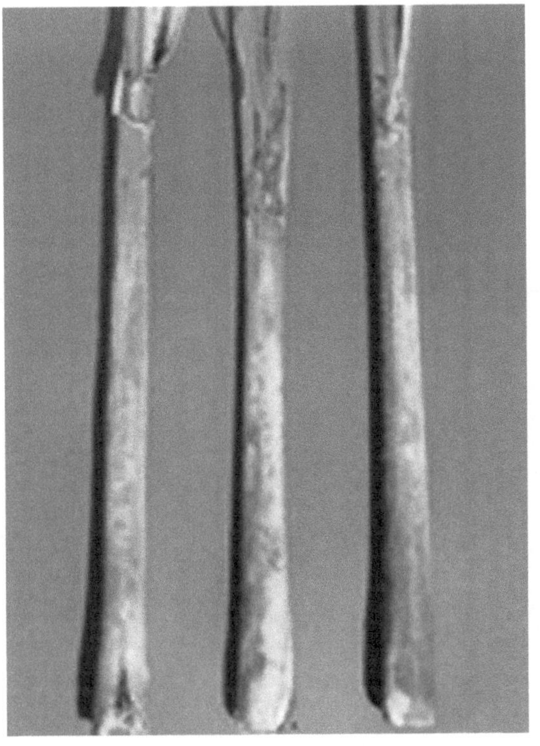

FIGURE 15.16 Sugarcane sheath rot disease. (Saumtally et al. 2000)

FIGURE 15.17 Red stripe disease observed in leaf. (Shan et al. 2017)

FIGURE 15.18 Leaf scorch. (Huang et al. 2018)

FIGURE 15.19. Leaf scald. (Huang et al. 2018)

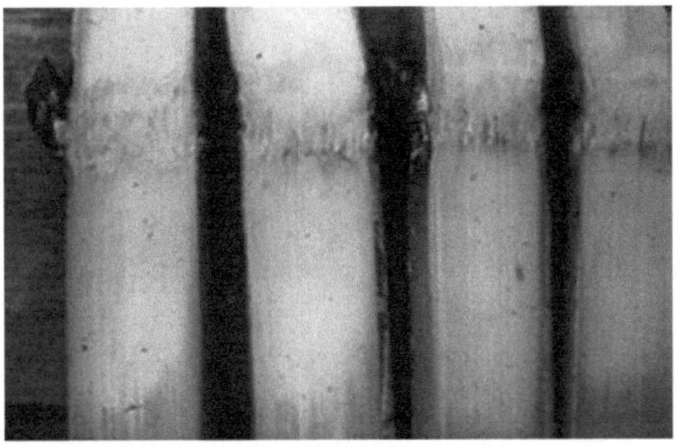

FIGURE 15.20. Internal tissue symptom of sugarcane ratoon stunt disease. (Agrios 2005)

FIGURE 15.21 Sugarcane mosaic disease. (Huang et al. 2018).

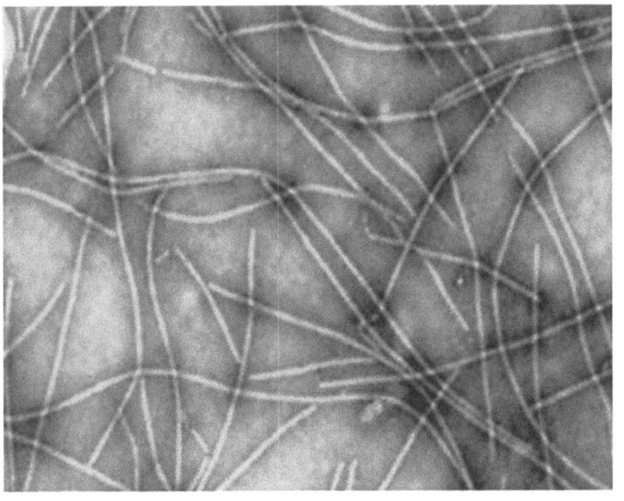

FIGURE 15.22 Sugarcane mosaic virus – flexuous thread virus in 80,000X. (Agrios 2005)

FIGURE 15.23 Streak mosaic. (Huang et al. 2018)

FIGURE 15.24 Yellow leaf syndrome. (Huang et al. 2018)

FIGURE 15.25 Leaf freckle/bacilliform disease. (Huang et al. 2018)

FIGURE 15.26 Maize streak virus disease. (Van Antwerpen et al. 2008)

FIGURE 15.27 White leaf. (Rao et al. 2012)

FIGURE 15.28 Green grassy shoot. (Rao et al. 2012)

FIGURE 15.29 Plants and leaves infected by the leaf yellow disease. (Rao et al. 2012)

15.3 ECO-FRIENDLY MANAGEMENT

Every year, loss of crop yield affecting the economy from the produce occurs as a result of biotic and abiotic stress conditions, despite the multiple defensive mechanisms created by plants to combat stress conditions and adapt. Preventing crop yield losses due to abiotic and biotic stressors is essential given the rapid global population expansion and the ensuing rise in food demand. Since 1979, numerous studies on plants' resistance to pathogens and herbivores have been conducted with the goal of understanding and controlling plant diseases. This is because a thorough understanding of plant defense mechanisms against pathogens and herbivores is needed to develop new strategies and enhance crop protection. Various biotic stresses and unfavorable environmental factors are encountered by plants. Plants react to these challenges in a variety of morphological, biochemical, and molecular ways, and there may be connections between their individual signaling pathways. Lots of pests and pathogens have been recorded since ancient times as the sources of biotic stressors in plants. The major pathogens that cause plant diseases include fungi, bacteria, nematodes, and viruses. Fungal pathogens can be classified into two types: biotrophs and necrotrophs, which use toxins to destroy their hosts' cells. Together with bacteria, they may infect various sections of the plant and produce symptoms including vascular wilts, leaf spots, and cankers. Although viruses seldom kill their hosts, other than causing local lesions there will be systemic harming which causes stunting, chlorosis, and deformities affecting many areas of the plant. Plant diseases put enormous biotic stress on plants, which results in farmers losing their crops and income besides spoiling the grains by producing toxins during storage. Different fungicides and a bactericidal molecule are being used by farmers to control the diseases, but using those resulted in environmental deterioration that is jeopardizing human health. Many plant diseases have evolved resistance to these substances, making it challenging to regulate plant health. To solve these issues, different eco-friendly approaches for management of the biotic stress are being employed by researchers. In sugarcane crops also, researchers have experimented with various eco-friendly management practices and the results are promising. This chapter is trying to epitomize the various eco-friendly approaches that are in vogue in sugarcane crops, with suitable examples.

15.3.1 Agronomic Practices

Initially, selection of disease-free planting material is the best way to prevent a disease. Healthy setts can be selected for cultivation. The period of planting should be chosen wisely, with an appropriate temperature and humidity. Better field irrigation management will be an advisable method to maintain proper moisture in the soil as well as to the crop. The waste after harvest can be burnt, as an act of killing the pathogen inoculum. Monitoring should be done regularly in nursery and commercial plots to avoid all major diseases.

Collateral hosts should not be grown near sugarcane fields, for example crops like sorghum when grown in the vicinity of sugarcane; sugarcane mosaic virus may spread through grasshoppers from sorghum to sugarcane (Anwar et al. 2010). Crop rotation is the best method for avoiding disease spread. Cultivating a different crop after harvesting sugarcane will be a better strategy to manage the disease incidence. The ratoons must be reduced in disease-prevailing areas and the rotation of non-infective crops *viz.* maize, rice, soyabean, alfalfa, etc., can be cultivated to avoid pathogen infection. Crop rotation reduces the source of pathogens and enhances soil fertility and soil structure (Huang 2018). Three-tier seed production, i.e., the level 1 to level 3 nursery system, can be used for the production of disease-free seed cane. This method increases the yield and sugar content in sugarcane, the ratooning period, thus leading to the economic benefit of sugarcane production.

At a temperature of 45–50°C and a depth of 5–6 cm, the radiant heat produced in the soil by sun radiation over a polythene sheet kills the insect and nematode pests. This method does not discharge any dangerous chemicals into the soil and is secure, practical, and affordable. It may be used with other techniques, like mulching and organic additives, to effectively control pests while growing crops. Damping off, root rot, stem rot, fruit rot, wilt, and blights caused by *Pythium* spp., *Phytophthora* spp., *Fusarium* spp., *S. rolfsii*, *R. solani*, *Sclerotinia sclerotiorum*, and *Verticillium* spp. are some of the major parasitic fungi and diseases that are effectively controlled by soil solarization. *Agrobacterium tumefaciens*, *Pythium* spp., *Macrophomina* spp., *Phytophthora* spp., *Verticillium* spp., *Rhizoctonia* spp., and other soil-borne fungal and bacterial diseases were effectively controlled by soil solarization (Chellemi and Mirusso 2006; Gelsomino et al. 2006). Using soil solarization technology, several of the microbial pathogens that cause wilt, root rot, and damping off diseases have reportedly been effectively controlled (Gill et al. 2017). Using numerous technologies, including mulching, push-pull technology, multiple crops, flowering/nectary plants, trap crops, cover crops, etc., many biotic pressures may be efficiently controlled. One of the agronomic strategies known as intercropping involves planting two or more crops in between the primary crop for increased yield, soil health, and financial gain.

Intercrops may include plants that are hosts or non-hosts of a certain pest. Some intercrops may reduce the number of pests that affect the main crop by attracting or repelling them. Allelopathic intercrops provide a good defense against soil-borne diseases and nematode pests. Intercrop root exudates successfully prevent the germination of spores, the development of pathogen mycelium, and the hatching of young nematodes. In the case of *Fusarium* spp., the allelopathic process has been well established in checking the pathogen (Hao et al. 2010).

15.3.2 Resistant Varieties

Although replacing susceptible varieties with resistant ones is the most common way to control the diseases, the breeding and selection of varieties with high resistance against a parasite is a huge task as the pathogens change their behavior in nature. Sugarcane varieties resistant to diseases can be cultivated to avoid biotic stress. Varietal resistance provides economically viable and complete management of the disease. Against smut disease, *Saccharum spontaneum* and *S. officinarum* genotypes have been found to provide resistance. Various clones have been discovered in *S. spontaneum* that are resistant to the smut disease (Viswanathan and Rao 2011). A few disease-resistant sugarcane varieties are listed in Table 15.2.

15.3.3 Heat Therapy

Setts are also treated with heat to inactivate or eliminate the pathogens. In thermotherapy, the differential effect of heat on the pathogen and the host is taken advantage of in freeing infected

TABLE 15.2
List of sugarcane disease resistant varieties

S.No.	Disease	Resistant Varieties	Moderately Resistant
1	Red rot	Co 90014, Co 0233, Co 86032, Co 86249, Co 87263, Co 87268, Co 94008, Co 98014, Co 2001-13, Co 0118, Co 0232, Co 0233, Co 0218, Co 05011, Co 0237, C0 0403	Co 89029, Co 99004, Co 2001-15, Co 0124, Co 0239,
2	Smut	Co 8371, Co 85004, Co 86032, Co 87025, Co 91010, Co 94008, Co 0403	Co 87044, Co 89029, Co 2001-13,
3	Wilt	Co 98014, Co 0232, Co 0233	

Source: ICAR-Sugarcane Breeding Institute, Coimbatore. (https://sugarcane.icar.gov.in/index.php/en/varieties)

stalks of grassy shoot, ratoon stunting, mosaic, leaf scald, downy mildew and smut. In this process, pathogens present inside the sett or on the surface are inactivated or eliminated at a temperature that is not lethal to host tissue. Heat therapy was not only found effective against certain seed-borne diseases but also against scale insects, mealy bugs and larvae of borers. Hot water treatment at 50°C for 2 hrs and moist hot air treatment at 54°C for 4 hrs was found effective. Hot water treatment at 52°C for 18 minutes is highly useful for controlling all sugarcane internally seed-borne diseases (Anwar et al. 2010). Heat therapy is one of the eco-friendly methods, where the pathogens are killed by heat while the sett/seed material tissues are undamaged. Grassy shoot disease (GSD) and ratoon stunting disease (RSD) can be prevented by aerated steam therapy (AST) and moist hot air therapy (MHAT). Any mistake in the treatment unit may result in damage to planting material. As an alternative to thermotherapy, meristem culture-derived seedlings can be used for seed nurseries producing disease-free seedlings. In the management of ratoon stunting disease, aerated steam therapy @ 50°C for 1–3 hrs or moist steam air @ 54°C for 4 hrs over the seed canes deactivates the pathogen's ability to infect. But care must be taken during treatment. *Sugarcane Mosaic Virus* can be managed by meristem culture alone or in combination with heat therapy (Balamuralikrishnan et al. 2002).

15.3.4 Biotechnological Tools

Genetic engineering is being used to improve the disease resistance against sugarcane yellow leaf syndrome (SYLS) (Huang et al. 2018). The resistance of the coat protein (cp) gene, the replicase gene, and the movement protein (MP) gene were developed and with the use of antisense RNA, the virus resistance properties were obtained. Somaclonal variation is one of the best methods for managing biotic stress in sugarcane *viz.* somaclones isolated with Fiji and downy mildew resistance (Manchanda et al. 2018), smut resistance (Dalvi et al. 2012), rust resistance (Oloriz et al. 2011), eye spot resistance (Manchanda et al. 2018), red rot (Shahid et al. 2012; Abdullah et al. 2013), sugarcane mosaic virus (Abdullah et al. 2013), improvement in yield (Doule et al. (2008) and sucrose content (Dalvi et al. 2012; Manchanda et al. 2018) are important examples. Somaclonal variation provides a variation that is induced because of mutagenesis and that variation could be more effective if it is successfully associated with selection on the cellular level and handling of large populations for screening (Hoy et al. 2003; Suprasanna et al. 2006).

Induced mutation through *in vitro* methods could be used for sugarcane disease management. Micropropagation is one of the *in vitro* methods for clonal multiplication of plants using shoot apical meristems as explant. Direct regeneration can be used in producing sugarcane plants. Apical/auxiliary meristem can be used or from immature leaf tissues (Geijskes et al. 2003; Lakshmanan et al. 2002) micropropagation can be done. The *in vitro* regeneration process for sugarcane that involves somatic embryogenesis has received the greatest research attention. Somatic embryogenesis has grown in popularity as an essential component of the genetic transformation system, while being first conceived as an alternate mechanism to regeneration. Numerous commercial sugarcane clones have been discovered to undergo somatic embryogenesis, which may be acquired either directly or indirectly from the leaf tissues. Embryogenic callus may be kept alive for many months without significantly losing its capacity to regenerate. Meristem culture has been utilized to effectively eradicate many diseases, including White Leaf Disease, Ratoon Stunting Disease, Sugarcane Yellow Leaf Virus (SCYLV), and Sugarcane Mosaic Virus (SCMV) (Kristini 2004; Tiwari et al. 2008). Meristem and callus cultures, together with heat treatment, were successful in creating pathogen-free stocks from plants infected with SCMV, downy mildew, and the Fiji disease virus (FDV) (Kristini 2004). Recent studies revealed that thin cell layer culture might be employed for direct plant regeneration to quickly produce disease-free plants from sugarcane afflicted with FDV and SCMV (Kristini 2004). The sweet potato tiny leaf (SPLL) phytoplasma was successfully eradicated from sweet potatoes using cryotherapy of the shoot tips. The Raspberry Bushy Dwarf Virus (RBDV), which may infect meristematic tissue, could not be eradicated by cryotherapy alone; however, RBDV could be eradicated by thermotherapy followed by cryotherapy. Cryotherapy may also be used to produce plants devoid of viruses and phytoplasma while also storing genetic material for the long term (Wang and Valkonen 2007). Additional research is required to discover and characterize antifungal genes implicated in disease resistance and to develop PCR diagnostic tools for phytoplasma linked to sugarcane yellow leaf syndrome and grassy shoot disease (GSD).

15.3.5 Biocontrol

Pseudomonas spp. strains when applied to sugarcane resulted in triggering induced systemic resistance (ISR) in the crop, against red rot disease. The bioagent also suppressed the soil-borne pathogen present in the rhizosphere (Nandita and Panigrahi 2017). *Trichoderma harzianum*, *T. viride* and *T. pseudokeningii* are found to be effective against sugarcane wilt (Viswanathan et al. 2009). *Bacillus xiamenensis* (PM14), a plant growth-promoting rhizobacteria (PGPR) promotes resistance to sugarcane against root fungal diseases. The microbe consists of 12 antibiotics which helps in preventing disease incidence (Xia et al. 2020). *Pseudomonas putida* NH-50 is used in the management of *Glomerella tucumanensis*, the pathogen causing red rot in sugarcane. *P. putida* produces pyoluteorin antibiotic which suppresses the infection due to the pathogen in the crop. The bioagent is applied through the soil drenching method (Hassan et al. 2011). Another biocontrol agent effective against red rot disease of sugarcane is *Trichoderma* (Ansari 2012). The mycelial growth and sporulation of *Colletotrichum falcatum* was effectively suppressed by *T. harzianum* (Vijai et al. 2008). Around 15 fluorescent *Pseudomonas* and *Trichoderma* isolates were antagonistic against *C. falcatum* under field conditions (Viswanathan and Malathi 2018). Lysis and mycoparasitism were the modes of action used by the antagonistic microbes to suppress the pathogen.

Pasteuria penetrans, a bacterial biocontrol agent has the ability to control *Meloidogyne* spp. that infects sugarcane (Bhuiyan et al. 2018). The crop gained induced systemic resistance when the *Pseudomonas* spp. strains are applied through sett treatment at the time of planting and by soil drenching at 60th and 120th day after planting (Viswanathan 2009). According to earlier research by Viswanathan (2018), *Pythium* spp. caused sugarcane seedling rot could be successfully controlled by *Pseudomonas* formulation. Additionally, *Pseudomonas* treatment in nursery trays or pots dramatically increased the seedlings' vigor. The use of biocontrol formulation in sugarcane seedling beds significantly decreased seedling rot, and the effectiveness of the biocontrol strategy in seedling nurseries decreased reliance on chemicals to manage seedling rot. Soil-borne plant pathogens can be managed by microbes that have biocontrol potential and colonizing in the rhizosphere. Biocontrol agents used for the management of major sugarcane diseases are given in Table 15.3.

TABLE 15.3
List of biocontrol agents used for the management of sugarcane diseases

S.No.	Disease	Biocontrol Agents	Reference
1	Red rot	*Trichoderma* spp.	Elamathi et al. 2016
		Pseudomonas spp., Plant Growth Promoting Bacteria (PGPR)	Viswanathan and Samiyappan 2008
2	Smut	*Trichoderma* spp.	Viswanathan et al. 2014
		Pseudomonas aeruginosa	Singh et al. 2021
3	Wilt	*Trichoderma harzianum, T. viride, T. pseudokoningii*	Viswanathan et al. 2012
4	Pineapple disease/Sett rot	*Trichoderma harzianum, T. pseudokoningii*	Rahman et al. 2009
5	Leaf scald	*Pantoea dispersa*	Zhang and Birch 1996

15.4 CONCLUSIONS

In order to reduce the reliance of farmers on pesticides and to increase their income through saving the cost due to the purchase of plant protection chemicals, researchers have been embarking on research to come up with pragmatic eco-friendly plant protection strategies. Sugarcane is one of the important cash crops that is being cultivated by a huge number of farmers who have been indiscriminately using plant protection chemicals against important diseases like red rot, grassy shoot disease, ratoon stunting disease, and the like, which culminates in environmental degradation. This important issue could be addressed by resorting to eco-friendly approaches or strategies. This chapter highlights the important literature related to agronomic practices and the selection of variety, which could be practiced by farmers at low or no cost. The review also pointed out biotechnological interventions, which have reached the pinnacle in managing diseases of sugarcane. Yet another important tool that this review brings to the fore is the biocontrol of sugarcane diseases. Plenty of literature supported the reduction in sugarcane diseases using biocontrol means. An array of microorganisms has been identified and their mode of action was demystified by various researchers. The exploitation of PGP microbes in the rhizosphere and other parts of the plants to mitigate biotic stress is another area that hogged the limelight in the recent past and which has also been highlighted in the review. Hence, this systematic review of the ecological management of sugarcane diseases will be very helpful for researchers working on sugarcane diseases and for students. Ultimately, these findings or strategies should reach farmers through effective extension strategies which would increase the income of sugarcane farmers.

15.5 FUTURE THRUST

- The biotechnological approach and importance of organic amendment of soil with substances like chitin that would improve the inducement of resistance, and further studies on this line would open up several possibilities.
- The application of endophytic bacteria in appropriate formulations and identifying appropriate methods of application are pivotal for effective disease management and more research endeavors are required to be taken in this area.

REFERENCES

Abdullah S, Khan FA, Iftikhar R, Raza MM, Aslam R, Hammad G, Ijaz U, Zafar MW, Ijaz U (2013) Detection of somaclonal variation in micropropagated plants of sugarcane and SCMV screening through ELISA. *J Agric Sci* 5(4):199–208.

Agrios GN (2005) *Plant Pathology*. Elsevier.

Ansari MI (2012) *Colletotrichum falcatum,* a causal organism of sugarcane red rot disease. *Trends Biosci* 5(2):95–96.

Anwar MS, Khan HWA, Muhammad M, Chattha AA, Amjad Z (2010) Integrated control strategies for sugarcane disease. *Pak Sugar J* 25(2):7–9.

Balamuralikrishnan M, Doraisamy S, Ganapathy T, Viswanathan R (2002) Combined effect of chemotherapy and meristem culture on sugarcane mosaic virus elimination in sugarcane. *Sugar Tech* 4(1):19–25.

Bhuiyan SA, Magarey RC, McNeil MD, Aitken KS (2021) Sugarcane smut, caused by *Sporisorium scitamineum,* a major disease of sugarcane: A contemporary review. *Phytopathology* 111(11):1905–1917.

Bhuiyan SA, Stirling GR, Garlick K, Anderson J, Wickramasinghe P, Wong E (2018) The bacterial biocontrol agent *Pasteuria penetrans* can help control root-knot nematode on sugarcane. *Proc Aus Soc Sugar Cane Tech* 40:304– 307.

Chellemi DO, Mirusso J (2006) Optimizing soil disinfestation procedures for fresh market tomato and pepper production. *Plant Dis* 90(5):668–674.

Dalvi SG, Vasekar VC, Yadav A et al (2012) Screening of promising sugarcane somaclones for agronomic traits, and smut resistance using PCR amplification of inter transcribed region (ITS) of *Sporisorium scitamineum*. *Sugar Tech* 14:68–75.

Doule RB, Kawar PG, Devarumath RM, Nerkar YS (2008) Field performance and RAPD analysis for assessment of genetic variation in sugarcane somaclones. *Indian J Genet* 68(3):301–306.

Duttamajumder SK (2008) *Red rot of sugarcane*. Indian Institute of Sugarcane Research.

Economic survey, 2021–2022.

Elamathi E, Malathi P, Viswanathan R, Sundar AR (2016) Potential of *Paenibacillus* and *Trichoderma* isolates against *Colletotrichum falcatum* causing red rot disease in sugarcane. *J Sugarcane Res* 6(2):1–12.

Gaur RK, Singh AP, Singh AK, Singh M, Raol GP, Mishra LC (2003) Serological identification of sugarcane yellow leaf luteovirus (SCYLV) from cane stalk juice. *Natl Acad Sci Lett* 26(5–6):143–145.

Gelsomino A, Badalucco L, Landi L, Cacco G (2006) Soil carbon, nitrogen and phosphorus dynamics as affected by solarization alone or combined with organic amendment. *Plant Soil* 279(1):307–325.

Geijskes RJ, Wang WLF, LiFang, Lakshmanan P (2003) Smartsett™ seedlings: tissue cultured seed plants for the Australian sugar industry.*cabidigitallibrary.org*.

Gill HK, Aujla IS, De Bellis L, Luvisi A (2017) The role of soil solarization in India: How an unnoticed practice could support pest control. *Front Plant Sci* 8:1515.

Hao WY, Ren LX, Ran W, Shen QR (2010) Allelopathic effects of root exudates from watermelon and rice plants on *Fusarium oxysporum* f. sp. *niveum*. *Plant Soil* 336(1):485–497.

Hassan MN, Afghan S, Hafeez FY (2011) Biological control of red rot in sugarcane by native pyoluteorin-producing *Pseudomonas putida* strain NH-50 under field conditions and its potential modes of action. *Pest Manag Sci* 67(9):1147–1154.

Hoy JW, Bischoff K, Milligan S, Gravois K (2003) Effect of tissue culture explant source on sugarcane yield components. *Euphytica* 129(2):237–240.

Huang YK, Li WF (2016) *Colored Atlas of Control on Diseases, Insect Pests and Weeds of Modern Sugarcane*. China Agricultural Press, pp 118–122.

Huang YK, Li WF, Zhang RY, Wang XY (2018) *Diagnosis and Control of Sugarcane Important Diseases. Color Illustration of Diagnosis and Control for Modern Sugarcane Diseases, Pests, and Weeds*. Springer, pp 1–103.

Kristini A (2004) The use of tissue culture to eliminate some important diseases in sugarcane (Doctoral dissertation, M. Sc. thesis, University of Queensland, Australia).

Lakshmanan P, Geijskes RJ, Elliott AR, Wang LF, McKeon MG, Swain RS, Borg Z, Berding N, Grof CPL, Smith GR (2002) A thin cell layer culture system for the rapid and high frequency direct regeneration of sugarcane and other monocot species. *Proc X Int Assoc Plant Tiss Cult Biotechnol*, Orlando.

Nandita J, Panigrahi MR (2017) Rhizobacteria and *Trichoderma* spp the potential biocontrol agents against sugarcane wilt pathogens. *Int J Res Eng Appl Sci* 7(10): 1–9.

Oloriz MI, Gil V, Rojas L,Veitía N, Höfte M, Jiménez E (2011) Selection and characterisation of sugarcane mutants with improved resistance to brown rust obtained by induced mutation. *Crop Pasture Sci* 62:1037–1044.

Rahman MA, Begum MF, Alam MF (2009) Screening of *Trichoderma* isolates as a biological control agent against *Ceratocystis paradoxa* causing pineapple disease of sugarcane. *Mycobiology* 37(4):277–285.

Rao GP, Mall S, Marcone C (2012) Recent biotechnological approaches in diagnosis and management of sugarcane phytoplasma diseases. *Funct Plant Sci Biotechnol* 6(2):19–29.

Ricaud C, Egan BT, Gillaspie AG, Hughes CG (2012) *Diseases of Sugarcane: Major Diseases*. Elsevier Science, p 352.

Rott P, Davis MJ (2000) *Leaf scald. In: A guide to sugarcane diseases* (eds) Rott P, Comstock JC, Croft BJ and Saumtally AS. CIRAD/ISSCT, Montpellier.

Saumtally AS (2000) *Sheath rot. In: A guide to sugarcane diseases* (eds) Rott P, Comstock JC, Croft BJ and Saumtally AS. CIRAD/ISSCT, Montpellier, 1770 180.

Shahid MR, Khan FA, Saeed A, Aslam M, Rasul F (2012) Development of somaclones in sugarcane genotype BF-162 and assessment of variability by random amplified polymerphic DNA (RAPD) and simple sequence repeats (SSR) markers in selected red rot resistant somaclones. *African J. Biotechnol* 11(15): 3502–3513.

Shan H, Li W, Huang Y, Wang X, Zhang R, Luo Z, Yin J (2017) First detection of sugarcane red stripe caused by *Acidovorax avenae* subsp. *avenae* in Yuanjiang, Yunnan, China. *Trop Plant Pathol* 42(2):137–141.

Singh P, Singh RK, Guo DJ, Sharma A, Singh RN, Li DP, Malviya MK, Song XP, Lakshmanan P, Yang LT, Li YR (2021) Whole genome analysis of sugarcane root-associated endophyte *Pseudomonas aeruginosa* B18—A plant growth-promoting bacterium with antagonistic potential against *Sporisorium scitamineum*. *Fron Microbiol* 12: 628376. doi: 10.3389/fmicb.2021.628376

Singh V, Srivastava SN, Lal RJ, Awasthi SK, Joshi BB (2008) Biological control of red rot disease of sugarcane through *Trichoderma harzianum* and *Trichoderma viride*. *Indian Phytopath* 61(4):486–491.

Sumael M, Lalusin A, Ocampo ET, Hernandez J (2020) Detection of downy mildew [*Peronosclerospora philippinensis* (W. Weston) CG Shaw] resistance in sugarcane (*Saccharum officinarum* L.) based on chlorophyll content and biomass. *J Agri Res Dev Ext Tech* 2(1):13–22.

Suprasanna P, Desai NS, Sapna G, Bapat VA (2006) Monitoring genetic fidelity in plants derived through direct somatic embryogenesis in sugarcane by RAPD analysis. *J New Seeds* 8(3). https:/doi.org.10.1300/J153v08n03_01

Tiwari AK, Yadav S, Tripathi S, Rastogi J, Bharti YP, Lal M, Sharma ML (2008) Performance of micropropagated, SCYLV infected and healthy plants of sugarcane. *Indian J Sugarcane Technol* 23(1&2):41–44.

Van Antwerpen T, McFarlane SA, Buchanan GF, Shepherd DN, Martin DP, Rybicki EP, Varsani A (2008) First report of Maize streak virus field infection of sugarcane in South Africa. *Plant Dis* 92(6):982–982.

Viswanathan R (2018) Changing scenario of sugarcane diseases in India since introduction of hybrid cane varieties: Path travelled for a century. *J Sugarcane Res* 8(1):1–35.

Viswanathan R, Malathi P (2018) Biocontrol strategies to manage fungal diseases in sugarcane. *Sugar Tech* 21(2):202–212.

Viswanathan R, Malathi P, Durai, AA, Prasanth CN, Scindiya M (2014) Sudden occurrence of wilt and pokkah boeng in sugarcane and status of resistance in the parental clones in national hybridization garden to these diseases. *J Sugarcane Res* 4(1):62–81.

Viswanathan R, Malathi P, Sundar AR (2009) *Final report of the network project on wilt of crops with special reference to cultural, morphological, molecular characterization and pathogenic variability of isolates in India*. Coimbatore: Sugarcane Breeding Institute.

Viswanathan R, Poongothai M, Malathi P, Sundar RA (2012) Sugarcane wilt: New insights into the pathogen identity, variability and pathogenicity. *Funct Plant Sci Biol* 6(2):30–39.

Viswanathan R, Premachandran MN, Balamuralikrishnan M, Jothi R (2003) A new stalk rot disease of sugarcane caused by *Phaeocytostroma sacchari* in India. *Sugar Tech* 5(1):61–64.

Viswanathan R, Rao GP (2011) Disease scenario and management of major sugarcane diseases in India. *Sugar Tech* 13(4):336–353.

Viswanathan R, Samiyappan R (2008) Bio-formulation of fluorescent *Pseudomonas* spp. induces systemic resistance against red rot disease and enhances commercial sugar yield in sugarcane. *Arch Phytopathol Pflanzenschutz* 41(5):377–388.

Vijai S, Srivastava SN, Ram Ji Lal, Awasthi SK, Joshi BB (2008) Biological control of red rot disease of sugarcane through *Trichoderma harzianum* and *Trichoderma viride*. *Indian Phytopath* 61(4):486–491.

Wang Q, Valkonen JPT (2007) Elimination of viruses and phytoplasma by cryotherapy of in vitro grown shoot tips: Analysis of all cases. *Adv Hortic Sci* 21(4):265–269.

Xia Y, Farooq MA, Javed MT, Kamran, MA, Mukhtar T, Ali J, Tabassum T, Rehman S, Munis, MFH, Sultan T, Chaudhary HJ (2020) Multi-stress tolerant PGPR *Bacillus xiamenensis* PM14 activating sugarcane (*Saccharum officinarum* L.) red rot disease resistance. *Plant Physiol Biochem* 151:640–649.

Yonzone R, Devi MS (2018) Red stripe/top rot disease of sugarcane: A review. *Int J Curr Microbiol Appl Sci* 7(1):1469–1478.

Zhang L, Birch RG (1996) Biocontrol of sugar cane leaf scald disease by an isolate of *Pantoea dispersa* which detoxifies albicidin phytotoxins. *Lett Appl Microbiol* 22(2):132–136.

16 Insect Pests of Sugarcane and Their Management

Arun Baitha, Singh M. R., Dinesh Singh, Srinivas K. and Hasan S. S.

16.1 INTRODUCTION

Sugarcane cultivation in India spans both subtropical and tropical regions, each presenting distinct agro-climatic conditions. Subtropical areas endure extreme weather and a brief growing season, while the tropics have warm climates and longer growing periods (Mazumder 2020). Productivity faces challenges from biotic factors such as pests, which can cause significant losses, estimated at 20% in cane yield and 15% in sugar recovery. While pests typically inflict more havoc in subtropical regions, tropical India also experiences pest severity under certain conditions, prompting the implementation of strategies to keep them in check. India's sugarcane crop contends with over 200 species of pests, with about two dozen insects identified as major threats. These include moth borers, termites, white grubs, scale insects, mealybugs, pyrilla, whiteflies, and woolly aphids, targeting the crop at various growth stages. Some pests, like stalk borers, internode borers, Plassey borers, Gurdaspur borers, green borers, scale insects, and woolly aphids, have spread unintentionally to new regions due to human activities. Over 200 different insect pests afflict sugarcane, with approximately two dozen deemed major pests. Natural enemies have been documented sporadically, with occasional reports of their significant role in controlling specific pests (David et al. 1986; Varma and Agnihotri 1993; Baitha and Varma 2016).

Pests of sugarcane can be grouped according to various criteria such as geographical distribution, taxonomic grouping, severity of damage caused, and feeding habit. Here, pests have been grouped by feeding habit as follows:

1. Pests that feed /bore on sugarcane – borers
2. Pests that suck the sap from the plant/leaf – sucking pests
3. Pests that feed on the subterranean parts of cane – soil pests

16.2 BORERS OF SUGARCANE

The sugarcane crop faces several discrete waves of attack of borers (in each brood) during its stay in the field. The borers of sugarcane pass through several generations (broods) in a calendar year. In the subtropical sugarcane belt, where borers pass the winter months in hibernation, the initial two broods at the beginning of the crop season (February–March) are very distinct (Mazumder 2020). In the sugarcane belt of the tropical south, due to mild winter, overlapping broods are common (at a given point in time, all four life stages remain available in the field).

Sugarcane is a year-round crop, consistently present in the fields. This uninterrupted availability provides an ideal habitat for most sugarcane borers, allowing them to thrive without needing to migrate to other plants or crops for sustenance and shelter. Thus, in the sugarcane agro-ecosystem,

DOI: 10.1201/9781032631578-16

TABLE 16.1
Distribution of sugarcane borers across India

Scientific name	Common name	Part attacked	Status	Distribution
Chilo infuscatellus Snellen	Shoot borer/Early shoot borer	Shoot as well as stalk	Major	Throughout India
Emmalocera depresella Swinhoe	Root borer	The underground portion of the cane	Minor	Andhra Pradesh, Karnataka, Gujarat, Maharashtra, West Bengal North India
Sesamia inferens Walker	Pink borer	Shoot as well as stalk	Minor	North India
Raphimetopus ablutellus Zeller	Green borer	Young shoot	Minor	North India
Chilo sacchariphagus indicus Kapur	Internode borer	Stalk	Major	Peninsular Occasionally subtropical India
Chilo auricilius Dudgeon	Stalk borer	Stalk	Major	Bihar, Uttar Pradesh, Haryana, Punjab, West Bengal
Acigona steniellus Hampson	Gurdaspur borer	Stalk	Major	North India
Chilo tumidicostalis Hampson	Plassey borer	Stalk borer	Minor	Bihar, Assam, Eastern U.P.
Scirpophaga excerptalis Walker	Top borer	Top portion during all stages of crop	Major	Subtropical and occasionally peninsular India

Sources: Srikant et al. 2014; Mazumder 2020, Anonymous 2022; Baitha et al. 2022a and Baitha et al. 2023.

the environment plays a pivotal role in the establishment and subsequent proliferation of borers. The crop is typically cultivated in regions surrounding sugar mills, where a plant crop is usually succeeded by one or two ratoon crops. This monoculture setup creates a stable agro-ecosystem conducive to the proliferation of insect pests. The dense sugarcane growth poses a significant challenge for chemical pest control. Consequently, alternative control methods have collected attention, with biological control emerging as a widely accepted approach. Technological advancements in biological control have progressed rapidly, leading to the adoption of several methods currently in practice. It involves the breeding and release of already-existing natural enemies in the environment, and the introduction of new natural enemies from the country of origin of the pest or from the pests of similar habits in other countries. The hardy nature of sugarcane crops means that they can withstand sizeable population densities of different borers. Therefore, the values of economic threshold and economic injury levels of major pests are too high.

Nine species of borers infest the crop in different sugarcane areas, singly or in combination with one or more species, every year. These are top borer (*Scirpophaga excerptalis* Walker), early shoot borer (*Chilo infuscatellus* Snellen), Stalk borer (*C.auricilius* Dudgeon), Plassey borer (*Chilo tumidicostalis* Hampson), internode borer (*C. sacchariphagus indicus* Kapur), Gurdaspur borer (*Acigona steniellus* Hampson), pink borer (*Sesamia inferens* Walker), root borer (*Emmalocera depressella* Swinhoe) and the green borer (*Raphimetopus ablutellus* Zeller) (Table 16.1).

16.2.1 Early Shoot Borer (*Chilo infuscatellus* Snellen)

The early shoot borer, prevalent in all sugarcane-growing regions of the country, primarily targets the crop during its initial growth phases. Its peak activity occurs during the summer months (March–June), characterized by elevated temperatures and low-to-moderate humidity. Borer activity decreases with the onset of the monsoon. It can co-occur with other borers like pink borer, green borer, and root borer. The larva enters the plant laterally by making entry holes at the base of the

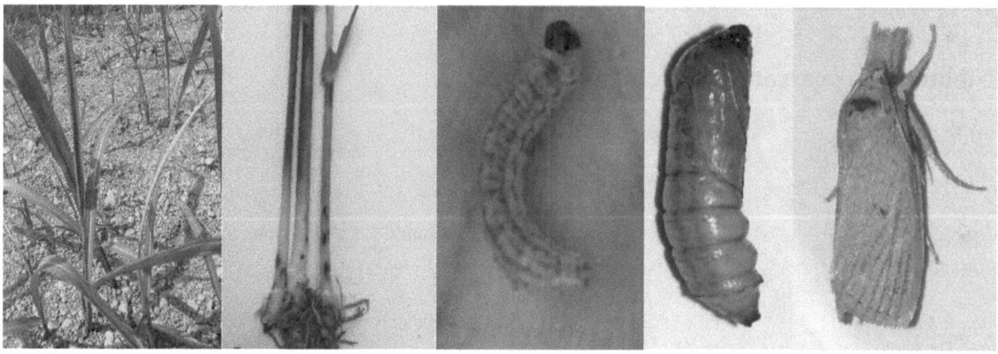

FIGURE 16.1 Early shoot borer.

FIGURE 16.2 Green affected cane (dead heart).

stalk and bores downwards or upwards killing the growing point. This would sever the central leaf spindle which dries up forming a dead heart that can be pulled out easily. The cut-off portion inside the bored plant decays and the dead heart emits an offensive odour when pulled out. Borer infestation during the germination phase kills the mother shoots resulting in the drying up of the entire clump creating gaps in the field. Shoot borer attack often induces compensatory tillering if the mother shoot is not damaged in the affected clump (Figure 16.1). A single larva can move between plants, causing damage to multiple ones. Mature larvae are dirty white with a dark brown head and five violet dorsal stripes.

16.2.2 Green Borer (*Raphimetopus ablutellus* Zeller)

It is a pest of very young shoots and is localized in Punjab, Haryana, U.P. and Bihar. The caterpillars of this borer are observed to occur in very small numbers in young sugarcane crops during April–June. It occurs in association with other species of sugarcane shoot borers, *Chilo infuscatellus*, *Sesamia inferens* and *Emallocera depresella* during summer. Due to the damage inflicted on the growing point of cane shoots, the central leaf whorl dries up (Figure 16.2). The attack of the species can be found only by digging the affected plants and examining the larvae.

The pest generally infests the young crop resulting in the drying up of the central whorl to form a dead heart, which when pullet out gives an offensive smell. It has been observed that temperature above 37°C with relative humidity raging between 20–40% is conducive for multiplication.

16.2.3 Pink Borer (*Sesamia inferens* Walker)

The pink borer has been observed attacking young sugarcane crops across the country. In Uttar Pradesh, its presence often coincides with that of the shoot borer (*Chilo infuscatellus* Snell). In most other states, it appears either alone or along with *C. infuscatellus*. Besides sugarcane, the pink borer has been documented attacking crops such as jowar, maize, paddy, bajra, wheat, sudan grass, narkal, johnson grass, and napier grass. In northern India, *S. inferens* typically targets sugarcane during the early stages of crop growth, resembling the behaviour of the shoot borer, *C. infuscatellus* (Figure 16.3). Two distinct types of attacks have been observed: primary and secondary. Primary attacks focus on late shoots or the top portion of the cane, initiated by newly hatched larvae entering the internodes through the bud point. Upon splitting open an affected internode longitudinally, numerous tunnels can be observed spiralling upwards, typically with one larva per tunnel, occasionally two. Normally, primary attacks are confined to a single internode without extending to adjacent ones. After completing their third instar, caterpillars create round exit holes through which they emerge, resulting in numerous exit holes on the internode's rind surface, corresponding to the number of caterpillars that fed on it.

FIGURE 16.3 Pink borer affected cane and larvae of pink borer.

A secondary attack occurs when mature caterpillars enter their solitary phase, with only one caterpillar entering a specific internode. In this scenario, the tunnel created is straight, with only two holes present on the internode's rind: the entrance and exit holes. Unlike in a primary attack, there are no minute holes. In a secondary attack, the tunnel may extend through two to three internodes.

16.2.4 Root Borer (*Emmalocera depresella* Swinhoe)

It is a significant pest in both the subtropical and tropical sugarcane belts of India, this pest attacks the underground stalk of the cane, with damage severity varying according to the crop's stage. In the early stages, typically May–June, it causes the formation of dead hearts, which cannot be easily pulled out and lack the offensive odour associated with shoot borers. Occasionally, one or two leaves near the central whorl also wither (Figure 16.4). Mature canes exhibit symptoms such as leaf yellowing. Damage detection requires uprooting the cane to inspect for larvae presence. First brood attacks affect tiller production, while subsequent broods progressively decrease cane length and weight. The larva feeds in an irregular semi-circular pattern and may leave the bored stem and enter into another underground stem of the same clump or adjoining shoots. Larvae hibernate from October to February. The life cycle is completed in 30–73 days depending on the season and region. The borer is active at high temperatures and moderate humidity levels. Generally, ratoon is more prone to its attack.

FIGURE 16.4 Root borer-infested sugar cane plants.

FIGURE 16.5 Top borer-infested sugarcane plants.

16.2.5 Top Borer (*Scirpohaga excerptalis* Walker)

Top borer poses a significant and consistent threat in subtropical regions but is only a minor and occasional nuisance in the tropics (Baitha et al. 2022a) (Figure 16.5). This pest can attack sugarcane at any stage of growth, typically leaving characteristic signs such as leaf mines, shot holes on open leaves, a weakened central shoot, stunted cane, and the sprouting of axial buds, resulting in the formation of a bunchy top. An early indication of an infestation is the tunnelling of the first larval instar into the midrib, creating a white streak that later turns reddish-brown, visible on the midrib's underside. The larva then tunnels into the central core through unfurled leaves, leaving rows of shot holes visible upon leaf unfolding. While feeding on the central core towards the growing point, the larva also nibbles on the surrounding inner leaf, causing it to dry up, atrophy, and turn dark brown, forming a dead heart when the fourth larval instar cuts the growing point. During the tillering phase, affected shoots may die, leading to the formation of side tillers. In the grand growth phase, crop growth is stunted, and the crown with the dead heart completely dries. In mature canes, the formation of dead hearts prompts lateral bud sprouting, giving rise to a 'bunchy top' appearance.

In subtropical conditions, the pest remains active from February to October, completing five broods and hibernating as larvae during the winter. Reduced rainfall in July impacts both the number and size of egg masses, while dry weather negatively affects hatchability. Abiotic factors are significant during the second brood, whereas biotic factors take precedence in the following broods.

16.2.6 Internode Borer, *Chilo sachhariphagus indicus* (Kapur)

The internode borer is a major pest of sugarcane in peninsular India with distribution in Tamil Nadu, Andhra Pradesh, Karnataka, Maharashtra, Bihar, Uttar Pradesh, and Haryana. The borer begins its activity soon after internode formation and the damage continues until harvest (Figure 16.6).

FIGURE 16.6 Sugarcane plants infested with internode borer.

The larva tunnels the inner tissues of the internode and pushes the frass to the exterior. The frass can be seen as a wet or dry mass hanging from the bore hole which can serve as an indicator of borer attack as well as the duration of attack. Heavy build-up of the borer occurs during July-December after the commencement of internode formation. Lodging, high dose of nitrogen, water-logged conditions and presence of a large number of water shoots favour pest build-up.

16.2.7 Stalk Borer (*Chilo auriLius* Dudgeon)

The stalk borer, a highly damaging pest of sugarcane in subtropical India, poses a significant threat in Uttar Pradesh, Haryana, Punjab, and Bihar (Avasthy 1981; Baitha et al. 2016) (Figure 16.7). Primarily feeding involves scraping the inner surface of these sheaths for approximately a week, completing two instars. Typically, two to six larvae infest a single leaf sheath, leading to its decay and subsequent drying of the entire leaf. As a result of feeding by the newly hatched larvae, leaves show distinct longitudinal orange-yellow streaks extending from tip to base on both sides of the midrib towards leaf margins. The grownup cane shows clear symptoms of infestation unless the leaf sheaths are removed when the boreholes may be seen clearly on the internodes. An infested cane may harbour 3–15 larvae at a time and the damaged internodes show symptoms of reddening which sometimes emit a rancid odour. Often in mature canes, three well-marked regions of borer infestation, *viz.* the lower, middle and top portions may be discerned during August–December. However, in epidemic conditions, larvae may be found in all the internodes of the infested cane during November–January. The borer completes five generations in Haryana, five to six in Uttar Pradesh, and seven in Bihar.

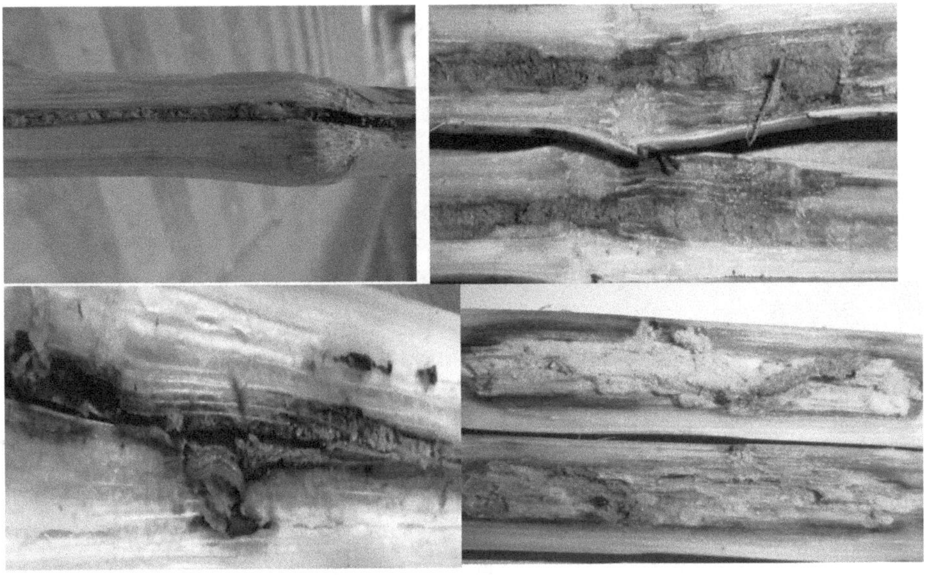

FIGURE 16.7 Stalk borer infestation in sugarcane crop.

16.2.8 Gurdaspur Borer (*Acigona steniellus* Hampson)

It is a major pest of sugarcane in Punjab, Haryana and western U.P. It is also reported from Sriganganagar in Rajasthan and some parts of Himachal Pradesh. The pest exhibits both gregarious and solitary phases. After hatching, larvae wander about for a few hours in search of a suitable site and enter the cane gregariously through a single hole. The entrance hole is always made on a sheathless internode and never below the third internode from the top. They feed gregariously in galleries in a spiral manner making minute punctures on the rind from within. Externally the tunnel appears as a dark spiral made up of a series of punctures. As many as 125 larvae may feed gregariously in one cane (Figure 16.8). After feeding two-thirds of the internode in a spiral manner, the larvae bore deeper and feed upward by making a straight tunnel. Third instar larvae start dispersing to other canes singly or in groups of two to three and in the later stages, one caterpillar is found infesting the internodes in each plant. Leaves wither due to larval feeding just below the rind but as the infestation advances and the central tunnel is formed, the entire whorl of leaves turns yellow and dries up. Due to the attack, canes fail to grow, damaged internodes become weak and infested stalks break off easily. Nodal buds in the remaining portions of canes are stimulated and side shoots are formed.

After the completion of two broods during June–September, more than 80% larval population comprising full grown third brood larvae move downwards to the below ground portion of the cane by November for hibernation. The pest flourishes well under moderate temperature and high humidity conditions. Early monsoon, heavy rains and water-logging also favour its multiplication.

16.2.9 Plassey borer (*Chilo tumidicostalis* Hampson)

Its name was drawn from its occurrence in an epidemic form at Plassey (West Bengal) in 1956 (Khanna et al. 1957). It is an important pest in some parts of West Bengal, Bihar and Assam (Baitha and Singh 2018; Mazumder 2020). The incidence of the borer can be easily detected by the presence of a dried crown of infested sugarcane. Borer attack occurs in two distinct forms. Primary infestation is caused by as many as 180 newly hatched larvae aggregating in the top three to five internodes. Due to such gregarious feeding, fresh red coloured frass is seen projecting out of the bore holes and

FIGURE 16.8 Gurdaspur borer infestation in sugarcane crop.

the top leaves of the canes ultimately dry up. Affected internodes break off easily and nodes adjacent to the infested internode develop sett roots which completely envelop the stalk; sprouting of nodal buds also occurs. In the case of secondary infestation, the grownup larvae disperse to either the lower healthy portion of the canes showing primary attack or the neighbouring canes. Larvae feed by tunnelling the internodes. One larva may bore into one to five internodes in a cane, but cane tops do not dry up in this form of damage (Figure 16.9).

There is a well-marked polymorphism and larvae are of four different types. However, a typical larva possesses four broad pinkish stripes present sub-dorsally and laterally in pairs. The pest gets activated from the end of February to the middle of November. All stages of this pest are simultaneously met with from April to October showing overlapping generations. Larvae hibernate from mid-November to February. The fourth and fifth broods overlap from February–June. Rainfall appears to favour its multiplication. The incidence is high in heavy soils and under waterlogged or flooded conditions.

FIGURE 16.9 Plassey borer infestation in sugarcane plant.

16.3 SUCKING PESTS OF SUGARCANE

16.3.1 Pyrilla (*Pyrilla perpusilla*)

Pyrilla poses a significant threat to sugarcane across many regions of the country, often reaching epidemic proportions in subtropical areas. Both adults and nymphs feed on phloem sap from leaves, leading to the development of sooty mould due to honeydew excretion (Baitha 2007). This direct and indirect damage adversely affects both sugar yield and quality. Nymph mortality is high during hot, dry weather, with surviving individuals maturing into adults by early July. The first and second broods are distinct, while subsequent broods overlap, with all stages of the pest present until December. The pest completes 4–5 generations in different parts of the country (Figure 16.10). Copious irrigation during hot months and high foliar density are conducive for the build-up of *Pyrilla*.

FIGURE 16.10 Pyrilla and its parasitoid on sugarcane.

16.3.2 Woolly aphid (*Ceratovacuna lanigera*)

The woolly aphid, previously confined to northeastern India since the late 1950s, emerged in epidemic proportions in peninsular India, specifically in Sangli (Maharashtra) and Belgaum (Karnataka) in 2002. Subsequently, it quickly proliferated in sugarcane-growing regions across Maharashtra, Karnataka, Tamil Nadu, Andhra Pradesh, Uttarakhand, Haryana, and Uttar Pradesh. These aphids infest plants by colonizing the underside of leaves (Baitha et al. 2021). The attack generally begins in the middle leaves and extends to the upper leaves. Within a leaf, colonies appear in the middle portion and extend to both ends. As the aphids grow and reach later instars, their body is covered with white waxy filaments giving the appearance of a white waxy coating to the infested leaves. A crop that has been attacked for about two months can be identified from a distance by the presence of white patches of plants contrasting with the green canopy background (Figure 16.11). The wax filaments that detach from the body of thousands of aphids drop down and cover the ground with a powdery coating giving it a white appearance which easily catches the attention of one's eye upon entering the field. The large volumes of honeydew excreted by the innumerable aphids settle on the upper surface of leaves located directly below the infested leaves. These leaves with the deposits not only gleam against bright light but also attract a variety of insects such as bees, wasps and flies, though regular attendance by symbiotic ants is seldom noticed. Sap sucking activity of huge colonies of nymphs and apterous adults causes stunted growth. The indirect damage includes the development of sooty mould on the honeydew deposits which interferes with photosynthesis; infested cane tops are believed to affect the health of cattle when fed to them. The pest generally begins colonizing a five-month or older crop though a younger crop is also known to support aphid attack in an epidemic situation. The pest remains almost until harvest and causes serious direct and indirect damage affecting both cane yield and quality.

The winged forms are responsible for plot-to-plot or long distance dispersal, the active first instar nymphs disperse to adjacent leaves or plants in the contiguous canopy in the morning hours. Another

FIGURE 16.11 Sugarcane woolly aphid infestation.

form of dispersal is through congregations of young nymphs on the leaf tip which are blown off by wind in a process called ballooning.

16.3.3 Whitefly (*Aleurolobus barodensis* and *Neomaskellia bergii*)

Whiteflies damage sugarcane by extracting large quantities of phloem sap from leaves which can result in greater than 50% yield reduction. The whitefly *Aleurolobus barodensis* is found in Andhra Pradesh, Bihar, Tamil Nadu, Haryana, Punjab, Gujarat, and Maharashtra. Significant clusters of *A. barodensis* nymphs feed on the sap from under the surface of the leaves, causing them to yellow (Figure 16.12). In severe infestations, leaves may exhibit a pinkish discoloration and gradually wither. Honeydew deposited on lower leaves encourages sooty mould growth which adversely affects photosynthesis. Whitefly infestation not only retards plant growth but also reduces sugar content in canes. *Neomaskellia bergii* is sporadic in occurrence and the colonies are restricted to small areas on leaves. The emergence of *A. barodensis* commences with the formation of a 'T' shaped split in the dorsum of the puparium. The entire life cycle spans from 25 to 47 days but extends to 119–130 days during winter. There are nine generations per year.

Drought during summer and dry spells during monsoon profoundly influence whitefly build-up. In semi-dry alkaline patches, a heavy incidence of the pest is observed due to physiological stress. Generally, crops subject to water-logging, especially in low-lying areas of eastern Uttar Pradesh,

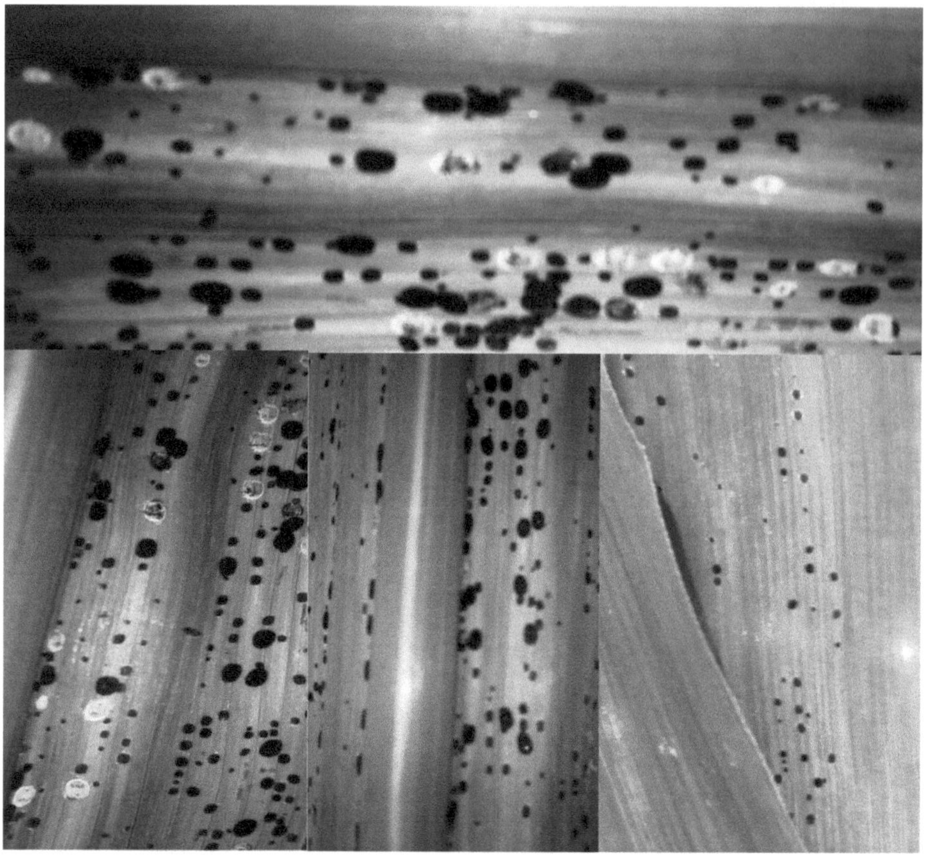

FIGURE 16.12 Whitefly (*Aleurolobus barodensis*) on the surface of sugarcane leaves.

Bihar, Tamil Nadu, Gujarat and Haryana, are heavily infested by whitefly. Water-logging leads to physiological starvation of the crop and is characterized by symptoms of nitrogen deficiency which promotes pest build-up. Similarly, the leaching of nitrogen due to heavy rains also promotes infestation levels. High levels of humidity generally favour its multiplication. In general, higher whitefly incidence is observed in neglected crops and unmanured ratoons as compared to manured ratoons. Varieties which have long and broad leaves are highly susceptible to the pest.

16.3.4 Black Bug (*Cavelerius sweeti* and *Dimorphopterus gibbus*)

Among the three black bug species infesting sugarcane, *Cavelerius sweeti* stands out as the primary pest of ratoons in north India, whereas *Dimorphopterus gibbus*, occurs sporadically. Nymphs and adults of *C. sweeti* inhabit leaf whorls and hide beneath the bases of leaves. Affected plant leaves exhibit a pale yellow colour with brown patches, and severe infestations lead to hole-riddled leaves as multiple bugs puncture the same spot. Gradual drying of leaf tips and margins occurs, potentially causing the entire plant to wither. This stunts plant growth and adversely affects juice quality as both nymphs and adults feed on sap, particularly impacting ratoon crops (Figure 16.13).

Black bug incidence is heavy in ratoon fields where trash is retained for a long time. Mild weather and large numbers of overwintering individuals are responsible for the outbreak of the pest during the pre-monsoon period. The loss due to bug damage is severe in unirrigated fields.

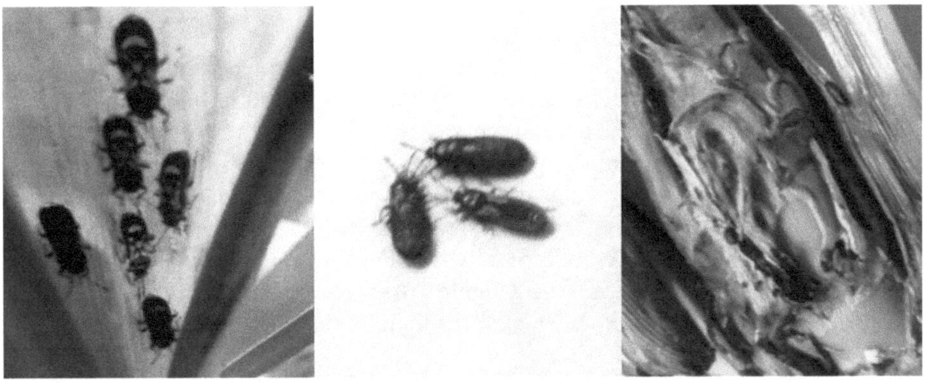

FIGURE 16.13 Sugarcane black bug infestation.

FIGURE 16.14 Scale insect appears on stem of sugarcane.

16.3.5 Scale Insect (*Melanaspis glomerata*)

Among the several species of scale insects that attack sugarcane, the armoured scale insect *Melanaspis glomerata* is economically important. The symptoms of attack in the field become evident only when heavy populations of the pest cover the entire cane, the encrustation giving a greyish-black appearance (Figure 16.14). When large colonies of different stages desap the cane, the crop loses general vigour, shrivels up and exhibits stunted growth with a reduction in intermodal length. Consequently, cane yield, percentage of juice extraction, juice quality and jaggery production suffer reduction.

Population growth is favoured by high temperatures and humidity from July to September. High humidity from September onward, encourages scale population growth and rainwater and high winds disperse crawlers to new areas.

16.3.6 Mealy bug

The mealy bugs derive their name from the whitish, powdery/filamentous mealy coating on their body. They are ubiquitous in all sugarcane-growing tracts of the world but seldom assume major pest status. Of the six species of mealy bugs that occur in India, adults of five species, including the common pink mealybug, are soft-bodied with well-developed legs. The adult males are delicate, alate insects and are not observed as commonly as the relatively large and apterous females. Depending on the species, severe infestation results in stunting of canes and yellowing of leaves. Loss of sap may even kill the young shoots or may result in a marked setback in cane growth, ultimately leading to the total drying of the crop (Figure 16.15. As a result of infestation, the stem gets covered with sticky honeydew which serves as a medium for the development of black sooty mould around the insect colonies.

The pink mealy bug *Saccharicoccus sacchari* is the most commonly occurring species throughout the sugarcane-growing areas of the country (Baitha et al. 2024). Their infestation starts as discrete colonies just below the node on the side opposite the bud, later spreading to the internodes as well. Germinating aerial buds also harbour the mealy bug. The adult females are distinctly pink in colour, with a round or dumb-bell-shaped, clearly segmented body. The yellow mealy bug *Kiritshenkella sacchari* commonly infests the nodal regions but it is also known to colonize leaf sheaths, internodes, and underground portions of stalk and roots as well. Adult females are flattened and covered by a white mealy secretion. Gregarious in habit, like the pink mealy bug, they are pale yellowish or light pink. *Dysmicoccus carens* generally infest the leaves in a severe form in potted plants under glasshouse conditions. Drought appears to be a major factor that influences mealybug populations. Rainfall dislodges populations and encourages entomopathogenic fungi. They are often found in large numbers on stunted canes. Varieties with loose clasping leaf sheaths harbour higher levels of mealybug populations than those with tight clasping leaf sheaths.

16.4 SOIL PEST

16.4.1 Termite, *Odontotermes obesus* and *Microtermes obesi*

Among the 13 termite species known to attack sugarcane from germination to harvest, the mound-forming *Odontotermes obesus* and the non-mound-forming *Microtermes obesi* stand out as significant.

FIGURE 16.15 Sugarcane mealybug.

FIGURE 16.16 Termites infested canes.

Termite infestation begins shortly after planting, affecting germinating setts and young shoots. Newly planted setts suffer the most serious damage as termites enter through their cut ends or the buds and feed on soft tissues replacing them with soil. The infested stools or shoots dry up soon after germination and these can be pulled out easily. Such shoots show no compensatory tillering. Occasionally 40–60% of the eye buds are destroyed leading to poor germination, gaps in the field and subsequent yield loss. In the ratoons, infestation occurs through the cut ends of the stubbles. In tillers, termites consume the inner tissues of the underground stem, filling them with soil. In mature canes suitable for milling, termites feed on the inner tissues, leaving the outer rind undisturbed. This creates cavities filled with moist soil containing termite galleries. Severely affected canes initially exhibit yellowing and drying of outer leaves, followed by inner leaves. Heavily damaged canes eventually collapse due to their weakened structure. Termite activity is observed in both loamy and heavy soils; areas of low rainfall or with poor irrigation potential are known to be more prone to damage (Figure 16.16).

16.4.2 White Grub (*Holotrichia consangunea*)

The white grub, comprising several species, is an important group of soil pest of sugarcane in India. Though inconspicuous late larval feeding activities, white grub has the potential to cause complete loss in sugarcane crops. In subtropical India, *Holotrichia consanguinea* has a life cycle similar to that of *H. serrata* but shorter in duration. In subtropical *Heteronychus sublaevis,* adults damage the underground portion of the stem during April and May while peak infestation is observed in August. The pest overwinters in adult and grub stages in earthen cells from October to February. Pupation occurs in March and adult emergence takes place in the following April. Adults feed on the underground portion of the stem leading to the drying of plants in a week after which the beetles move to other shoots (Figure 16.17).

Beetles emerge at dusk after first rains during April-June and congregate on preferred host trees like neem and *Ailanthus* where they mate, feed on the leaves and return to soil before day-break. Eggs are laid in the soil in sugarcane fields, fallow lands or grassy bunds along water channels. First instar grubs hatching feed on roots of grasses or organic matter initially and switch to sugarcane roots. After a short one-month period in each of the first two instars, third-instar grubs feed voraciously on sugarcane roots for three to four months. Pupation takes place in earthen cells from September and the adults emerging in early summer showers in the following year stimulate them

FIGURE 16.17 Sugarcane white grub.

to emerge out of the soil. *Holotrichia serrata* thrives in moist sandy soil as well as black cotton and clayey soils. Grubs can endure drought conditions by staying dormant inside earthen cells for over three months across various growth stages. High moisture is detrimental to the grub and heavy rains and waterlogged conditions force the grubs to come to the soil surface.

16.5 MANAGEMENT OF INSECT PESTS OF SUGARCANE

Insect pest management has been recognised as one of the important inputs of sugarcane production technology. In general, management of insect pests in sugarcane, owing to long duration of the crop and multiple pest complex, has combined two or more practices, largely due to their failure to achieve the desired objective singly (Srikant et al. 2016).

16.5.1 Early Shoot Borer

- Early planting should be followed in endemic areas to avoid synchrony between the susceptible crop stage and peak borer activity period (April–June).
- Two to three rounds of light earthing-up and trash mulching operations should be given during the early stages of crop growth. Trash mulching conserves soil moisture which is inimical to the borer.
- Frequent light irrigations and trash mulching should be practiced instead of heavy irrigation at long intervals.
- Release of egg parasitoid, *Trichogramma chilonis* Ishii @ 50,000/ha at 7 days interval in southern India and also release of *Sturmiopsis inferens* @ 125 gravid female/ha in Coastal Tamil Nadu.
- Soil application of Fipronil 0.3%GR @ 25–3 kg/ha or Chlorantraliniprole 0.4GR @ 18.75 kg/ha or Thiamethoxam 75 SG @ 160g/ha in 500–1000 lit at the time of planting in furrows over the cane setts before covering them with soil have been recommended in different states.

16.5.2 Green Borer

- Provide light earthings 2 to 3 earthings given during the early stages of the crop are beneficial.
- Timely cutting of attacked shoots at or just below ground level ensures the destruction of most of the caterpillars preset within the borer larva in the attacked shoot can also be pulled by first pulling out the dead heart.

16.5.3 Root Borer

- Mechanical destruction of affected shoots along with borer larvae in the pre-monsoon period reduces pest incidence.
- Give profuse and timely irrigation to ratoon and plant crops and remove borer-infested shoots 3–5 cm below the ground level and kill them at weekly intervals.
- Deep harvesting of cane meant for rationing helps in the reduction of this borer.
- Collection and destruction of moths using light traps also minimize the perpetuation of the borer.
- Egg parasitoid *T. chilonis* is released at a rate of 50,000 adults per hectare, with intervals of 10 days from July to October.
- Soil application of Fipronil 0.3G @ 25–33 kg/ha at the time of planting in furrows over the cane setts before covering them with soil.

16.5.4 Top Borer

- Autumn planting is to be followed as it escapes the borer attack in the ensuing winter months in subtropical India.
- Setting up of sex pheromone traps @ 50 numbers/ha for the first two broods and operated up to 30 days for each brood in subtropical states reduces moth population and brings down the severity of subsequent broods. However, trap catches appear to decline from the third brood onwards making its use uneconomical. In tropical India, traps can be set up in hot spots with lure changes decided by the extent of borer occurrence.
- Harvesting of crops (particularly those having high incidence) by the middle of February (before moth emergence) from the over-wintering brood in subtropical India causes a considerable reduction in the borer incidence due to the reduction of initial incidence of first brood during March and April.
- Avoidance of frequent irrigation during the peak period of moth emergence during summer protects the crop from heavy borer incidence.
- Suitable drainage to avoid water-logging, application of higher dosage of nitrogenous fertilizers and growing of resistant varieties with higher spindle length provide considerable protection of the crop.
- Release of *Isotima javensis* @ 125/ha from July to August manages them effectively.
- Release of pupal parasitoid, *T.howardi* @ 5000/ha and temperature tolerant strain of *Trichogramma japonicum* @ 100000 adults/ha released against II brood to IV brood of top borer.
- Application of carbofuran @ 66 kg/ha. or Chlorantraliprole 0.4GR @ 18.5 kg/ha or Chlorantraliprole 18.5% SC @ 375 ml/ha in 1000 lit of water in second week of June.

16.5.5 Internode Borer

- Removal of water/late shoots during September and October reduce incidence.
- Setting up sex pheromone traps @ 25 per hectare beginning fourth or fifth month age of the crop with three to four lure changes at 30–45 day intervals reduces moth population.
- Removal of trash with a view to destroy larvae and pupae is suggested to manage the pest.
- Release of egg parasitoid, *T. chilonis* @ 50,000/ha with intervals of 10 days from July to October.
- Release of larval parasitoid, *Cotesia flavipes* @ 500 females/ha with intervals of 7 days from July to November.
- Releasing of pupal parasitoid, *T.howardi* @ 5000 adults /ha in monthly intervals from July to October.

16.5.6 Stalk Borer

- Remove dry leaves and late shoots from the standing crop from October onwards and avoid taking seed canes from the heavily infested cane field.
- Water-logging and lodging of canes should be avoided.
- All water and late shoots should be harvested along with the harvesting of the crop.
- Release of egg parasitoid, *T. chilonis* @ 50,000/ha with intervals of 10 days from July to October.
- Release of larval parasitoid, *Cotesia flavipes* @ 500 females/ha with intervals of 7 days from July to November.

16.5.7 Gurdaspur Borer

- Affected canes with five to six top internodes harbouring borer larvae in the gregarious phase should be cut and removed regularly. The operation should be carried out on a campaign basis from the second fortnight of July to the end of September.
- Stubbles harbouring hibernating larvae, easily identified by the presence of round tunnel holes at the cut ends, should be uprooted and kill the larvae.
- Release of egg parasitoid, *T. chilonis* @ 50,000/ha at 10 days interval from July to October.
- Release of larval parasitoid, *C. flavipes* @500 females/ha at 7-day intervals from July to November.

16.5.8 Plassey Borer

- Collection of moths in light traps to minimize pest population.
- Collection and destruction of egg masses and removal and destruction of cane tops, infested by primary infestation from mid-February to mid-July.
- Release of egg parasitoid, *T. chilonis* @ 50,000/ha at 10 days interval from July to October.
- Release of larval parasitoid, *C. flavipes* @ 500 females/ha at 7 days interval from July to November.

16.5.9 Pyrilla

- Cane trash should be completely removed from the field after harvesting to be utilized as compost. This operation should be carried out until the middle of March to destroy unhatched egg masses and overwintering nymphs.
- Removal and destruction of the white-coloured puffy pyrilla egg masses or egg-bearing sugarcane leaves themselves at regular intervals minimizes build-up of populations.
- Egg masses or cocoons of the parasitoid collected from sugarcane fields can be colonized in target sugarcane fields at a dosage of 10,000 cocoons or 10,000,000 eggs per hectare. Leaf bits containing egg masses or cocoons are stapled on the under surface of the leaves in the center of infested field if the parasitoid inoculum is small. If a large consignment is available, it is distributed uniformly in the field.

16.5.10 Woolly Aphid

- Complete avoidance of transport of seed material or green tops as fodder from affected areas to minimize the risk of introduction into a new area.
- Destruction of affected leaves in the early stages of infestation and priority for the harvest of affected cane to prevent the spread of the pest in recently invaded areas.

- Wider spacing and paired row planting, and wrapping and propping operations render the crop canopy more amenable to monitoring and chemical control operations.
- Treated fields should be monitored and spot application of insecticides continued (spraying of Chlorpyrifos 20EC @1.5 l/ha in 500–1000 lit of water) if predators have not established.
- Mass multiplication of *Dipha* by tray rearing method and field release at dosages up to 1,000 cocoons per ha. Leaf bits bearing the cocoons can be inserted in the leaf axils of the standing cane.

16.5.11 White Fly

- The proper water management of crops to avoid extremes of dry conditions or water-logging should be followed. This should be coupled with nitrogen management, especially in ratoon crops, to minimize the pest.
- Detrashing and destruction of leaves bearing whitefly puparia before the emergence of the adults which can be determined by the presence of 'T' shaped split in the dorsum of the puparia.
- Redistribution of the nymphal puparial parasitoid *Amitus minervae* from high-activity fields to parasitoid-free fields holds the key to successful biological control.
- Spraying of Imidacloprid 17.80 SL@1.0–1.5 l/ha in 1000 lit of water.

16.5.12 Black Bug

- Removing of trash and crop residues after harvest, by the end of April and irrigating the fields reduce black bug infestation.
- Application of chlorpyriphos 20EC @ 750 ml/ha in 500–1000 lit directed towards the leaf whorls to target all stages.

16.5.13 Scale Insect

- Prevention of water-logging reduces humidity and helps minimize scale populations.
- Detrashing of crop two or three times beginning with internode formation would expose the colonizing sites as well as the crawlers that move upwards to settle underneath the tightly clasping leaf sheath.
- Wide-row spacing, wrapping and propping, and detrashing help maintain crop hygiene and facilitate insecticide application in endemic areas.
- Spraying of Imidacloprid 17.80 SL@1.0–1.5 l/ha in 1000 lit of water reduce the incidence.

16.5.14 Mealy Bug

- Use of pest-free seed cane is an essential pre-requisite for the prevention of the spread of the mealy bug.
- The removal of trash and stubble for effective control and to prevent the carryover of the pest into new fields
- In grownup crop showing considerable incidence of mealy bugs, detrashing is essential to expose the colonies to abiotic (high temperature and sunlight) and biotic (natural enemies) stresses.
- Detrashing is the most effective practice that can easily disturb and dislodge fully developed females. This practice is incidentally useful against other sucking pests like scale insect which too grow under the shelter of leaf sheaths.

- Avoidance of overdose of nitrogen and repeated ratoons in areas prone to the pest should be ensured. Ratoons can act as the inoculums for the next season crop since the mealy bugs can multiply in the underground portions of the plant.
- Manage the water stress and general health of the crop to reduce mealybug incidence.
- Hot water treatment of seed cane at 50°C for 3 hours, to prevent diseases of sugarcane, completely controls carryover of the mealy bug.
- Seed treatment with Imidacloprid 17.8% SC @ 0.75 ml/lit of water.
- In cases of severe infestation, Imidacloprid 17.80% SL @ 1.0–1.5 l/ha in 1000 lit of water should be sprayed after detrashing. Insecticide should be directed towards lower and freshly exposed internodes that harbour colonies.
- Indiscriminate use of insecticides should be avoided if activity of predators is observed in the field.

16.5.15 Termites

- Location and destruction of termite colonies in those species that build conspicuous mounds by breaking open the nest and removing the queen termite.
- Although secondary reproductive often replace the destroyed queen, the time delay ensures a reduction in overall colony build-up.
- Ploughing or disturbing high field bunds would expose and destroy colonies of subterranean termites.
- Clearing of crop residues and other debris in the crop ecosystem deprives the subterranean termites of food and reduces the activity of foraging workers.
- Application of Chlorpyriphos 20EC @ 6.25 lit/ha or Bifenthrin 10EC @ 1 lit/ha in 500 lit of water at the time of planting in furrows over the cane setts before covering them with soil.

16.5.16 White Grub

- Large-scale mechanical collection of beetles congregating on the most preferred neem trees during mass emergence, starting from the night of first showers and continuing for a week, reduces beetle population in the habitat progressively over a few years.
- Alternatively, kerosene vapour (Petromax) lamps may be placed near the trees to attract the dislodged beetles. The beetles that hit the light source and drop down should be collected in double-layered synthetic cement or fertilizer bags.
- Mechanical collection yields better results when carried out on a community basis with the active participation of both sugarcane and non-sugarcane farmers. Incentives may be offered for the beetles.
- Repeated deep ploughing at land preparation in February, i.e. well before summer showers, particularly in farms with neem trees nearby or fields around beetle-intense trees, exposes quiescent adults to desiccation and vertebrate predation.
- Ploughing and flooding for 24–48 h of similar fields in February, when planting is taken up during the pest activity period (June–September), would be useful in minimizing grub populations of different stages.
- Puddling and rotation with paddy can have a similar longer-lasting negative effect on grub populations. Rotation with the non-preferred sunflower is also advisable to break the continuous availability of sugarcane.
- Application of Fipronil 40% + Imidacloprid 40%WG @ 437.5–500 g/ha (for Holotrichia consanguinea) in 1000–1250 lit of water or Acephate 50 % + Imidacloprid 01.80 % SP @2.5lit in 500 lit of water.

16.6 FUTURE THRUST

The insect pests of sugarcane in India host numerous natural enemies. Multiplication of these natural enemies in different isolated geographical sectors over a period of time under the influence of natural selection pressure led to the development of new strains possessing wide adaptability to abiotic stresses. Selection of potential species from such new strains and their introduction into new areas may prove effective against the target borers. The sugarcane agro-ecosystem of the country has huge reservoir for bio-agents possessing different attributes. These appears to be ample scope for the identification of promising bio-agents within the county.

The semi-perennial nature of sugarcane habitats fosters natural biological control and offers an ideal setting for applied biological control methods. A primary strategy for controlling sugarcane pests through biological means is to prioritize the preservation of natural biological control mechanisms by avoiding disruptions to the system. Research and development efforts should focus on identifying, introducing, colonizing, and establishing major natural enemies, particularly in cases of accidental introductions or invasions from other countries.

The use of *Trichogramma* in protecting sugarcane from the stem borer is more a necessity than preference, since the crop and the pest concerned are such that conventional chemical control methods are neither useful nor possible. Realizing the significance and advances made in this field, several cooperative and private sugar factories in different states, established biological control laboratories to multiply natural enemies of pests specific to the region. *Trichogramma chilonis* heads the list of natural enemies with a nationwide distribution and most laboratories set up for its production. Several cooperative sugar mills and private factories initiated laboratories to multiply *T. chilonis* with financial aid from the state government. Most of the failures or ineffectiveness of the mass-bred (on factitious hosts) and released individuals in controlling the target pest may mostly be ascribed to the artificial laboratory condition in which the *Trichogramma* species are produced and to which the parasitoids become adapted.

Biocontrol technology can complement integrated pest management (IPM) strategies against sugarcane pests by curbing their carryover population in endemic areas. Sugarcane, cultivated as an industrial crop near sugar mills under expert supervision, offers an opportunity to establish laboratories in each sector for breeding potential bio-agents. This infrastructure supports the development of biological control-based IPM strategies, serving as a vital research resource in sugarcane cultivation regions.

REFERENCES

Anonymous (2022) *Annual Report, All India Coordinated Research Project on Sugarcane*, ICAR-IISR, Lucknow.

Avasthy PN (1981) Stalk borer, *Chilo auricilius* Dudg management in sugarcane. *Proc Nat Symp Stalk Borer* 31–42.

Baitha A (2007) Spatial distribution of *Pyrilla perpusilla* walker and their parasitoids. *Insect Environ* 13(3): 103–104.

Baitha A, Lal RJ, Maurya BL, Nigam R (2016) Bio-intensive management of internode borer (*Chilo sachhariphagus indicus* Kapur) and stalk borer (*Chilo auricilius* Dudgeon). *Int J Agril Inv* 1 (1): 80–83.

Baitha A, Singh BD (2018) Gregarious phase of Plassey borer, *Chilo tumidicostalis* in Sugarcane. *Ann Pl Protec Sci* 26(1): 219–220.

Baitha A, Tripathi GM, Singh MR, Roy S (2021) Status of invasive woolly aphid management on sugarcane in India. *Hexapoda* 28(1–2): 94–102.

Baitha A, Singh MR, Roy S (2022a) Identification of insect-pests of sugarcane and their control measures. In: *Compendium: Winter School on Recent Approaches for Doubling Farmers Income in Sugarcane Based Cropping System (December 08-28, 2022)*, Shukla et al. (Eds.), ICAR-IISR, Lucknow, pp. 295–308.

Baitha A, Jaiswal A, Kumar A (2022b) Distribution pattern of protective silken discs weaved by larvae of sugarcane top borer *Scirpophaga excerptalis* (Walker). *Indian J Entomol.* DOI:5958/IJE.2021.143 (Online)

Baitha A, Singh MR, Roy S, Singh D, Srinivas K, Goswami SK, Singh S, Raj C (2023) Biological control of sugarcane borers. In: *Sustainable Sugarcane Production and Utilization: Issues and Initiatives (Souvenir-Cum-Edited Book)*, Kumar N, Singh AK, Kamat DN, Kumar A, Minnatullah Md, Kumar A, Singh SN, Jha (Eds.), pp. 99–122. Dr. Rajendra Prasad Central Agricultural University, Pusa, Samastipur, Bihar.

Baitha A, Singh MR, Roy S, Singh D, Srinivas K, Kumar A (2024) Varietal preference of the sugarcane pink mealy bug, *Saccharicoccus sacchari* (Cockerell) (Homoptera: Pseudococcidae). In: *National Conference on Plant Health for Food Security: Threats and Promises, 1-3 February 2024*, Viswanathan et al. (Eds.), ICAR-Indian Institute of Sugarcane Research, Lucknow, pp. 204–205.

Baitha A, Varma A (2016) Field population dynamics of *Trichogramma chilonis* on eggs of sugarcane internode borer. *Ann Pl Protec Sci* 24(1): 86–88.

David H, Easwaramoorthy S, Jayanti R (1986) *Sugarcane Entomology in India*. Sugarcane Breeding Institute, Coimbatore, p. 564.

Khanna KL, Nigam LN, Puri VD (1957) Stem borer, *Chilo tumidicostalis* Hampson, a serious pest of sugarcane in Bihar. *Proc Indian Acad Sci (B)* 46: 75–95.

Mazumder SKD (2020) *Moth Borer of Sugarcane*. Daya Publishing House, New Delhi, p. 639.

Srikanth J, Easwaramoorthy S, Jalali SK (2016) A 100 years of biological control of sugarcane pests in India: review and perspective. *CAB Rev* 11 (013): 1–32.

Varma A, Agnihotri VP (1993) Microbial control of insect pests of sugarcane in India. In: *Fungal Ecology and Biotechnology*, Rai B (Eds.), pp. 47–54. Rastogi publication, India.

17 Pathogen-induced Biotic Stresses in Tomato Crops and Their Management Strategies

Anshu Kumar, Sabyasachi Mukhopadhyay, Krishnendu Kundu, Susmita Pati, Ashis Roy Barman, Subhendu Jash, Anindita Patwari, Subhas Patar, Ujjwal Sarkar, Dipasree Sadhukhan, Sujit Kumar Ray and Subrata Dutta

17.1 INTRODUCTION

The tomato (*Solanum lycopersicum* L.), the second most important vegetable crop in the world belonging to the Solanaceae family, originated in the South American Andes. The tomato is a useful source of dietary minerals, lycopene, vitamins (vitamins A, C and E), carotene, flavonoids, folic acid, and potassium ions (Willcox et al. 2003).

Tomato plants face several biotic and abiotic stresses which create a high impact on production worldwide. Some of the most common fungal diseases of tomato which causes huge losses in the production scenario are early blight (*Alternaria solani*), late blight (*Phytophthora infestans*), Septoria leaf spot (*Septoria lycopersici*), target leaf spot (*Corynespora cassiicola*), damping off (*Pythium aphanidermatum*) and wilts like Fusarium wilt (*Fusarium oxysporum* f. sp. *lycopersici*) and Verticillium wilt (*Verticillium dahliae*; *Verticillium albo-atrum*). Major bacterial diseases of tomato causing high growth reduction and/or mortality, degraded fruit quality and high yield loss include bacterial wilt (*Ralstonia solanacearum*), bacterial canker (*Clavibacter michiganensis* subsp. *michiganensis*) and bacterial speck (*Pseudomonas syringae* pv. *tomato*). Tomato plants get easily infected by many viruses resulting in huge damage in tomato production. Major viral diseases of tomato include spotted wilt caused by GBNV (Tospo virus), tomato fern leaf caused by Cucumber mosaic virus (CMV), tomato mosaic disease caused by Tomato mosaic virus (ToMV) and tomato yellow leaf curl caused by tomato yellow leaf curl virus (TYLCV). The overview of different biotic and abiotic stresses is presented in Figure 17. 1 and a list of economically important diseases of tomatoes and losses caused by them throughout the world are presented in Table 17.1.

17.2 DETECTION AND DIAGNOSIS OF PLANT DISEASES

Detection and diagnosis of plant diseases and their causal agents is one of the most fundamental steps for developing effective and holistic disease management strategies. The easiest way to identify the causal agents is to look for the symptoms they produce on the host plant. But this identification of causal agent through symptomatology does not always hold true in the case of a viral infection. Symptoms produced by a viral agent may be similar to those of other viral agents resulting in ambiguity in the identification of the exact causal agent of the disease.

DOI: 10.1201/9781032631578-17

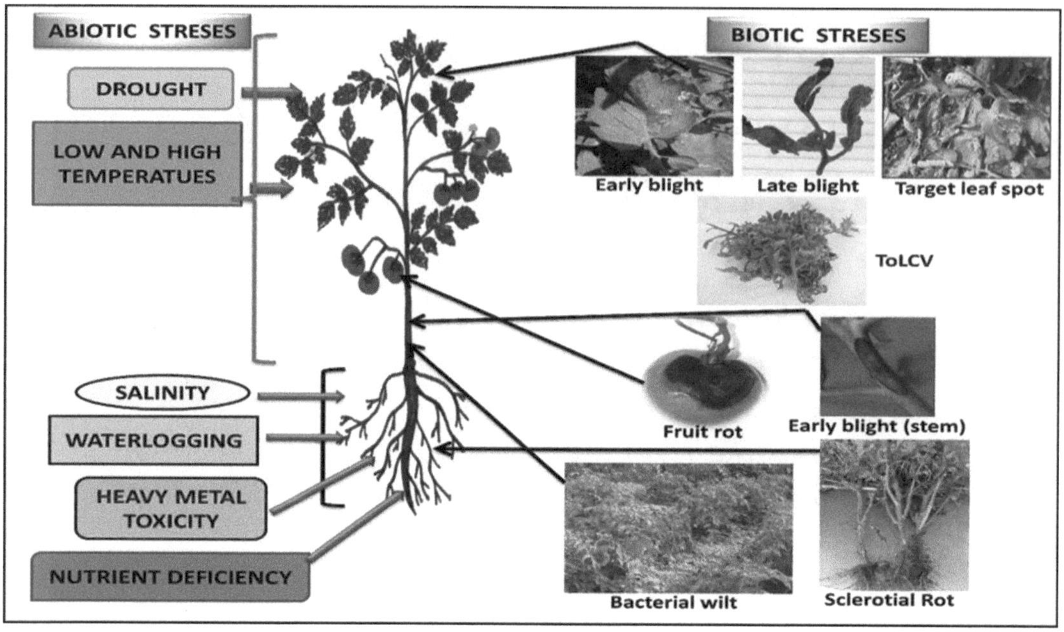

FIGURE 17.1 Overview of different stresses in tomato.

The progressive development of diagnostic tools and techniques indicates enormous development in terms of accuracy, ease and specificity. Earlier laboratory detection techniques were simply based on diffusion tests and hybridization techniques. Later, enzyme-linked immunosorbent assay (ELISA) and different variants of ELISA were established as effective techniques of detection. ELISA-kit was an efficient detection tool for the diagnosis of even two closely related species of viruses like TYLCV and tomato yellow leaf curl Sardinia virus (TYLCSV) (Dalmon et al. 2000). But with the advancement of polymerase chain reaction (PCR)-based screening, the overall trend of detection strategy has been transformed entirely. Multiplex RT-PCR was employed for the species-specific diagnosis of tobacco mosaic virus (TMV), and tomato mottle mosaic virus (ToMMV), *Xanthomonas* etc. (Sui et al. 2017). RT-PCR was designed against three wilt-causing fungi, *Fusarium solani*, *Rhizoctonia solani* and *Verticillium* sp., and the oomycete *Pythium ultimun* (Lievens et al. 2006). RT-PCR was also used to differentiate *Potato spindle tuber viroid* (PSTVd) and *Tomato chlorotic dwarf viroid* (TCDVd) (Matsushita et al. 2010). Rapid and simple detection of TYLCV has been performed by using loop-mediated isothermal amplification (LAMP) reaction with the formation of white precipitate indicating the presence of the virus (Fukuta et al. 2003). Furthermore, tomato germplasm was successfully screened against Tomato leaf curl New Delhi virus (ToLCNDV) using a chip graft inoculation assay (Akhtar et al. 2019).

Ralstonia solanacearum is grouped into five races based on its host range and five biovars depending on its ability to produce acid by utilizing different carbohydrate sources (Denny and Hayward 2001; Denny 2006). Molecular aptamers can be designed and used as a diagnostic tool for this race 3 biovar 2 (Champoiseau et al. 2009). PCR-based detection can be done with *R. solanacearum-specific* primers Y2 (5'-CCCACTGCTGCCTCCCGTAGGAGT-3') and OLI1 (5'GGGGGTAGCTTGCTACCTGCC3') as illustrated by Singh et al. 2010. ITS1 (TCCGTAGGTGAACCTGCGG) and ITS4 (TCCTCCGCTTATTGATATGC) primers are used for the characterization of *Fusarium* isolates (White et al. 1990). Universal primer, namely, uni-F (ATCATCTTGTGCCAACTTCAG) and uni-R (GTTTGTGATCTTTGAGTTGCCA) are employed to amplify all the three races, race 1, 2 and 3 of *Fusarium oxysporum* f. sp. *lycopersici* infecting tomato. Race-specific primer, sp13-F (GTCAGTCCATTGGCTCTCTC) and sp13-R

TABLE 17.1
List of economically important diseases, their characteristics symptoms, epidemiology and dissemination of pathogens causing important diseases of tomatoes

Name of the disease	Causal organism	Percent yield loss (Country)	Symptom	Favourable conditions	Survival and spread of the pathogen	References
Early blight	*Alternaria solani*	Up to 48.85 % (India)	Concentric ring or target board on leaf lamina, spots irregularly distributed.	Medium to high temperatures *i.e.*, 24–29°C high relative humidity (more than 80%) and intermittent rainfall.	In infected crop debris or soil for many years. Secondary spread occurs through water, air, agricultural implements or by other animals including human beings.	Chaerani and Voorrips 2006
Late blight	*Phytophthora infestans*	Up to 100 % (India)	Small, irregular, greyish, water-soaked spots on the adaxial surface of the leaves. Greyish white pathogenic growth on abaxial surface of the affected leaves. The foliage turns dark, withers and dries up within a few days. In case of fruit infection, irregular brown blotches appear which transform into foul-smelling masses and complete deterioration of the fruits occurs.	Cool nights and warm days and prolong leaf wetness and intermittent rainfall are congenial for the late blight disease. Temperature above 30°C is not favourable for the disease development.	Being seed-borne in nature, primary infection can happen through infected tomato seedlings. The secondary infection generally occurs through wind in favourable conditions.	Gleason and Edmunds 2006
Septoria leaf spot	*Septoria lycopersici*	Up to 100% (USA)	Appearance of light-colored greyish leaf spots with black or dark brown margin. Under favourable condition, numerous fungal fruiting bodies (pycnidia) are observed as tiny black specks at the centres of these spots. Lesions coalesce and entire leaf turns yellow, wither and eventually fall off.	High relative humidity, moist weather with intermittent showers and temperatures between 20–25°C is congenial for the disease.	Resting structure pycnidia can survive on different solanaceous weeds hosts e.g. *Solanum nigrum*, *S. carolinense*, *Datura stramonium* etc Secondary spread occurs through rain splash, wind or by mechanical means.	Sanoubar and Barbanti 2017

(continued)

TABLE 17.1 (Continued)
List of economically important diseases, their characteristics symptoms, epidemiology and dissemination of pathogens causing important diseases of tomatoes

Name of the disease	Causal organism	Percent yield loss (Country)	Symptom	Favourable conditions	Survival and spread of the pathogen	References
Fusarium wilt	*Fusarium oxysporum* f. sp. *lycopersici*	Up to 70 % (India)	Leaf yellowing and drooping of the plant; "yellow flag" symptoms; red-brown discoloration of the vascular tissue.	Infection is more in warm condition (28°C) predominated by dry weather condition and soil with acidic pH (5–5.6). Excessive application of micronutrients, phosphorous and nitrogen accelerates the infestation.	Chlamydospores in soil.	Shankar et al. 2014
Verticillium wilt	*Verticillium dahlia* and *Verticillium albo-atrum*	Up to 50% (Greece)	V-shaped lesions; light-tan or reddish discoloration of vascular and pith tissues can be observed. The discoloration is typically lighter in color than that of *Fusarium* wilt.	Temperature of 21–25°C is favourable for disease development. Fungi can enter the host through root wounds caused by interculture operations or by nematode feeding.	Produces micro sclerotia in soil and hypha in infected crop residue.	Suresh 2017
Target spot	*Corynespora cassiicola*	50–70 % (Sri Lanka)	Small, pinpoint, water-soaked spots appeared on leaves which later become enlarged (up to 2 cm diameter) and irregular with pale brown centre along with dark blackish or brownish rings surrounded by yellow halo. On the fruits small, light brown, slightly sunken lesions appear which enlarge in size and coalesce resulting in large areas of sunken, necrotic tissue.	Disease may occur between temperatures of 20–32°C; however, 24–25°C temperature coupled with 80–85% relative humidity.	Sporulates on moist crop debris or on dead tomato leaves. Spores remain viable up to two years. Spores are disseminated by wind or wind-blown rain or by wind-blown soil particles.	Shankar et al. 2014

Damping off	*Pythium aphanidermatum, Phytophthora spp,* and *Rhizoctonia solani*	>60% (India)	Seedlings rot at collar region. Root rot or collar rot causes stunting of the plant in extreme disease condition.	High relative humidity, moist damp soil, cloudy weather, low temperature (below 22–24°C) for a few days and poor aeration within the foliage.	Produce oospore in soil and spread through irrigation water or by agricultural implements.	Satyagopal et al. 2014
Bacterial canker	*Clavibacter michiganensis* subsp. *michiganensis*	Up to 84% (USA)	Vascular wilt, leaf spots and fruit spots are the typical symptoms. On fruits, slightly raised yellow to brown spots surrounded by a white halo are known as **"bird's eye spots"**.	Moderate temperature *i.e.*, 18–24°C and more than 80% relative humidity is favourable for disease development. Symptoms are more severe in sandy soils.	Seed-borne. Survive in infected crop debris, weeds, volunteer tomato plants and inside the seeds for several years.	Sharabani et al. 2014
Bacterial speck	*Pseudomonas syringae* pv. *tomato* (race 0 & 1)	Up to 75 % (Turkey)	Small, dark brown to black irregular lesions surrounded by a yellow halo appear on leaves. Lesions on the fruits are usually superficial, 1 mm in diameter and speck-like which may change to sunken lesion.	Cool rainy weather with 13–25°C temperature and sprinkler irrigation may lead to disease development.	The bacteria are seed borne. Dispersion mainly takes place through contaminated seeds, but long-distance spread occurs through splashing rain or by sprinkler irrigation.	Shankar et al. 2014; Suresh 2017
Bacterial wilt	*Ralstonia pseudosolanacearum, Ralstonia solanacearum*	60–90 % (India)	Yellowing, drooping and wilting of the plant. The vascular system shows yellow to light-brown discolouration which later turns dark brown, and the stem become completely hollow. Oozing out from the stem is the diagnostic feature of the disease.	Warm temperature (29–35°C) and high soil moisture.	The bacteria can persist in soil where it infects roots through natural wounds. Secondary spread occurs by rainwater or surface irrigation, or by intercultural operation.	Satyagopal et al. 2014; Suresh 2017
Spotted wilt disease of tomato	Tomato spotted wilt virus (TSWV)	Up to 42.1 % (Turkey)	Small, black, round dots on younger leaves appear which later turn bronzed, become dark brown or black in colour and withers along the leaf lamina. Several spots merge together and become necrotic with purple or violet colouration known as **"leaf bronzing"**.	Presence of infected alternate hosts in adjacent fields can pose a threat to newly planted crops. Temperature above 22°C is favourable for the hatching of egg of thrips which results in explosive spread of the vector and virus.	Circulative and persistent transmission by thrips species such as *Thrips tabaci, Frankliniella schultzi* and *Frankliniella occidentalis.*	Davino et al. 2018

(continued)

TABLE 17.1 (Continued)
List of economically important diseases, their characteristics symptoms, epidemiology and dissemination of pathogens causing important diseases of tomatoes

Name of the disease	Causal organism	Percent yield loss (Country)	Symptom	Favourable conditions	Survival and spread of the pathogen	References
Tomato leaf curl	Tomato leaf curl virus (TLCV)	Up to 80% (Pakistan)	Yellowing upward curling, distortion, reduce size of leaf and stunted plant growth.	Neither seed nor sap transmissible. Dodder transmission is reported in India.	Vector-Whitefly (*Bemisia tabaci*).	Sugano et al. 2011
Tomato fern leaf	Cucumber mosaic virus (CMV)	Up to 20% (India)	Young leaves become narrow and tendril-like known as **"shoestring"**. Severely infected plants are stunted, bushy in appearance and produce no marketable fruit.	CMV mainly found in temperate climate. It cannot survive in extreme dry condition. Optimum temperature for symptomatology is 25–30°C.	Non-persistent transmission by green peach aphid (*Myzus persicae*). Transmission by seeds.	Lecoq and Desbiez 2012
Tomato mosaic disease	Tomato mosaic virus (ToMV)	Up to 18% (India)	Mosaic, mottling, blistering and puckering of leaves. Yellow-colored necrotic patches might be seen on green or ripe fruits. Occasionally tomato mosaic virus is accompanied with potato virus X and causes severe damage resulting in the development of **"double streak"** symptom.	The virus may survive in hostile environmental conditions and can persist in infected plant debris for 2 years or in moist soil for at least 1 month. It can also survive in greenhouse conditions for long periods of time.	Active in seeds up to ten years. Seed-borne in nature, the virus can also be spread mechanically by contaminated hands, clothing or tools used in intercultural operations.	Shankar et al. 2014

(TCCTTGACACCATCACAGAG) are used for the identification of race 1 and 3 while primer, sp23-F (CCTCTTGTCTTTGTCTCACGA) and sp23-R (GCAACAGGTCGTGGGAAAA) is specific for race 2 and 3 as described by Hirano and Arie (2006). Polygalacturonase gene (pg1) is responsible for cell wall degradation and determines the virulent nature of *Fusarium* isolates which can be amplified using primer Endo F (CCAGAGTGCCGATACCGATT) and Endo R (AAGTGTTGGTAGGATAGTTG) (Hirano and Arie 2006). Similarly, fumonisin, a mycotoxin produced by *Fusarium* can be identified by a fumonisin-specific primer, FUM1 F (AGTCGTGCTGACGTCGAG) and FUM1 R (TGCTGCTGTCGCATCATA) (Ramana et al. 2011).

17.3 DISEASE MANAGEMENT STRATEGIES

Formulating a disease management strategy and needs-based disease forecast is not an easy task and needs a detailed understanding of the causal agents, weather parameters in which the crop is cultivated and dynamics of disease development in order to forecast disease outbreaks. This chapter is intended to give the reader a quick review of various strategies of plant disease management including cultural, physical, biological, and chemical, the concept of integrated disease management (IDM) and the recent trends in disease management.

17.3.1 Cultural Methods

17.3.1.1 Crop Rotation

Crop rotation can be defined as a "system of growing different kinds of crops in recurrent succession on the same land". Practices like monocropping increase this initial inoculum load build-up. Crop rotation reduces or eradicates the primary inoculums. Non-host crops or crops that have some adverse effect on the established pathogen population or crops that induce the population of beneficial microorganisms must be incorporated in cropping sequences for a predetermined number of seasons as per the survival ability of the pathogen in the soil in the absence of its host. It should be noted that modifications of the cropping system alone may not be sufficient to manage many soil-borne diseases because many pathogens possess a wide host range or produce durable resting structures to avoid harsh climatic conditions. Therefore, these agronomic measures have to be integrated with other disease management strategies for effective management of the pathogens.

"Prevention is better than cure" is the slogan rightly scripted for the management of viruses like ToLCV. Crop rotation including non-host crops of the vector should be followed to restrict the build-up of a large population of vectors. If feasible, the field should be left fallow to aid in the management of the vectors, as they may suffer in the absence of a host. Nurseries should be prepared in a protective environment, particularly under insect-proof nets to ensure minimum infestation of vectors. Field inspection is a must to look for the presence of whiteflies. Reflective plastic mulches or yellow plastic mulches repel while sticky traps made of cardboard with yellow color painted on them can attract and kill the insects. The other adjacent crops should also be monitored for the presence of whiteflies. Plants that show the presence of whiteflies should be bagged to prevent the flight of whiteflies to healthy non-infected plants. Furthermore, the virus-infected plants may be uprooted and burnt.

17.3.1.2 Intercropping

Intercropping means the cultivation of two or more crops simultaneously within the same field maintaining proper row-to-row and plant-to-plant distances. Moreover, any crop with an allelopathic effect on soil-borne pathogens as an intercrop with the primary crop can reduce the disease incidence in the current and onward seasons (Figure 17.2). Cultural practices (intercropping with marigold, mulching and stacking) when integrated with fungicides reduced the percent disease index and increased the yield. Intercropping tomatoes with soybean (*Glycine max*) or sesame (*Sesamum*

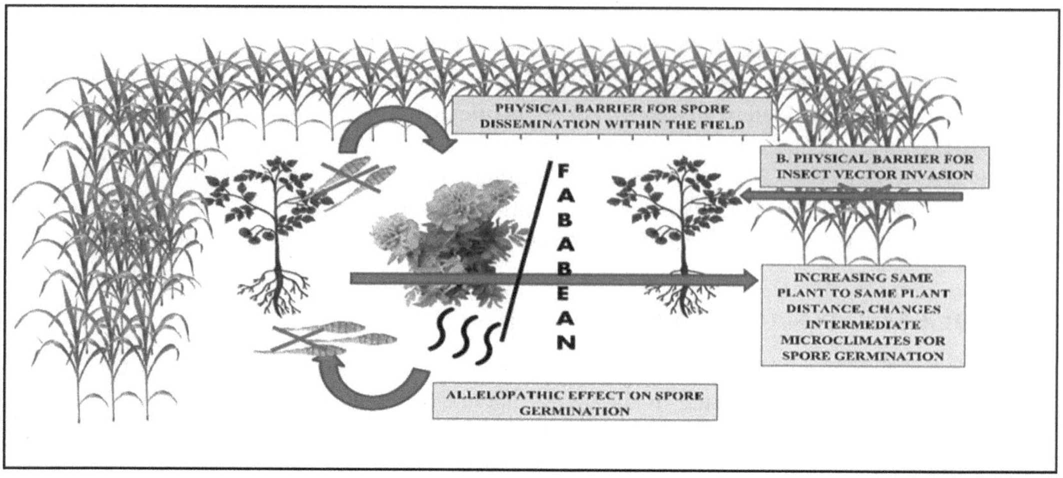

FIGURE 17.2 Illustration on the effect of intercropping and border crop on different diseases of tomato.

indicum), with sanitation, limited late blight development, but taller intercrops suppressed tomato growth and production (Tumwine et al. 2002).

17.3.1.3 Biofumigation

Biofumigation is based on the incorporation of fresh plant mass into the soil, which will release several volatile substances like isothiocyanates to suppress soil-borne pests. Dahlin et al. (2020) focused on the role of allyl isothiocyanate (AITC) in controlling the root knot nematode (RKN) *Meloidogyne hapla* by supplementing AITC with *Brassica juncea* cv. terrafit as biofumigant to manage RKN and obtained significant result.

Healthy plants particularly of the Brassicaceae family contain Myrosinase and Glucosinolate molecules in separate isolated areas of the same cell and when these two compounds interact during chopping, they produce unstable aglucones that spontaneously rearrange to a variety of reactive toxic compounds such as different isothiocyanates which have the potential to control soil-borne-pest and diseases. Gao et al. (2020) investigated isopropyl isothiocyanate derived from *Apium graveolens* on the mycelial growth of *Fusarium oxysporum*. The majority of the research is based on Brassicales plant biofumigation but there is a need to elucidate non-brassicales plant-based biofumigations in the near future. Some examples of brassicales and non-brassicales plant-source-based biofumigations are given in Table 17.2.

17.3.1.4 Organic Amendments and Its Role in Disease Management

The use of organic amendments could be a potential way to overcome the ill effects of chemical fertilizers, which helps to maintain the sustainability of soil ecology, environment and crop productivity. Compost, wood chips, charcoal, biochar, animal manure, straw, husk and sewage sludge are some of the examples of organic amendments. The ability to reduce soil-borne plant diseases using organic amendments such as compost, animal manure and organic industrial by-products has been well documented (Cheuk et al. 2005) as presented in Table 17.3.

Mostly, the application of some opportunistic microorganisms in soil *viz.*, plant growth-promoting rhizobacteria (PGPRs) such as *Bacillus, Pseudomonas*, etc., actinomycetes like *Streptomyces* and plant-growth-promoting fungi (PGPFs) such as *Trichoderma* and *Gliocladium* increases the efficacy of organic amendments. El-Tantawy (2009) observed that after application of goat manure and farm-yard manure on tomato plants as well as chitosan and aminofort spraying significantly increased all vegetative parameters compared to control treatments. They also found that the chitosan application increased the yield.

TABLE 17.2
Successful examples of pathogen suppression in tomatoes by utilizing different biofumigants

Pathogen	Disease of tomato	Sources of glucosinolates/active chemical compounds	Possible/ responsible active chemical	References
Alternaria solani	Early blight of tomato	*Moringa oleifera*	Sinigrin	Mvumi et al. 2018
Fusarium oxysporum f. sp. *lycopersici*	Fusarium wilt	Clove Mint, Oregano Spearmint	Essential oil	Selim et al. 2020
F. oxysporum f. sp. *lycopersici*	Fusarium wilt	Cruciferous plant tissue	Allylisothiocyanate	Ren et al. 2018
F. oxysporum f. sp. *lycopersici*	Fusarium wilt	Thyme	Essential oil	Ben-Jabeur et al. 2015
Pythium aphanidermatum	Damping off	Cruciferous plant tissue	Allylisothiocyanate	Ren et al. 2018
Phytophthora sp.	Late blight	Cruciferous plant tissue	Allylisothiocyanate	Ren et al. 2018
Ralstonia solanacearum	Bacterial wilt	Commercially available Thymol (derived mainly from *Thymus* sp. Plant Palmarosa)	Essential oil	Ji et al. 2005
Ralstonia solanacearum	Bacterial wilt	Indian mustard (*Brassica juncea*)	Isothiocyanate	Pontes et al. 2019
Ralstonia solanacearum	Bacterial wilt	Thymol Palmarosa Lemongrass	Essential oil	Pradhanang et al. 2003
Ralstonia solanacearum	Bacterial wilt	Clove	Essential oil	Huang et al. 2010
Ralstonia solanacearum	Bacterial wilt	Palmarosa	Essential oil	Deberdt et al. 2018

TABLE 17.3
Organic amendments and their impact on tomato disease management

Organic amendments	Disease	Findings	Researchers'
Compost-amended container media	Damping-off (*Rhizoctonia solani)*	Reduce damping off disease by 70%.	Krause et al. 2001
Decomposing fresh green organic manure	Soil-borne pathogens	Increases soil microbial activity.	Cardoso et al. 2006
Dhaincha leaf (25 q /ha) as green manure	Bacterial wilt	Reduce bacterial wilt disease severity by 42.3% in contrast to vermicompost (18.03% disease reduction).	Ghosh et al. 2015
Wheat straw biochar	Bacterial wilt	Reduce disease incidence by 61–78%. Biochar enhances plant growth while changing the rhizosphere's organic acid and amino acid composition.	Tian et al. 2021
Biochar amendment	Bacterial disease of tomato	Reduce pathogen population by 70–80% and utilize the plant-derived carbon in alkaline calcareous soils.	Liao et al. 2021
Tobacco stem biochar	Bacterial wilt	Upgrades soil physical properties, reduces pathogen population by reducing disease incidence and controls the bacterial wilt disease.	Li et al. 2022
Organic soil additives, compost and digestate, poultry manure	Soil borne pathogens	Compost and digestate (80%) perform better than poultry manure (50%).	Rizzo et al. 2022

The major constraint of using organic amendments is the emission of NH_3 which may have a negative impact on the environment and cause human and animal health hazards through eutrophication and nitrate contamination of ground waters and soil acidification. But overall, organic amendments provide an enriched source of nutrients by improving soil health and maintaining sustainable crop productivity. It improves the yield of crops by enhancing soil-beneficial microbial activity, plant nutrient mobility and availability and enhancement of soil fertility and productivity. Different soil amendments alone or in combination with the recommended dose of fertilizer or antagonistic biocontrol strains resulted in considerably higher yields in different crops in varying agro-climatic zones.

17.4 RESISTANT VARIETIES

Resistance of the host plant is the most effective strategy for disease management, whereas the heritable genetic potential of the plant is the ability to prevent reproduction, dissemination and disease generation potential of the virulent pathogen and thus reducing crop losses (Mitchell et al. 2016). Resistant cultivars are a great panacea and are the most trusted way of managing TYLCV disease (Gilreath et al. 2000). Arka Rakshak and Arka Samrat are the resistant varieties released by the Indian Institute of Horticultural Research (IIHR), Bangalore, against tomato leaf curl, bacterial wilt and early blight, whereas Arka Ananya is resistant to tomato leaf curl and bacterial wilt. Similarly, IIHR-Bangalore releases Arka Alok (BWR-5), Arka Shreshta, Arka Abha (BWR-1) and Arka Abhijit (BRH 2) resistant varieties to manage bacterial wilt of tomato. The College of Horticulture, Kerala Agricultural University, Vellanikkara, also developed resistant varieties like Shakthi, Mukthi and Angha for bacterial wilt. Hisar Anmol (H24) was developed by HAU, Hisar, against the tomato leaf curl virus. Kashi Amul, Kashi Adarsh, Kashi Aman, Kashi Amrit and Kashi Vishesh are the notified varieties released by the Indian Institute of Vegetable Research (IIVR), Varanasi, which is resistant against tomato leaf curl.

17.5 PHYSICAL METHODS

17.5.1 Soil Solarization

Soil solarization can be defined as a method of soil disinfestation by using transparent polythene mulch to cover the soil during the hot summer months, thereby reducing the soil-borne pathogens population (Katan and Gamliel 2014). Soil-borne fungal and bacterial pathogens can survive in soil for long periods of time and therefore management of these pathogens is a very problematic task. The environmentally beneficial practice of soil solarization was initially created to combat soil-borne pathogens. In this technique, solar energy is used to kill the fungal or bacterial pathogens present in the soil. Solarization decontaminates the soil thereby reducing pathogen inoculum significantly. In soil solarization, the field is covered with a transparent air-proof polythene sheet after land preparation for a few days to accumulate solar energy which will increase the soil temperature up to 45–60°C in the topsoil zone. Thus, under this high soil temperature, most of the pathogenic resting structures will desiccate and be killed. When the crop is sown after that, there will be reduced or no pathogenic infection. By using this technique several soil-borne diseases of tomatoes have been demonstrated to be managed successfully (Table 17.4).

17.6 GRAFTING TECHNIQUES FOR THE MANAGEMENT OF SOIL-BORNE PATHOGENS

Grafting is an asexual propagation technique in which two living plant components (the rootstock and the scion) are joined together to form a genetically composite organism that behaves as a single plant (Mudge et al. 2009). Arwiyanto et al. (2015) first reported experiments on grafting of commercial

TABLE 17.4
Successful management of tomato diseases by soil solarization

Disease	Pathogen	Comment	References
Fusarium wilt	*Fusarium oxysporum* f. sp. *lycopersici*	Reduced pathogen population by 86%.	Barakat and AL-Masri 2012
Damping off	*Pythium* sp.	Reduced disease incidence by >83% through soil solarization combined with biocontrol agents.	Jayaraj and Radhakrishnan 2008
Natural infection	*Pseudomonas fluorescens*	Solarization treatments increased the yield up to 45% while limiting bacterial infections by approximately 90%.	Castello et al. 2017
Bacterial spot	*Xanthomonas campestris* pv. *vesicatoria*	Soil solarization reduced the population of *Xanthomonas campestris* pv. *vesicatoria* from 10.68 to 8.79 CFU g^{-1}.	Misrak et al. 2014

TABLE 17.5
Successful examples of grafting in tomatoes for controlling tomato diseases

Diseases for which grafting used successfully	Rootstock	Scion	Reference
Verticillium wilt, Bacterial wilt, Fusarium wilt	*S. sisymbriifolium*, *S. torvum*, *S. intergrifolium* and *S. toxicurium*	Tomato	Kawaguchi et al. 2008
Bacterial wilt	Cheong Gang RST-04-106-T and Shield	Red Mountain tomato	Suchoff et al. 2019
Fusarium wilt, Bacterial wilt	CRA66, Hawaii7996 and Maxifort	German Johnson	Rivard et al. 2008
Bacterial wilt	RST-04-106-T, BHN 1054, BHN 998, BHN 1053, JakKung and Cheong Gang	BHN-602	Freeman et al. 2010
Bacterial wilt	BHN 998, BHN 1054, and RST-04-106-T	BHN 602	Kunwar et al. 2015
Bacterial wilt	Eggplant rootstock (KER-DC-117)	Karma 444	Pandey et al. 2020

cultivars with readily available rootstocks, which is considerably easier to do and has been shown to be effective in lowering the prevalence of bacterial wilt in many countries. Tomato grafting has been used successfully to combat various soil-borne diseases (Table 17.5). This method, along with maintaining high fruit quality, also helps to overcome various abiotic stresses like excess moisture, high salinity, and soil temperature extremes and also allows the extension of the growing season.

17.7 BIOCONTROL AGENTS

Biological management offers the possibility of increasing crop production with existing resources while avoiding the concern of pathogen resistance to chemicals and providing a reasonably pollution-free and risk-free solution. There is a plethora of information about the successful management of plant diseases through application of biocontrol agents (Table 17.6). *Pseudomonas*, *Bacillus*, *Serratia* and *Trichoderma* are some of the bacterial and fungal agents which are commonly applied as biocontrol agents (Mazzola et al. 2017). These biocontrol agents not only manage disease but additionally improve overall plant development functioning as PGPR.

TABLE 17.6
Biocontrol agents for the management of tomato disease

BCA	Target pathogen	Method of application	Mechanism of action	Reference
P. fluorescens isolate Pf1	*F. oxysporum* f. sp. *lycopersici*	Seeds treatment with talc-based bio formulation at 10 g / kg of seed followed by soil application @ 2.5 kg/ha	Accumulation of phenolics, activation of Phenylalanine ammonia-lyase (PAL), β-1,3-glucanase and induction of PO1 isoform, PPO1 and PPO2 isoforms and Chi2 isoform.	Ramamoorthy et al. 2002
Bacillus pumilis, Pantoea agglomerans	*Alternaria solani*	Foliar application of 0.5× 10^8 cfu/ml bacterial suspension	Antifungal compounds (subtiline, bacitracin and bacilomycin) *in vitro* mechanism.	Yazici et al. 2011
Serratia sp.	*R. solanacearum*	Soil application	Competition.	Guo et al. (2004)
Pseudomonas marginalis, P. fluorescens, P. putida, P. corrugata	*Pythium ultimum* or *P. aphanidermatum*.	Seed treatment with 4×10^7 bacteria ml^{-1}	Antibiotic compounds.	Gravel et al. 2005
Pseudomonas spp.	*C. michiganensis* subsp. *michiganensis*	Seed treatment with pure culture @ 1×10^8 cfu/ ml	-	Umesha 2006
	Ralstonia solanacearum	Seed treatment with 1×10^8 cfu ml^{-1}	ISR	Vanitha et al. 2009
Pseudomonas fluorescens SS101	*Phytophthora infestans*	Seed treatment with bacterial suspensions @10^9 cfu ml^{-1}	Production of antibiotic *i.e.*, cyclic lipopeptide massetolide A.	Tran et al. 2007
T. harzianum T22	*P. ultimum*	Seed treatment with the conidial suspension at the rate of 20 μl g^{-1} to deposit 2 × 10^7 cfu per gram of seed	Competition and mycoparasitism.	Mastouri et al. 2010
T. harzianum, T. viride, T. atroviride T. longibrachiaum	*F. oxysporum* f. sp. *lycopersici*	Seed treatment with 1×10^{12} cfu per 100 kg seed	Competition, cell wall degrading enzymes and mycoparasitism.	Enespa and Dwivedi 2014; Hafez et al. 2013; Marzano et al. 2013
Staphylococcus epidermidis (Endophytes)	*R. solanacearum*	Seedling root dip method with 10^7-10^8 cfu/ml for 12-14 h	Antibiosis and competition.	Nawangsih et al. 2011
P. macerans	*A. solani*	foliar application of 0.5 x 10^8 cfu/ml bacterial suspension	Antifungal metabolites, competition for nutrients and space.	Yazici et al. 2011
Paenibacillus lentimorbus,	*A. solani*	Foliar spray @ 1.8 × 10^9 cfu/ml	Cell wall degradation,	Khan et al. 2012
Streptomyces rubrolavendulae	*P. infestans*	Inoculated on peat moss @ 10^6 cfu/g	Cell wall degrading enzymes; chitinase and cellulase activities.	Loliam et al. 2012

TABLE 17.6 (Continued)
Biocontrol agents for the management of tomato disease

BCA	Target pathogen	Method of application	Mechanism of action	Reference
Ralstonia pickettii QL-A6	*R. solanacearum*	Stem injection @ 10^5 cfu	Shoot and root colonization; competition for resources.	Wei et al. 2013
Bacillus pumilus	*F. oxysporum* f. sp. *lycopersici*	Soil drenching with 10 ml bacterial suspension having 1.5×10^7 cfu/ml	Siderophore production, chitinase activity.	Heidarzadeh and Baghaee-Ravari 2015
Pseudomonas veronii	*Pseudomonas syringae* pv. *tomato* (race 0 & 1)	Seed treatment @ 1×10^8 cfu ml^{-1}	Antibacterial compounds.	Romero et al. 2016
Bacillus cereus	*F. oxysporum* f. sp. *lycopersici*	Soil drenching with 25 ml bacterial suspension having 10^8 cells/ml	Proteolytic and chitinolytic activity.	Abdallah et al. 2016
Alcaligenes faecalis	*F. oxysporum* f. sp. *lycopersici*	Soil drenching with 25 ml of bacterial suspension having 10^8 cells/ml	Chitinolytic activity.	Abdallah et al. 2016
Pseudomonas putida	*C. michiganensis* subsp. *michiganensis*	Seedling root dip method @10^9 cfu/ml for 2 h at room temperature	Production of phenolic compounds.	Aksoy et al. 2017
P.aeruginosa PM12	*F. oxysporum* f. sp. *lycopersici*	Soil drenching @ 10 mM with 3-hydroxy-5-methoxy benzene methanol (HMB)	ISR by HMB.	Fatima and Anjum 2017
Pythium oligandrum	*Rhizoctonia solani*	Soil drenching	Antibiosis and mycoparasitism.	Abdeljalil et al. 2021
Aspergillus spp. (*Aspergillus alabamensis*+ *A. tubingensis*+ and *A. oryzae*)	*F. oxysporum* f. sp. *lycopersici*	Soil treatment	Production of phenolic compounds and increased activity of oxidative enzymes.	

Transcriptome analysis revealed that Phenazine-1-carboxylic acid is responsible for altering the gene expression in *Phytophthora infestans* when *Pseudomonas* strains are used as a biocontrol agent (Roquigny et al. 2018). *Bacillus subtilis*-based commercial biocontrol products such as Companion and Serenade can also be used to suppress the growth of *Phytophthora infestans* (Elliott et al. 2009). Endophytic bacteria potentially reduce the growth of pathogens. The fast colonization ability of endophytic bacteria may prevent the build-up of pathogens in the xylem vessel. Nawangsih et al. (2011) for the first time found *Staphylococcus epidermidis* as an endophyte of tomato. Inoculation with non-pathogenic strains of the pathogens also demonstrated control from pathogenic agents. *Trichoderma harzianum* and *Streptomyces griseoviridis* are effective biocontrol agents for *Verticilium* wilt disease of tomato pathogen. Bacterial canker is also managed by the application of vesicular arbuscular mycorrhiza (VAM) and inoculation of avirulent mutants of *Clavibacter michiganensis* subsp. *michiganensis*. Actinomycetes can be used for controlling different soil-borne diseases including late blight of tomato (Shahidi et al. 2006). Seed treatment by the ethyl acetate extracts of *Steptomyces* spp. isolates significantly reduced disease severity (Abdel-Aziz et al. 2021).

Trichoderma has been established as an effective biocontrol agent for various plant diseases. The reason for the high level of success of the utilization of *Trihcoderma* is its strong antagonistic behavior towards other fungi and easy multiplication in the laboratory. According to Naher et al. (2014), the mode of action entails a number of mechanisms, including mycoparasitism, competition, antibiotic synthesis and the ability to induce systemic resistance. Al-Surhanee (2022) showed that the combination of *T. harzianum* and salicylic acid was found to be a commercially viable tool for improving defenses against *Fusarium* wilt of tomato.

17.8 CURRENT TRENDS IN PLANT DISEASE MANAGEMENT PRACTICES OF TOMATO

Plant disease management plays an important role in reducing losses caused by various pathogens. Several earnest attempts have been made to cultivate disease-resistant cultivars that not only safeguard but significantly boost the production of crops. But plant-pathogen interaction is dynamic in nature and the pathogen keeps recasting itself with respect to the changing environment. So, to cope with turbulence in the behavior of pathogens, the disease management strategies need to ameliorate with superior tools and techniques. There is a need to upgrade the available resources with more precise tools to manage the most stentorious disease of tomato, like TYLCV. It is the most investigated virus of all time due to its devastating influence on overall tomato yield. The widespread nature and divergent host range offer arduous challenges in the management of this stubborn viral disease. Management practices require the integration of traditional and non-traditional tactics, ranging from quarantine to typical IPM procedures, as well as advances in molecular approaches such as transgenics, genome editing, and RNA interference (RNAi) technology.

This section will give a glance at current trends in plant disease management of tomatoes exploring the immense potential of molecular approaches comprising marker-assisted breeding, gene pyramiding, eco-friendly techniques of virus-induced gene silencing (VIGS), the transgenic approach and application of antibody production in plants for management of virus. Plant pathogens have been combated using molecular techniques since the introduction of molecular tools for disease management. This section will also deal with the treasure of RNA biology which consists of RNAi, spray-induced gene silencing (SIGS) and artificial micro-RNA mediated gene silencing technique. The potential of clustered, regularly interspaced, short palindromic repeats (CRISPR-Cas 9) and assisting molecular chaperones to combat plant disease is also discussed. This section also discusses nanoparticles as an emerging technique for managing tomato disease. The concept of artificial intelligence, which has the potential to revolutionize plant protection, is also addressed.

17.9 MOLECULAR APPROACHES

Conventional breeding techniques such as hybridization, backcross, composite breeding, pure line selection and recurrent selection are time-consuming, expensive and inconvenient. Furthermore, these time-consuming approaches could not cope with the breakdown of resistance caused by rapidly changing microorganisms. As a result, breeders have redesigned breeding approaches to pave their way into the new world of molecular biology which consists of tools like marker-assisted selection (MAS), gene pyramiding, RNA-mediated gene silencing, VIGS and others to explore the molecular nature of the pathogen and to combat disease in less time. Molecular approaches provide a strong vision, especially for the management of viral diseases.

Markers are readily employed for the rapid detection of the resistance gene in the pathogen. In a study, markers were developed and/or evaluated in the identification of the target resistance gene *i.e.*, *Ve1* and *Ve2* genes responsible for resistance against *Verticillium dahlia*; *Tm1*, *Tm2*, and *Tm2*^{2} genes, which confer ToMV resistance; *I* and *I2* genes linked showing resistance against *Fusarium oxysporum* f. sp. *lycopersici* (Arens et al. 2010). There is substantial genetic diversity among different isolates of *Fusarium oxysporum* f. sp. *lycopersici* derived from various geographical regions in India, as validated by the random amplification of polymorphic DNA (RAPD) pattern (Manikandan et al. 2018), and this genetic variability may be correlated with different levels of pathogenicity among *Fusarium oxysporum* f. sp. *lycopersici* (Sharma et al. 2014). TOMSatdb (Tomato Microsatellite database) is the first database of microasatellites present in the entire genome which allows the web access of up to more than 1.4 million DNA markers (Iquebal et al. 2013). Mapping of the resistance genes and marker-assisted breeding of TYLCV contributed to the development of novel tomato varieties. Polymorphic DNA markers have been used to map the *Ty-1* gene and *Ty-3* (Ji et al. 2007) on chromosome 6; *Ty-4* on chromosome 3 (Ji et al. 2009) and *Ty-5* on chromosome 4 (Anbinder et al. 2009) from different wild *Solanum* species. Molecular markers were developed for *Ty-2* and *Ty-3* genes for screening of germplasm against TYLCV (Kim et al. 2020). Marker-assisted backcrossing for combining *Ty-1, Ty-2, Ty-3, Ph-2* and *Ph-3* genes followed by phenotypic selection resulted in the development of multiple disease-resistant lines providing resistance against leaf curl and late blight disease of tomato (Kumar et al. 2022).

Breeding approaches such as gene pyramiding are often employed for the development of resistance cultivars by incorporating resistant genes in tetraploid cultivated varieties of tomato from wild solanum species such as *S. cheesmaniae*, *S. chilense*, *S. habrochaites*, *S. peruvianum* and *S. pimpinellifolium*. TY20, the first commercially available tolerant cultivar derived from *S. peruvianum*, postponed the occurrence of disease symptoms and the development of viral DNA (Pilowski and Cohen 1990). TYLCV-resistant cultivars were often mildly affected by the virus and maintained the optimum yield of tomatoes. Resistant cultivars maintained a pool of dominant and semi-dominant genes which function in different ways to execute resistance against TYLCV (Anbinder et al. 2009; Ji et al. 2007, 2009). Gene pyramiding of *Ty-1* and *Ty2*; *Ty-3, Ty-4* and *Ty-5; Ty-6* and *Ty-5* significantly lowered the TYLV disease severity while *Ty-3* and *Ty-4* provided resistance against tomato mosaic virus (Hutton et al. 2015).

VIGS, which is based on the post-transcriptional gene silencing (PTGS) technique, is an excellent tool for gene functional study in plants. Plants overcome viral infection by introducing double-stranded RNA (ds RNA) accomplishing the PTGS to kill the invading viral DNA. VIGS is a technique in which viral DNA is modified to introduce the target gene fragments, thus functioning as vectors and VIGS vectors carrying the specific gene to induce PTGS in the host plant. VIGS is successfully employed to study the metabolic and developmental process, analyzing different aspects of biotic and abiotic stresses in tomatoes, etc. The technique of VIGS unfolds the governing pathway and gene responsible for bacterial infection and may be exploited to limit the further spread of the pathogen. Coronatine, a non-host-selective phytotoxin released by pathovars of *Pseudomonas syringae* interferes with JA-Ile signaling which operates the normal defense mechanism of plants leading to cell death and chlorosis. VIGS is employed to induce silencing of *SGT1* resulting in reduction of symptoms of chlorosis suggesting that *SGT1* is behind coronatine mediated-signaling and development of disease symptom by *P. syringae* pv. tomato (Uppalapati et al. 2011). Another experiment identified *SlALC1* as the COR response gene and subsequent silencing through VIGS produces a hypersensitive response and checks further invasion of the pathogen, *Pseudomonas syringae* pv. tomato DC3000 (*Pst* DC3000) (Wangdi et al. 2010). Furthermore, *SlSKIP1b,* a component of the spliceosome, plays a significant role in providing resistance against *Pseudomonas syringae* pv. tomato possibly by regulating the accumulation of reactive oxygen species (ROS) as well as the expression of disease resistance genes as found during VIGS (Zhang et al. 2020). In another study, VIGS showed that salicylic acid, ethylene and mitogen-activated protein kinases (MAPK)-related defense signaling regulate many layers of disease resistance towards the ingression of *Ralstonia solanacearum* (Chen et al. 2009). DNA and RNA viruses are reconstructed to serve as a vector for VIGS (Becker and Lange 2010). A new vector for VIGS has been developed through modified DNAβ (DNAmβ) of tomato yellow leaf curl China virus (TYLCCNV) which activates gene silencing in tomato (Tao and Zhou 2004; Cai et al. 2007). Moreover, new information on the mechanism of TYLCV infection is best revealed through tobacco rattle virus-mediated gene silencing in tomatoes. Similarly, the apple latent spherical virus was also used as a potent vector for VIGS. VIGS defines the role of genes that confer resistance to viruses. Through VIGS it was demonstrated that the Permease1-like protein gene expression is involved in resistant cultivar against TYLCV (Eybishtz et al. 2009). Apart from this, virus-induced gene silencing and allele mining were employed to study allelic variation in the *Ty-1/Ty-3* gene in a few *S. chilense* accessions which led to the discovery of a number of allele-specific polymorphisms (Yan et al. 2018). Thus, VIGS has a lot of promise in terms of identifying gene function to manage biotic stressors in a subtle manner.

Transgenic approaches possess immense potential against viruses, but they are commercially less exploited due to low acceptability in developing nations like India. One such concept is pathogen-derived resistance which is a promising asset against TYLCV. *CP* and truncated *Rep* gene of TYLCV have been incorporated to confer resistance in host plants. Another strategy is RNA-i which involves inhibition of gene expression through degradation of specific m-RNA by dsRNA, is gaining popularity for the identification of pathogenicity genes and exploiting these genes for the development of transgenics. The role of two pathogenicity genes, *FOW2* and *chsV* in *Fusarium oxysporum* and *F. solani* are well illustrated by RNA-i (Shanmugam et al. 2017). Further, an RNA-i construct was used to target ornithine decarboxylase, a key polyamine biosynthesis gene to develop a transgenic line expressing small interfering RNAs (siRNAs) and exhibited moderate-to-high resistance against *F. oxysporum* (Singh et al. 2020). In another experiment, RNAi-mediated suppression of two genes, *PEX6* and *GAS1* in *Fusarium oxysporum* f. sp. *lycopersici* resulted in enhanced resistance against Fusarium wilt of tomato (Tetorya et al. 2021). Nanosheets were utilized as a delivery system for topical spray of dsRNA to target three key genes, *FoCYP51*, *FoChs1* and *FoEF2* to provide resistance against *F. oxysporum* f. sp. *radicis-lycopersici* (Mosa and Youssef 2021). RNA-i is widely employed for the management of TYLCV. Incorporating dsRNA sequences from TYLCV-IR, Oman's V1 and V2 regions resulted in a 75% reduction in infection (Ammara et al. 2015). A unique bi-viral RNAi construct was made by combining C1, C2 and C3 sections of the tomato leaf curl Taiwan virus with the DNA-A genome of tomato yellow leaf curl Thailand virus to exhibit resistance against these two different begomoviral species infecting tomato (Chen et al. 2016). AC1 and AC4 genes of the tomato leaf curl virus were silenced through RNA-i to defend the respective pathogen (Ramesh et al. 2007).

SIGS is a novel approach in which dsRNA/siRNA are sprayed on plants to induce gene silencing (Cagliari et al. 2019). SIGS target genes are involved in *Phytophthora infestans* infection such as *PiGPB1*, *PiHmp1 PiCut3* and *PiEndo3* to mitigate late blight of potato (Kalyandurg et al. 2021). Artificial microRNAs (amiRNAs)-mediated gene silencing is another robust tool to manage TYLCV (Tiwari et al. 2014).

The usage of peptide aptamers is another comparable noble strategy for management of plant diseases. Peptide aptamers are developed to bind specifically with tep in order to disrupt the replication cycles of TYLCV and tomato mottle virus (ToMoV) conferring resistance to the host.

Traditional tomato plant breeding is typically arduous and time-consuming. CRISPR/Cas9 system is a promising technology, particularly in terms of quickly introducing increased crop features for effective tomato cultivation. They are successfully used in genome editing to provide protection against different categories of pathogens including bacteria, fungi and viruses. The toxin coronatine produced by the bacteria *Pseudomonas syringae* pv. *tomato* (*Pto*) DC3000 interferes with the salicylic acid defense pathway and induces stomata reopening which allows active colonization of the bacteria. CRISPR-Cas9-mediated gene editing of SlJAZ2, a co-receptor of coronatine, prevented this reopening and bacterial entry (Ortigosa et al. 2019). Similarly, CRISPR-Cas9 was employed to target small deletions of the *SlDMR6-1* gene which causes frameshift mutation and premature termination of the protein resulting in mutants with broad-spectrum resistance to pathogens including *P. syringae*, *P. capsici* and *Xanthomonas* spp. (Thomazella et al. 2016). CRISPR-Cas9-mediated knock-down of the *Solyc08g075770* gene demonstrated resistance to Fusarium wilt. This technique was used in the silencing of two micro-RNAs, miR482b and miR482c which further caused expression changes in other miRNAs, implying miRNA cross-regulation to show resistance against *P. infestans*. Molecular chaperones are often employed to restrict virus infection. The endosymbiont of whitefly / aphids produce GroEL, a molecular chaperone that binds to coat the protein or read through the protein of the virus and facilitates circulative transmission of plant viruses. Moreover, the interaction of GroEL with TYLCV has proven revolutionary for the generation of transgenic tomatoes. It was suggested that the manipulation of the GroEL-TYLCV complex prevented virus replication and movement and rendered the plants resistant.

17.10 APPLICATION OF NANOPARTICLE

The indiscriminate use of pesticides causes adverse effects on the environment through bioaccumulation and biomagnification. Moreover, the heavy residual effect of pesticides hampers the beneficial organisms present in the soil and the accumulation of contaminants in plants and food products occurs. To circumvent and improve these specific difficulties associated with biopesticides, nanoformulations of biopesticides possess enormous benefits in terms of specificity, improved shelf life, slow-release, better survivability of the PGPR, etc. Nanomaterials allow the slow release of constituents into the environment. Nanoparticles are becoming popular as a viable method for controlling biotic stresses (Table 17.7). Nanotechnology is effectively used to detect (Frewer et al. 2011) and protect plants from biotic stresses. Nanopesticides are required in much smaller concentrations compared to the doses of conventional pesticides for effective action. Plant disease management can be achieved by silver nanoparticles, copper nanoparticles and zinc nanoparticles.

Chitosan nanoparticles induce protection against Fusarium crown/root rot of tomatoes (Kashyap et al. 2015). Many researchers have reported that metal nanoparticles in their oxide form (CuO, MgO and ZnO) comprehensively control a wide range of foliar, air and soil-borne plant pathogens such as *P. infestans* (late blight of tomato), *Fusarium solani* (Fusarium root rot of tomato), *F. oxysporum* f.sp. *radicis-lycopersici* (crown root rot of tomato) and plant pathogenic bacteria such as *R. solanacearum*, *P. aeruginosa*, etc.

From Table 17.7 it is obvious that nanoparticles have a huge role in combating biotic stresses in tomatoes. Although, the application of nanoparticles against biotic stress is still in its early phase and in-depth research is needed towards its actual fate in the environment and ecological niche of the soil-crop system. Moreover, the toxicological data must be thoroughly analyzed for its safe application in agro-ecosystems.

17.11 ARTIFICIAL INTELLIGENCE

Plants go through several biochemical and physiological reaction cascades to protect themselves before observable symptoms appear as a result of host-pathogen interaction in the host plant. These changed responses of the host plant like increased/decreased rate of photosynthesis, transpiration, modified morphological pattern *viz.*, leaf shape, tissue color, leaf angle, etc., can easily be identified by available modern sensors like RGB (red, green, and blue), multispectral, hyperspectral, thermal, chlorophyll fluorescence, 3D sensors (Mahlein 2016). At an early stage of symptom development, modern sensors recognize the primary disease foci from where disease initiation takes place in the field in a heterogeneous pattern and gives us the opportunity to manage the disease by using site-specific targeted pesticides if needed. Hence, applications of this sensor-based method will ultimately lower the cost of plant protection with minimum ecological harm. These sensors can be mounted on different vehicles/instruments like microscopes, mobiles, cameras, drones, airplanes, satellites, etc., based on needs (Mahlein 2016). After snapping large amounts of images, data analysis is carried

TABLE 17.7
Effect of nanomaterial for management of different biotic stresses in tomato

Agent	Effect in disease management	Dose/applications	References
Nanosized Ag–silica	Upregulation of SAR marker gene (*PR1*, *PR2* and *PR5*). 1.6 times lower average cfu/cm^2 leaf of *Pseudomonas syringae* pv. *tomato* was detected in treated plant of *Arabidopsis* as compared to control	Foliar spray (10 ppm) on the *Arabidopsis* plants	Chu et al. 2012
DNA-directed silver nanoparticles grown on graphene oxide	Reduced disease severity of bacterial spot caused by *Xanthomonas perforans*	At 16 ppm showed antibacterial capability in culture and 100 ppm control disease severity	Ocsoy et al. 2013
Al_2O_3 NPs. mesoporous alumina sphere (MAS) nanoparticles	35.70 to 78.57 % mycelial inhibition of *Fusarium oxysporium in vitro* condition. 20 times increase in survivability of Tomato plant inoculated with *Fusarium oxysporium* treated with MAS nanoparticles under greenhouse condition	400 mg/l concentration *in vitro* condition. Root dip treatment @ 400 mg/l	Shenashen et al. 2017
CuNP + Potassium silicate	Reduced disease severity of *Clavibacter michiganensis*	50mgL^{-1} CuNPs + 460mgL^{-1} Si	Cumplido-Nájera et al. 2019
MgONPs	Reduced disease severity of *Xanthomonas perforans*	Foliar spray at 200 μg/ml	Liao et al. 2019
Sulfur nanoparticles of size 30 nm	Decrease the disease incidence of *Fusarium oxysporum* f. sp. *lycopersici* by 47.6%	Foliar application with 1 mg/plant	Cao et al. 2021
Cu and quaternary ammonium (Quat) loaded onto the silica core (hybrid nanoparticle)	Inhibit the growth of *Xanthomonas perforans*	4 μg mL^{-1} (Cu): 1 μg mL^{-1} (Quat)	Ozcan et al. 2021
Chitosan hydrochloride-cellulose nanocrystals-high amylose starch Nanostructured microparticles	Inhibit bacterial growth *in vitro* and reduce disease severity of *Pseudomonas syringae* pv. *tomato*	*In vitro* and foliar application at 1% w/v	Schiavi et al. 2021

out utilizing an algorithm-based image categorization model (Table 17.8). A general cascade of processing methods such as image pre-processing, picture segmentation and new feature vector extraction are utilized for image analysis-based detection of plant diseases (Ghorai et al. 2021).

17.12 USE OF CHEMICALS

Nonjudicial uses of chemical fungicides not only cause environmental hazards but also develop resistance to the pathogen. Fungicide applications should be based on weather-forecasting models (Table 17.9). As a consequence of the actual seasonal application of fungicides, this will help in lowering the fungicidal load. Historical disease data, if available, can be very beneficial and be utilized for the development of prediction models and early forewarning to circumvent yield losses. In order to generate disease prediction and forewarning systems, it is very necessary to record the periodic disease as well as weather data at specific intervals for a long-term basis on a fixed plot.

TABLE 17.8
Some successful examples of different algorithm-based models for plant disease detection

Disease name	Algorithm-based model that deployed	Process of data collection	Reference
Early blight	ELM model	Hyperspectral imaging	Xie et al. 2015
	SVM	RGB images of tomato leaf	Mokhtar et al. 2015a
	SVM and random forest	RGB images of tomato leaf	Govardhan et al. 2019
	CNN	PlantVillage dataset of 14,903 images of tomato leaves	Kumar and Vani 2019
Late blight	ELM model	Hyperspectral imaging	Xie et al. 2015
	MATLAB, Classification Tree	RGB images of tomato leaf	Sabrol and Satish 2016c
	Decision Tree	RGB images of tomato leaf	Sabrol and Kumar 2016b
	Statistical-model	PlantVillage image dataset	Hlaing et al. 2017
	SVM	PlantVillage image dataset	Hlaing et al. 2018
	SVM and random forest	Images of tomato leaf	Govardhan et al. 2019
Septoria leaf spot	MATLAB, Classification Tree	RGB images of tomato leaf	Sabrol and Satish 2016c
	Decision Tree	RGB images of tomato leaf	Sabrol and Kumar 2016b
	Statistical-model	PlantVillage image dataset	Hlaing et al. 2017
	SVM	PlantVillage image dataset	Hlaing et al. 2018
	SVM and random forest	RGB images of tomato leaf	Govardhan et al. 2019
	KNN and PNN	RGB images of tomato leaf	Balakrishna et al. 2019
	CNN	PlantVillage dataset of 14,903 images of tomato leaves	Kumar and Vani 2019
Fusarium wilt	Naïve Bayes approach	RGB images of tomato leaf	Dhaya 2020
Verticillium wilt	KNN and PNN	RGB images of tomato leaf	Balakrishna et al. 2019
Target spot	Statistical-model	PlantVillage image dataset	Hlaing et al. 2017
	SVM	PlantVillage image dataset	Hlaing et al. 2018
	SVM and random forest	RGB images of tomato leaf	Govardhan et al. 2019
	CNN	PlantVillage dataset of 14,903 images of tomato leaves	Kumar and Vani 2019
Bacterial canker	MATLAB, Classification Tree	RGB images of tomato leaf	Sabrol and Satish 2016c
	Decision Tree	RGB images of tomato leaf	Sabrol and Kumar 2016b
Spotted wilt disease of tomato	SVM	Infected tomato leaf images	Mokhtar et al. 2015b
Tomato yellow leaf curl	MATLAB, Classification Tree	RGB images of tomato leaf	Sabrol and Satish 2016c
	Decision Tree	RGB images of tomato leaf	Sabrol and Kumar 2016b
	MLBPNN (MultiLayer Feed Forward Back Propagation Neural Network	RGB images of tomato leaf	Sabrol et al. 2016a
	SVM	Infected tomato leaf images	Mokhtar et al. 2015b
	Hyperspectral Image analysis by using ENVI and MATLAB	Hyperspectral Imaging	Lu et al. 2017
	SVM and random forest	RGB images of tomato leaf	Govardhan et al. 2019
	CNN	PlantVillage dataset of 14,903 images of tomato leaves	Kumar and Vani 2019
Tomato mosaic disease	SVM RF+ MDC+ RF	RGB images of tomato leaf	Annabel and Muthulakshmi 2019
	SVM and random forest	RGB images of tomato leaf	Govardhan et al. 2019

Moreover, new fungicide molecules are emerging day by day to manage crop diseases and holistic utilization of various disease management strategies coupled with forewarning system-based chemical management may pave the way for sustainable eco-friendly management of tomato diseases. The list of fungicides used for various tomato disease management is prepared based on the list of registered pesticides (CIBRC), generated by the Directorate of Plant Protection, Quarantine & Storage, Department of Agriculture & Farmers Welfare (2024)" to "Major Uses of Pesticides (2024) and is shown in Table 17.10.

TABLE 17.9
Weather forecasting models developed for prediction of different diseases of tomato

Name of the disease	Name of the model/Statistical approach	Country	References
Early blight	FAST	Pennsylvania	Madden et al. 1978
Early blight	CUFAST	New York	Zitter and Sandlan 1990
Early blight, Septoria leaf spot	TOMCAST	Ontario	Gleason et al. 1995
Late blight	SIMCAST	North America	Fry et al. 1983
Early blight, late blight	Extreme Learning Machine (ELM) classifier model	China	Xie et al. 2015
Yellow leaf curl disease	Receiver Operator Characteristic (ROC) curve analysis	China	Lu et al. 2017

TABLE 17.10
List of chemicals, their applicable doses and target diseases for management

Chemical composition	Dose	Applied against
Azoxystrobin 23% SC	200 ml in 200 lit of water/acre	Early blight, late blight
Captan 50% WP	1000 g in 300–400 lit of water/acre	Early blight, late blight
Captan 75% WP	1000 g in 400 lit of water/acre for soil drenching	Damping off in nursery
Captan 75% WS	20–30 g/ kg seeds	Damping off in nursery
Copper oxychloride 50% WP	1000 g in 300–400 lit of water/acre	Early blight, late blight, leaf spot
Iprodione 50% WP	600 g in 200 lit of water/acre	Early blight
Kasugamycin 3% SL	400–500 ml in 240 lit water/acre	Early blight
Mancozeb 35% SC	0.5% or 1000 g in 200 lit water/acre	Early blight, late blight
Mancozeb 75% WG	400 g in 200 lit of water/acre	Early blight
Pyraclostrobin 20% WG	150–200 g in 200 lit of water/acre	Early blight
Ziram 80% WP	600–800 g in 300-400 lit of water/acre	Early blight
Famoxadone 16.6% + Cymoxanil 22.1% SC	200 g in 200 lit of water/acre	Early blight, late blight
Thifluzamid 24% SC	200 ml in 200 lit of water/acre	Early blight
Metiram 55% + Pyraclostrobin 5% WG	600–700 g in 200 lit of water/acre	Early blight
Kresoxim-methyl 18% + Mancozeb 54% WP	555 gm in 200 lit water/acre	Early blight, late blight
Boscalid 25.2% +Pyraclostrobin 12.8% WG	200 ml in 200 lit of water/acre	Early blight
Mancozeb 40% + Azoxystrobin 7% OS	600 g in 200 lit water/acre	Early blight, late blight
Cymoxanil 8% + Mancozeb 64% WP	600 g in 200–300 lit of water/acre	Late blight

TABLE 17.10 (Continued)
List of chemicals, their applicable doses and target diseases for management

Chemical composition	Dose	Applied against
Metalaxyl M 3.3% + Chlorothalonil 33.1% SC	0.2% or 400 g in 200 lit of water/acre	Late blight, early blight
Chlorothalonil 40.0% + Difenoconazole 4.0 SC	400 ml in 200 lit of water/acre	Late blight, early blight
Ametoctradin 27% + Dimethomorph 20.27% SC	320–400 ml in 200 lit water/ acre	Late blight
Azoxystrobin 18.2% + Difenoconazole 11.4% SC	200 ml in 200 lit water/ acre	Late blight, early blight
Tebuconazole 50% + Trifloxystrobin 25% WG	140 g in 200 lit of water/acre	Early blight
Valifenalate 6 % + Mancozeb 60 % WG	2000–2500 g in 500 lit water / ha	Early blight, late blight
Cyazofamid 34.5% SC	80 ml in 200 lit of water/acre	Late blight
Metiram 75% WG	1000 g in 200–300 lit of water/acre	Early blight
Metalaxyl M 31.8% ES	2 gm as seed dresser	Damping off
Mandipropamid 23.4% SC	0.08% or 0.8 ml/Ltr	Late blight
Mancozeb 75% WP	600–800 g in 300 lit of water/acre	Septoria leaf spot, buckeye rot, late blight (Prophylaxis)
Flubendiamide 7.5% + Kresoxim-Methyl 37.5% SC	267 ml in 200 lit of water/acre	Late blight, early blight
Fluxapyroxad 250 g/l + Pyraclostrobin 250 g/l SC	80–100 ml in 200 lit of water/acre	Septoria leaf spot, early blight
Streptomycin sulfate 9% + Tetracycline hydrochloride 1% SP (Streptocycline)	40–100 ppm. Two sprays, 1 after the appearance of first true leaves but before transplanting (seed beds) and 2nd after transplanting in main field	Bacterial leaf spot
Diafenthiuron 50 % WP	240 ml in 200 lit of water/acre	Insect vector management for viral diseases
Imidacloprid 70 % WG	20 g in 200 lit of water/acre	Insect vector management for viral diseases
Imidacloprid 17.8 SL	60–70 ml in 200 lit of water/acre	Insect vector management for viral diseases
Oxydemeton-methyl 25 % EC	400 ml in 200–400 lit of water/acre	Insect vector management for viral diseases
Thiamethoxam 25 WG	80 g in 200 lit of water/acre	Insect vector management for viral diseases
Fenazquin 10% + Bifenthrin 4% EC	500 ml in 200 lit of water/acre	Insect vector management for viral diseases
Oxydemeton-methyl 25% EC	400 ml in 200–400 lit of water/acre	Insect vector management for viral diseases
Acetamiprid 20% + Chlorantraniliprole 20% WG	60 g in 200 lit of water/acre	Insect vector management for viral diseases
Spiromesifen 22.9% SC	250 ml in 200 lit of water/acre	Insect vector management for viral diseases
Propargite 50 % + Bifenthrin 5 % w/w SE	440–460 ml in 200 lit of water/acre	Insect vector management for viral diseases

* Prepared based on the list of registered pesticides (CIBRC), Directorate of Plant Protection, Quarantine & Storage, Department of Agriculture & Farmers Welfare (2024).

17.13 CONCLUSION

Food, nutritional and livelihood security are the major issues that must be addressed in agricultural research, education and planning. Biotic stresses are the major bottleneck in harnessing optimum harvest. The diseases described above are the prime causes behind the poor productivity of tomatoes. Chemical input-based intensive agricultural systems not only have adverse effects on the environment and ecosystem but also enhance the attack of different biotic stresses. Agro-ecosystem-based holistic exploitation of different plant disease management approaches coupled with modern technological breakthroughs has opened new possibilities for plant disease management that lead to effective, environmentally benign and sustainable tomato disease management.

REFERENCES

Abdallah RA, Mokni-Tlili S, Nefzi A, Jabnoun-Khiareddine H, Daami-Remadi M (2016) Biocontrol of *Fusarium* wilt and growth promotion of tomato plants using endophytic bacteria isolated from *Nicotiana glauca* organs. *Biol Control* 97: 80–88.

Abdel-Aziz MS, Ghareeb MA, Ahmed AA, Rashad EM, Eslam, El-Sawy ER, Saad IM, Ghoneem KM (2021) Ethyl acetate extract of *Streptomyces* spp. isolated from Egyptian soil for management of *Fusarium oxysporum*: the causing agent of wilt disease of tomato. *Microbiol Res* 169: 59–65.

Abdeljalil NO, Vallance J, Gerbore J, Yacoub A, Daami-Remadi M, Rey P (2021) Combining potential oomycete and bacterial biocontrol agents as a tool to fight tomato *Rhizoctonia* root rot. *Biol Control* 155: 104521.

Akhtar KP, Akram A, Ullah N, Saleem MY, Saeed M (2019) Evaluation of *Solanum* species for resistance to tomato leaf curl New Delhi virus using chip grafting assay. *Scientia Horticult* 256: 108646.

Aksoy HM, Kaya Y, Ozturk M, Secgin Z, Onder H, Okumus A (2017) *Pseudomonas putida*-induced response in phenolic profile of tomato seedlings (*Solanum lycopersicum* L.,) infected by *Clavibacter michiganensis* subsp. *michiganensis*. *Biol Control* 105: 6–12.

Al-Surhanee AA (2022) Protective role of antifusarial eco-friendly agents (*Trichoderma* and salicylic acid) to improve resistance performance of tomato plants. *S J Biol Sci* 29: 2933–2941.

Ammara U, Mansoor S, Saeed M, Amin I, Briddon RW, Al-Sadi AM (2015) RNA interference-based resistance in transgenic tomato plants against *Tomato yellow leaf curl virus-Oman* (TYLCV-OM) and its associated betasatellite. *Virol J* 12: 1–12.

Anbinder I, Reuveni M, Azari R, Paran I, Nahon S, Shlomo H, Chen L, Lapidot M, Levin I (2009) Molecular dissection of tomato leaf curl virus resistance in tomato line TY172 derived from *Solanum peruvianum*. *Theor App Gen* 119: 519–530.

Annabel LSP, Muthulakshmi V (2019) AI-powered image-based tomato leaf disease detection. In: *2019 Third International Conference on I-SMAC (IoT in Social, Mobile, Analytics and Cloud)*, Palladam, India. pp. 506–551. doi: 10.1109/I-SMAC47947.2019.9032621

Arens P, Mansilla C, Deinum D, Cavellini L, Moretti A, Rolland S, van der Schoot H, Calvache D, Ponz F, Collonnier C et al (2010) Development and evaluation of robust molecular markers linked to disease resistance in tomato for distinctness, uniformity and stability testing. *Theor Appl Genet* 120: 655–664.

Arwiyanto T, Nurcahyanti SD, Indradewa D, Widada J (2015) Grafting local commercial tomato cultivars with H-7996 and eg-203 to suppress bacterial wilt (*Ralstonia solanacearum*) in Indonesia. *Acta Hortic* 1069: 173–178.

Balakrishna K, Rao M (2019) Tomato plant leaves disease classification using KNN and PNN. *IJCVIP* 9: 51–63.

Barakat RM, AL-Masri MI (2012) Enhanced soil solarization against *Fusarium oxysporum* f. sp. *lycopersici* in the uplands. *Int J Agron* 368654. https://doi.org/10.1155/2012/368654

Becker A, Lange M (2010) VIGS-genomics goes functional. *TPl Sci* 15: 1–4.

Ben-Jabeur M, Ghabri E, Myriam M, Hamada W (2015) Thyme essential oil as a defense inducer of tomato against gray mold and *Fusarium* wilt. *Plant Phy Biochem* 94: 35–40.

Cagliari D, Dias NP, Galdeano DM, Dos Santos EÁ, Smagghe G, Zotti MJ (2019) Management of pest insects and plant diseases by non-transformative RNAi. *Front Plant Sci* 10: 1319.

Cai XZ, Wang CC, Xu YP, Xu QF, Zheng Z, Zhou XP (2007) Efficient gene silencing induction in tomato by a viral satellite DNA vector. *Vir Res* 125: 169–175.

Cao X, Wang C, Luo X, Yue L, White JC, Elmer W, Dhankher OP, Wang Z, Xing B (2021) Elemental sulfur nanoparticles enhance disease resistance in tomatoes. *ACS Nano* 15: 11817–11827.

Cardoso SC, Soares AC, Brito AD, Laranjeira FF, Ledo CA, Santos AP (2006) Control of tomato bacterial wilt through the incorporation of aerial part of pigeon pea and crotalaria to soil. *Sum Phytopathol* 32: 27–33.

Castello I, D'Emilio A, Raviv M et al (2017) Soil solarization as a sustainable solution to control tomato Pseudomonads infections in greenhouses. *Agron Sustain Dev* 37: 59.

Chaerani R, Voorrips RE (2006) Tomato early blight (*Alternaria solani*): the pathogen, genetics, and breeding for resistance. *J Gen Pl Pathol* 72: 335–347.

Champoiseau P, Jones JB, Allen C, Momol T (2009) Description and strategies for best management of *Ralstonia solanacearum* Race 3 biovar 2 as a potential incitant of bacterial wilt of tomato. In: *Proceedings of the 24th Annual Tomato Disease Workshop Held on, November 3–5*. State College, Pennsylvania, pp. 1–35.

Chen HM, Lin CY, Tsai WS, Kenyon L, Chan MT, Yen JY, Chang SY, de la Peña R, Schafleitner R (2016) Resistance to viral yellow leaf curl in tomato through RNAi targeting two begomovirus species strains. *J Plant Biochem Biotechnol* 25 : 199–207.

Chen YY, Lin YM, Chao TC, Wang JF, Liu AC, Ho FI, Cheng CP (2009) Virus-induced gene silencing reveals the involvement of ethylene-, salicylic acid- and mitogen-activated protein kinase related defense pathways in the resistance of tomato to bacterial wilt. *Physiol Plant* 136: 324–335.

Cheuk W, Victor LK, Copeman R, Joliffe P, Fraser B (2005) Disease suppression on greenhouse tomatoes using plant waste compost. *J Environ Sci Health A* 40: 449–461.

Chu H, Kim HJ, Kim JS, Kim MS, Yoon BD, Park HJ, Kim CY (2012) A nanosized Ag-silica hybrid complex prepared by γ-irradiation activates the defense response in *Arabidopsis*. *Radiat Phys Chem* 81:180–184.

Cumplido-Nájeraa CF, González-Moralesb S, Ortega-Ortízc H, Cadenas-Pliegoc G, Benavides-Mendozaa A, Juárez-Maldonadod A (2019) The application of copper nanoparticles and potassium silicate stimulate the tolerance to *Clavibacter michiganensis* in tomato plants. *Sci Horti* 245: 82–89.

Dahlin P, Hallmann J (2020) New insights on the role of allyl isothiocyanate in controlling the root knot nematode *Meloidogyne hapla*. *Plants* 9(5): 603.

Dalmon A, Cailly M, David C (2000) Comparison of serological and molecular techniques for detection of tomato yellow leaf curl begomovirus. *EPPO Bull* 30: 457–462.

Davino SW, Panno S, Parrella G, Davino M, Cocuzza GEM, Rapisarda C, Caruso AG, Carpino C (2018) *Viruses*. In Davino SW (ed) *Tomato Diseases. Viruses and Soil Borne Fungi*. Edizioni L'Informatore Agrario s.r.l., Verona, Italy, pp. 11–137.

Deberdt P, Davezies I, Coranson-Beaudu R, Jestin A (2018) Efficacy of leaf oil from *Pimenta racemosa* var. *racemosa* in controlling bacterial wilt of tomato. *Plant Dis* 102: 124–131.

Denny TP, (2006) Plant pathogenic *Ralstonia* species. In: Gnanamanickam SS (ed.), *Plant –Associated Bacteria*. Springer Publishing, Dordrecht, The Netherlands, pp. 573–644.

Denny TP, Hayward AC (2001) *Ralstonia* In: Jones JB, Chun W (eds.), *Laboratory Guide for Identification of Plant Pathogenic Bacteria*. APS Press, St Paul, pp. 154–174.

Dhaya R (2020) Flawless Identification of *Fusarium oxysporum* in tomato plant leaves by machine learning algorithm. *JIIP* 2: 194–201.

Elliott M, Shamoun SF, Sumampong G, James D, Masri S, Varga A (2009) Evaluation of several commercial biocontrol products on European and North American populations of *Phytophthora ramorum*. *Biocont Sci Technol* 19: 1007–1021.

El-Tantaway EM (2009) Behavior of tomato plants as affected by spraying with chitosan and aminofort as natural stimulator substances under application of soil organic amendments. *Pak J Biol Sci* 12(17):1164–1173.

Enespa DS, Dwivedi SK (2014) Effectiveness of some antagonistic fungi and botanicals against *Fusarium solani* and *Fusarium oxysporum* f.sp. *lycopersici* infecting brinjal and tomato plants. *AJ Plant Pathol* 8: 18–25.

Eybishtz A, Peretz Y, Sade D, Akad F, Czosnek H (2009) Silencing of a single gene in tomato plants resistant to tomato yellow leaf curl virus renders them susceptible to the virus. *Plant Mol Biol* 71: 157–171.

Fatima S, Anjum T (2017) Identification of a potential ISR determinant from *Pseudomonas aeruginosa* PM12 against *Fusarium* wilt in tomato. *Front Plant Sci* 8: 848.

Freeman JH, McAvoy T, Rideout S, Wimer A (2010) Utilization of grafted tomato seedlings for open field production. *HortSci* 45: 499.

Frewer LJ, Norde W, Fischer A, Kampers F (eds.) (2011) *Nanotechnology in the Agri-Food Sector: Implications for the Future*. Wiley, Weinheim.

Fry WE, Apple AE, Bruhn JA (1983) Evaluation of potato late blight forecasts modified to incorporate host resistance and fungicide weathering. *Phytopathology* 73:1054–1059.

Fukuta S, Kato S, Yoshida K, Mizukami Y, Ishida A, Ueda J, Ishimoto Y (2003) Detection of tomato yellow leaf curl virus by loop-mediated isothermal amplification reaction. *J Virol Methods* 112(1–2): 35–40.

Gao X, Li K, Ma Z, Zou H, Jin H, Wang J (2020) Cucumber Fusarium wilt resistance induced by intercropping with celery differs from that induced by the cucumber genotype and is related to sulfur-containing allelochemicals. *Sci Hortic* 271: 109475.

Ghorai AK, Mukhopadhyay S, Kundu S, Mandal SN, Barman AR, De Roy M, Dutta S (2021) Image processing based detection of diseases and nutrient deficiencies in plants. *SATSA Mukhapatra* 25(1): 1–24.

Ghosh PP, Dutta S, Chattopadhyay A (2015) Integration of organic and inorganic amendments with native bioagents for bio-intensive management of vascular bacterial wilt on eggplant (*Solanum melongena*). *Indian Phytopath* 68: 32–38.

Gilreath P, Shuler, K, Polston J, Sherwood T, McAvoy G, Stansly P et al (2000) Tomato yellow leaf curl virus resistant tomato variety trials. *Proc Fla State Hort Soc* 113: 190–193.

Gleason ML, Macnab AA, Pitblado RE, Ricker MD, East DA, Latin RX (1995) Disease warning systems for processing tomatoes in eastern North America: are we there yet? *Plant Dis* 79: 113–121.

Gleason ML, Edmunds BA (2006) *Tomato Diseases and Disorders*. Department of Plant Pathology, Iowa State University of Science and Technology, Ames, Iowa.

Gomez-Rodriguez O, Zavaleta-Mejia E, Gonzalez-Hernandez AV, Livera-Munoz M, Cardenas Soriano E (2007) Physiological and morphological adaptations in tomato intercropped with *Tagetes erecta* and *Amaranthus hypochondriacus*. *Rev Fitotec Mex* 30(4): 421–428.

Govardhan M, Veena M (2019) Diagnosis of tomato plant diseases using random forest. In: *Global Conference Advance Technol (GCAT)*, Bangalore, India, pp. 1–5. doi: 10.1109/GCAT47503.2019.8978431

Gravel V, Martinez C, Antoun H, Tweddell RJ (2005) Antagonist microorganisms with the ability to control *Pythium* damping-off of tomato seeds in rock wool. *Bio Cont* 50: 771–786.

Guo JH, Qi HY, Guo YH, Ge HL, Gong LY, Zhang LX, Sun PH (2004) Biocontrol of tomato wilt by plant growth-promoting rhizobacteria. *Biol Contr* 29: 66–72.

Hafez EE, Meghad A, Elsalam HA, Ahmed SA (2013) Biological and molecular studies on *Trichoderma viride*–plant pathogenic fungi interactions. *J Wor App Sci* 21: 1821–1828.

Heidarzadeh N, Baghaee-Ravari S (2015) Application of *Bacillus pumilus* as a potential biocontrol agent of *Fusarium* wilt of tomato. *Arch Phytopathol Plant Protect* 48(13–16): 841–849.

Hirano Y, Arie T (2006) PCR-based differentiation of *Fusarium oxysporum* f. sp. *lycopersici* and *radicis-lycopersici* and races of *F. oxysporum* f. sp. *lycopersici*. *J Gen Plant Pathol* 72: 273–283.

Hlaing CS, Zaw SMM (2017) Model-based statistical features for mobile phone image of tomato plant disease classification. In: *2017 18th International Conference on Parallel and Distributed Computing, Applications and Technologies (PDCAT)*, , Taipei, Taiwan, pp. 223–229. doi: 10.1109/PDCAT.2017.00044

Hlaing CS, Zaw SMM (2018) Tomato plant diseases classification using statistical texture feature and color feature. In: *IEEE/ ACIS 17th International Conference on Computer and Information Science (ICIS), Singapore*, pp. 439–444. doi: 10.1109/ICIS.2018.8466483

Huang Q, Lakshman DK (2010) Effect of clove oil on plant pathogenic bacteria and bacterial wilt of tomato and geranium. *J Plant Pathol* 92: 701–707.

Hutton SF, Scott JW, Shekasteband R, Levin I, Lapidot M (2015) Combinations of Ty resistance genes generally provide more effective control against begomoviruses than do single genes. *Acta Hortic* 1069: 59–64.

Iquebal MA, Arora V, Verma N, Rai A, Kumar D (2013) First whole genome-based microsatellite DNA marker database of tomato for mapping and variety identification. *BMC Plant Biol* 13: 197. https://doi.org/10.1186/1471-2229-13-197

Jayaraj J, Radhakrishnan NV (2008) Enhanced activity of introduced biocontrol agents in solarized soils and its implications on the integrated control of tomato damping-off caused by *Pythium* spp. *Plant Soil* 304: 189–197.

Ji P, Momol MT, Olson SM, Pradhanang PM, Jones JB (2005) Evaluation of thymol as biofumigant for control of bacterial wilt of tomato under field conditions. *Plant Dis* 89(5): 497–500.

Ji Y, Schuster DJ, Scott JW (2007) *Ty-3*, a begomovirus resistance locus near the *Tomato yellow leaf curl virus* resistance locus *Ty-1* on chromosome 6 of tomato. *Mol Breed* 20(3): 271–284.

Ji Y, Scott JW, Schuster DJ, Maxwell DP (2009) Molecular mapping of Ty-4, a new tomato yellow leaf curl virus resistance locus on chromosome 3 of tomato. *J Am Soc Hortic Sci* 134: 281–288.

Kalyandurg PB, Sundararajan P, Dubey M, Ghadamgahi F, Zahid MA, Whisson S, Vetukuri RR (2021) Spray-induced gene silencing as a potential tool to control potato late blight disease. *Phytopathology* 11(12): 2168–2175. https://doi.org/10.1094/PHYTO-02-21-0054-SC

Kashyap PL, Xiang X, Heiden P (2015) Chitosan nanoparticle based delivery systems for sustainable agriculture. *Int J Boil Macromol* 77: 36–51.

Katan J, Gamliel A (2014) Plant health management: soil solarization. *Encyclopedia Agri Food Syst* 4: 60–471.

Kawaguchi M, Taji A, Backhouse D, Oda M (2008) Anatomy and physiology of graft incompatibility in solanaceous plants. *J Horti Sci Biotechnol* 83(5): 581–588.

Khan N, Mishra A, Nautiyal CS (2012) *Paenibacillus lentimorbus* B-30488r controls early blight disease in tomato by inducing host resistance associated gene expression and inhibiting *Alternaria solani*. *Biol Cont* 62: 65–74.

Kim M, Park Y, Lee J, Sim S (2020) Development of molecular markers for Ty-2 and Ty-3 selection in tomato breeding. *Sci Horticult* 265: 109230. https://doi.org/10.1016/j.scienta.2020.109230

Krause SM, Madden LV, Hoitink HAJ (2001) Effect of potting mix microbial carrying capacity on biological control of *Rhizoctonia* damping off of radish and Rhizoctonia crown and root rot of poinsettia. *Phytopathology* 91: 1116–1123.

Kumar A, Jindal SK, Dhaliwal MS et al (2022) Development of multiple disease-resistant tomato lines through marker assisted breeding and their evaluation for horticultural traits. *Indian Phytopath* 75: 47–55.

Kumar A, Vani M (2019) Image based tomato leaf disease detection. In: *2019 10th International Conference on Computing, Communication and Networking Technologies (ICCCNT)*, Kanpur, India, pp. 1–6. doi: 10.1109/ICCCNT45670.2019.8944692

Kunwar S, Paret ML, Olson SM,Ritchie L, Rich JR, Freeman JH, McAvoy T (2015) Grafting rootstocks with resistance to *Ralstonia solanacearum* against *Meloidogyne incognita* in tomato production. *Plant Dis* 99: 119–124.

Lecoq H, Desbiez C (2012) Viruses of cucurbit crops in the Mediterranean region: An ever-changing picture. *Adv Virus Res* 84: 67–126.

Li C, Ahmed W, Li D, Yu L, Xu L, Xu T, Zhao Z (2022) Biochar suppresses bacterial wilt disease of flue-cured tobacco by improving soil health and functional diversity of rhizosphere microorganisms. *Appl Soil Ecol* 171: 104314.

Liao H, Zheng C, Long J, Guzmán I (2021) Effects of biochar amendment on tomato rhizosphere bacterial communities and their utilization of plant-derived carbon in a calcareous soil. *Geoderma* 396: 115082 https://doi.org/10.1016/j.geoderma.2021.115082.

Liao YY, Strayer-Scherer AL, White J, Mukherjee A, De la Torre-Roche R, Ritchie L et al (2019) Nano-magnesium oxide: A novel bactericide against copper-tolerant *Xanthomonas perforans* causing tomato bacterial spot. *Phytopathology* 109(1): 52–62.

Lievens B, Brouwer M, Vanachter AC, Cammue BP, Thomma BP (2006) Real-time PCR for detection and quantification of fungal and oomycete tomato pathogens in plant and soil samples. *Plant Sci* 171(1): 155–165.

Loliam B, Morinaga T, Chaiyanan S (2012) Biocontrol of *Phytophthora infestans*, fungal pathogen of seedling damping off disease in economic plant nursery. *Psyche* 324317:1–6.

Lu J, Zhou M, Gao Y, Jiang H (2017) Using hyperspectral imaging to discriminate yellow leaf curl disease in tomato leaves. *Precision Agric* 19(3): 379–394.

Madden L, Pennypacker SP, MacNab AA (1978) FAST, a forecast system for *Alternaria solani* on tomato. *Phytopathology* 68(9):1354–1358.

Mahlein AK (2016) Plant disease detection by imaging sensors–parallels and specific demands for precision agriculture and plant phenotyping. *Plant Dis* 100(2): 241–251.

Major Uses of Pesticides (2024) Department of Agriculture & Farmers Welfare, Directorate of Plant Protection, Quarantine & Storage, Central Insecticide Board Registration Committee, N.H.-IV, Faridabad-121001 (Haryana). URL: fungicides_mup_as_on_30.11.2024_0.pdf

Manikandan R, Harish S, Karthikeyan G, Raguchander T (2018) Comparative proteomic analysis of different isolates of *Fusarium oxysporum* f. sp. *lycopersici* to exploit the differentially expressed proteins responsible for virulence on tomato plants. *Front Microbiol* 9: 420. https://doi.org/10.3389/fmicb.2018.00420

Marzano M, Gallo A, Altomare C (2013) Improvement of biocontrol efficacy of *Trichoderma harzianum* vs. *Fusarium oxysporum* f. sp. *lycopersici* through UV-induced tolerance to fusaric acid. *Biol Control* 67: 397–408.

Mastouri F, Björkman T, Harman GE (2010) Seed treatment with *Trichoderma harzianum* alleviates biotic, abiotic, and physiological stresses in germinating seeds and seedlings. *Phytopathology* 100: 1213–1221.

Matsushita Y, Usugi T, Tsuda S (2010) Development of a multiplex RT-PCR detection and identification system for *Potato spindle tuber viroid* and *Tomato chlorotic dwarf viroid*. *Eur J Plant Pathol* 128: 165–170.

Mazzola M, Freilich S (2017) Prospects for biological soil-borne disease control: application of indigenous versus synthetic microbiomes. *Phytopathology* 107: 256–263.

Misrak K, Amare A, Dechassa N (2014) Evaluation of soil solarization and bio-fumigation for the management of bacterial spot of tomato. *Afr J Food Agric Nutr Dev* 14(4): 8998–9015.

Mitchell C, Brennan RM, Graham J, Karley AJ (2016) Plant defense against herbivorous pests: exploiting resistance and tolerance traits for sustainable crop protection. *Front Plant Sci* 7: 1132. https://doi.org/10.3389/fpls.2016.01132

Mokhtar U, Ali MAS, Hassanien AE, Hefny H (2015b) Identifying two of tomatoes leaf viruses using support vector machine. In: Mandal, J., Satapathy, S., Kumar Sanyal, M., Sarkar, P., Mukhopadhyay, A. (eds) *Information Systems Design and Intelligent Applications,* vol 339. Springer, New Delhi. https://doi.org/10.1007/978-81-322-2250-7_77

Mosa MA, Youssef K (2021) Topical delivery of host induced RNAi silencing by layered double hydroxide nanosheets: an efficient tool to decipher pathogenicity gene function of *Fusarium* crown and root rot in Tomato. *Physiol Mol Plant Pathol* 115: 101684.

Mudge K, Janick J, Scofield S, Goldschmidt EE (2009) A history of grafting. *Hort Rev* 35: 437–493.

Mvumi C, Ngadze E, Marais D, duToit ES, Kugara J (2018) Determination and quantification of sinigringlucosinolates in *Alternaria solani* susceptible tomato (*Solanum lycopersicum*) leaves treated with moringa (*Moringa oleifera*) leaf extract. *Arch Phytopat Plant Prot* 51(7–8): 432–444. https://doi.org/10.1080/03235408.2018.1478610

Naher L, Yusuf UK, Ismail A, Hossain K (2014) *Trichoderma* spp.: A biocontrol agent for sustainable management of plant diseases. *Pak J Bot* 46(4): 1489–1493.

Nawangsih AA, Damayanti I, Wiyono S, Kartika JG (2011) Selection and characterization of endophytic bacteria as biocontrol agents of tomato bacterial wilt disease. *HAYATI J Biosci* 18(2): 66–70.

Ocsoy I, Paret ML, Ocsoy MA, Kunwar S, Chen T, You M, Tan W (2013) Nanotechnology in plant disease management: DNA-directed silver nanoparticles on graphene oxide as an antibacterial against *Xanthomonas perforans*. *ACS Nano* 7(10): 8972–8980.

Ortigosa A, Ibanez SG, Leonhardt N, Solano R (2019) Design of a bacterial speck resistant tomato by CRISPR/Cas9-mediated editing of SlJAZ2. *Biotechnol J* 17: 665–673.

Ozcan A, Young M et al (2021) Copper-fixed quat: A hybrid nanoparticle for application as a locally systemic pesticide (LSP) to manage bacterial spot disease of tomato. *Nanoscale Adv* 3: 1473–1483.

Pandey S, Koirala U, Acharya P, Shrestha S (2020) Management of bacterial wilt using grafting technique in tomato (*Ralstonia solanacearum*). *World* 8(4): 129–133.

Pilowski M, Cohen S (1990) Tolerance to tomato yellow leaf curl virus derived from *Lycopersicon esculentum. Plant Dis* 74: 248–250.

Pontes NC, Yamada JK, Fujinawa MF, Dhingra OD, de Oliveira, JR (2019) Soil fumigation with mustard essential oil to control bacterial wilt in tomato. *Eur J Plant Pathol* 155: 435–444.

Pradhanang PM, Momol MT, Olson SM, Jones JB (2003) Effects of plant essential oils on *Ralstonia solanacearum* population density and bacterial wilt incidence in tomato. *Plant Dis* 87: 423–427.

Ramana MV, Balakrishna K, Murali HS, Batra HV (2011) Multiplex PCR-based strategy to detect contamination with mycotoxigenic *Fusarium* species in rice and finger millet collected from southern India. *J Sci Food Agric* 91: 1666–1673. doi: 10.1002/jsfa.4365

Ramamoorthy V, Raguchander T, Samiyappan R (2002) Induction of defense-related proteins in tomato roots treated with *Pseudomonas fluorescens* Pf1 and *Fusarium oxysporum* f. sp. *lycopersici*. *Plant Soil* 239: 55–68.

Ramesh SV, Mishra AK, Praveen S (2007) Hairpin RNA-mediated strategies for silencing of *Tomato leaf curl virus* AC1 and AC4 genes for effective resistance in plants. *Oligonucleotides* 17: 251–257.

Ren Z, Li Y, Fang W, Yan D, Huang B, Zhu J, Cao A (2018) Evaluation of allylisothiocyanate as a soil fumigant against soil-borne diseases in commercial tomato (*Lycopersicon esculentum* Mill.) production in China. *Pest Manag Sci*. https://doi.org/10.1002/ps.4911

Rivard CL, Louws FJ (2008) Grafting to manage soilborne diseases in heirloom tomato production. *HortSci* 43: 2104–2111.

Rizzo PF, Young BJ, Viso NP, Carbajal J, Laura Martínez LE, Riera NI,Bres PA, Beily ME, Barbaro L, Farber M, Zubillaga MS, Crespo DC (2022) Integral approach for the evaluation of poultry manure, compost, and digestate: amendment characterization, mineralization, and effects on soil and intensive crops. *Waste Manage* 139: 124–135.

Romero FM, Marina M, Pieckenstain FL (2016) Novel components of leaf bacterial communities of field-grown tomato plants and their potential for plant growth promotion and biocontrol of tomato diseases. *Res Microbiol* 167(3): 222–233.

Roquigny R, Novinscak A, Arseneault T, Joly DL, Filion M (2018) Transcriptome alteration in *Phytophthora infestans* in response to phenazine-1-carboxylic acid production by *Pseudomonas fluorescens* strain LBUM223. *BMC Genomics* 19(1): 474–489.

Sabrol H, Kumar S (2016a) Fuzzy and neural network based tomato plant disease classification using natural outdoor images. *Indian J Sci Technol* 9(44): 1–8.

Sabrol H, Kumar S (2016b) Intensity based feature extraction for tomato plant disease recognition by classification using decision tree. *Int J Comput Sci Inf Secur* 14(9): 622–626.

Sabrol H, Satish K (2016c) Tomato plant disease classification in digital images using classification tree. In: *2016 International Conference on Commun and Signal Processing*, Melmaruvathur, India, pp. 1242–1246. doi: 10.1109/ICCSP.2016.7754351

Sanoubar R, Barbanti L (2017) Fungal diseases on tomato plant under greenhouse condition. *Eur J Biol Res* 7: 299–308.

Satyagopal K, Sushil SN, Jeyakumar P, Shankar G, Sharma OP, Sain SK, Saha S (2014) *AESA Based IPM Package for Tomato*. National Institute of Plant Health Management, Hyderabad, AP.

Schiavi D, Balbi R, Giovagnoli S, Camaioni E, Botticella E, Sestili F, Balestra GM (2021) A green nanostructured pesticide to control tomato bacterial speck disease. *Nanomaterials* 11(7): 1852. https://doi.org/10.3390/nano11071852

Selim EMI, Ammar MM, Amer GA, Awad HM (2020) Effect of some plant extracts, plant oils and *Trichoderma spp.* on tomato *Fusarium* wilt disease. *Menoufia J Plant Protec* 5: 155–167.

Shahidi BH, Barkhordar B, Pakgohar N, Aghighi S, Biglary S, Rashid Farrokhi P, Aminaii M, Mahdavi MJ, Aghelizadeh A (2006) Biological control of *Phytophthora drechsleri* Tucker, the causal agent of pistachio gummosis, under greenhouse conditions by use of actinomycetes. *Plant Pathol J* 5: 20–23.

Shankar R, Harsha S, Bhandary R (2014) *A Practical Guide to Identification and control of Tomato Diseases*. P 24 TROPICA SEEDS PVT LTD, Bangalore India.

Shanmugam V, Sharma V, Bharti P et al (2017) RNAi induced silencing of pathogenicity genes of *Fusarium* spp. for vascular wilt management in tomato. *Ann Microbiol* 67: 359–369.

Sharabani G, Manulis-Sasson S, Chalupowicz L, Borenstein M, Shulhani R, Lofthouse M, Sofer M, Frenkel O, Dror O, Shtienberg D (2014) Temperature at the early stages of *Clavibacter michiganensis* subsp. *michiganensis* infection affects bacterial canker development and virulence gene expression. *Plant Pathol* 63: 1119–1129.

Sharma M, Nagavardhini A, Thudi M, Ghosh R, Pande S, Varshney RK (2014) Development of DArT markers and assessment of diversity in *Fusarium oxysporum* f. sp. *ciceris,* wilt pathogen of chickpea (*Cicer arietinum* L.). *BMC Genomics* 15: 454.

Shenashen M, Derbalah A, Hamza A, Mohamedd A, El Saftya S (2017) Antifungal activity of fabricated mesoporous alumina nanoparticles against root rot disease of tomato caused by *Fusarium oxysporium*. *Pest Manag Sci* 73: 1121–1126.

Singh N, Mukherjee SK, Rajam MV (2020) Silencing of the ornithine decarboxylase gene of *Fusarium oxysporum* f. sp. *lycopersici* by host-Induced RNAi confers resistance to *Fusarium* wilt in tomato. *Plant Mol Biol Rep* 38: 419–429.

Singh D, Sinha S, Yadav DK, Sharma JP, Srivastava DK, Lal HC, Jaiswal RK (2010) Characterization of biovar/races of *Ralstonia solanacearum*, the incitant of bacterial wilt in solanaceous crops. *Indian Phytopath* 63(3): 261–265.

Suchoff DH, Louws FJ, Gunter CC (2019) Yield and disease resistance for three bacterial wilt-resistant tomato rootstocks. *HortTech* 29: 330–337.

Sugano J, Melzer M, Pant A, Radovich T, Fukuda S, Migita S, et al (2011) Field evaluations of tomato yellow leaf curl virus-resistant varieties for commercial production. *Plant Dis PD* 78: 1–4.

Sui X, Zheng Y, Li R, Padmanabhan C, Tian T, Groth-Helms D, Keinath AP, Fei Z, Wu Z, Ling KS (2017) Molecular and biological characterization of tomato mottle mosaic virus and development of RT-PCR detection. *Plant Dis* 101: 704–711.

Suresh LM et al (2017) Tomato disease field guide. *Seminis Vegetable Seeds*. https://seminis-us.com/resources/disease-guides/tomatoes/

Tao XR, Zhou XP (2004) A modified viral satellite DNA that suppresses gene expression in plants. *Plant J* 38: 850–860.

Tetorya M, Rajam MV (2021) RNAi-mediated silencing of *PEX6* and *GAS1* genes of *Fusarium oxysporum* f. sp. *lycopersici* confers resistance against *Fusarium* wilt in tomato. *3 Biotech* 11: 443. https://doi.org/10.1007/s13205-021-02973-8

Thomazella DPT, Brail Q, Dahlbeck D, Staskawicz B (2016) CRISPR-Cas9 mediated mutagenesis of a DMR6 ortholog in tomato confers broad-spectrum disease resistance. *Bio Rxiv* 064824.https://doi.org/10.1101/064824

Tian J, Rao S, Gao Y, Lu Y, Cai K (2021) Wheat straw biochar amendment suppresses tomato bacterial wilt caused by *Ralstonia solanacearum*: potential effects of rhizosphere organic acids and amino acids. *J Integ Agri* 20(9): 2450–2462.

Tiwari M, Sharma D, Trivedi PK (2014) Artificial micro RNA mediated gene silencing in plants: Progress and perspectives. *Plant Mol Biol* 86: 1–18.

Tran HTT, Ficke A, Asiimwe T, Hofte M, Raaijmakers JM (2007) Role of the cyclic lipopeptide surfactant massetolide A in biological control of *Phytophthora infestans* and colonization of tomato plants by *Pseudomonas fluorescens*. *New Phytol* 175: 731–742.

Tumwine J, Frinking HD Amp, Jeger M J (2002) Integrating cultural control methods for tomato late blight (Phytophthora infestans) in Uganda. *Ann Appl Biol* 141(3): 225–236.

Umesha S (2006) Occurrence of bacterial canker in tomato fields of Karnataka and effect of biological seed treatment on disease incidence. *Crop Prot* 25: 375–381.

Uppalapati SR, Ishiga Y, Ryu CM (2011) SGT1 contributes to coronatine signaling and *Pseudomonas syringae* pv. tomato disease symptom development in tomato and *Arabidopsis*. *New Phytol* 189: 83–93.

Vanitha SC, Niranjana SR, Mortensen CN, et al (2009) Bacterial wilt of tomato in Karnataka and its management by *Pseudomonas fluorescens*. *Bio Control* 54: 685–695.

Wangdi T, Uppalapati SR, Nagaraj S (2010) A virus-induced gene silencing screen identifies a role for thylakoid formation1 in *Pseudomonas syringae* pv. *tomato* symptom development in tomato and Arabidopsis. *Plant Physiol* 152: 281–292.

Wei Z, Huang J, Tan S, Mei X, Shen Q, Xu Y (2013) The congeneric strain *Ralstonia pickettii* QL-A6 of *Ralstonia solanacearum* as an effective biocontrol agent for bacterial wilt of tomato. *Biol Control* 65(2): 278–285.

White TJ, Bruns T, Lee S, Taylor JW (1990) Amplifcation and direct sequencing of fungal ribosomal RNA genes for phylogenetics. In: Innis MA, Gelfand DH, Sninsky JJ, White TJ (eds.), *PCR Protocols: A Guide to Methods and Applications*. Academic Press, New York, pp. 315–322.

Willcox JK, Catignani GL, Lazarus S (2003) Tomatoes and cardiovascular health. *Crit Rev Food Sci Nutr* 43: 1–18.

Xie C, Shao Y, Li X et al (2015) Detection of early blight and late blight diseases on tomato leaves using hyperspectral imaging. *Sci Rep* 5: 16564. https://doi.org/10.1038/srep16564

Yan Z, Pérez-de-Castro A, Díez MJ, Hutton SF, Visser RGF, Wolters, AMA et al (2018) Resistance to tomato yellow leaf curl virus in tomato germplasm. *Front Plant Sci* 9:1198. https://doi.org/10.3389/fpls.2018.01198

Yazici S, Yanar Y, Karaman I (2011) Evaluation of bacteria for biological control of early blight disease of tomato. *Afr J Biotechnol* 10(9): 1573–1577.

Zhang H, Yin L, Song F, Jiang M (2020) SKIP silencing decreased disease resistance against *Botrytis cinerea* and *Pseudomonas syringae*pv. tomato DC3000 in tomato. *Front Plant Sci* 11. https://doi.org/10.3389/fpls.2020.593267

Zitter TA, Sandlan KP (1990) CU-FAST-beta test version: Cornell University forecaster of Alternaria solani on tomato. *Plant Pathol Extension Rep* 90–1, Cornell University, Ithaca, NY.

18 Integrated Management of Brinjal (*Solanum melongena* L.) Diseases

Mudasir Bhat, Ali Anwar, Zakir Amin, Roohi Jan, Zohra Shabir, Shivanshi Gupta and Raja Junaid Rashid

18.1 INTRODUCTION

Brinjal (*Solanum melongena* L.) is also known as aubergine or eggplant and usually finds a place as a "poor man's crop". It is a versatile crop adapted to different agro-climatic regions and can be grown throughout the year, but in Kashmir, it is grown only during the summer from June to October. It is also known as eggplant due to the shape of the fruit of some varieties which are white and shaped similarly to chicken eggs (Santos et al. 2010). In Jammu and Kashmir, it is cultivated over an area of 2.02 thousand hectares with an annual production of 45.24 thousand MT and productivity of 22.39 tonnes per hectare (Saxena and Gandhi 2014). It is valued for its medicinal properties and has a decholestrolizing property primarily due to the presence of polyunsaturated fatty acids (linoleic and lenolenic) present in the flesh and seeds of fruit in higher amounts. In terms of nutrition, eggplant is among the healthiest vegetables due to its high amount of vitamins, minerals, and bioactive substances that are beneficial for human well-being (Raigón et al. 2008; Docimo et al. 2016). It also has a very low calorie value. According to Cao et al. (1996), eggplant ranks among the finest 10 vegetables on account of its ability to absorb oxygen radicals. For instance, chlorogenic acid in the flesh of the fruit (Stommel et al. 2015) and anthocyanins in the skin of the fruit (Mennella et al. 2012) are two phenolic chemicals that have been linked to the bioactive qualities of eggplant (Plazas et al. 2013). Anthocyanins and phenolic acids both provide a variety of advantages for human well-being (Plazas et al. 2013; Braga et al. 2016). Additionally, eggplants have been reported to contain a reasonable number of antioxidants, antimutagenic, antimicrobials, antivirals and to be completely devoid of cholesterol. In the areas experiencing high temperature and humidity during the rainy season, the production of brinjal suffers immensely due to the attack of disease and insect pests which became a regular feature and an important limiting factor in its production. Growing eggplant substantially enhances the livelihood of small-holders owing to its fruit-bearing ability and multiple harvestings on a weekly basis. However, a number of pathogenic diseases caused by fungi like damping off in the nursery and fungal fruit rot caused by *Phomopsis vexans*, or bacteria like bacterial wilt (*Ralstonia solanacearum*), or viral diseases like necrotic leaf virus and other causal organisms substantially limit the sustainable production each year (Ramesh 2008; Singh et al. 2014; Pagoch et al. 2015;). Brinjal is thus subjected to many fungal, bacterial, mycoplasmal and viral diseases which are explained as follows.

DOI: 10.1201/9781032631578-18

18.2 BACTERIAL DISEASES

18.2.1 Bacterial Wilt

Bacterial wilt is caused by *Ralstonia solanacearum* (Smith) Yabuuchi et al., primarily a soil-borne disease of wide distribution in the tropics, subtropics and warm-temperate regions of the world (Singh and Sharma 2014; Ramesh et al. 2016). In India, the disease has been reported from West Bengal. The pathogen also attacks other solanaceous crops such as potatoes, tomatoes, chillies and other cultivated plants which include castor, groundnut, banana, ginger and many wild plants. The disease is often associated with nematode infestation. It survives long term in water and soil, and latently infected plants and can be transmitted by soil, water, equipment and infected plant materials (Hayward 1991; Rahman et al. 2010). The pathogen invades through wounds in roots, lateral root emergence points or stomata and colonizes the intercellular spaces of the root cortex and vascular parenchyma, eventually entering the xylem vessels (Vasse et al. 1995). In the xylem vessels, the pathogen dissolves the cell walls and produces highly polymerized polysaccharides that increase the viscosity of the xylem and result in plugging. Blocking of vessels by bacteria is the major cause of wilting.

18.2.2 Symptoms

Wilting, stunting, yellowing of the foliage and finally collapse of the entire plant are the characteristic symptoms of the disease. Lower leaves may droop first before wilting occurs. The vascular system becomes brown. Bacterial ooze comes out from the affected parts. Plants that show wilting symptoms at noontime will recover at night, but die soon.

18.2.3 Favorable Conditions

All soil types, including sandy and clay soils, are susceptible to the disease. The frequency of BW is primarily concentrated in humid coastal regions and on acidic soils (soil pH <7.0). Moisture and high temperatures are advantageous. Root knot nematode presence predisposes the plant and expedites the onset of disease (Ramesh 2008).

18.2.4 Causal Organism

The bacterium is non-acid fast, non-spore-forming, non-capsulated and motile by a polar flagellum. The bacterium produces acid but no gas in dextrose, sucrose, lactose and glycerol. Starch is hydrolyzed with a slight liquefaction of gelatin. The soil bacterium *Ralstonia solanacearum*, formerly known as *Pseudomonas solanacearum*, is responsible for bacterial wilt in brinjal. There are five distinct races of the pathogen, and each one infects a different kind of plant. Based on the results of the biochemical tests, the *R. solanacearum* strains are divided into five biovars. Race 1 and biovar 3 strains predominately produce wilt disease in brinjal. A wide variety of hosts are present in Race 1, including solanaceous plants like tomato and chilli. Race 1 strains from Goa differ in terms of brinjal pathogenicity and genetic variability (Ramesh 2008).

18.2.5 Disease Cycle

The bacterium infects banana, chillies, fennel, ginger, potato, radish, tomato, etc.; the bacterium, though a non-spore former, is found to be alive and viable for more than 16 months under laboratory conditions. The pathogen is found to be alive in the infected plant debris for about 10 months. The presence of root knot nematode, *Meloidogyne javanica* increases the wilt incidence.

18.2.6 Disease Management

Crop rotation with cruciferous vegetables such as cauliflower helps in reducing the disease incidence. Crop rotation with okra, tomato, and potato should be avoided. Removal and destruction of the affected plant parts and using disease-resistant varieties help to reduce the disease incidence. Before sowing, the seeds should be dipped in a solution of Streptocycline (1 g/40 litres of water) for 30 minutes. Spray copper fungicides to control the disease (2% Bordeaux mixture.). The disease is more prevalent in the presence of root knot nematodes, so control of these nematodes will suppress the spreading of the disease (Ramesh 2008).

18.3 FUNGAL DISEASES

18.3.1 Verticillium Wilt

18.3.1.1 Symptoms

The disease attacks young plants as well as mature plants. The infected young plants show dwarfing and stunting due to the shortening of the internodes (Sheu et al. 2023). Such plants do not flower or fruit. Infection after the flowering stage results in the development of distorted floral buds and fruits. The affected fruits finally drop off. The infected leaves show the presence of irregularly scattered necrotic pale-yellow spots over the leaf lamina. Later on, these spots coalesce resulting in complete wilting of the leaves. The roots of the affected plants are split open longitudinally; a characteristic dark brown discolouration if the xylem vessels are observed (Figure 18.1). (Ramesh et al. 2016)

18.3.1.2 Causal Organism

Verticillium wilt caused by *Verticillium dahlia* Kleb. is a soil-borne fungus that colonizes the vascular tissues of plants. There are many strains of *V. dahliae* which are categorized into vegetative compatibility groups (VCG). These groups comprise strains that can exchange genetic material via anastomosis. Each VCG affects a few or only one host and the virulence of the pathogen varies by host. While individual *V. dahliae* strains are relatively host-specific, as a species it has a wide range.

FIGURE 18.1 Symptoms of *Verticillium* wilt disease in brinjal.

V. dahliae has a broad host range, affecting over 300 plant species causing vascular discoloration and wilt of many economically important crops. Some susceptible crops include brussels sprouts, cabbage, eggplant, cucumbers, mint, pepper, potatoes, pumpkin, spinach, tomato, watermelon, honeydew, and cantaloupe. Of these, tomato, potato, and eggplant have resistant or tolerant varieties. Microsclerotia can be seen under a lens as small black structures in the vasculature of living and dead plants. This feature can be used to distinguish *V. dahliae* from *V. albo-atrum*, the other verticillium wilt pathogen. Microsclerotia produced by *V. dahliae* may survive under field conditions for up to 14 years in the absence of a host. The microsclerotia germinate in the vicinity of host roots and cause infection. Verticillium wilt is favoured by cool air and soil temperatures (Sheu et al. 2023).

18.3.1.3 Disease Management

Soil fumigation of *V. dahliae* using chloropicrin or metam sodium is effective. Soil application and foliar application with Benlate (0.1%) is effective in reducing the wilt disease. Because of the longevity of microsclerotia and the broad host range of *V. dahliae,* crop rotation is usually not a feasible option for the control of verticillium wilt in many crops. However, rotations with broccoli, corn, wheat, barley, sorghum or safflower for a period of at least 2 years (the longer the rotation, the better) can reduce inoculum. There have been reports of a few biofertilizers showing antagonistic effects on the *Verticillium* pathogen (Singh and Sharma 2013).

18.3.2 Fusarium Wilt

18.3.2.1 Symptoms

In *Fusarium* wilt, the underground stems become dry and brown as a result of cortical decay while roots may have a soft and water-soaked appearance. Stunted growth, withering of immature fruits, yellowing of lower leaves, drooping of the apical portion, browning of vascular bundles and ultimate drying of the whole plant. When the roots and stems are cut diagonally, reddish-brown streaks may be seen in the vascular tissues. The plant may produce fewer or no fruits. Wilting of seedlings is also a common characteristic of the diseases.

18.3.2.2 Disease Management

Scientists have recently started to comprehend the negative effects of the persistent and extensive usage of chemical fungicides that endanger human interests. One of the options is that fungi might be better biocontrol agents. Presently, different trichoderma species are employed to control various plant pathogens (Verma et al. 2007; Najar et al. 2011). Some higher plant parts and their composition have already been reported to control the disease. Due to their lower phytotoxicity, fungitoxicants have been said to have effective properties (Kuri et al. 2011).

18.3.3 Cercospora Leaf Spot

Cercospora melongenae is found primarily in warm climates close to the equator. In the US it is prevalent in Hawaii, and could potentially be found in the southern Continental US (Srivastava and Nelson 2012). *C. melongenae* is a fungal plant pathogen that causes leaf spot on eggplant. It is a deuteromycete fungus that is primarily confined to eggplant species (Yadav et al. 2019). This plant pathogen only attacks the leaves of eggplants and not the fruit. It is fairly common among the fungi that infect community gardens and home gardens of eggplant (Srivastava and Nelson 2012).

18.3.3.1 Symptoms

Symptoms begin to show on the underside of older, lower leaves first due to the proximity to the soil. They are the first leaves to receive the inoculum. Lesions can be found on leaves, stems and

FIGURE 18.2 Symptoms of Cercospora spot on the leaf of brinjal.

petioles. The lesions appear small and circular at first and later progress to angular and more irregular shapes. The lesions are usually 4–10 mm in diameter, and have light to dark tan concentric spots (Figure 18.2). Spots appear on the upper leaf surface first and eventually progress to the lower side of the leaf. Upon observation in the field with a handheld scope, signs of conidiophores can be seen in the stomata of infected tissue.

18.3.3.2 Causal Organism

The leaf spot disease in eggplant is caused by *Cercospora solani-melongenae* Chupp., *C. melanagana* Welles and *C. solani* in. Recently, the fungus *Paracercospora egenula* as the primary cercosporoid has also been reported to cause leaf spot in eggplants in Hawaii. Although it was previously reported as *C. solani-melongenae*, based on an interception at Oahu, Hawaii, in 1946. Deighton (1979) listed *C. solani-melongenae* as a heterotypic synonym of *Paracercospora egenula*. Conidia of *C. melongenae* typically have a truncate base with a somewhat thickened convex scar and a sub-acute apex (Kumar et al. 2020). The conidia are typically 40–150 microns long, 3–4.5 microns wide, and 4–14 septate. Later in the disease cycle, excessive sporulation can be readily observed in the centre of the lesions (Srivastava and Nelson 2012). Eventually, the lesions will dry up and, in some cases, fall out. The fungi do not affect the fruit; however, yield can be impacted due to loss of photosynthetic material.

18.3.3.3 Disease Cycle

The fungal pathogen exists wholly in its asexual stage; even when grown in culture in the lab, no sexual stage is seen. This means that this disease has an imperfect life cycle, making it of the deuteromycota type. *C. melongenae* overwinters in conidiophores which asexually produce conidia. It survives the winter on lesions of previously infected fruit, plant debris, or simply in the soil (Yadav et al. 2019). This fungus can survive up to a year in the soil or plant debris as conidiophores. Conidia are released in the spring when the fungus is ready to infect its host, the eggplant. The conidia disperse by air or wind, or by rain-splash, and are able to infect a suitable host this way

(Kumar et al. 2020). Conidia infect an eggplant host by landing on the lower surface of young leaves and entering the leaf through stomata or other natural openings of the plant or wounds on the plant. From infection, there appear to be visible lesions on the leaves. The germ tube of the fungus can enter both closed and open stomata. Multiple germ tubes from various conidia can enter the same stomata (Agrios 2005). This fungus has the best infection rates among young leaves and has a reduced infection rate in mature leaves. It is thought that the reduced infection rate of mature leaves is due to the saprophytic microflora on the leaves and the anti-fungal compounds produced by the mature leaf. This fungus produces secondary conidia within the same season; this furthers the disease cycle and allows the fungus to infect more plants. Mycelium rapidly develops. Under near-perfect conditions, mycelium can develop within twelve hours of infection. The mycelium protects the next generation of conidia and ensures survival through the winter via conidiophores, and the cycle starts again the next season.

Fungi in the genus *Cercospora* produce the plant toxin cercosporin, which causes the leaf spot appearance. The fungal spores are not harmed by the production of this toxin because they produce pyridoxine which neutralizes the reaction (Agrios 2005). While the toxin damages the host cells, conidia are produced and are able to infect the current host or spread to other susceptible hosts in the area.

18.3.3.4 Epidemiology

Cercospora melongenae enters through breaks in the plant surface on the young leaf host of eggplants, typically through lesions caused by the fungi. Fruiting bodies of the fungus are overall larger when the fungus is able to proliferate during a heavy rain period versus a dry season. The fungus favours wet or moist leaves and high relative humidity for optimal infection. This fungus produces larger fruiting structures during the rainy season than during the dry season, indicating the need for a moist or wet environment (Yadav et al. 2019). Essentially, the amount of moisture in the air is correlated with the sizes of the fruiting structures. In almost all cases and especially in severe cases, signs and symptoms of *C. melongenae* have already appeared by the beginning of the dry season (or the end of the wet or rainy season). Because *C. melongenae* is a polycyclic disease and heavily dependent on a moist or wet environment, an extremely heavy rainy season is conducive for larger and greater conidia production and dispersal. What is interesting to note is that in less susceptible eggplant varieties, such as the Siamese variety, signs and symptoms of *C. melongenae* can be unseen and not appear until at least four weeks after the start of the dry season.

18.3.3.5 Disease Management

Prevention: Proper growing methods, crop rotation, and sanitary procedures are the most effective methods of prevention of *C. melongenae* infection. Leaf spot is a common disease found in home gardens and preventing its spread can be done by preventing excessive moisture and humidity accumulation. Ways to reduce relative humidity and dampness include weed control, irrigating in the morning, avoiding overhead sprinkler irrigation, increasing aeration at the base of plants, and using covers to minimize dampness. Crop rotation is also important, as well as the inclusion of diverse intercropping with many other vegetables. Being crop-specific, diversity will help slow down the spread of *C. melongenae*. Pant Samrat variety is resistant to the leaf spot so the disease can be managed by growing resistant varieties.

Chemical control: There are many fungicides which control *C. melongenae* including oils with no added pesticides, and some with. The most commonly recommended is a Bordeaux mixture, which only needs to be applied once every two weeks. Neem oil, octanoic acid and copper salt solutions, and various concentrations of chlorothalonil are all recommended fungicides, as well (Srivastava and Nelson 2012).

18.3.4 Alternaria Leaf Spot

18.3.4.1 Symptoms

Cracks appear in leaf spot. The two species of *Alternaria* occur commonly, causing the characteristic leaf spots with concentric rings. The spots are mostly irregular, 4–8 mm in diameter and may coalesce to cover large areas of the leaf blade. Severely affected leaves may drop off. *A. melongenae*also infects the fruits causing large deep-seated spots. The infected fruits turn yellow and drop off prematurely (Shafique et al. 2021).

18.3.4.2 Causal Organism

The disease is caused by *Alternaria melongenae* Rangs and Sams and *A. solani* Ell. and Mart. Mycelium is septate, branched, light brown to dark brown. It is inter- and intra-cellular. Conidiophores emerge through stomata and are dark coloured. Conidia are single-celled, muriform, beaked and produced in chains. The conidia are with 5–10 transverse septa and a few longitudinal or oblique septa (Shafique et al. 2021).

18.3.4.3 Disease Cycle

The disease is spread by wind-borne conidia.

18.3.4.4 Disease Management

Removal and destruction of affected plant parts and spraying 1 per cent Bordeaux mixture or 2.0 g copper oxychloride or 2.5 g zineb per litre of water effectively controls leaf spots.

18.3.5 Damping Off

Damping off is considered a destructive disease responsible for poor germination and poor stand of seedlings in nursery beds. Brinjal damping off incited by *Pythium ultimum* Trow. is one of the most important and destructive diseases of brinjal, which can cause mortality up to 90% (pre-emergence seed rot and post-emergence seedling mortality).

18.3.5.1 Symptom

Two types of symptoms are observed during the raising of the nursery plants.

Pre-emergence damping off: The pre-emergence damping off results in seed and seedling rot before these emerge out of the soil.

Post-emergence damping off: The post-emergence damping off phase is characterized by infection of the young, juvenile tissues of the collar at the ground level. The infected tissues become soft and become water-soaked. The collar portion rots and ultimately the seedlings collapse and die. Sudden collapsing of the seedlings occurs in the seed bed. The seedlings are attacked at the collar region and the attacked seedlings are toppled down (Sanjeevkumar 2020). The disease spreads through fungi present in the soil. The disease spreads through fungi present in the soil.

18.3.5.2 Causal Organism

Many fungi prevalent in soils can cause damping off. *Fursarium* spp., *Pythium* spp., and *Phytophthora* spp. are most active in cool, wet soils whereas, *Cylindrocladium* spp. and *Rhizoctonia* spp. are more common in warm, wet soils. The damping off in brinjal is caused by *Pythium* spp., including *P. aphanidermatum*, *P. irregulare* and *P. ultimum* Trow.

18.3.5.3 Disease Cycle

The pathogen is soil-borne with a wide host range and almost worldwide distribution. High soil, moisture and moderate temperature along with high humidity especially in the rainy season lead to the development of the disease (Sanjeevkumar 2020).

18.3.5.4 Disease Management

Being soil-borne, *P. ultimum* is very difficult to manage with fungicides alone and uneconomical. Healthy seeds should be selected for sowing. The seeds should be treated with Thiram @ 2 g/kg of seed before sowing. Continuous raising of the nursery in the same plot should be avoided. The top-soil 2 of the nursery should be treated with Thiram @ 5 g/m area of the soil and the nursery should be drenched with the same chemical @ 2 g/litre of water at fortnightly intervals.

18.3.6 Collar Rot

The host range of *S. rolfsii* is very wide and includes not only many important horticultural and agronomic crops but many of noneconomic importance. The soil-borne plant pathogenic fungus *S. rolfsii* attacks more than 500 species of plants belonging to over 100 families. The pathogen causes a great economic loss on various crops. On several other hosts like chilli, tomato, groundnut, etc. it causes a collar rot of the plants.

18.3.6.1 Symptoms

The disease occasionally occurs in a serious form. The lower portion of the stem is affected by the soil-borne inoculum (sclerotia) (Jadon and Tiwari 2011). Decortications is the main symptom. Exposure and necrosis of underlying tissues may lead to the collapse of the plant. Near the ground surface on the stem may be seen the mycelia and sclerotia. Lack of plant vigour, accumulation of water around the stem and mechanical injuries help in development of this disease.

18.3.6.2 Causal Organism

The disease is caused by *Sclerotium rolfsii* Sacc. Choudhury (1967) states that foot rot disease of eggplant/brinjal is caused by *Sclerotium* sp. Sharma et al. (2002) reported variability among 26 isolates of *S. rolfsii* collected from various hosts/soil samples and localities in India. The pathogen, *S. rolfsii* secrete oxalic acid and tissue-degrading enzymes such as cellulose during infection (Song et al. 2024). Massive mycelial growth on plant tissues produces large quantities of oxalic acid that facilitate penetration of hyphae into tissue. Oxalic acid and tissue-degrading enzymes work together to break down cell walls, resulting in tissue maceration. The pathogens then absorb nutrients from the macerated tissue (Zhu et al. 2024). Maceration interrupts the transport of water and nutrients in plant tissues, thereby causing wilting, yellowing, and necrosis.

18.3.6.3 Disease Cycle

Sclerotia are the primary structures that enable *S. rolfsii* to survive periods of unfavourable conditions. Sclerotia are formed from masses of hyphal strands. Mature sclerotia include an outer layer (rind), an underlying cortical layer, and a central medullary region comprised of loosely interwoven hyphae (Shanmugaraj et al. 2024). Sclerotia of *S. rolfsii* survived from 2 months to 7 years in field soil depending on experimental conditions (Sailaja Rani et al. 2024). Temperature and moisture may interact in influencing the survival of *S. rolfsii* under field conditions.

18.3.6.4 Disease Management

The pathogen overwinters in or under the soil and plant debris, therefore it is often difficult to destroy all the inoculum by fungicide application. However, both systemic and non-systemic fungicides in the group of triazole, strobilurin and dithiocarbamates have shown potential against *S. rolfsii*. The inclusion of non-host crop plants is an option to reduce the pathogen inoculum in soil by crop rotation. Seed treatment with 4 g of *Trichoderma viride* formulation per kg seed will help in reducing the disease even by the mycoparasite (Anwar et al. 2008; Jadon and Tiwari 2011). Spraying with Mancozeb @ 2 g/litre of water. Collection and destruction of diseased parts and portions of the plant (Suryawanshi et al. 2015).

18.3.7 Phomopsis Blight and Fruit Rot

Fruit rot is the most destructive stage of the disease, as it damages the fruits partially or completely in the fields or during transit. The disease on stems and leaves results in a reduction of fruit size and weight as well as loss of plants. In India, the yield losses due to fruit rot ranged from 10 to 20% in the Punjab and Delhi (Singh et al. 2024). In an advanced stage of the disease, seed quality is also adversely affected, and infected seeds become discoloured, with poor germinability and reduced seed viability (Tripathi et al. 2024). The disease was first reported in Gujarat in 1914 and later in other parts of India. The causal pathogen causes seedling blight, leaf spots and fruit rot (Manda et al. 2020).

18.3.7.1 Symptoms

The symptoms range from poor germination and seedling blight to fruit rot. Post-emergence damping off of seedlings results from infection of the stem just above the soil surface. The symptoms on leaves are more prominent during the early stages of plant growth. Irregular spots result from coalescence. After transplanting, leaves coming into contact with the soil may become infected directly or develop leaf spot due to infection by conidia. Lesions on the petiole or the lower part of the midrib can result in the death of the entire leaf (Manda et al. 2020). The diseased plant bears smaller leaves, and the axillary buds are often killed. When stem girdling occurs, the shoot above the infected area wilts and dries up and the plant may be toppled by the wind (Sherf and MacNab 1986; Manda et al. 2020). Pycnidia develop readily in lesions on young stems, but rarely on older ones. On the fruits the symptoms appear first as minute sunken greyish spots with a brownish halo, which later enlarge and coalesce, producing concentric rings of yellow and brown zones (Mahadevakumar et al. 2017). If the infection enters the fruits through the calyx, the whole fruit may become mummified due to dry rot (Mahadevakumar et al. 2017). Rot may appear in fruit, in transit after harvest (Sherf and MacNab 1986) (Figure 18.3).

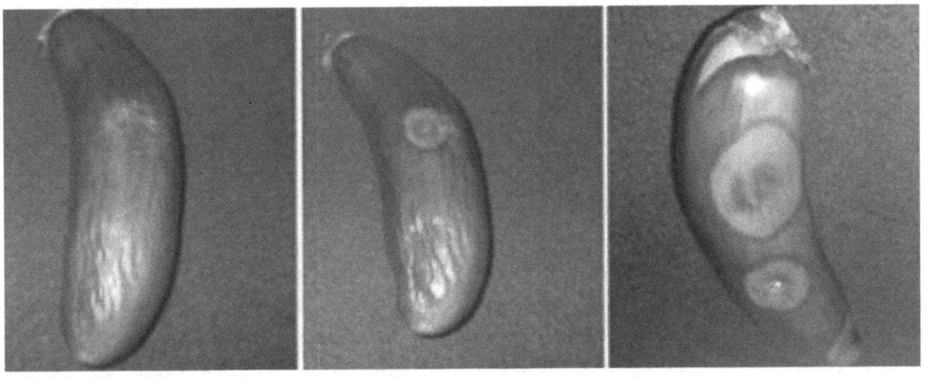

FIGURE 18.3 Symptoms of Phomopsis rot on the fruits of brinjal.

18.3.7.2 Causal Organism

The disease caused by *Phomopsis vexans* (Sacc. & Syd.) Harter is a pynicidial anamorph with a teleomorph in the genus *Diaporthe*. Conidia germinate after 6 hours and penetration occurs after 12 hours. In tissue, the spread of the fungus is both intercellular and intracellular. Seedlings and young stems are highly susceptible. Mature tissue exhibits hypertrophy and hyperplasia below the infected region, preventing further spread of the fungus. Pycnidia with or without beak are found in the affected tissue. They are globose or irregular. Conidiophores in the pycnidium are hyaline, simple or branched (Mahadevakumar et al. 2017).

18.3.7.3 Epidemiology

P. vexans requires hot and humid conditions for infection and disease development. Spore germination is optimal at 27°C, and pycnidial formation is greatest between 30 and 35°C. The optimum relative humidity for disease development is 55% RH and above, and the optimum temperature for fungal growth is 28°C. Fruit rot was maximal at 30°C and 50% RH in the growth chamber (Islam and Sitansu 1990); temperatures of 5, 10 and 40°C were unfavourable for disease development in inoculated, detached fruit (Islam et al. 2009).

18.3.7.4 Disease Management

Physical and chemical treatment: It is appropriate to collect the seeds from healthy plants (disease-free) and use them for sowing. Hot water treatment of seeds at 50°C for 30 min or 56°C for 15 min can be used to control the disease incidence, as the fungus is sensitive to high temperatures (Singh et al. 2014). Seeds should be dipped in hot water at 50°C for 30 min. Spraying with captan 0.2% in the nursery and field at 7–10-day intervals controls the disease. Spraying the crop in the field with zineb 0.2% or Bordeaux mixture 0.8% is effective in controlling Phomopsis blight (Islam et al. 2009). Several chemicals have been tested and recommended for seed treatment with bavistin (1 g kg–1 seed) or seedling treatment (0.05% solution) for 30 minutes or foliar spray (0.05%) at 10 to 15-day intervals for control of P. *vexans* effectively (Pandey 2010). Propiconazole 25 EC, followed by carbendazim 50 WP, copper oxychloride 50 WP, and metalaxyl + mancozeb 68 WP in decreasing order of their performance, emerged as the most effective fungitoxicant, showing the greatest disease control (Anwar et al. 2017).

Biological control: Antagonistic *Pseudomonas fluorescens* and *Trichoderma harzianum* seed treatment and spray treatment were found to be effective against *P. vexans* (Srinivas et al. 2005).

18.3.8 Phytophthora Rot

Phytophthora fruit rot was first reported in the United States in Colorado and California in the late 1930s. It apparently occurred sporadically in most of the United States except California until the 1980s, when the incidence of both fruit rot and crown rot increased notably in Florida, Georgia, Michigan, and the northeastern states. The increase followed a hurricane in some areas.

18.3.8.1 Symptoms

The initial symptoms of phytophthora fruit rot are a water-soaked or depressed spot. The underside of the fruit, where it is in contact with the ground, is often affected first. Symptoms also develop on the upper surface of the fruit, especially following rain or irrigation which provides splashing water for fungal dispersal (Kumar 2012). The fungus produces white, yeastlike growth that contains many sporangia, especially under moist conditions. This growth does not become very thick. Affected fruit can be completely covered by sporangia (Chowdappa et al. 2016). Infections can progress rapidly until the fruit collapses. Symptoms also develop rapidly after harvest.

18.3.8.2 Causal Organism

The disease is caused by *Phytophthora nicotianae* Breda de Haan.

18.3.8.3 Disease Cycle

The fungus survives in soil between crops for at least 2 years, according to observations of disease occurrence in northern states. Soil moisture conditions are important for disease initiation. Sporangia form when the soil is at field capacity (within 24 h, under controlled conditions), and they release zoospores when the soil is saturated (5–6 h, under controlled conditions) (Kumar 2012; Chowdappa et al. 2016). Subsequently, the disease can develop rapidly, because sporangia on affected fruit are produced and dispersed under a wider range of conditions.

18.3.8.4 Disease Management

Rotation with non-host crops is recommended. Other hosts are pepper, tomato, eggplant, cocoa, and macadamia. It is not known how long *Phytophthora* can survive in soil as oospores, which are believed to be the primary overwintering propagule or as chlamydospores. An important control strategy is to manage soil moisture by selecting well-drained fields, avoiding low-lying areas, subsoiling, preparing dome-shaped raised beds for non-vining crops, and not over-irrigating.

Fumigation and a preventive fungicide program initiated at the start of fruit formation may provide effective disease suppression when combined with these cultural practices. Movement in soil on equipment is probably an important means by which *Phytophthora* has been spread between fields and may account for disease occurrence in fields with no history of susceptible crops.

18.4 VIRAL DISEASES

18.4.1 Tobacco Mosaic Virus (TMV)

18.4.1.1 Symptoms

Mosaic mottling of leaves and stunting of plants are the characteristic symptoms of potato virus Y. Mosaic symptoms are mild in early stages but later become severe. Infected leaves are deformed, small and leathery. Very few fruits are produced on infected plants (He et al. 2019). The important symptom produced by the tobacco mosaic virus is conspicuous mottling of leaves. Leaves also develop blisters in advanced cases. Severely infected leaves become small and misshapen (Hidayat et al. 2019). Plants infected early remain stunted. PVY is easily sap-transmitted. It is transmitted in the field through aphids, *Aphis gossypii* and *Myzus persicae* and perpetuates on weed hosts like *Solanum nigrum* and *S. xanthocarpum*. TMV is transmitted by sap, contaminated implements and clothes, soil debris and hands of labour (He et al. 2019; Hidayat et al. 2019).

18.4.1.2 Disease Management

The disease incidence can be minimized by reducing the population of aphids, removing and destroying infected plants and eradicating susceptible weed hosts. In the nursery, aphids can be controlled by the application of Carbofuran (1 kg a.i./ha) in the nursery bed at the time of sowing seeds followed by 2–3 foliar sprays of Phosphamidon (0.05%) at an interval of 10 days. Spraying Phosphamidon (0.05%) at 10-day intervals starting from 15–20 days after transplanting effectively controls the aphids in the field. Avoid planting of cucumber, pepper, tobacco, and tomato near brinjal seed beds and fields. Wash hands with soap and water before working in seed beds (Zhong et al. 2015).

18.5 DISEASES CAUSED BY PHYTOPLASMAS

Wall less prokaryotes known as phytoplasmas, which are parasitic on both plants and insects, comprise a sizable monophyletic group within the class Mollicutes (Maheshwari et al. 2017). The phytoplasma infections that affect eggplant crops can be terrifying in that they cause substantial damage to the crop all over the world and are frequently reported from the newer areas (Rao and Kumar 2017).

18.5.1 Little Leaf of Brinjal

This disease of brinjal was reported from India in 1938 and as far as known it occurs only in India and Sri Lanka. In almost all the states of the country, it has become a serious problem facing brinjal cultivation. The yield loss is 100 per cent in the diseased plants.

18.5.2 Symptoms

The characteristic symptom is the smallness of the leaves. The petioles are so short and the leaves appear to be sticking to be stem. Such leaves are narrow, soft, smooth and yellow. Newly formed leaves are much shorter. The internodes of the stem are also shortened. Axillary buds get enlarged but their petioles and leaves remain shortened (Rao and Kumar 2017). This gives the plant a bushy appearance Figure 18.4). Mostly, there is no flowering but if flowers are formed they remain green. Fruiting is rare (Vandana et al. 2015).

FIGURE 18.4 Symptoms of little leaf of brinjal.

18.5.3 Causal Organism

Little leaf was first considered a disease caused by a virus. In 1969 it was attributed to a mycoplasma-like organism, closely related to aster-yellows and curly top. It is a sap-transmissible disease (Kumar et al. 2012). The organism has been transmitted to *Datura*, tomato and tobacco. It occurs in nature on *Datura fastuosa* and *Vinca rosea*. Natural transmission is through a vector, *Cestius phycytis* (*Eutettix phycytis*) while *Empoasca devastans* is a less effective vector. The perennation of the organism is through its weed hosts.

18.5.4 Disease Cycle

The disease is transmitted by leaf hoppers, *Hishimonas phycitis* and *Empoasca devastans* and grafting. *E. devastans* is a less effective vector (Majumdar and Das 2020). Perennation of the virus is through the weed host. This disease has a very wide host range.

18.5.5 Disease Management

The severity of the disease can be reduced by the destruction of affected plants and the spraying of insecticides. New crops should be planted only when diseased plants in the field and its neighbourhood have been removed (Rakesh and Chaturvedi 2019). Spraying of insecticide such as Methyldemeton 25EC @ 2.0 ml/lit., Dimethoate 30EC 2.0 ml/lit. or Malathioan 50 EC @ 2.0 ml /lit. are found to manage insect vectors. Although mycoplasmas are reported to be suppressed by tetracyclines field application of this method has not yet been recommended.

18.6 CONCLUSIONS

Solanum melongena, sometimes known as brinjal or eggplant, is a vegetable crop that is grown extensively around the world, including the Indian subcontinent. Wilt, damping off, soft rot, root rot, and other fungal, viral and phytoplasma diseases affect this crop causing severe economic and yield losses. It is essential to have knowledge about the occurrence and treatment of fungal diseases as they are frequently encountered and result in substantial financial losses. After a thorough search in the literature, it may be concluded that the main fungi that cause fungal infections in brinjal include Alternaria leaf spot, damping off, collar rot, fruit rot, and leaf spot. Among the bacterial diseases, a significant production constraint in the brinjal industry is bacterial wilt, and there is currently no single viable management strategy for the disease. The diseases caused by virus and phytoplasmas may also incur severe losses and there is a need to keep track of the advancement of various diseases in an area keeping into consideration the prevailing environmental conditions and infection load. Recent studies have shown various effective strategies including organic additives, soil solarization, and biological control agents that could be used to control various diseases in brinjal. However, to manage diseases and prevent economic loss in an area, integrated disease management strategies can be helpful.

REFERENCES

Agrawal MK, Fageria MS, Dhaka RS (2000) Breeding for multiple disease resistance in vegetables: a review. *Agricultural Rev* 21(2): 125–128.

Agrios GN (2005) *Plant pathology*. Elsevier.

Anwar A, Bhat G, Bhat K (2008) Mycoparasitic behaviour of certain bioagents against sheath blight pathogen (*Rhizoctonia solani*) of rice. *Indian J Mycol Plant Pathol* 38: 135–137.

Anwar A, Bhat M, Ganaie N, Ambardar V, Hassan MG (2017) Prevalence and management through relative performance of organic mulches and fungi toxicants of noxious Phomopsis fruit rot (*Phomopsis vexans* Sacc. & Syd.) Harter, in brinjal ecology of Kashmir. *The Pharma Innov J* 6: 318–323.

Bhat M (2017) *Variability in isolates of Phomopsis vexans (Sacc. & Syd.) Harter causing blight disease in brinjal (Solanum melongena L.) and its integrated management in Kashmir*". SKUAST Kashmir.

Braga PC, Scalzo RL, Dal Sasso M, Lattuada N, Greco V, Fibiani M (2016) Characterization and antioxidant activity of semi-purified extracts and pure delphinidin-glycosides from eggplant peel (*Solanum melongena* L.). *J Functional Foods* 20: 411–421.

Cao G, Sofic E, Prior RL (1996) Antioxidant capacity of tea and common vegetables. *J Agri Food Chem* 44(11): 3426–3431.

Choudhury B (1967). Vegetables. Sixth revised edition. National book trust, 56–57.

Chowdappa P, Kumar BN, Kumar SM, Madhura S, Bhargavi BR, Lakshmi MJ (2016) Population structure of *Phytophthora nicotianae* reveals host-specific lineages on brinjal, ridge gourd, and tomato in South India. *Phytopathology* 106: 1553–1562.

Docimo T, Francese G, Ruggiero A, Batelli G, De Palma M, Bassolino L, Toppino L, Rotino GL, Mennella G, Tucci M (2016) Phenylpropanoids accumulation in eggplant fruit: characterization of biosynthetic genes and regulation by a MYB transcription factor. *Front Plant Sci* 6: 1233.

Hayward A (1991) Biology and epidemiology of bacterial wilt caused by *Pseudomonas solanacearum*. *Annu Rev Phytopathol* 29: 65–87.

He Y, Huang K, Yao L, Li W, Liu C, Chen L, Chen H, Chen S, Wang J (2019) First report of tobacco mosaic virus infecting dutch eggplant (*Solanum aculeatissimum*) in China. *Plant Dis* 103: 2973.

Hidayat SH, Wiyono S, Damayanti TA (2019) Characteristic of Tobacco mosaic virus isolated from cucumber and tobacco collected from East Java, Indonesia. *Biodiversitas J Biol Diversity* 20.

Islam M, Meah M, Islam M, Islam M, Faruq A (2009) Studies on some physiological aspects of Phomopsis vexans causing Phomopsis blight and fruit rot of eggplant. *J Sher-e-Bangla Agric Univ* 3: 96–103.

Islam S, Sitansu P (1990) Variabilities among isolates of *Phomopsis vexans*. *Environ Ecol* 8: 315–319.

Jadon KS, Tiwari P (2011) Pathogen physiology and management of Brinjal collar rot caused by *Sclerotium rolfsii*. *Ann Plant Protec Sci* 19: 113–117.

Kumar J, Gunapati S, Singh S, Lalit A, Sharma N, Tuli R (2012) First report of a '*Candidatus* Phytoplasma asteris'(16SrI group) associated with little leaf disease of *Solanum melongena* (brinjal) in India. *New Disease Rep* 26: 2044–0588.

Kumar N, Kumar A, Chand G, Biswas S (2020) Diseases of brinjals (eggplant) and their management. In: Chand G, Akhtar MN, Kumar S (eds), *Diseases of fruits and vegetable crops: recent management approaches*, vol. 270. Apple Academic Press, New York.

Kumar S (2012) Factors affecting the development of fruit rots of brinjal. *Plant Dis Res* 27: 39–44.

Kuri S, Islam R, Mondal U (2011) Antifungal potentiality of some botanical extracts against important seedborne fungal pathogen associated with brinjal seeds, Solanum melongena L. *J Agri Technol* 7: 1139–1153.

Mahadevakumar S, Amruthavalli C, Sridhar K, Janardhana G (2017) Prevalence, incidence and molecular characterization of *Phomopsis vexans* (*Diaporthe vexans*) causing leaf blight and fruit rot disease of brinjal in Karnataka (India). *Plant Pathol Quar* 7: 29–46.

Maheshwari M, Kumar M, Rao GP (2017) Identification and characterization of phytoplasma associated with brinjal little leaf, phyllody and witches' broom disease in India. *Indian Phytopath* 70: 258–261.

Majumdar S, Das BK (2020) Studies on little leaf of brinjal and morphotaxonomy of the leafhopper species associated from Bengal. *J Entomol Zool* 8: 514–521.

Manda RR, Addanki VA, Srivastava S (2020) Phomopsis blight of *Solanum melongena*–brinjal/eggplant. *Plant Cell Biotechnol Mol Biol* 17: 7–12.

Mennella G, Lo Scalzo R, Fibiani M, D'Alessandro A, Francese G, Toppino L, Acciarri N, de Almeida AE, Rotino GL (2012) Chemical and bioactive quality traits during fruit ripening in eggplant (*S. melongena* L.) and allied species. *J Agri and Food Chem* 60: 11821–11831.

Najar A, Anwar A, Masoodi L, Khar M (2011) Evaluation of native biocontrol agents against *Fusarium solani* f. sp. *melongenae* causing wilt disease of brinjal in Kashmir. *J Phytol* 3: 31–34.

Pagoch K, Srivastava J, Singh AK (2015) Damping-off disease of seedlings in Solanaceous vegetables: current status and disease management. In: Awasthi LP (ed), *Recent Advances in the Diagnosis and Management of Plant Diseases*, pp. 35–46. Springer, New Delhi. https://doi.org/10.1007/978-81-322-2571-3_4

Pandey A (2010) Studies on fungal diseases of eggplant in relation to statistical analysis and making of a disease calendar. *Recent Res Sci Technol* 2(9): 01–03.

Plazas M, Andujar I, Vilanova S, Hurtado M, Gramazio P, Herraiz FJ, Prohens J (2013) Breeding for chlorogenic acid content in eggplant: interest and prospects. *Notulae Botanicae Horti Agrobotanici Cluj-Napoca* 41: 26–35.

Plazas M, Prohens J, Cuñat AN, Vilanova S, Gramazio P, Herraiz FJ, Andújar I (2014) Reducing capacity, chlorogenic acid content and biological activity in a collection of scarlet (*Solanum aethiopicum*) and gboma (*S. macrocarpon*) eggplants. *Int J Mol Sci* 15: 17221–17241.

Rahman M, Islam M, Rahman T, Meah M (2010) Biochemical characterization of *Ralstonia solanacerum* causing bacterial wilt of brinjal in Bangladesh. *Progr Agri* 21: 9–19.

Raigón MD, Prohens J, Muñoz-Falcón JE, Nuez F (2008) Comparison of eggplant landraces and commercial varieties for fruit content of phenolics, minerals, dry matter and protein. *J Food Comp Anal* 21: 370–376.

Rakesh P, Chaturvedi A (2019) Effective management strategy of little leaf disease in brinjal (*Solanum melongena* L.). *J Exper Biol Agri Sci* 7: 295–300.

Ramesh R (2008) Bacterial wilt in brinjal and its management. *Technical Bulletin No: 10, ICAR Research Complex for Goa (ICAR)*, Ela, Goa, India.

Ramesh R, Achari G, Asolkar T, Dsouza M, Singh N (2016) Management of bacterial wilt of brinjal using wild brinjal (*Solanum torvum*) as root stock. *Indian Phytopath* 69: 260–265.

Rao G, Kumar M (2017) World status of phytoplasma diseases associated with eggplant. *Crop Protec* 96: 22–29.

Sailaja Rani J, Reddi Kumar M, Khayum Ahammed S, Rao KS, Ravindra Reddy B, Sarada Jayalakshmi R (2024) Cultural and morphological characterization of *Streptomyces* and interaction study with *Sclerotium rolfsii* by SEM. *J Sci Res Rep* 30: 176–192.

Sanjeevkumar K (2020) Chapter-1 general studies of damping off (*Pythium aphanidermatum*) of brinjal. In: *Current research and Innovation Plant Pathology 1*. Edited by Singh HK, © AkiNik Publications, pp. 1–13.

Santos B, Stall W, Zhang S, Webb S, Smith S (2010) Eggplant production in Florida. *Vegetable Prod Handbook* 10: 107.

Saxena M, Gandhi CP (2014) *Indian horticulture database-2014*. National Horticulture Borad.

Shafique MS, Amrao L, Saeed S, Ahmed MZ, Ghuffar S, Anwaar HA, Sheikh UAA, Khan MA, Qadir, A, Abdullah A (2021) Occurrence of leaf spot caused by *Alternaria alternata* on eggplant (*Solanum melongena*) in Pakistan. *Plant Dis* 105: 1224–1224.

Shanmugaraj C, Kamil D, Gogoi R, Das A, Kumar GS, Hussain Z, Singh PK, Iquebal M (2024) Characterizing genetic diversity of *Sclerotium rolfsii* isolates by biomapping of mycelial compatibility groupings and multilocus sequence analysis. *World J Microbiol Biotechnol* 40: 251.

Sharma BK, Singh UP, Singh KP (2002) Variability in Indian isolates of *Sclerotium rolfsii*. *Mycologia* 94(6): 1051–1058.

Sherf AF, MacNab AA (1986) *Vegetable diseases and their control*. John Wiley & Sons.

Sheu ZM, Chiu MH, Chang JH, Oliva R (2023) First report of Verticillium wilt Caused by *Verticillium dahliae* in grafted tomato in Taiwan. *Plant Dis* 107(7): 2236.

Singh BK, Singh S, Yadav SM (2014) Some important plant pathogenic disease of brinjal (*Solanum melongena* L.) and their management. *Plant Pathol J (Faisalabad)* 13: 208–213.

Singh D, Sharma P (2014) Diseases of tomato and brinjal. In: *Diseases of vegetable crops: diagnosis and management*. Edited by Singh D, Chowdappa P, Sharma P, Indian Phytopathological Society, Division of Plant Pathology, Indian Agricultural Research Institute Pusa, pp. 59–142.

Singh N, Sharma C (2013) Antagonistic effect of biofertilizers against seed born mycoflora of brinjal (*Solanum melongena*). *Biotech Today* 3: 24–26.

Singh P, Singh R, Khilari K, Mishra P, Singh H (2024) Effect of different culture media on growth and establishment of *Phomopsis vexans* Inciting fruit rot of brinjal (*Solanum melongena* L.). *J Adva Biol Biotechnol* 27: 58–64.

Song D, Wen R, Chen K, Wen F, Xing D, Wu Z (2024) Application of thifluzamide to stem rot in peppers: infection and control mechanisms of *Sclerotium rolfsii*. *Pesticide Biochem Physiol* 200: 105846.

Srinivas C, Niranjana S, Shetty H (2005) Effect of bioagents and fungicides against *Phomoposis vexans* and on seed quality of brinjal. *Crop Improvement* 32: 95–101.

Srivastava S, Nelson S (2012) *Cercospora leaf spot of eggplant*. College of Tropical Agriculture and Human Resources (CTAHR), pp. 1–5.

Stommel JR, Whitaker BD, Haynes KG, Prohens J (2015) Genotype × environment interactions in eggplant for fruit phenolic acid content. *Euphytica* 205: 823–836.

Suryawanshi A, Borgaonkar A, Kuldhar D, Dey U (2015) Integrated management of collar rot (*Sclerotium rolfsii*) of brinjal (*Solanum melongena*). *Indian Phytopath* 68: 189–195.

Tripathi A, Maurya S, Pandey K, Behera T (2024) Global scenario of vegetable fungal diseases. *Vegetable Sci* 51: 54–65.

Vandana Y, Mahadevakumar S, Janardhana G, Amruthavalli C, Sreenivasa M (2015) Molecular detection of *Candidatus* phytoplasma trifolii associated with little leaf of brinjal from Kerala state of Southern India. *Interl J Life Sci* 9: 109–112.

Vasse J, Frey P, Trigalet A (1995) Microscopic studies of intercellular infection and protoxylem invasion of tomato roots by *Pseudomonas solanacearum*. *Mol Plant Microbe Intera* 8: 241–251.

Verma M, Brar SK, Tyagi R, Surampalli RN, Valero J (2007) Antagonistic fungi, *Trichoderma* spp.: Panoply of biological control. *Biochem Engineering J* 37: 1–20.

Yadav F, Kumar R, Yadav A, Yadav M (2019) Effective management of cercospora leaf spot disease of brinjal by fungicides in Ayodhya UP. *J Medicinal Plants* 7: 134–135.

Zhong Y, Peng JJ, Chen ZZ, Xie H, Luo D, Dai JR, Yan F, Wang JG, Dong HZ, Chen SY (2015) Dry mycelium of *Penicillium chrysogenum* activates defense responses and restricts the spread of Tobacco mosaic virus in tobacco. *Physiol Mol Plant Pathol* 92: 28–37.

Zhu Y, Wu C, Deng Y, Yuan W, Zhang T, Lu J (2024) Recent advances in virulence of a broad host range plant pathogen *Sclerotinia sclerotiorum*: a mini-review. *Front Microbiol* 15: 1424130.

19 Biotic Stress Management Strategies for Cruciferous Vegetable Crops

Savitha A. S., Renuka M., Ajithkumar K. and Yenjerappa S. T.

19.1 INTRODUCTION

Cruciferous vegetables belonging to the Brassicaceae family are valuable exports and important staples of the human diet in many countries as they are low in calories, providing essential vitamins, minerals and fibre. The plant name derives from the Latin *crucifer* ("cross-bearing"), a reference to the symmetrical arrangement of four petals in the flowers of several species in this family. Globally, crucifers are popular temperate vegetables and occupy an important place among vegetables. These are rich in important organic compounds: vitamin C, vitamin A, soluble fibre glucosinolates, comprised of glucose and amino acids containing nitrogen and sulphur which imparts pungent aromas and spicy taste. Among the brassicas, cabbage and cauliflower are grown on a commercial scale, whereas knol khol, sprouting broccoli and brussels sprouts are grown scattered on a small scale. These crops are prone to various diseases, which results in severe yield loss in terms of quantity and quality, which in turn affects the seed industry and commercial vegetable production. Important diseases of crucifers found worldwide are black rot, curd rot, club root, black spot, wilt, watery soft rot, wire stem and cyst nematode (Sharma et al. 2018). Hence, the chapter is designed to provide sufficient information related to the symptomatology, aetiology, epidemiology and management aspects of the crucifer major diseases.

19.2 DISEASES OF CRUCIFEROUS VEGETABLE CROPS

19.2.1 Leaf Spot and Blight

19.2.1.1 Disease Occurrence, Distribution and Importance

Alternaria brassicae and *A. brassicicola* (Schwein) are the necrotrophs affecting the majority of cole crops (cabbage, cauliflower, broccoli, canola and mustard) across the globe. The disease is known by other names such as black spot of cabbage, brown spot of cauliflower, *Alternaria* storage rot, *Alternaria* sooty spot and curd blight. It is distributed across the tropical and sub-tropical regions that prevailed with warm and moist conditions. *A. brassicae* and *A. brassicicola* were first noticed in cabbage in 1909 and 1924 in the USA and Holland, respectively. Blight can cause a reduction in cabbage seed yield to an extent of 59 per cent, while under severe cases it results in complete yield loss (Bhere et al. 2020).

19.2.1.2 Symptoms

Initially, the pathogen infects the seedling and forms small, circular, dark brown necrotic spots on the stem portion leading to the damping off, while on adult plants the symptoms are noticed on the

DOI: 10.1201/9781032631578-19

leaf, leaf stalk, flower stalk and seed head. On the leaf surface, initially, small circular dark spots are seen. Later, the infected leaf area becomes papery and the centre will fall off leading to a shot hole symptom. In the case of leaf blight, the individual spots merge with one another resulting in large, blighted areas with a target pattern of concentric rings surrounded by a yellow halo, whereas the corresponding lower surface consists of a black spore mass. Whenever the warm and wet weather prevails the whole leaf becomes blighted, dried and defoliates prematurely (Nowicki et al. 2012).

19.2.1.3 Causal Organism

Alternaria brassicae and *A. brassicicola* form a thick, dark brown to black colony on cultural media. Microscopically, the mycelium is initially septate, highly branched and unpigmented and later develops olive-grey to grey-black pigmentation at maturity. The mycelium further differentiates and forms olive-brown septate conidiophore (100–200 μm) which bears chains of 8–10 dark brown, smooth-walled and muriform conidia of size 60 x 14 μm. Pramila et al. (2014) studied the molecular diversity of 10 isolates of *A. brassicae* using random amplification of polymorphic DNA (RAPD) and constructed the dendrogram and found that isolates were categorized into two clusters with a 15 per cent similarity coefficient. Group I was composed of seven isolates namely VR, DV, P7, LM. P10, KR and ND with 18 per cent similarity (82 % dissimilarity), while group II was composed of only three isolates namely JD, KA and AS with only 24 per cent similarity (76 % dissimilarity).

19.2.1.4 Disease Cycle

Both pathogens are saprophytes that survive in the infected crop debris and weed host in the form of dormant mycelia. They even survive in infected seeds as externally and internally seed-borne. When the temperature of 25–30°C and 70–100 per cent relative humidity prevails, the pathogen undergoes asexual reproduction and forms enormous amounts of muriform conidia within a week which is easily dispersed by wind and rain splash to neighbouring plants and starts a new disease cycle (Mamgain et al. 2013).

19.2.1.5 Disease Management

Cultural Practices: Removal and destruction of crop debris containing the harvest to restrict the sources of primary inoculum. Deep summer ploughing should be done during May and June to destroy the resting structures of the pathogen. Go for crop rotation with non-host crops for at least three years. Early sowing of certified crucifer seeds was done before 20th October to protect the crop from *Alternaria* blight under North Indian conditions. Hot water treatment at 50 °C for 30 min, helps to minimize the disease incidence. Intercropping of broccoli with fennel (1:2) was followed to reduce disease severity. Five rows of oats were used as a barrier crop all around the cruciferous field to decrease the disease intensity by up to 66.2 per cent. Avoid overhead irrigation during head development and application of excessive nitrogen fertilizers which facilitates the pathogen growth. Foliar spray with nutrients such as boron, potassium and zinc was helpful for the management of the blight disease in oilseed rape.

Host Resistance: Three genotypes of mustard such as PR-8988, PR-9024 and Kranti have partial resistance against the *Alternaria* leaf blight and showed the lowest blight severity. PHR-2, PAB 9511, PAB 9534, EC 399301, EC399299, EC399313 and JMM 915 genotypes of Indian mustard were tolerant to disease. Cabbage varieties such as CKD-1104, CKD-1105, Snow drift, IIHR series (1, 36, 49, 51, 54, 55, 60, 61, 64, 67, 82, 89, 91, 142, 158, 166, 192, 217, 218 and 222), while four cauliflower lines, *i.e.* IIHR 73-3-20, IIHR 250-4-4-16-27, IIHR 264-3 and IIHR 392 were utilized for cultivation as they showed resistance response to the disease.

Biological Control: Treat the seedling with *Trichoderma harzianum* as well as soil application of *T. harzianum* enriched vermicompost had recorded maximum disease control (62 %) and enhanced yield in cauliflower. *Bacillus subtilis* was also found efficient in managing *Alternaria* leaf spot of cabbage with a minimum disease index (10.8 %) over the control (79.4 %). Khalse et al. (2017) found that the minimum disease was noticed in eucalyptus leaf extract at 10 % (25.60 %) followed by datura leaf extract at 10 % (28.13 %) as compared to the untreated control (34.05 %).

Chemical Control: The seed production plots should be sprayed at full bloom, pod set and pre-harvest stage with captan (0.2 %) or copper oxychloride (0.5 %) for the disease management. Foliar spray with chlorothalonil (0.2 %) and macozeb (0.2 %) were considered as best fungicides to manage *Alternaria* blight in crucifers (Bharat and Gupta 2011).

19.2.2 Club Root of Cabbage

19.2.2.1 Disease Occurrence, Distribution and Importance

Club root is an economically important disease in crucifers, known to occur in America, Australia, Germany, Canada and India. This is the oldest disease dating back to the 4th century AD when Pallidus reported the development of spongy-type roots on the surface of radish, turnip and rape. This is a bound disease of cabbage because of its occurrence in almost all the cultivating areas of crucifers and results in 10–15 per cent yield loss across the world.

19.2.2.2 Symptoms

The initial above-ground symptoms are reduced development of the plant, flagging of leaves and wilting of the entire plant on hot sunny days, as if the plant is suffering from water deficiency though ample moisture is available in the soil. As the pathogen enters through root hairs, it mainly targets these roots resulting in hypertrophy because of the secretion of excess auxin and cytokinin, which results in the production of "club"-shaped galls on the root system. Clubs may be present only on the entire main root or the lateral root or both. They may appear as tumours as in radishes and even dark decomposed spots in the root system (Gahatraj et al. 2019).

19.2.2.3 Causal Organism

Plasmodiophora brassicae (Woronin) is a soil-borne obligate parasite that produces subspherical to spherical resting spores of 3 μm size and the surface is covered by spines. These resting spores release pyriform primary zoospores of 2.8–5.9 μm long and biflagellate. Inside the host root hairs, the pathogen forms primary plasmodia, and undergoes several nuclear divisions synchronously followed by cleaving resulting in the formation of zoosporangia. The clusters in root hairs are formed by the zoosporangia and sometimes in epidermal cells. Each zoosporangium releases 4–16 secondary zoospores, which further enlarge and form secondary plasmodium.

Differential cultivars and RAPD primers were employed to assess the extent of genetic diversity among nine single-gall populations of *P. brassicae* and 37 single-spore isolates (SSI) and categorised all the pathogenic isolates into eight pathotypes and each isolate was associated with unique molecular genotype.

19.2.2.4 Disease Cycle

The pathogen survives as a resting spore in soil (17–18 years) or crop debris. The second stage *i.e.* root infection starts with the release of primary zoospores from the resting spore which swims to the root surface and penetrates through the cell wall. Further, the pathogen forms primary plasmodia inside the root hair, which turns into zoosporangia consisting of 4–16 secondary zoospores which are released into the soil ready for the penetration of the cortical tissue of the main root system, a process that is called cortical infection. Later, the pathogen forms secondary plasmodia and connects with cellular hypertrophy leading to gall formation of tissues. Finally, these plasmodia develop into a new generation of resting spores that release into the soil as overwintering structures.

The temperature of 25 °C coupled with 5.2–6.6 soil pH and 60–70 per cent soil moisture always favours the pathogen.

19.2.2.5 Disease Management

Cultural Practices: Removal and destruction of infected roots after harvesting of economic plant parts. This practice gradually reduces spore load in the soil. Prevention can be done by applying agricultural lime in fields with an acidic soil reaction, to increase the soil pH to 7 which will in turn inhibit the germination of resting spores. Some plants, such as lettuce, spinach and Italian ryegrass, may be rotated to minimize the root hair infection and clubroot disease severity. Radish and tori can be used as trap crops to reduce the club root; even intercropping with flax helps in the reduction of club root because the hydrogen cyanide produced by the roots of flax has an inhibitory action on the club root pathogen. Over-irrigation and water-logging situations should be strictly avoided. Application of borax has reduced club root as it inhibits both the primary and secondary stages of infection.

Host Resistance: *Brassica rapa* offers a great source of qualitative and quantitative resistance to club root, where at least 17 loci were identified, of which Crr1, Crr2, Crr3, Crr4, CRa, CRb, CRc and CRk are involved majorly in providing resistance (Peng et al. 2018).

Biological Control: The antagonistic microorganism such as *Trichoderma* reduced club root incidence by 45 per cent along with the production of new roots allowing continuous development of crops and resulting in higher crop yield (Cuevas et al. 2011) and *P. fluorescens* is a ubiquitous bacteria in agricultural soils that is well suited to managing soil-borne pathogens and acts as a growth-promoting agent. *Streptomyces graminifaciens* G11 and *P. fluorescens* H237 had reduced club formation and disease severity values by 15–20 per cent compared to the control under glasshouse conditions.

Chemical Control: Protozoicides such as fluazinam, cyazofamid, flusulfamide, procymidone and prohexadione-calcium are efficient in minimizing the club root incidence. Even seed treatment with carbendazim at 10 g kg^{-1} of seeds or soil drenching with carbendazim at 2.5 kg ha^{-1} or seedling dip in carbendazim solution at 1-2 gl^{-1} for two min had delimited club root pathogen growth. Soil fumigation with metham sodium and dazomet helps deplete the pathogen load in the soil (Donald and Porter 2014).

19.2.3 *Fusarium* Wilt

Fusarium wilt is distributed worldwide, causing a major threat to cabbage production as well as severe economic losses in crucifers. The first report of the disease was from the USA and thereafter spread across the globe by infecting cabbage, cauliflower, broccoli, brussels sprouts, Chinese cabbage, collard, kale, kohlrabi, mustards, radishes, rape, rutabaga, seakale, turnips and watercress (Khafagi et al. 2020).

19.2.3.1 Symptoms

As the pathogen is a soil inhabitant, soon after seedling emergence it enters through root hairs, assimilates in vascular vessels, leads to blockage of the vessels and prevents the transport of water and food. When such infected stem portions are cut open, a characteristic brown discolouration of the xylem vessel is noticed in the stem. The above-ground symptoms include yellowing of leaves, which turn brown, curled and drop prematurely. Even the adult plant shows stunting, yellowing and twisting on one side, abnormal head development and dies rapidly in warm soil at 27 °C (Liu et al. 2019).

19.2.3.2 Causal Organism

Fusarium oxysporum f. sp. *conglutinans* (Wollenw) produces white fluffy mycelia, the hyphae were filamentous, hyaline and septate, which further differentiates to produce conidiophore-bearing microconidia and macroconidia. Microconidia are microscopic, small, hyaline, single-celled, oval to elliptical or slightly curved. Whereas macroconidia are large, slightly curved, septate, hyaline spores. Even, the pathogen forms round thick-walled chlamydospores that survive for longer periods in the soil being resistant to unfavourable environmental conditions (Ling et al. 2022). Liu et al. (2019) evaluated the genome level diversity of race 1 and 2 strains of the pathogen by next-generation sequencing and genome assembly. They discovered that the race 2 strain has abundant single-nucleotide polymorphisms and 645 insertion–deletions (InDels) compared with the race 1 strain, which showed low genetic variability.

19.2.3.3 Disease Cycle

The fungus survives as dormant mycelia, macroconidia and chlamydospore in soil, debris and roots of weeds for many years. The pathogen invades the plant through the young rootlets or wounds in the older roots at transplanting time or later. Then it enters into the xylem and then progresses up to leaves through the stem. After sufficient assimilation, the mycelium differentiates to form conidiophore-bearing microconidia and macroconidia and helps in dissemination through infected transplants, infested soil clinging to transplants, farm equipment, plant refuse, animals, surface-drainage water and air currents. The fungus is highly favoured by soil temperatures of 32 to 35 °C (Okungbowa and Shittu 2012).

19.2.3.4 Disease Management

Cultural Practices: Sowing of healthy and certified seeds in well-drained disinfested fields. Crop rotation with lettuce, capsicum and tomato for several years can prevent inoculum build-up in soil. Prevent the spread of infested soil carried on equipment, tools, feet and irrigation water to clean fields. In the nursery, the seedling trays can be decontaminated by dipping them in household bleach for 20 min. The seedlings can be grown in sterilized soil mixes and the nursery areas should be well drained.

Host Resistance: Efficient management occurs only through the utility of resistant cultivars of cabbage such as Bajrang, BSS-44, H-113, Sri Ganesh Gol and Sudha.

Biological Control: Weed-Max and Oligo-X algae were found effective in reducing wilt and enhancing the yield of cabbage because they reduced the activities of polygalacturonase (PG) and cellulase (CX) enzymes as well as increased total phenols and total chlorophyll in cabbage (Khafagi et al. 2020).

Chemical Control: Root dipping of cabbage in Thiophanate-methyl (topsin M 70 %) at 1 gl^{-1} suspension for 30 min and soil drenching with the same chemical had reduced wilt severity (14.65 %) compared to control (43.30 %) and enhanced cabbage head diameter and yield.

19.2.4 Damping Off and Wire Stem

The worldwide distribution of disease is noticed, primarily seen in areas having moderate temperature and high soil moisture. All crucifers are susceptible to wire stem due to the polyphagous nature of the fungus.

19.2.4.1 Symptoms

The pathogen also initiates damping off in the nursery, and after transplanting it shows wire stem disease in the mature crop. The infection initiates on the stem as a dark water-soaked spot extending

1–2 cm below the soil line. Later, stems become wiry as the lesion girdles the stem and is slender at the point of the lesion, hence it is called a "wire stem". If such infected seedlings are transplanted to the field they grow poorly, are wilted, stunted and may eventually die. These infected plants usually mature late and fail to produce a marketable head. Once head formation starts, the lower side of head leaves which are near the soil, develop sunken, black and sharply elliptical lesions parallel to the side of the midrib. Lesions may dry out and become papery brown in appearance, which is covered with a sparse, web-like mycelium. Eventually, a general black decay develops near the leaf base, leading to bottom rot (Ahuja and Sharma 2014).

19.2.4.2 Causal Organism

The pathogen *Rhizoctonia solani* (Kuhn) (TS: *Thanetophorus cucumeris*) produces brown septate mycelium of 12 µm in diameter with right-angle branching near the distal septum in young hyphae. Under unfavourable conditions the mycelia aggregates and forms irregular, brown to black sclerotia, and also the pathogen produces both terminal and intercalary, barrel-shaped chlamydospores. Whenever the favourable environment prevails, the pathogen undergoes sexual reproduction resulting in the formation of a thin, hymenial layer having basidia on the plant surface near to soil level. These basidia are barrel-shaped, clavate and have four sterigmata on which four hyaline and ellipsoidal basidiospores (6-14 × 4-8 µm) are formed (Chauhan et al. 2000).

Erper et al. (2021) evaluated 132 *R. solani* isolates recovered from red cabbage plants with root rot and wire stem symptoms and carried PCR amplification of ITS primers and found that 124 isolates of *R.* solani were multinucleate while eight were binucleate. The most prevalent anastomosis group (AG) was AG 4 (84 %), which was subdivided into AG 4 HG-I (81 %) and AG 4 HG-III (3 %), followed by AG 5 (10 %) and AG-A (6 %), respectively.

19.2.4.3 Disease Cycle

The pathogen survives saprophytically in the soil as dormant mycelium or sclerotia, which acts as a primary source of inoculum infecting the seedlings and older plants. Dissemination of sclerotia across the field is achieved through rainfall and irrigation water. Wind also helps in further disease by dispersing the basidiospores to new fields. The intensity of the disease is high in sandy soils compared to clay soils and favoured by 25 °C (Chauhan et al. 2000).

19.2.4.4 Disease Management

Cultural Practices: Careful monitoring of disease, removal and destruction of infected plants. Good hygiene measures are required in propagation nurseries as *Rhizoctonia* is often coupled with reused trays which are unclean and disinfected. Composts should be kept covered prior to use so that they do not get contaminated. In the field, crop rotation with non-brassica crops should be beneficial as they reduce the pathogen load. Application of powders of commercial cellulose, rice stubbles or biomass of water hyacinth with ammonium nitrate was effective in reducing damping off.

Biological Control: A number of biofungicides are available for disease management, such as "Serenade ASO" (*Bacillus subtilis*, strain QST 713) and "Prestop" (*Gliocladium catenulatum* strain M17223) can be applied as a root drench during drilling, planting or transplanting. The seedling dip with *T. harzianum* has significantly reduced the *Rhizoctonia* infection.

Chemical Control: Preventive measures include treating seeds with thiram at 2–3 g kg^1 of seeds. Drenching the soil around the affected seedling with mancozeb 45 (0.2 %) or carbendazim (0.1 %) may possibly reduce the secondary spread of the disease.

19.2.5 Sclerotinia Rot/White Mould

The fungus causes serious losses in the field, storage, transit and also in the market if humid conditions persist. The pathogen is aggressive with a polyphagous nature and infects over 360 species of plants and is widely distributed in relatively cool, moist areas. In the Cruciferae family alone, its infection is recorded in 18 genera and 32 species (Sharma et al. 2015).

19.2.5.1 Symptoms

Initially, the infections occur on the collar region of the stem just above ground level as small, circular areas with tan and water-soaked patches, later covered with fungal growth which is a white cottony structure. The affected tissue disintegrates to become soft and watery as the disease progresses. Eventually, fungus growth covers the cabbage head and produces large, black, seeds resembling structures called sclerotia on the infected tissue (Sharma et al. 2015).

19.2.5.2 Causal Organism

Sclerotinia sclerotiorum [(Lib.) de Bary] forms white, aerial mycelia, whereas the reverse side is a salmon buff colour. Microscopically, the hypha was hyaline, branched, and multinucleate. The mycelia aggregates and forms cylindrical to globose sclerotia having a black outer layer, whereas the inner cortex has white. Apothecia are individual to many which are borne from a sclerotium with tan and amber coloured. The matured apothecia measures 4–6 mm in diameter with a length of 3–5 mm. It consists of cylindrical asci having eight hyaline single-celled and ellipsoid uniseriate ascospores. The asci are scattered with sterile structures called paraphyses, which help in the discharge of ascospores (Prova et al. 2018). The genetic diversity among the 16 isolates of *S. sclerotiorum* was analysed by using RAPD markers (OPA-14, OPA-16, OPA-17 and OPA-20). Categorization based on RAPD fingerprint data revealed the formation of two groups with 10 independent branches (0.70 similarity) (Kapatia et al. 2016).

19.2.5.3 Disease Cycle

The pathogen overwinters as sclerotia in soil and infected seeds for the period of seven years and upon favourable conditions develops into mycelia or a tan, mushroom-like-structured sexual fruiting body called an apothecia containing ascus with ascospores which are released forcibly and dispersed by the wind. The ascospores, after landing on a nearby plant surface, demand food and a thin film of moisture for germination and infection. Although the fungus can grow at a wide range of temperatures, the moist conditions support the disease development in the field (Sharma et al. 2015).

19.2.5.4 Disease Management

Cultural Practices: It is always advisable to plant the crop with the wind direction to facilitate free air movement within the plants which will reduce sclerotia germination. Crop rotation with non-susceptible crops such as cereals has reduced the disease incidence by more than 60 per cent. Intercropping with onion or garlic (2:1) significantly reduced the disease. Soil mulching by sunflower inflorescence residues or pine needles and even soil amendment such as sunflower cake, mustard cake, gypsum and bean straw minimized the disease intensity through the enhancement of antagonistic microorganisms in the soil. The application of poultry manure (37.43 %) and goat manure (36.36 %) was found highly promising in managing the disease incidence (Jajoriya et al. 2022). Soil solarization for nine weeks during the summer is often more effective in preventing white mould fungal growth on cabbage. At most, care should be taken to avoid mechanical damage to heads at the time of harvesting which facilitates adequate room for fungal infection.

Biological Control: Garlic and neem extract applied to cabbage seedlings reduced *Sclerotinia* rot by up to 48.44 and 42.94 per cent, respectively. *Trichoderma* species efficiently managed the disease by inhibiting the pathogen growth and destroying the sclerotia.

Chemical Control: Application of carbendazim (500 ppm), thiophenate methyl (100 ppm) and mancozeb (100 ppm) were efficient in reducing *Sclerotinia* rot. Combined application of *B. amyloliquefaciens* isolates B15 and tebuconazole + trifloxystrobin was highly effective with the minimum disease incidence (Krishnamoorthy et al. 2017).

19.2.6 Black Rot

Black rot is the most destructive and yield-limiting disease of *Brassicaceae* members, especially cabbage and cauliflower and was reported for the first time from Kentucky, USA, later disseminated to all over the world, including Australia, Europe, Asia and Africa.

19.2.6.1 Symptoms

The pathogen infects crops from the seedling stage; whenever the bacteria-contaminated seeds are sown, the pathogen enters through the epicotyl of a young developing seedling, reaches the cotyledonary leaves, and shows as shrivelling, blackening and dropping. On the foliage, conspicuous symptoms of V or U-shaped chlorotic and necrotic lesions appear from the margin of the leaf as the pathogen enters through hydathodes (see Figure 19.1). Later, the midrib of the leaf turns black, veins become darkened and the entire leaf shows rotting and dropping. Further, bacteria enter into the vascular system and become systemic; as a result, the plants show stunting and even the heads will deteriorate rapidly after harvest (Liu et al. 2016).

19.2.6.2 Causal Organism

Xanthomonas campestris pv. *campestris* (Pammel) Dowson is a Gram-ve, rod-shaped, monotrichous and aerobic bacteria. On nutrient agar media it a produces circular, smooth, glossy, yellow mucoid colony with a raised centre (Vincente and Holub 2012). Race profiling of the isolates of *Xcc* revealed that races 1, 4 and 6 are prevailing, while race 1 followed by race 4 are dominant in India (Singh et al. 2016).

19.2.6.3 Disease Cycle

The pathogen survives in debris inside the soil for up to 2 years, and infected seeds also serve as initial inoculums for the infection. Whenever such seeds are sown, the bacterium affects the emerging seedling; later, the pathogen spreads systemically inside the host system eventually leading to the

FIGURE 19.1 V-shaped necrotic lesions on cabbage.

collapse of the seedling. Foliar infection is caused by the invasion of bacteria through hydathodes or leaf wounds. In temperatures of 25–30 °C with sufficient rainfall or heavy dew prevails, bacteria can be seen in guttation droplets and are easily spread by rain, water splashes and mechanical equipment to the surrounding healthy plants.

19.2.6.4 Disease Management

Cultural Practices: Clean cultivation and sanitation of the field. Avoid cultivating the crops in areas where crucifers have been continuously grown for the past two years. Hot water treatment at 50 °C for 30 min minimizes the bacterial load on the seeds. Crop rotation with non-cruciferous crops at least for three years both in the nursery and the main field is practiced in order to starve the bacterium. Covering the soil surface with grass mulch will reduce the extent of splashing of the infested soil and avoid the secondary dispersal of the bacterium. Maintenance of high soil moisture after harvesting helps to reduce the inoculum load of the bacterium.

Host Resistance: Resistant cultivars are the cheapest and best measure of managing black rot. For example, cabbage (IIHR 54, 55, 103, 104, 106, 380, 386, 416, 431, 466 and 472) and cauliflower (Avans, Iglory, Pusa Kea, Kt-2, Punjab katki, IIHR 73–56 and IIHR 250-4-1-11-28) are the resistant sources.

Biological Control: The combined foliar application of *P. fluorescens* + *Bacillus* was found fruitful in minimizing disease incidence and severity in cauliflower. *P. fluorescens* strain TO7 producing 2,4-diacetylphloroglucinol (2,4-DAPG) was treated by the root dip method in cabbage had reduced the number of lesions, vein discolouration, and necrosis and chlorotic lesions in the infected plants. Fontana et al. (2021) revealed that 0.5 and 0.1 mg ml^{-1} of methanolic, hydroalcoholic and hydroalcoholic maltodextrin of *Madia glomerata* extracts has bacteriostatic and bactericidal activities by changing the membrane permeability and biofilm formation of *Xcc*. Another bacteria, "*Acinetobacter lactucae* QL-1", effectively suppressed *Xcc* virulence through quorum quenching in radish, and co-inoculation of *Xcc* and QL-1 considerably decreased the incidence and severity of black rot (Ye et al. 2019).

Chemical control: Seed treatment with streptocycline (100 ppm) followed by a seedling dip in the same chemical for 15 min and foliar spray with streptocycline (200 ppm) at an interval of 10 days was effective in reducing the disease incidence. Pre-treatment of *Xcc*-contaminated cabbage seeds with CAC-717 significantly reduced *Xcc* cell numbers recovered from the seeds compared with distilled water (Sanna et al. 2022). Foliar application of abiotic elicitors acibenzolar-S-methyl at 350 µg/ml was found very effective in managing the disease (Neelam and Singh 2022).

19.2.7 Curd Rot/Soft Rot

Soft rot is considered a destructive disease of vegetables and is distributed in temperate and tropical zones. The bacterium infects a variety of plants including carrot, crucifers, potato, cucumber, onion, tomato, lettuce, iris, *etc.* The disease can be noticed commonly in the field, in transit, in storage and during marketing, resulting in great economic losses (Umunna and Austin 2016).

19.2.7.1 Symptoms

Initially, small, water-soaked spots appear on the centre of the curd, which later rapidly enlarge and become sunken and soft. Interior tissues beneath the spots become mushy and discoloured, with the discolouration ranging anywhere from cream to black. Within a few days, the affected plant part may collapse. Seepage of internal content from affected areas and the emission of a strong, disagreeable offensive odour is common. Rot exceeding 50 per cent of the curd surface may lead to the death of the plant.

19.2.7.2 Causal Organism

Erwinia carotovora subsp. *caratovora* (Winslow) is a G-ve, rod-shaped bacterium that lives alone or aggregates into pairs and chains. Non-spore forming and peritrichously flagellated, facultatively anaerobic, mesophilic bacterium requires a temperature range from 27–30 °C. The bacterial colonies are greyish-white to creamy-white, smooth, round, glistening and slightly raised and visible after 24 hours of isolation. The genetic variability among 87 strains of the soft rot pathogen *E. carotovora* subsp. *carotovora* differentiated strains into A and B groups. Group A contains typical Ecc strains which provided negative reactions in testing the production of reducing substances from sucrose and acids from α-methyl glucoside, whereas group B contains atypical Ecc strains and the properties of this group were similar to those of *E. carotovora* subsp. *atroseptica*.

19.2.7.3 Disease Cycle

The pathogen survives on dead and decaying matter in the soil. The bacteria disseminates in water and infects the plants through wounds or natural openings. They feed and multiply at first on the liquid released by the broken cells on the wound surface, and then they produce increased amounts of pectolytic enzymes which could break the pectic substance of the middle lamella resulting in maceration of the tissue. Bacteria continue to move and multiply in the intercellular space. As a result, cells are plasmolyzed or collapsed and die in the intercellular spaces, while their enzymes advance ahead of them and prepare the tissue for further invasion. The pathogen is generally spread by irrigation water and rain splash. Disease development is favoured by warm temperatures of 25–30 °C, humid conditions or following periods of wet weather accompanied by free moisture on the plant surface (Umunna and Austin 2016).

19.2.7.4 Disease Management

Cultural Practices: Careful removal and destruction of infected plant parts. Proper drainage facilities should be allocated to avoid bacterial development. Alternation of crucifer crops with other crops such as corn, cereals and grasses. Avoid injuries to the plants when cultivating, handling during and after harvest, packing and storing. Avoid frequent irrigation during curd development. The storage room should be kept clean and thoroughly sprayed from ceiling to floor with sodium hypochlorite (1000 to 1900 ppm) disinfectant and disinfect tools by dipping in 70 per cent ethanol. In storage, use buffering materials such as straw or paper to prevent injuries to the head.

Host Resistance: The hybrids of cauliflower as such Carillon × Sel-12, Pyramis × K-1 and Pyramis × Carillon showed resistance to soft rot pathogen. Shruthy (2018) found that four varieties of cauliflower such as NS 60N, Himpriya 60, G 45 and Himshort also exhibited resistance to the soft rot.

Biological Control: Salem and Abd El-Shafea (2018) tested the efficacy of *B. subtilis, P. fluorescens, P. aeruginosa* and *Streptomyces* spp. under *in vitro* conditions against *E. carotovora* pv. *carotovora*. Among all these four bioagents, *Streptomyces* spp. showed the highest inhibition zone of 40 mm followed by *P. fluorescens* and *B. subtilis*.

Chemical Control: An effective control of soft rot disease can be achieved through the prevention of bacteria in the host. This is done by the application of copper-based fungicides as such Greencop WP, Trinity Gold 452 WP, Colonizer 440 WP and Biodistinction Xtra 700 CS, while Pyramid 700 WP suppresses the bacterial activity within the host system and the optimizer boosts the immunity of the crop and manages the disease. The application of bactericide streptocycline (100 ppm) in combination with copper oxychloride could manage the soft rot.

19.2.8 Cabbage Cyst Nematode

19.2.8.1 Disease Occurrence, Distribution and Importance

The cabbage cyst nematode was first noticed on the roots of cabbage in the eastern part of Turkey in 1963. Further, it is seen in Europe, the USA, Australia, Iran, Azerbaijan and Russia. The nematode infects only crucifers such as brussels sprouts, cabbage, cauliflower, broccoli, radish, kohlrabi and rapeseed.

19.2.8.2 Symptoms

Above-ground symptoms include chlorosis, curling and red spots seen on the leaves. Later, the whole plant shows stunting, while below-ground symptoms include invaded roots branching profusely, while the tap roots remain small. Some plants may die prematurely and if the plant survives it produces loose small heads and discoloured root systems. The best diagnostic symptom, however, is the appearance of the pearly white, tan or reddish bodies of females dotting the root surface.

19.2.8.3 Causal Organism

The most infective stage of *Heterodera cruciferae* (Franklin) is the J_2 stage, which is cylindrical in shape, with a slightly offset anterior part and a tapering round tail tip. The stylet is strong with basal knobs shallowly concave anteriorly. The female nematodes are initially white and later turn brown and are swollen and lemon-shaped with a projecting neck containing an oesophagus and vulvar cone structure. The cyst wall is dark brown in colour and has a zig-zag pattern. They generally enclose most of their eggs to protect against adverse environmental conditions and predation. Males are vermiform, with a rounded tail less than 1/5th of the body width in length. The body adopts a curved position after heat relaxation, with the posterior part often twisted 90 or 180° (Mokrini et al. 2017). The genetic diversity of *H. cruciferae* has been analysed by Toktay et al. (2022) using the sequences of ribosomal DNA region (rDNA-ITS) and cytochrome oxidase subunit 1 (mtDNA-COI). They noticed the differences between 13 geographic populations of *H. cruciferae* and revealed genetic variability within nematode populations.

19.2.8.4 Disease Cycle

H. cruciferae is an amphimictic, sedentary endoparasite having various stages including the egg, four juvenile stages and an adult stage. The J_2 stage invades just behind the growing apex of the root tip and next passes through cells towards the stele where they initiate the development of a cluster of multinucleate-feeding specialized feeding cells called a syncytium. This J_2 then go through a moult to the third stage (J_3). The syncytium provides food for the development and maturation of the juveniles to the adult stage. The strong sexual dimorphism was seen after the fourth stage (J_4). The adult males become vermiform and leave the root, whereas the females swell into a white lemon-shaped body that protrudes out of the surface of the root. After mating, the eggs produced by females are kept within their bodies. When all eggs have formed, the female dies and becomes a cyst, detached from the root, and survives in the soil. The eggs contain a single first-stage juvenile (J_1), which moults inside the egg to become a second-stage juvenile (J_2), and these are released outside by hatching the egg. The nematode survives three or four months under low soil temperature (5–15 °C) and high soil moisture (Mokrini et al. 2017).

19.2.8.5 Disease Management

Cultural Practices: In severe infestations, long crop rotations of more than 6 years are useful as the pathogens persist in soil and have a limited host range. Soil fumigation, soil solarization, early planting and destruction of weed hosts can be useful.

Host Resistance: Resistant cultivars will not allow the nematodes to produce functional feeding sites in the host after invasion and to develop subsequently as reproducing females. Mennan and Aydinli (2012) evaluated white head cabbage cultivars and found that cultivars such as 524 C, 166 T, 519 C, 542 C, 508 C, 115 T and 530 T showed high resistance without any females in the roots and soils.

Biological Control: *P. fluorescens* strain 6 was found effective in suppressing the growth and egg-hatching of *H. cruciferae.* Further, it also reduced the performance of the J_1 and J_2 juveniles.

Chemical Control: The nematicide such as 1,3-dichloropropene, 1,2-dichloropropane, 1,2-dibromo-3-chloropropane and ethyl 4-(methylthio)-m-tolyl isopropyl phosphoramidate are applied for managing *H. cruciferae* in brussels sprouts and cabbage, they even enhanced the marketable yield and significantly reduced the number of cysts extracted from the soil after the crop had been harvested (Avato et al. 2013).

19.3 CONCLUSIONS

Cruciferous vegetables are extremely perishable goods, particularly in the post-harvest, storage, transit, retail and consumer phases. Numerous bacterial, nematode and fungal pathogens cause infection on these crops. Therefore, it is essential to implement effective disease management strategies, which primarily focus on preventing the illness from entering a field and keeping disease pressure low in case a disease is present. An integrated approach consisting of cultural methods, host-plant resistance, and biological and chemical strategies is employed in managing the black rot, club root, curd rot, black spot, wilt, watery soft rot, wire stem and cyst nematode. This chapter also focused on disease importance, distribution, symptoms, aetiology, epidemiology, and molecular variability of the pathogens, providing basic information regarding the subject.

REFERENCES

Ahuja DB, Sharma P (2014) Diseases of cole crops. In: Singh D, Chowdappa P, Sharma P (eds), *Diseases of vegetable crops: diagnosis and management*, pp. 263–299. Indian Phytopathological Society, New Delhi.

Avato P, D'Addabbo T, Leonetti P, Argentieri MP (2013) Nematicidal potential of Brassicaceae. *Phytochem Rev* 12: 791–802.

Bharat NK, Gupta BK (2011) Occurrence and management of pod blight of cabbage seed crop under arid temperate conditions. *Plant Dis Res* 26: 92–93.

Bhere DT, Solanke KM, Subhadarshini A, Tiwari S, Narode MK (2020) *In-vivo* management of *Alternaria* leaf spot of cabbage (*Alternaria brassicicola*). *Int J Curr Microbiol App Sci* 9(8): 2553–2559.

Chauhan RS, Maheshwari SK, Gandhi SK (2000) Effect of soil type and plant age on stem rot disease (*Rhizoctonia solani*) of cauliflower (*Brassica oleracea* var. *botrytis*). *Agr Sci Digest* 20: 58–59.

Cuevas VC, Lagman CA, Cuevas AC (2011) Potential impacts of the use of *Trichoderma* spp. on farmers' profit in the field control of club root disease of crucifers caused by *Plasmodiophora brassicae* Wor. *Philippine Agr Sci* 94(2): 171–178.

Donald EC, Porter IJ (2014) Clubroot in Australia: the history and impact of *Plasmodiophora brassicae* in Brassica crops and research efforts directed towards its control. *Canadian J Plant Pathol* 36: 66–84.

Erper I, Ozer G, Kalendar R, Avci S, Yildirim E, Alkan M, Turkkan M (2021) Genetic diversity and pathogenicity of *Rhizoctonia* spp. isolates associated with red cabbage in Samsun (Turkey). *J Fungi* 7: 1–20.

Fontana R, Caproni A, Buzzi R, Sicurella M, Buratto M, Salvatori F, Pappadà M, Manfredini S, Baldisserotto A, Marconi P (2021) Effects of *Moringa oleifera* leaf extracts on *Xanthomonas campestris* pv. *campestris*. *Microorganisms* 9: 1–17.

Gahatraj S, Shrestha SM, Devkota TR, Rai HH (2019) A review on club root of crucifers: Symptoms, life-cycle of the pathogen, factors affecting severity and management stratergies. *Arch Agr Environ Sci* 4: 342–349.

Jajoriya A, Ahir AA, Meena R, Sharma D (2022) Management studies on *Sclerotinia sclerotiorum* (Lib.) de Bary, causing *Sclerotinia* rot of cauliflower through plant extracts, and amendments. *The Pharma Innov J* 11: 600–603.

Kapatia A, Sharma GM, Khan A, Kulshrestha S (2016) Isolation and analysis of genetic diversity amongst *Sclerotinina sclerotiarum* isolates infecting cauliflower and pea. *Indian J Biotechnol* 15: 589–595.

Khafagi EM, Abd El-Syed MHF, Elwan SE (2020) Controlling cabbage *Fusarium* wilt (yellows) using topsin M and some commercial bio-fertilizer products. *Plant Prot Pathol Res* 47: 519–530.

Khalse DK, Lal AA, Simon S (2017) Efficacy of bio-agents and plant extracts against the *Alternaria* leaf spot of cabbage (*Alternaria brassicae*). *J Pharmacognosy Phytochem* 6: 1980–1982.

Krishnamoorthy KK, Sankaralingam A, Nakkeeran S (2017) Management of head rot of cabbage caused by *Sclerotinia sclerotiorum* through combined application of fungicides and biocontrol *Bacillus amyloliquefaciens*. *Int J Chem Studies* 5: 401–404.

Ling J, Dong X, Ping X, Li Y, Yang Y, Zhao J, Lu X, Xie B, Mao Z (2022) Genetic diversity and population structure of *Fusarium oxysporum* f. sp. *conglutinans* race 1 in northern China samples. *J Fungi* 8: 1–12.

Liu K, Garrett C, Fadamiro H, Kloepper JW (2016) Antagonism of black rot in cabbage by mixtures of plant growth-promoting rhizobacteria (PGPR). *Biocontrol* 61(5): 605–613.

Liu X, Xing M, Kong C, Fang Z, Yang L, Zhang Y, Wang Y, Ling J, Yang Y, Lv H (2019) Genetic diversity, virulence, race profiling and comparative genomic analysis of the *Fusarium oxysporum* f.sp. *conglutinans* stains infecting cabbages in China. *Frontiers Microbiol* 10: 1–14.

Mamgain A, Roychowdhury R, Tah J (2013) *Alternaria* pathogenicity and its strategic controls. *Res J Biol* 1: 1–9.

Mennan S, Aydinli G (2012) Screening resistance level of Brassicaceae plants to cabbage cyst nematode, *Heterodera cruciferae* Franklin, 1945 (Tylenchida: Heteroderidae). *Turk Entomol Magazine* 36: 3–10.

Mokrini F, Viaene N, Waeyenberge L, Afechtal M, Essarioui A, Meons M (2017) The importance, biology and management of cereal cyst nematodes (*Heterodera* spp.). *Rev Mar Sci Agron Vet* 5: 414–421.

Neelam G, Singh D (2022) Efficacy evaluation of abiotic elicitors for the management of black rot disease of cauliflower incited by *Xanthomonas campestris* pv. *campestris*. *Int J Environ Climate Change* 1210: 1023–1030.

Nowicki M, Nowakowska M, Niezgoda A, Kozik EU (2012) *Alternaria* black spot of crucifers: Symptoms, importance of disease, and perspectives of resistance breeding. *Veg Crops Res Bull* 76: 5–19.

Okungbowa FI, Shittu HO (2012) *Fusarium* wilts: an overview. *Environ Res J* 6: 84–102.

Peng L, Zhou L, Li Q, Wei D, Ren X, Song H, Mei J, Si J, Qian W (2018) Identification of quantitative trait loci for club root resistance in *Brassica oleracea* with the use of *Brassica* SNP microarray. *Front Plant Sci* 9: 1–8.

Pramila, GP, Tasleem M, Taj G, Mal R, Kumar AM (2014) Morphological, cultural, pathogenic and molecular variability amongst Indian mustard isolates of *Alternaria brassicae* in Uttarakhand. *African J Biotechnol* 13: 441–448.

Prova A, Akanda AM, Islam S, Hossain MM (2018) Characterization of *Sclerotinia sclerotiorum*, an emerging fungal pathogen causing blight in hyacinth bean (*Lablab purpureus*). *Plant Pathol J* 34: 367–380.

Salem EA, Abd El-Shafea YM (2018) Biological control of potato soft rot caused by *Erwinia carotovora* subsp. *carotovora*. *Egypt J Biol Pest Control* 28: 1–5.

Sanna M, Gilardi G, Gullino ML, Mezzalama M (2022) Evaluation of physical and chemical disinfection methods of *Brassica oleracea* seeds naturally contaminated with *Xanthomonas campestris* pv. *campestris*. *J Plant Dis Prot* 5: 1145–1152.

Sharma A, Rathore JP, Ali A, Qadri I, Hussain SM, Angmo T (2018) Major diseases and pathogen ecology of cabbage. *The Pharma Innov J* 7: 574–578.

Sharma P, Meena PD, Verma PR, Saharan GS, Mehta N, Singh D, Kumar A (2015) *Sclerotinia sclerotiorum* (Lib.) de Bary causing *Sclerotinia* rot in oilseed brassicas: a review. *J Oilseed Brassica* 6: 1–44.

Shruthy ON (2018) Screening cauliflower varieties for incidence of physiological disorders, pests and diseases. *IOSR J Agr Vet Sci* 11: 22–25.

Singh D, Rathaur PS, Vicente JG (2016) Characterization, genetic diversity and distribution of *Xanthomonas campestris* pv. *campestris* races causing black rot disease in cruciferous crops of India. *Plant Pathol* 65: 1411–1418.

Toktay H, Akyol BG, Evlice E, Imren M (2022) Molecular characterization of *Heterodera cruciferae* Franklin, 1945 from cabbage fields in Nigde province, Turkey. *Mol Biol Rep* 49: 1–6.

Umunna OE, Austin AA (2016) An overview of characterization and identification of soft rot bacterium *Erwinia* in some vegetable crops. *Greener J Biol Sci* 6: 46–55.

Vicente JG, Holub EB (2012) *Xanthomonas campestris* pv. *campestris* (cause of black rot of crucifers) in the genomic era is still a worldwide threat to brassica crops. *Mol Plant Pathol* 14(1): 2–18.

Ye T, Zhou T, Fan XH, Bhatt P, Zhang LH, Chen SH (2019) *Acinetobacter lactucae* strain QL-1, a novel quorum quenching candidate against bacterial pathogen *Xanthomonas campestris* pv. *campestris*. *Front Microbiol* 10: 1–11.

20 Important Plant Parasitic Nematodes Infecting Vegetable Crops and Their Management

Bhagawati B., Pranjal Pratim Neog and Uday Kurulkar

20.1 INTRODUCTION

Vegetables, being a rich and cheap source of vitamins and minerals, occupy an important place in the food basket of Indian consumers, provide food and nutrition security, generate income and employment and also earn foreign exchange. India is the second-largest producer of vegetables in the world, occupying 12 percent of the total area and producing 10 percent of the vegetable production in the world. Vegetable crops are attacked by various pests and diseases. Among them, plant parasitic nematodes (PPNs) constitute one of the major causes of crop losses all over the world. Sasser and Freckman (1987) have indicated annual crop losses due to PPNs on a worldwide basis that exceeds $100 billion. Large numbers of PPNs have been found to be associated with vegetables in India but pathogenicity has not been proved for all. The major PPNs infecting vegetable crops in India are reviewed in this chapter.

20.2 MAJOR NEMATODE PESTS

20.2.1 Root-knot Nematode

Root-knot nematodes (RKNs), known scientifically as *Meloidogyne* spp., pose significant challenges to vegetable cultivation, especially in the tropical and subtropical regions of developing countries. Among various plant parasitic nematodes, RKNs are particularly destructive, affecting a wide range of vegetable crops including legumes, cucurbits, and solanaceous plants, as well as root and bulb crops such as onions, garlic, carrots, and radishes. They directly harm crops and also exacerbate damage by facilitating interactions with soil-borne fungi, bacteria, and viruses. The presence of RKNs in vegetable crops dates back to the 19th century when Berkeley (1855) first reported the disease on glasshouse-grown cucumbers in England, initially describing it as 'Vibrios'. Since then, RKNs have been identified in various plants across different regions. The most prevalent species of RKNs include *M. incognita*, *M. javanica*, *M. arenaria*, and *M. hapla*, which collectively constitute more than 95 per cent of RKN populations worldwide. Characteristic symptoms of RKN infestation include the formation of root galls or knotted roots, easily identifiable without magnification. Virtually all vegetable crops are susceptible to RKN infestation, making it a significant limiting factor for vegetable cultivation. Studies have demonstrated substantial yield losses attributable to RKNs, with losses of up to 91%, 46%, and 27% in okra, tomatoes, and eggplants, respectively.

DOI: 10.1201/9781032631578-20

20.2.1.1 Symptoms

RKN infection manifests distinct disease symptoms on the underground root system, commonly recognized as 'root galls' or 'knotted roots'. The size of these galls varies depending on the nematode species and host plant. For instance, cucurbits typically develop large galls, while chili plants exhibit smaller galls. Notably, infections by *M. hapla* generally result in smaller galls compared to *M. incognita* and *M. javanica*. Moreover, the extent of infection influences gall size, with heavy infestations leading to the formation of larger or multiple galls, including secondary galls due to hypertrophy or hyperplasia of infected cells (Goswami 2002). In specific crops like carrots and potatoes, RKNs induce distinct symptoms such as forking in carrots and tubercle formation on potato tubers. Above-ground symptoms are typically non-specific, including yellowing, stunting, mineral deficiencies, wilting during hot periods, chlorosis, premature leaf shedding, and overall poor plant appearance. RKNs also interact with various soil-borne fungi, bacteria, and viruses, exacerbating crop damage. These interactions are observed across a range of crops, including vegetables, fibers, pulses, and plantation crops. Notably, one common interaction involves the breakdown of disease resistance, leading to the wilting of previously healthy plants. A significant example of such interaction is the association between root-knot nematodes and *Ralstonia solanacearum* resulting in bacterial wilt in tomato, eggplant, and potato crops.

20.2.1.2 Etiology

RKNs are soil-dwelling endoparasites. Their second-stage juveniles (J2) are infectious and develop from eggs laid in a gelatinous matrix secreted by rectal glands. These eggs undergo changes, leading to the emergence of the infective J_2 stage. Hatching occurs freely in water or moist soil, but adverse conditions can delay it to ensure prolonged survival. Pre-parasitic J_2 juveniles move through the soil, equipped with glycogen and glycolipid reserves for months of survival without a host. They feed on underground plant parts, penetrating newly formed tissues like the pericycle. As they feed, they secrete hydrolytic enzymes to liquefy and digest cell contents extracorporeally. Post-infection, juveniles feed for weeks, molting rapidly before maturing into adult males or females. Adult males elongate and emerge from roots, while females remain attached, producing eggs continuously, often exceeding 1000. The life cycle's duration varies depending on host and temperature, ranging from 3 weeks to several months. Under optimal conditions, the cycle typically lasts 3–4 weeks.

20.2.1.3 Epidemiology

General distribution and spread of root-knot nematodes over the field or from one field to another are affected by farm machinery, irrigation water, the feet of animals, seedling plants, and nursery stocks or any other agency by which infested soil and /or plant material is moved from one point to another. Spread over long distances occurs through the movement of nursery stocks, seedling plants, tubers, rhizomes, and other plant materials, especially when these planting materials are transplanted in uninfested soil. The infective stage larvae can remain in the soil freely for quite a considerable period and on the availability of the host, the population builds up. The eggs and larvae may remain alive in the remains of the roots of the harvested crop or on the weed host.

20.2.2 Reniform Nematode

Reniform nematode (*Rotylenchulus reniformis*) is polyphagous, known to feed and multiply on a wide range of field, horticultural and plantation crops. Its host range extends to more than 115 plant species. In India, it is considered a pest of great significance after root-knot nematode. The adult female of reniform nematode (*Rotylenchulus reniformis*) is an obligate, sedentary semiendoparasite of a large number of vegetable crops. Some of the vegetable crops in which the nematode feeds and causes noticeable injury are tomato, okra, eggplant, cowpea and bean, etc.). The common name 'reniform' was derived from the kidney-shaped mature female. It has a worldwide distribution and is receiving importance as a national pest of crops. A significant loss in yield of cowpea by 15.80 percent at initial inoculums of 220 pre-adult of *R. reniformis* per 100 g soil was obtained as compared to carbofuran (Furadon 3G) treated soil (Roy et al. 2008).

20.2.2.1 Symptoms

The symptoms of damage to crops are non-specific on the above-ground or even in the below-ground parts necessitating a close observation to confirm their presence and damage. It feeds on cortical tissue, phloem, and pericycles. The infected plants generally show stunted growth, yellowing of leaves but root injury with epidermal cell necrosis is more characteristic of this disease. Severely infected plants exhibit drying of leaf margins and burning of the tip resulting in the drying and brittling of leaves that are shed off. The axillary buds on the main stem may develop into tiny leaves which also fall down early. The rootlets turn dark brown and necrotic that results in greatly reduces the yield. Apart from directly damaging plant roots, root-knot nematodes can form disease complexes in collaboration with other pathogens such as *Fusarium* spp., *Verticillium* spp., *Sclerotium rolfsii*, and *Rhizoctonia solani*. Additionally, there are reports of root-knot nematodes parasitizing the bacterial nodules formed by *Rhizobium* species.

20.2.2.2 Etiology

Rotylenchulus reniformis is a sedentary semiendoparasite of roots. The young female is the only infective stage. The nematode head is located near the stellar region, while the rest of the body remains outside. Within a week's time, the posterior half of the body becomes swollen. Eggs are deposited in an eggmass surrounding the body on the root surface. The mature female typically produces between 40 and 80 eggs, with reproductive capacity varying depending on the host plant. The initial molt occurs within the eggs, and hatching occurs in water without the need for root exudates. Juveniles develop to the young stage without feeding on host tissue, undergoing three consecutive molts rapidly. Upon infecting roots, young females orient themselves perpendicular to the root's longitudinal axis, with their posterior portion protruding outside the root. Once they establish a feeding site, they transform into kidney-shaped females, with the posterior portion remaining external. Egg-laying typically commences within 7–10 days after invasion, with eggs deposited into a gelatinous matrix secreted by specialized cells around the vagina. Each egg mass can contain 30–200 eggs. The entire life cycle is completed within 3 to 4 weeks, contingent on temperature and host suitability. While males are non-parasitic, reproduction primarily occurs through amphimixis in most populations.

20.2.2.3 Epidemiology

The nematode exhibits remarkable survival abilities in air-dried soil over extended periods. Retaining molted cuticles from previous stages is a unique adaptation facilitating its soil survival. Young females, males, and fourth-stage juveniles can enter a coiled anhydrobiotic state with encrusted cuticles, enabling prolonged survival in soil. The nematode's survival is inversely correlated with soil moisture loss rates, with alternating drying and wetting cycles resulting in significant population declines. Immature females freely inhabit the soil and can be transported between fields via irrigation water, nursery stocks, planting materials, and farm implements. Larvae remain viable for up to seven months in dry soils.

20.2.3 Lesion Nematodes

Lesion nematode (*Pratylenchus* spp.) is one of the major plant parasitic nematodes infecting vegetable crops. The vernacular name of 'root lesion nematode' is derived from the discolored patches (lesions) that develop on roots. It has a very wide host range including important vegetable crops. Among vegetables, tomato is an important host for this nematode.

20.2.3.1 Symptoms

The above-ground symptoms caused by the nematode are non-specific in nature. The nematode usually infects the roots, rhizomes, or tubers and multiply. Infection of this nematode causes dark red-brown lesions due to necrosis of the infected cells. Root lesion is the most characteristic symptom. The lesions initially appear as small elongated, water-soaked spots which soon turn brown to almost black. Other symptoms associated with the infection of this nematode are loss of primary roots, decay of roots, reduced size of blossoms, etc. Normal functioning of infested roots is badly impaired due to secondary infection by other soil-borne fungi and bacteria resulting in rotting and decay of roots. The affected area may be sloughed off and the overall root system reduced. The above-ground symptoms resemble those due to malnutrition or water deficiency. In the initial stages, the infestation can be noticed as tiny patches of stunted growth and yellowing of leaves. Stunting becomes more pronounced as growth advances. Leaves may turn yellowish-brown and wilt during the hot period of the day. In severe cases, plants may be killed (Walia and Kanwar 2002).

20.2.3.2 Etiology

Pratylenchus spp. are migratory endoparasites of roots, capable of infecting host roots at all stages of their life cycle. Eggs are laid individually within the root cortical parenchyma, but reproduction also occurs in the soil. Migration within the root can occur inter- or intracellularly. Initially, most of the nematode population resides in the soil, but as the crop grows, they predominantly invade the roots to multiply. Compared to sedentary endoparasites, their multiplication rates are lower. On average, a single female lays between 10–25 eggs, with reproduction occurring through sexual or parthenogenetic means. At crop maturity, lesion nematodes may exit the roots and return to the soil or persist within root remnants, primarily as adults or in the egg stage (Walia and Kanwar 2002).

20.2.3.3 Epidemiology

The biology of lesion nematodes is influenced by a variety of edaphic factors. Temperature plays an important role in governing the life cycle duration. On tomato, *Pratylenchus alleni* and *Pratylenchus scribneri* completed the life cycle at 20°C temperature. Several species of lesion nematodes are known to survive for long periods in the absence of host or unfavorable environmental conditions. Adult and 4th stage larvae are the main overwintering stages of *Pratylenchus penetrans*, although many eggs are also capable of surviving winters in roots. Root lesion nematodes are also capable of surviving the winter in soil without a host.

20.2.4 Cyst Nematode

Various species of *Heterodera* and *Globodera* are popularly known as the cyst nematodes infecting vegetable crops. The important species are *Heterodera carotae* mainly infecting carrots, *H. cruciferae* infecting cabbage, *H. goettingiana* infecting peas, *H. schachtii* infecting beet, broccoli, cabbage, cauliflower, etc. *Globodera rostochiensis* and *G. pallida* are serious nematode pests of potatoes in some southern states of India like Tamil Nadu, Karnataka, and Kerala. The nematode is popularly known as 'golden nematode' and has been recognized as one of the major crop protection problems of the world. In India, potato cyst nematode is known from 1961 when Jones detected this nematode at Nilgiri Hills in Tamil Nadu. However the losses due to potato cyst nematodes depend on the initial inoculum levels. They are reported to be responsible for inflicting average losses of 9 percent of global potato production, which amounts to about 40 million tonnes (Krishna Prasad 1995)

20.2.4.1 Symptoms

Detecting disease symptoms at low nematode infestations can be challenging, as symptoms become prominent only with population buildup in the soil. Initially, affected areas may consist of small patches of underperforming plants, with temporary wilting during the day, stunted growth, unhealthy yellowish foliage, and poorly developed root systems. In potatoes, infestation by cyst nematodes can lead to reduced tuber size and quantity, resulting in a gradual decline in crop yield over time. Root development may be significantly hindered, with reductions in size and the proliferation of excessive secondary roots. Hypertrophied, multinucleate syncytia form in the roots without inducing hyperplastic reactions. Initially, slight root swelling may occur, but the most evident sign of infection is the appearance of white, pearl-like females on the roots, which later turn brown and develop into cysts.

20.2.4.2 Etiology

Cyst nematodes (*Heterodera* spp. and *Globodera* spp.) are sedentary endoparasites of roots. The second-stage larvae are the infective stage which enters the roots near the tip and cuts through the cell walls leaving a row of ruptured cells. They rest with their heads towards the stele and begin feeding either on cortex, pericycle, or endodermal cells. The cell is enlarged with breakdown resulting in the formation of a syncytial transfer cell occurs due to the partial dissolution of cell walls of several adjacent cells with dense granular cytoplasm. The nematode continues to feed on the transfer cell until its development is complete. It undergoes a series of molts, through third and fourth-stage larvae to adults. In the fourth stage, females are saccate and enclosed in the third stage cuticle. The female dies and the cuticle turn to form a tough, leathery cyst. During harvest, the cysts are detached from the roots and become free in the soil where they tide over unfavorable conditions.

20.2.4.3 Epidemiology

The cysts can remain viable in the soil without the host even up to 30 years. They can be spread through the displacement of soil along with seed material for a very long distance. Movement of farm machinery and tools helps to carry the infested soil along containing the cysts. The cysts being very light in weight can be easily carried by wind to a long distance.

20.3 MANAGEMENT OF PLANT PARASITIC NEMATODES

Plant parasitic nematodes (PPNs) exhibit high reproductive rates and possess unique survival mechanisms, making their management challenging within intensive crop cultivation systems. To mitigate their impact, growers are advised to maintain nematode populations below economically damaging levels through various available tactics. Young, tender seedlings of various crops are particularly susceptible to nematode attacks, although older plants often develop some degree of tolerance. Tailored to farmer preferences and conditions, the following management strategies are recommended for controlling nematode pests in vegetable crops:

20.3.1 Cultural Control

Cultural practices encompass a range of techniques such as crop rotation, fallowing, flooding, sanitation, summer plowing, mulching, optimizing plant spacing, timing of sowing, employing trap crops and antagonistic plants, and incorporating soil organic amendments. An effective approach to the use of trap crops is to plant crops that are highly susceptible to invasion by the nematode but resistant to the development of larvae to adults. In this case, the crop can be harvested or used as a cover crop and turned under as green manure. Crotalaria is a crop that has been successfully used for this purpose in reducing the population of certain species of root-knot nematodes. In recent years considerable interest has been focused on plants with roots containing nematicidal substances. These toxic substances found in *Tagetes* species (α-terthienyle) and *Asparagus officinalis* (Asparagusic acid) reduce the population level of some nematode species. Intercropping of marigold (*Tagetes* sp.) with brinjal plants was very effective in reducing nematode population, egg masses, and galling of *M. javanica* (Dhangar et al. 2002).

Rabbing is a cultural method for the management of plant parasitic nematodes. It is generally practiced in nurseries. Paddy husk was found to be the most effective rabbing material in reducing the population of root-knot nematode (Ahmed and Choudhury 2004a).

Summer plowing 2–3 times during the hottest period of the year helps to expose nematodes to the drying action of sun and wind and reduce the population of plant parasitic nematodes in soil. A drastic reduction of plant parasitic nematodes with a significant increase in the yield of maize with three deep plowings has been reported. Out of many cultural practices used for the management of plant parasitic nematodes, crop rotation is very common. This is an age-old practice but still practicable as well as profitable under certain sets of conditions. The rotational sequence of potato-French bean-pea-potato in a four-year rotation had decreased the cyst populations by 98 to 99 % with 90% increased yields of potato. The crop sequence, French bean-ahu rice-sesamum-French bean was found to be effective in reducing the root galls, eggmasses and nematode population in soil (Ahmed and Choudhury 2004b).

Soil amendments using green manure, compost, and various oil cakes like neem, mahua, mustard, groundnut, cotton, linseed, karanj, and sawdust have shown efficacy in reducing nematode populations in soil. Extracts from numerous cultivated and weed plants have also demonstrated toxicity to root-knot nematodes, with chopped plant parts incorporated into the soil resulting in population reduction (Budhram and Baheti 2006; Neog et al. 2012).

20.3.2 Biological Control

Plant parasitic nematodes (PPNs) face numerous adversaries in nature, ranging from fungi and bacteria to predatory nematodes. In recent years, leveraging bio-agents has emerged as a successful strategy in managing nematode populations. Despite some inherent constraints, biological control presents itself as a cost-effective and environmentally sustainable approach to mitigating the damage caused by plant parasitic nematodes. As a component of integrated nematode management, biological suppression of plant parasitic nematodes is well documented. The arbuscular mycorrhizal fungus *Glomus fasciculatum* increased the resistance in tomato plants against reniform nematode infestation. Inoculation of tomato transplant on seed bed with the fungus significantly reduced the juvenile penetration and development on mycorrhiza-treated plants over the control plants (Sitaramaiah and Sikora 1982). Less gall production by root-knot nematode on tomato roots was observed due to colonization by *G. fasciculatum*. Egg parasitic fungus *Paecilomyces lilacinus* has been reported to be pathogenic to reniform nematode infecting eggplant. Soil application of *P .lilacinus* was found to be effective against root-knot nematode infecting several vegetables (Khan and Goswami 2000). Sharma (2002) reported that antagonistic fungi *T. harzianum* and *T. viride* were found to be effective in reducing the population of root-knot nematode on tomatoes. Several workers have reported *Trichoderma* species to be the most effective bioagent against root-knot nematodes in various crops, wherein apart from antibiosis, egg parasitic behavior of the fungus has also been recorded (Annapurna et al. 2018; Bharali et al. 2019; Kurulkar et al. 2019). In recent years, there has been significant attention and research focused on *Pasturia penetrans*, a bacterial parasite of nematodes. Numerous greenhouse and microplot tests have consistently demonstrated the efficacy of *P. penetrans* in suppressing nematode-induced root galling and egg production, leading to enhanced plant growth. Additionally, *P. penetrans* has been shown to reduce the populations of *P. scribneri* recovered from both soil and bean roots. Remarkably, even small amounts of soil containing spores of this bacterium have proven effective in transmitting the organism to previously uninfested sites.

20.3.3 Regulatory Control

Legal enforcement to prevent pests and pathogens from spreading to new areas, or to prevent them from multiplying and spreading further in case they have already found entry, can be helpful, particularly in the case of endoparasitic nematodes. Numerous attempts have been made to prevent the introduction of nematodes into countries or provinces by means of quarantine. The government of India has also legislated an act, known as the Destructive Insect Pest Act 1914 (DIP Act 1914), which regulates the entry or movement of plant material into the country. Besides the central legislation, domestic regulatory measures are also possible, as has been done in the case of golden cyst nematode of potatoes, restricting the movement of potato seed material from the infested fields of Tamil Nadu to other areas of the country.

20.3.4 Resistant Varieties

Plant resistance plays an important role in the integrated management of nematode diseases, however, resistant varieties of vegetable crops are very few in number and many of them are not acceptable to the farmers for their suitability. Several varieties/cultivars of different vegetable crops screened by various workers from time to time exhibited resistant or tolerant reactions to root-knot nematodes (Yadav and Chand 2003).

20.3.5 Physical Control

Physiotherapy methods, such as hot water treatment of infected plant materials in their dormant state, particularly tubers and bulbs, have proven to be valuable in reducing the spread and damage caused by nematodes to crops. Such practices hold particular importance for various crops. Additionally, soil treatment involving heat for a brief period effectively eradicates a significant portion of nematode populations. However, techniques such as electricity, radiation, X-rays, and ultrasonics, which also include heat, are typically viable only for capital-intensive crops or certified seed production programs. Furthermore, Srivastava (2004) highlighted the efficacy of soil solarization in reducing galls on brinjal roots and suppressing nematode populations in the soil.

20.3.6 Chemical Control

Chemical control through the application of nematicides has historically been the most effective method of nematode management. However, many potent nematicides have been withdrawn from the global market. Currently, only a few insecticides with nematicidal properties are available to farmers. However, their high doses and associated costs often render their use cost-ineffective and leave behind significant pesticide residues on harvested crops. Despite these challenges, judicious application of chemical nematicides can significantly reduce doses and costs.

Various methods such as bare-root dip treatment, seed treatment, and nursery bed treatment have proven effective in protecting young seedlings from nematode attacks. Seed treatment with nematicides has shown effectiveness against root-knot nematodes in several vegetable crops (Bhagawati and Phukan 1990).

Furthermore, foliar application of systemic pesticide oxamyl on tomatoes shows promise against *R. reniformis*. A single application of this chemical can block nematode entry into roots. In regions infested by the golden nematode, such as the potato areas of Nilgiris, treatment with dasanit has been found effective in reducing nematode populations. Other chemicals like mocap, aldicarb, carbofuran, and fensulfothion have also shown effectiveness against golden cyst nematodes in potatoes under field conditions. Soil application of phorate, carbofuran, aldicarb, etc. has been effective in reducing populations of root-knot and reniform nematodes infecting different hosts (Singh 2006).

20.3.7 Integrated Control

The development of pest management programs for effective control requires the careful selection of control techniques that minimize environmental impact while maintaining economic feasibility. Various methods such as resistant cultivars, pesticides, sanitary and cultural practices, and biocontrol agents can be utilized to manage pests (Stephan et al. 1988; Novaretti et al. 1989). Integrated management strategies aim to prevent the overabundance of any single pest population and reduce the development of pest resistance. Flexibility is crucial in integrated pest management systems, as recommendations must adapt to changing pest populations. Factors considered in system development include the species and race of pests, availability of resistant host plants, longevity of pests and crops, cropping systems, and regional climate. In tropical and subtropical conditions, nematicide use can be limited to nursery beds in transplanted crops and as seed dressers in both transplanted and direct-seeded crops. These treatments can be combined with summer plowing, organic manuring, supplementary fertilizers, resistant varieties, biocontrol agents, botanical pesticides, and crop rotation with non-host crops.

Nematicides with organic amendments or farmyard manures have proven effective against root-knot nematodes (Bhattacharya and Goswami 1988. Soil amendments with neem and groundnut cake, along with antagonistic fungus *Trichoderma viride*, have been effective against disease complexes caused by nematodes and fungi on okra (Singh et al. 2003). The integration of various agents like *T. viride*, *G. fasciculatum*, and mustard cake can effectively manage disease complexes in okra (Saikia and Bhagawati 2014). Combining carbofuran with endospore-forming bacteria *Pasteuria penetrans* has shown effectiveness against root-knot and cyst nematodes (Singh and Dhawan 1998). Neem products combined with fungal biocontrol agents have also been effective against nematodes infecting eggplant (Mojumder et al. 2002). The integration of summer plowing followed by seed treatment with carbosulfan has reduced root-knot nematode populations in okra (Jain and Gupta 1996). Biocontrol agents like *V. chlamydosporium* with neem cake and carbofuran have been effective in reducing nematode infestation in tomatoes (Gopinatha et al. 2002). Similarly, applications of *P. lilacinus*, *Trichoderma harzianum* with neem cake, and marigold have been effective against root-knot nematodes (Verma et al. 2005). Combining mycorrhizal fungus, *Glomus fasciculatum*, with carbofuran has also been effective in reducing nematode populations in brinjal (Saikia and Borah 2008).

20.4 FUTURE APPROACHES AND CONCLUSION

Plant parasitic nematodes constitute one of the major limiting factors for the cultivation of vegetable crops. The changes in agricultural situations have tremendous effects on the emergence of new nematode problems in India. Plant parasitic nematodes are mostly neglected and considered as low-priority factors for crop production and protection in India due to their hidden nature and inconspicuous symptoms. Therefore, the economic significance of nematodes in agriculture is very often underestimated and their damage potential is not recognized by plant protection specialists, scientists, and administrators. As an important component of integrated pest management, nematode pathogens cannot be ignored and they could be tackled with the intelligent planning of nematode suppressive crop sequences, summer plowing, organic manuring, clean cultivation, adjusting sowing time, water and irrigation management and sensible use of nematicides. The increasing concern about the ill effects of chemical pesticides on the environment has driven the recent research interests in the use of several alternative strategies like botanicals, biopesticides, and cropping system research for the management of insect pests and diseases including nematodes. Some of the successes have been obtained for managing plant parasitic nematodes with neem-based formulations, fungal formulations of *P. lilacinus*, bacterial formulations of *Pasteuria penetrans*, and *Pseudomonas fluorescens*. However, the wide application of bio-agents in field-scale is still limited probably due to their inconsistent efficacy in different agro-ecological situations. The integration of two or more methods can be explored on the basis of their compatibility, economic viability, availability and acceptability to the farmers. Developing holistic approaches for managing the nematodes as well as the disease complexes caused by them interacting with other pathogens like fungi, bacteria, and viruses will be the major future areas of research thrust.

REFERENCES

Ahmed JA, Choudhury BN (2004a) Effect of rabbing on *Meloidogyne incognita* in French bean. *Ann Plant protect Sci* 12: 156–158.

Ahmed JA, Choudhury BN (2004b) Effect of rabbing on *Meloidogyne incognita* in French bean. *Ann Plant Protect Sci* 12: 118–120.

Annapurna M, Bhagawati B, Kurulkar U (2018) *In-vitro* efficacy of native fungal bioagents against *Meloidogyne incognita. Int J Curr Microbiol App Sci* 7(11): 396–410.

Berkeley MJ (1855) Vibrio forming excrescences on the roots of cucumber plants. *Gard Chron* 14: 200.

Bhagawati B, Goswami BK, Singh CS (2000) Management of disease complex of tomato caused by *Meloidogyne incognita* and *Fusarium oxysporum* f. sp. *lycopersici* through bioagent. *Indian J Nematol* 30(1): 16–22.

Bhagawati B, Phukan PN (1990) Chemical control of *Meloidogyne incognita* on pea. *Indian J Nematol* 20: 79–83.

Bharali A, Bhagawati B, Kurulkar U (2019) Nematicidal toxicity of native antagonists against *Meloidogyne incognita. Int J Curr Microbiol App Sci* 8(2): 1385–1398.

Bhattacharya D, Goswami BK (1988) Effect of oil-cakes used alone and in combination with aldicarb on *Meloidogyne incognita* infecting tomato. *Nematol Medit* 16: 143–145.

Budhram BR, Baheti BL (2006) Management of reniform nematode, *Rotylenchulus reniformis* on cowpea (*Vigna unguiculata* L.) with plant products. *J Nematol* 36: 140–142.

Dhangar DS, Gupta DC, Jain RK (2002) Studies on intercropping of marigold (*Tagetes* sp.) with brinjal on plant growth and population of root-knot nematode (*Meloidogyne javanica*). *Indian J Nematol* 32: 220–221.

Gopinatha KV, Nanjegowda D, Nagesh M (2002) Management of root-knot nematode, *Meloidogyne incognita* on tomato using bio-agent *Verticillium chlamydosporium*, neem cake, marigold and carbofuran. *Indian J Nematol* 32: 179–181.

Goswami BK (2002) Control of major species of the root-knot nematode. In: Alam MM, Sharma N (eds.) *Nematode Control in Crops*. International Book Distributing Co., Lucknow, India, pp. 361–378.

Jain RK, Gupta DC (1996) Integrated management of root-knot nematode (*Meloidogyne javanica*) through summer ploughing and nematicides in okra. *Indian J Nematol* 26: 121–122.

Khan MR, Goswami BK (2000) Effects of different doses of *Paecilomyces lilacinus* isolate 6 on *Meloidogyne incognit* infecting tomato. *Indian J Nematol* 30: 5–7.

Krishna Prasad KS (1995) Nematode problems of potato. In: Swarup G, Dasgupta DR, Gill JS (eds.) *Nematode Pest Management–An Appraisal of Ecofriendly Approaches*. Nematological Society of India, New Delhi, pp. 177–185.

Kurulkar U, Bhagawati B, Neog PP (2019) Biological control of *Meloidogyne incognita* by *Trichoderma harzianum. Int J Curr Microbiol App Sci* 8(2): 2176–2188.

Mojumder V, Dhawan SC, Pankaj, Singh J (2002) Compatibility of neem products and bioagents for the management of *Meloidogyne incognita* and *Rotylenchulus reniformis* infecting eggplant. *Indian J Nematol* 32: 197–198.

Neog PP, Chauhan MK, Deka SN (2012) Incidence and ecofriendly management of root-knot nematode in Assam. *Ann Plant Protect Sci* 20: 267–268.

Novaretti WRT, Carderan JO, Strabelli J, Amorim E (1989) Effect of compost application alone or in combination with nematicides and mineral fertilizers, on nematode control and productivity of sugarcane. *Nematol Brasileria* 13: 93–107.

Roy K, Mukhopadhyay AK, Pramanik A (2008) Assessment of yield loss due to *Rotylenchulus reniformis* Linford and Oliveira, 1940 in cowpea, *Vigna unguiculata* (L.) Walp. *Ann Plant Protect Sci* 16: 192–194.

Saikia D, Bhabesh B (2014) Management of *Meloidogyne incognita* and *Rhizoctonia solani* complex on 0okra through integration of *Trichoderma viride, Glomus fasciculatum* and mustard cake. *Indian J Nematol* 44(2): 139–143.

Saikia J, Borah A (2008) Comparative efficacy of fungal and bacterial bioagents with nematicides against *Meloidogyne incognita*. *Indian J Nematol* 38: 165–167.

Sasser JN, Freckman DW (1987) A world perspective on nematology: the role of Society. In: Veech JA, Dickson DW (eds.), *Vistas on Nematology*, pp. 7–14. Society of Nematologist, Hyattsville, Maryland.

Sharma R (2002) Effect of *Trichoderma* spp. against root-knot nematode *Meloidogyne incognita* on tomato. *Indian J Nematol* 32: 227–228.

Singh B, Dhawan SC (1998) Compatibility of *Pasteuria penetrans* with carbofuran in controlling *Heterodera cajani* on cowpea. *Indian J Nematol* 28: 110–113.

Singh CL, Singh S, Goswami BK (2003) Effect of cakes with *Trichoderma viride* for management of disease complex caused by *Rhizoctonia bataticola* and *Meloidogyne* incognita on okra. *Ann Plant Protec Sci* 11: 178–180.

Singh VK (2006) Management of root-knot nematode, *Meloidogyne incognita* infecting cauliflower. *Indian J Nematol* 36: 127.

Sitaramaiah K, Sikora RA (1982) Effect of the mycorrhizal fungus *Glomus fasciculatum* on the host parasitic relationship of *Rotylenchulus reniformis* in tomato. *Nematologica* 28: 412–419.

Srivastava SS (2004) Integration of soil solarization in the management of *Meloidogyne incognita* on brinjal. *Ann Plant Protect Sci* 12: 471–472.

Stephan ZA, Al-Mammoovry IK, Antoon BG (1988) The efficacy of nematicides, solar heating and the fungus, *Paecilomyces lilacinus* in controlling root-knot nematode, *Meloidogyne javanica* in Iraq. *Iraq J Agric Sci Zanco* 6: 69–76.

Verma AC, Singh HK, Khan MN (2005) Management of root-knot nematode, *Meloidogyne incognita* through antagonistic approaches in pointed gourd. *Indian J Nematol* 35: 78–79.

Walia RK, Kanwar RS (2002) Control strategies for root-lesion nematodes. In: Alam MM, Sharma N (eds.) *Nematode Control in Crops*. International Book Distributing Co., Lucknow, India, pp. 391–414.

Yadav BD, Chand R (2003) Field screening of tomato, pea, cowpea, coriander, fenugreek, cluster bean and bottle gourd germplasm against *Meloidogyne incognita*. *Ann Plant Protect Sci* 11: 349–351.

21 Nematode Problems in Tea and Their Management

Pranjal Pratim Neog, Bhabesh Bhagawati and Uday Kurulkar

21.1 INTRODUCTION

Tea, scientifically known as *Camellia sinensis* (L.) O. Kuntze, is a member of the Theaceae family and stands as a cornerstone in the realm of plantation agriculture, thriving not only in India but across various regions worldwide. Renowned as one of the most beloved and affordable beverages globally, tea boasts major production hubs in countries like India, China, Kenya, Sri Lanka, Turkey, and Vietnam. India, having one of the oldest organized tea industries globally, boasts a robust network encompassing tea producers, retailers, distributors, auctioneers, exporters, and packers. Its roots trace back to the region near the confluence of latitude 29°N and longitude 98°S, close to the source of the Irrewaddi River. Within India, tea cultivation flourishes predominantly in regions such as Assam, West Bengal, Kerala, Tamil Nadu, and Karnataka.

The tea crop faces numerous challenges, including attacks from various pests and diseases such as fungi, bacteria, viruses, and nematodes. Within these threats, plant parasitic nematodes (PPNs) emerge as particularly significant adversaries, capable of inflicting considerable losses on tea production. Various workers recorded a considerable number of PPNs from the rhizosphere of nursery, young and mature tea grown in diverse parts of the world (Sivapalan 1972; Basu and Roy 1979; Phukan and Sanwal 1979; Mukherjee and Dasgupta 1981; Sivapalan and Gnanapragasam 1991; Kepenekci and Akgul 1999; Orisajo 2012, 2013; Kibet et al. 2013; Mirghasemi et al. 2014). Out of several PPNs associated with tea, root-knot nematode, *Meloidogyne* spp., and root lesion nematode, *Pratylenchus* spp., are considered to be the most destructive nematode pests of nursery, young, and mature tea plants (Basu and Roy 1976; Rang et al. 1984).

21.2 MAJOR NEMATODE PESTS

21.2.1 Root-knot Nematode

The initial observation of root-knot nematode (RKN) infestation in young tea plants was documented in southern India, where a significant number of tea seedlings were found to be affected by this nematode (Barber, 1901). Subsequently, various researchers reported the presence of root-knot nematodes in the rhizosphere of tea plants (Basu and Roy 1976). Additionally, Rang et al. (1984) reported the distribution of RKNs specifically *Meloidogyne* spp., on tea seedlings across 18 different countries.

They also recorded 100% infection and 50% death of seedlings in seriously infested areas of tea plantations by *M. incognita*, *M. arenaria*, *M. javanica,* and *M. thamesi*. Among RKNs *M. incognita*

DOI: 10.1201/9781032631578-21

and *M. hapla* are considered to be economically important pests of nursery and young tea. In nursery conditions, both seedlings and vegetative propagated clonal plants are severely damaged by these species. It is observed that its infestation is mostly confined to seedlings in nurseries for about 3–5 months from the sowing of seeds. *M. incognita, M. javanica, M. thamesi* are economically important species causing severe damage to nursery seedlings less than one year old, except *M. bravicauda* (Sivapalan 1972). The occurrence of *M. bravicauda* and *M. javanica* was also recorded from tea seedlings in Madras state (Seshadri and Kumaraswami 1963; Thirugnanam and Rangaswami 1967). *M. incognita* and *M. javanica* are found to be the most harmful RKN species on nursery tea while *M. bravicauda* on mature tea.

21.2.1.1 Symptoms

Tea plants grown both from seeds and vegetative propagation are severely damaged by root-knot nematode. Root-knot nematode attacks are mainly confined to seedlings 3–5 months old causing galls on the feeder as well as tap roots and decreasing the height of tea seedlings, and number of leaves (Dutta 1995; Kalita 2000, Bora and Neog 2006a). Occasionally the collar region also gets infested where such galls are evident. Usually, tea seedlings infested with root-knot nematode are not always killed outright and symptoms of attack can be mistaken for drought, mineral deficiencies, lack of organic constituents, water logging, and so on. In the field, patches of stunted, wilted, chlorotic, or generally unhealthy growth of tea seedlings are observed. These are caused by atrophied root symptoms due to rotting, galling, or by killing the fine tender roots and lateral rootlets. Diseased seedlings look yellowing and unhealthy during the rainy period and suddenly wilt and die when a period of hot, dry weather prevails towards the end of it.

Root-knot nematode, *M. bravicauda* causes severe galling on the roots of mature tea bushes with resultant decline. The infected bushes grow more slowly, and the leaves tend to become smaller, dull, and yellow in appearance. The roots show the characteristic swelling and pitting that is typical symptoms of *M. bravicauda* (Sivapalan 1972)

21.2.2 Root Lesion Nematode

Various species of root lesion nematodes have been identified as threats to tea plants across different regions worldwide, with notable attention focused on *Pratylenchusloosi* and *P. brachyurus*. These nematodes primarily target young tea plants, causing significant damage (Sivapalan 1972; Basu and Roy 1976). Furthermore, *P. brachyurus* and *P. curvitatus* have been documented as pathogenic to mature tea plants, emphasizing the broad spectrum of nematode threats to tea cultivation (Sivapalan 1972; Basu and Roy 1976).

21.2.2.1 Symptoms

Tea bushes affected by root lesion nematodes exhibit visible symptoms above ground, often characterized by the presence of unthrifty tea plants with weakened foliage maintenance. The leaves appear pale yellowish and dull mainly due to nutritional imbalance caused by an altered rate of uptake of essential nutrients. The heavily infected plants show a tendency of premature flowering and fruiting (Gnanapragasam 2002). The infected plants are sometimes found to have small shoots and internodes and the leaves at the plucking table remain as tufts with few buds giving rise to pluckable shoots. Examination of the roots of such infected plants shows a marked reduction in the feeder root system. The feeder roots of plants infected with *P. loosi* appear brownish in colour and dark brown necrotic patches appear on both the feeder roots as well as the storage roots. Peeling pieces of the bark or cortex from the junction between the dying and healthy tissues and examination of the undersurfaces of the peeled strips would reveal large numbers of nematodes. *P. brachyurus* attacks mainly the feeder roots of young tea and occasionally the tap roots forming reddish-brown lesions (Gnanapragasam 2002).

21.3 MANAGEMENT OF PLANT PARASITIC NEMATODES

In the past decade, significant research efforts have been devoted to exploring diverse control methods aimed at reducing yield losses attributed to plant parasitic nematodes. While the use of nematicides has proven highly effective in managing nematode pests across various crops, including tea, frequent application raises concerns about potential residual effects, phytotoxicity, and risks to human and livestock health. A number of other methods like physical, cultural, biological and use of resistant clones are being experimented for effective and economic management of nematode pests of tea by several workers (Rao 1976; Basu 1978; Sivapalan 1980; Gnanapragasam 1982; Dutta 1995; Neog and Bora 1999a, 1999b, 1999c; Bora and Neog 2006b; Kalita and Bora 2006; Baruah 2011).

Cultural methods: Application of organic matter in nematode-infested soil is presently considered to be an alternative, easy, cheap, and efficient method of nematode management. Several workers emphasized the beneficial effect of the application of organic matter for the suppression of PPNs both in nurseries as well as in the main field. Visser (1959) reported that the incorporation of organic matter into nematode-infested soil is known to suppress the incidence of PPNs in tea. Significant reductions of gall formation have also been reported on tea seedlings grown on RKN-infested soil by dressing with oil cakes of mustard, sesame, and groundnut each at the rate of 0.5, 1.0 and 2.0 g in 100 g of soil and the effects were found to be more pronounced at comparatively high rates of application (Anonymous 1974a). Application of different organic amendments *viz.* mustard oil cake, neem cake, sawdust, decaffeinated tea waste at the rate of 10–30 g per kg soil before planting of tea seedlings were found to be effective in increasing the plant growth parameters and reducing the number of gall, eggmasses and population of *M. incognita* in soil (Dutta 1995; Neog and Bora 1999a; Bora and Neog 2006b; Kalita and Bora 2006). Organic amendments *viz.*, neem cake, mustard oilcake, jatropa cake and vermicompost each with two doses viz. 15 g and 30 g per kg of soil were found to be effective in increasing the plant growth parameters of tea seedlings and reducing the number of galls, egg masses, collar rot incidence, and population of *M. incognita* in soil (Baruh 2011).

The application of castor oil cake has been shown to achieve maximum suppression of the population of *P. loosi*, while simultaneously promoting the growth of tea plants under field conditions. Similarly, other oil cakes such as mahua, neem, and karanji have demonstrated equal effectiveness in this regard. However, various organic amendments, including poultry manure, coconut poonac, and tea waste, did not exhibit significant suppression of nematode populations. Nonetheless, plants treated with these organic amendments showed notable improvements in growth parameters, thereby enhancing tolerance to nematode attacks (Sivapalan 1980; Gnanapragasam 1987, 1991).

In both young and mature tea plantations, the application of potassium-enriched fertilizers has proven effective in controlling the population growth of root lesion nematodes. *P. loosi* was managed by doubling the doses of potassium fertilization from 554 mg of K_2O to 1108 mg per plant per year. At the higher level of 'K' application, the nematode population was significantly suppressed (Gnanapragasam 1982).

Visser and Vythilingham (1959) reported that two species of marigold, viz., *Tageteserecta* and *T. patula*, considerably reduced the multiplication of *M. javanica* and *P. coffeae* in the soil. Several studies have highlighted the positive impact of planting marigolds in vacant areas of tea gardens, particularly in reducing the population of PPNs (Hutchinsion 1961; Sivapalan 1972). Planting of *Vetiveria zizanioides* and *Eragrostis curvula* has been found to reduce the nematode population (Gnanapragasam 1981, 1985). Therefore, before the cultivation of tea plants, these crops may be cultivated to reduce nematode infestation and also in vacant areas of tea gardens, these crops should be cultivated.

Physical Methods: Management of plant parasitic nematodes through physical methods is mainly confined to nursery soil. The practice of spreading nursery soil over galvanized sheets to a depth of 4 cm and heating it to 60–67°C for 4–5 minutes before planting has proven effective in reducing RKN infestations and promoting seedling growth in nurseries. (Rao1976; Basu 1978; Neog and Bora 1999c). Neog and Bora (1999c) reported that steam sterilization of soil at 15 lb pressure per sq. inch for half an hour, heat treatment of soil by hot pan for 3 h at 60°C and sun drying of soil continuously for 10 days are some of the effective methods of management of root-knot nematode and thereby producing healthy tea seedlings. Out of all these, steam sterilization of soil at 15 lb pressure per sq. inch was found to be most effective and it provides nematode-free conditions to the seedlings throughout the observing period.

Biological Method: PPNs have many natural enemies including fungi, bacteria, and predacious nematodes. Bio-agents have emerged as a highly effective option in nematode management in recent years. Despite its several limitations, biological control of plant parasitic nematode is a cost-effective and eco-friendly method. As a component of integrated nematode management, biological suppression of plant parasitic nematodes is well documented.

Biopesticides like botanicals (plant extracts) and nematopathogenic microorganisms (bacteria and fungi) can be a potential tool for controlling PPNs because they are considered environmentally safe, biodegradable, economical, and renewable. It also encourages the use of biopesticides as a component of integrated pest management of PPNs in tea (Paull 2014). Fungal antagonists viz., *Trichoderma viride*, *T. harzianum* @ 4 g, *Pochonia chlamydosporia* @ 15 g and VAM fungus, *Glomus fasciculatum* @ 300 spores per kg of soil were found to be effective in increasing the plant growth parameters of tea seedlings and reducing the number of galls, egg masses, collar rot incidence and root-knot nematode *M. incognita* population in soil (Baruah 2011; Gogoi 2017).

Resistant/tolerant clones and seed stock: Utilizing resistant or tolerant clones stands out as one of the most cost-effective long-term strategies for managing nematode-infected tea crops. This approach not only eliminates the need for additional nematode control expenses but also mitigates the risk of environmental pollution. Very limited works have so far been done in this direction. Eight Tocklai-released tea clones viz., TV1, TV2, TV9, TV10, TV14, TV17, and TV18, were tested for their reactions to *M. incognita* and it was found that barring TV9 and TV14 the rest of the clones exhibited varying degrees of resistance to *M. incognita* (Anonymous 1972). It was also observed that out of four Tocklai-released tea clones tested against *M. incognita*, TV13, and TV15 were found to be slightly more susceptible than TV1 and TV18 (Anonymous, 1974b). Neog and Bora (1999b) also reported that out of ten Tocklai-released tea clones, two clones viz., TV1 and TV18 were found to be moderately resistant while the rest were found to be susceptible to *M. incognita* in varying degrees. Apart from screening tea clones, tea seedlings raised from tea seed stocks were also screened for their reactions to *M. incognita* and it was found that the number of galls per plant was high in stocks 462, 463, and 464 and low in stocks 491 and 493, while stock 461 occupied an intermediate position (Anonymous 1981). Gnanapragasam et al. (1985) reported that six tea clones were free from galling six years after planting in an area cleared of old seedling tea heavily infested with *M.brevicauda*. However, after 10 years, clone TRI 2141 showed a root-knot index of 3.24, while the rest of the clones had indices of less than 0.6.

Chemical method: Previously, fumigant nematicides were deemed effective in managing plant parasitic nematodes. DD and EDB were particularly noteworthy as pre-plant treatments, while DBCP served effectively as a post-plant treatment against RKNs affecting nursery tea seedlings, leading to a notable decrease in the incidence of root galls (Ichinohe 1965). Soil fumigant methyl bromide @ 0.5, 1.0, and 2.0 lb per square foot successfully reduced the infestation caused by

PPNs at the nursery site and increased seedling growth (Sivapalan 1969). The use of some other nematicides such as nemagon, hexanema, and vapam at the nursery site has proved effective against the nematode pest of tea (Mukerjea 1966; Rao 1970). Later, due to several disadvantages relating to the application of such chemicals, the non-fumigant granular organophosphate and carbamate group of nematicides were applied for the nematode management program. Some granular nematicides, *viz.*, carbofuran, phorate, diazinon, aldicarb, and ethoprophos were found to be effective when tried at different doses against root-knot nematode on the nursery and young tea, which resulted in good plant growth with less gall index (Mukerjea 1966; Rao 1970; Basu and Gope 1985; Dutta 1995). Similarly, population *P. loosi* per g of roots was found to be reduced when susceptible tea clone TRI 2024 was planted in nematode-infested soil along with incorporation of some systemic nematicide *viz.*, aldicarb @ 7 and 14 g per plant, carbofuran @ 3.5 g and 7 g per plant and fenamiphos 3.5 g per plant in the planting hole. Treatment with these nematicides also increases the shoot and root weight of the plants six months after planting (Sivapalan et al. 1980).

Among the various systemic granular nematicides, carbofuran (Furadan 3G) has shown efficacy in reducing RKN populations and enhancing seedling growth. A pre-plant application of 3–5 g per seedling in nursery settings and 3–5 kg per mu in young tea planting sites has yielded positive results (Rang et al. 1984; Neog and Bora 1999a). Additionally, Kalita and Bora (2006) found that combining carbofuran (Furadan 3G) at 2 g per seedling with 5 g of either neem cake or mustard oil cake was more effective in promoting tea seedling growth and reducing RKN infestations compared to individual applications. Moreover, fenamiphos 5 percent granules at 7 g per plant have demonstrated effectiveness in suppressing nematode infestations by up to 60 percent while enhancing tea plant growth (Gnanapragasam et al. 1987). However, it's important to note that many of these chemicals have faced bans or restrictions.

21.4 FUTURE PROSPECTS

Plant parasitic nematodes (PPNs) pose a significant challenge in tea cultivation, representing a major limiting factor. Addressing the losses incurred due to these nematodes requires a multifaceted approach, as no single method may suffice on its own. Therefore, different management strategies should be integrated and applied in such a way as to maintain the nematode population levels below the damage threshold level. Since the use of most of the chemicals has either been banned or restricted for their adverse effect on the environment and health hazards, trust should be given to the use of non-chemical means for managing the PPNs in tea. Efforts should be made to explore and exploitation of native fungal or bacterial antagonists of plant parasitic nematodes infecting tea that are eco-friendly.

REFERENCES

Anonymous (1972) Clonal susceptibility to eelworms. *Annual Scientific Report, 1971-72*, TRA, Tocklai Experimental Station, pp. 53–54. Tocklai Experimental Station, Jorhat, Assam, India.

Anonymous (1974a) Organic dressing and nematode pathogenecity. *Annual Scientific Report, 1973-74*, TRA, Tocklai Experimental Station, pp. 47–48. Tocklai Experimental Station, Jorhat, Assam, India.

Anonymous (1974b) Clonal resistance. *Annual Scientific Report, 1973-74*, TRA, Tocklai Experimental Station, pp. 47–48. Tocklai Experimental Station, Jorhat, Assam, India.

Anonymous (1981) Susceptibility of the seedlings of different tea seed stocks to *Meloidogyne incognita* (Kofoid and White) chitwood. *Annual Scientific Report, 1980-81*, TRA, Tocklai Experimental Station, p. 46. Tocklai Experimental Station, Jorhat, Assam, India.

Barber CA (1901) A tea eelworm disease in South India. *Trop Agricult* 21: 239.

Baruah A (2011) Management of *Meloidogyne incognita* and *Rhizoctonia solani* disease complex in tea seedling. *M.Sc. (Agri.) Thesis*, Assam Agricultural University, Jorhat, Assam, p. 672.

Basu SD (1978) Heat the soil and kill the nematode. *Two and a Bud* 25(1): 42.

Basu SD, Gope B (1985) Control of root-knot nematode *Meloidogyne incognita* (Kofoidand White) Chitwood on tea seedling with some systemic nematicides. *Two and a Bud* 32(1–2): 32–35.

Basu SD, Roy SK (1979) Sample survey on nematodes in doors. *Two and a Bud* 26(2): 69–70.

Bora BC, Neog PP (2006a) Pathogenicity of *Meloidogyne incognita* on tea. *Ann Plant Protect Sci* 14(2): 521–522.

Bora BC, Neog PP (2006b) Effect of oil cakes for management of *Meloidogyne incognita*in tea. *Ann Plant Protect Sci* 14(2): 522–523.

Dutta P (1995) Pathogenicity and management of *Meloidogyne incognita* on tea seedlings. *M. Sc. (Agri.) Thesis*, Assam Agricultural University, Jorhat, Assam.

Gnanapragasam NC (1981) The influence of cultivating *Eragrostis curvula* in nematode infested soils and on subsequent build-up of populations in replanted soil. *Tea Quarterly* 50: 160–162.

Gnanapragasam NC (1982) Effect of potassium fertilization and soil temperature on the incidence and pathogenicity of the root-lesion nematode, *Pratylenchus loosi* Loof on tea. *Tea Quarterly* 51: 169–174.

Gnanapragasam NC (1985) *Report of the Tea Research Institute of Sri Lanka 1985*, pp. 82–96. Tea Research Institute of Srilanka, St.Coombs, Talawakelle, Srilanka.

Gnanapragasam NC (1987) *Report of the Tea Research Institute of Sri Lanka 1987*, pp. 70–83. Tea Research Institute of Srilanka, St.Coombs, Talawakelle, Srilanka

Gnanapragasam NC (1991, August) Influence of soil amendments in reducing pathogenicity to tea bythe root lesion nematode, *Pratylenchus loosi*. In: *Proceeding of the International Symposium Tea Science*, Shizuoka, Japan, pp. 684–688.

Gnanapragasam NC (2002) Nematode pathogenic to tea and their control. In: Alam MM, Sharma N (eds.), *Nematode Control in Crops*, pp. 133–147. International Book Distributing Company, Lucknow.

Gnanapragasam NC, Jebamalai AS, Anpalagan VT (1987) Control of root lesion nematodes *Pratylenchus loosi* in contaminated tea nursery plants. *Sri Lanka J Tea Sci* 56: 95–97.

Gnanapragasam NC, Sivapalan P, Jebamalai AS, Dharmasena WAM (1985) Tolerance/ susceptibility of six different clones to *Meloidgyne bravicauda* the root-knot nematode of mature tea. *Int Nematol Net Newsletter* 2: 9–10.

Gogoi HJ (2017) Occurrence and distribution of plant parasitic nematodes associated with tea gardens in Dibrugarh district of Assam. *M.Sc. (Agri) Thesis*, Assam Agricultural University, Jorhat, Assam.

Hutchinson MT (1961) Developments in the control of the meadow nematode. *Tea Quarterly* 32: 129–132.

Ichinohe M (1965) Nematodes menace perennial field crops. *Biokemia* 9: 10–13.

Kalita S (2000) Pathogenicity and management of *Meloidogyne incognita* on tea seedlings. *M.Sc. (Agri) Thesis*, Assam Agricultural University, Jorhat, Assam.

Kalita S, Bora BC (2006) Effect of nematicides and organic amendments in the management of *Meloidogyne incognita* in nursery tea. *Indian J Nematol* 36: 148–149.

Kepenekci I, Akgul HC (1999) Plant parasitic nematodes associated with tea (*Camellia sinensis* L.) in Rize region, Turkey. *Pakistan J Nematol* 17(2): 181–184.

Kibet TK, Kinyua J, Mamati EG, Onguso J (2013) Diversity and distribution of soil nematodes in Ngere tea catchment area of Murang'a country, Kenya. *African J Agril Res* 8(18): 1986–1989.

Mirghasemi SN, Seraji A, Jamali S, Choshali AH (2014) Reported some new species of plant parasitic nematodes from rhizosphere of tea plantation in Iran. *Intl J Biosci* 5(9): 37–46.

Mukerjea TD (1963, August 22–23) Chemical control of root-knot nematodes parasitic on tea seedlings in North-East India. In: *All India Nematology Symposium*, New Delhi. Nematological Society of India.

Mukerjea TD (1966) Chemical control of root-knot nematodes parasitic on tea seedlings in North-East India. *Trop Agric Trin* 43(4): 335–340.

Mukherjee B, Dasgupta MK (1981) New records of plant parasitic nematodeinfestation and their significance in decline in tea plantations of West Bengal, India. *Curr Sci* 50: 824–825.

Neog PP, Bora BC (1999a) Organic amendments for the management of *Meloidogyne incognita* in tea nursery. *JASS* 12(2): 203–206.

Neog PP, Bora BC (1999b) Screening of certain tea clones against *Meloidogyne incognita*. *Indian J Nematol* 29(1): 92–93.

Neog PP, Bora BC (1999c) Management of root-knot nematode (*Meloidogyne incognita*) in tea nursery with physical treatments. *Indian J Nematol* 29(1): 101–103.

Orisajo SB (2012) Distribution of plant-parasitic nematodes associated with tea in Nigeria. *World J Agri Sci* 8(5): 459–463.

Orisajo SB (2013) Effect of plant parasitic nematodes associated with tea in Nigeria. *African J Basic Appl Sci* 5(4): 184–187.
Paull SK (2014) Biopesticides: A potential tool for the management of plant parasitic nematodes in tea. *Tea J Bangladesh* 43: 24–33.
Phukan PN, Sanwal KC (1979) Taxonomic studies on nematodes from Assam, India. *Indian J Nematol* 9: 20–26.
Rang WZ, Juan J, Zhang KS (1984) Study of tea seedling disease caused by root-knot nematode (*Meloidogyne* sp.). *Acta Phytopathol Sin* 14: 225–232.
Rao GN (1970) Tea pests in Southern India and their control. *Pest Artic News Summs* 16 (4): 667–672.
Rao GN (1976) Control of nematode in nursery soil. *Planters Chronicle* 71: 257–259.
Seshadri AR, Kumaraswami T (1963) A preliminary report of the occurrence of root-knot nematode (*Meloidogyne* spp.) in some important crop plants in Madras State. *Madras Agricult J* 50: 98–99.
Sivapalan P (1969) Further development in the control of the root lesion nematode (*Pratylenchus loosi*) in tea nurseries. *Tea Quarterly* 40(2–3): 111–114.
Sivapalan P (1972) Nematode pest of tea. In: Webster JM (ed.), *Economic Nematology*, Academic Press, New York and London, pp. 285–311.
Sivapalan P (1980) *Report of the Tea Research Institute of Sri Lanka*, pp. 83–96. Tea Research Institute of Srilanka, St.Coombs, Talawakelle, Srilanka
Sivapalan P, Gnanapragasam NC (1991) Plant parasitic nematodes of tea. In: Luc M, Sikora RA, Bridge J (eds.), *Plant Parasitic Nematodes in Subtropical Agriculture*, Commonwealth Agricultural Bureau, London, pp. 404–430.
Sivapalan P, Gnanapragasam NC, Jebamalai AS (1980) Chemical control of root lesion nematode, *Pratylenchusloosi* in new tea plantings. *Tea Quarterly* 49(2): 30–36.
Thirungnanam M, Rangaswami G (1967) A report on the natural occurrence of root-knot nematode (*Meloidogyne* spp.) on some plants in Madras State. *Madras Agri J* 54: 278–282.
Visser T (1959) Observation on the prevalence and control of parasitic eelworms in tea. *Tea Quarterly* 30(2): 96–107.
Visser T, Vythilingam MK (1959) The effect of marigold and some other crops on the *Pratylenchus* and *Meloidogyne* population in tea soil. *Tea Quarterly* 30: 30–38.

22 Plant Nematodes and Their Management Strategies in Major Plantation Crops

Rashid Pervez and Rajkumar

22.1 INTRODUCTION

Plantation crops are high-value commercial crops that are extremely important to many developing and industrialized nations' agricultural economies and export trade. The business features of farming these crops have significant economic relevance as the World Trade Organization (WTO) places more focus on agriculture. The plantation-grown crops include cashews, coconuts, areca nuts, oil palms, rubber, cocoa, tea, coffee, rubber, and cocoa. Except for coconut, areca nut, cocoa, and coffee, which are all grown either as monocultures or as multiple crops, most of these plantation crops are grown as monocultures. Major plantation crops growing states include Kerala, Tamil Nadu, Karnataka, and Andhra Pradesh, as well as Odisha, Maharashtra, Gujarat, Assam, Tamil Nadu, West Bengal, and Maharashtra (Figure 22.1). The other crops are generally restricted to the tropical region, while tea, cardamom, and coffee demand subtropical climates, frequently under shady settings.

Plantation crops generate foreign exchange. Both primary products and byproducts are in high demand domestically and in several ancillary industries, in addition to having export potential. Plantation crops contribute 4.8% of all exports and 27% of all agricultural commodity income. Less than 3% of the total cultivated area is taken up by plantations (i.e., 1.82 percent of all cropland, or 4 million hectares out of 143.00 million ha, or roughly 2.3%). India is the world's top producer of several plantation crops, including cocoa, cashew, areca nut, tea, and rubber. One of the main factors limiting the quantity and quality of these crops is plant nematodes. This chapter focuses on the main plant nematodes that infest coconut, areca nut, cashew, and cocoa and their management strategies.

22.2 PLANT NEMATODE PROBLEMS IN MAJOR PLANTATION CROPS

22.2.1 Coconut Crop

One of the most exquisite trees in the world, coconut palms are cultivated in over 90 tropical nations. In addition to providing food, water, and shelter, it also serves as a source of raw materials for numerous businesses that are closely related to both home and commercial life. This wonder palm's entire plant is beneficial to humanity in some way. Because of this, the palm has come to be known as Kalpvriksh (Tree of Heaven). Coconuts are primarily farmed along India's coastlines. Major coconut-growing states include Kerala, Tamil Nadu, Karnataka, and Andhra Pradesh, as well as Odisha, Maharashtra, Goa, Assam, Puducherry, Lakshadweep, and the Andaman & Nicobar Islands. Additionally, the North Eastern area, Gujarat, Madhya Pradesh, and Bihar have also gained momentum in their cultivation.

With a well-developed protrusable stylet utilized to pierce the plant tissues for food intake and toxin release, plant nematodes are obligate root parasites. The burrowing nematode (*Radopholus*

DOI: 10.1201/9781032631578-22

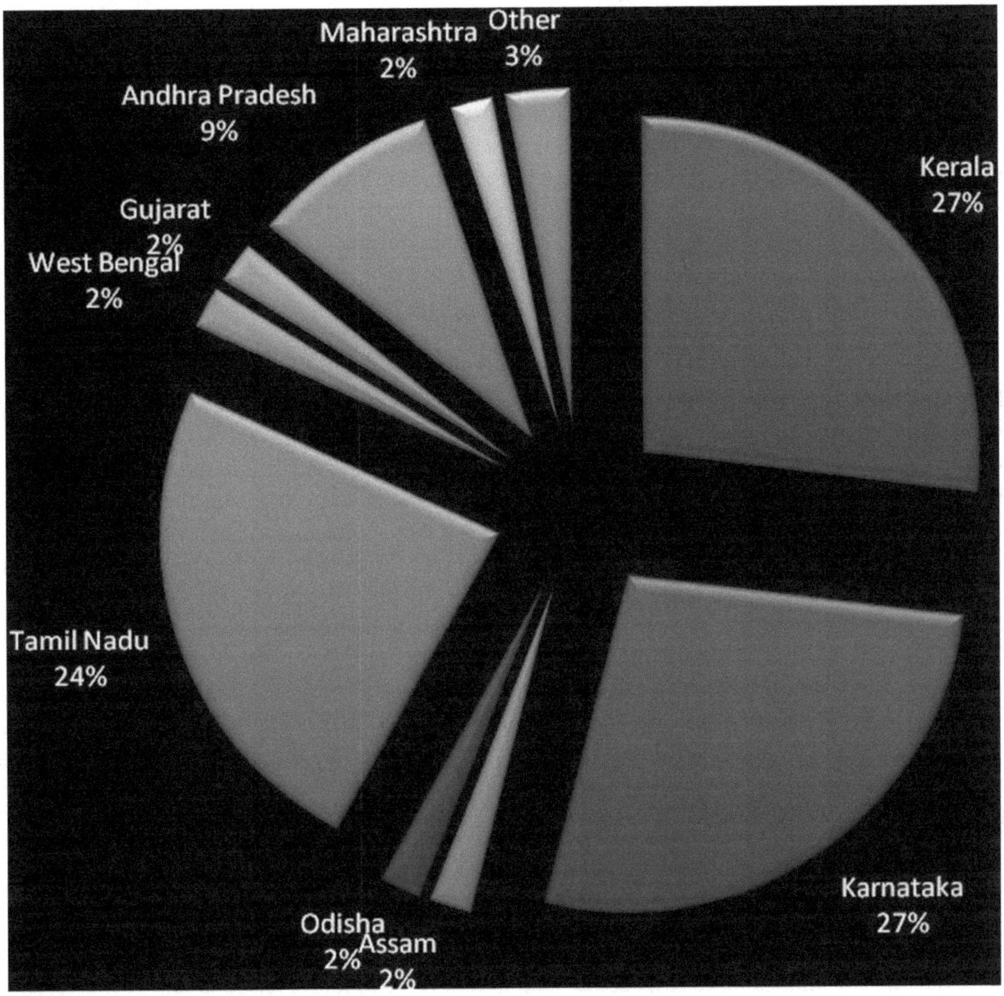

FIGURE 22.1 Plantation Crops Producing Major States. (www.kerenvis.nic.in/Database/Crops_1732.aspx)

similis), lesion nematode (*Pratylenchus coffeae*), and root-knot nematode (*Meloidogyne* spp.) infests intercrops cultivated in interspaces of coconut all over the world, and red ring nematode (*Bursaphelenchus cocophilus*) causes red ring disease. Nematode-infested palms are more vulnerable to bacterial and fungal diseases, which makes the disease situation worse. Higher nematode populations during the seedling stage result in severe root lesions and seedling death. *Scuttellonem brachyurum*, *Rotylenchulus reniformis*, and *Helicotylenchus elegans* were also discovered in coconut plantations in West Bengal (Rama and Dasgupta 2000), as well as *Aphelenchus* sp., *Aphelenchoides* sp., *Atylenchus* sp., *Caloosia* sp., *Criconemoides* sp., *Dolichodor* (Koshy et al. 2002; Rajkumar et al. 2019, 2020).

i. *Radopholus similis*

Radopholus similis is one of the most significant root pathogens in plantation crops, causing more damage to coconut seedlings than to mature palms. A widespread and extensive occurrence that results in 30% yield losses in the coconut cropping system has been documented in Kerala, Tamil Nadu, and Karnataka.

The nematode species known as the burrowing nematode is *Radopholus similis* (Cobb 1893) Thorne 1949. It is a plant parasite that causes damage to a variety of agricultural crops. It is a particularly significant coconut pest. Due to their migratory nature, these endoparasites have the ability to generate several localized necrotic patches across an infected root system. While secondary symptoms including wilting, weakening structure, and stunting may result from the root damage these parasites cause, *R. similis* signs are typically limited to the roots. Both adults and juveniles have a vermiform body form. Adults exhibit sexual dimorphism. The male features a sharp, curving spicule encased in a sac, a knob-like head, and a poorly developed stylet. The male is 500 to 600 μm in length, while the female is about 550 to 880 μm long. The female is around 550 to 880 μm long, whereas the male is 500 to 600 μm long. The female's stylet is fully formed. The long, tapering tails of both males and females have rounded or indented ends. Since *R. similis* burrows in the roots of its host plant, it is a burrowing nematode.

Parasites such as these are endoparasites, a term that describes how they get nutrition. Rather than residing outside the plant and obtaining nutrients through other means, they sit inside the plant and drain nutrients from the cytoplasm of the nearby cells. Additionally, they are migratory endoparasites, which means they can travel throughout the host by entering the roots. Since they finish laying their eggs inside their hosts, only females can infect roots. Their ability to reproduce asexually or sexually allows them to exist in three different forms: female, hermaphrodite, and male. Because stylet allows an organism to spread throughout its host and infect other hosts when its present host runs out, individuals at every stage of the life cycle possess stylet.

They frequently live in the parenchyma, where the females deposit three to five eggs every day in the tissues they occupy. Typically, eggs hatch in 5–10 days, mature into adults in 10–13 days, and become gravid in 2 days after they are laid. This all adds up to a life cycle of 20–25 days from egg to gravid adult. At 25 °C, the nematode takes around 21 days to complete its life cycle. Juveniles and females ingest inside roots, particularly in the vicinity of the tips. Males do not feed because of their feeble stylets. Every day, females lay two to six eggs.

ii. *Pratylenchus coffeae*

A genus of parasites usually referred to as lesion nematodes is *Pratylenchus coffeae* (Goodey 1951). Their roots become injured because they are parasitic on plants. Instead of staying in one location like a cyst or root-knot nematodes, lesion nematodes are migratory endoparasites that feed and reproduce inside the root and migrate around. In most cases, they merely attack the root cortex. Secondary infections can also result from the lesions serving as entry points for harmful fungi and bacteria. The plant droops, turns chlorotic, becomes stunted, and frequently dies above ground. When plants wilt and die, a crop field could appear irregular. There can be large yield losses due to root lesion nematodes.

The stylet morphology is an essential characteristic that sets *P. coffeae* apart. The lip area is low, anteriorly flattened, and either not at all or very slightly offset from the contour of the body. Esophageal glands moderately overlap the ventral side of the intestine. Poor development of the esophageal-intestinal valve. Phasmids are situated in the middle or slightly behind the tail. The posterior branch of the female genital tract is reduced to a post-uterine sac, making it monovarial. The female is thin. The tail is round at the end, rarely pointed, and two to three times the diameter of the anal body.

Temperature, aeration, pH, organic matter, mineral content, and soil moisture all have an impact on *P. coffeae* abundance. They are more prevalent in areas with weed infestations and sandy soils. *P. coffeae* is tolerant of a broad variety of environmental circumstances. For breeding and underground migration, moist temperate soils are ideal, but they can also survive in warm, dry conditions. They remain dormant in extremely dry conditions until moisture levels rise and plant growth resumes. When soil temperatures fall below 15 °C, the majority of lesion nematodes remain dormant, and they don't become fully active until they reach 20 °C. Single eggs are laid by female nematodes in the soil or in the roots. By parthenogenesis, they create viable eggs without the need for fertilization. There are not many males in many species.

iii. *Bursaphelenchus cocophilus*

The coconut red ring disease is brought on by the nematode *Bursaphelenchus cocophilus*. *Rhadinaphelenchus cocophilus* is an alternate scientific name that has been used to identify it in literature. Its distinctive symptom gave rise to the common name "red ring nematode". *B. cocophilus* is one of the nematode plant diseases that are most frequently banned and/or inspected worldwide.

This nematode is distinguished by its well-developed metacorpus from J_2 to adult, as well as its short stylet, which measures 11–15 µm in adults and is usually 1 mm in length. The vulva, which is located two-thirds of the length of the body in females, has a vulval flap. Females have a round tail and a long post-uterine sac. Males have a spade-shaped bursa, unique spicules, and seven papillae in the tail region.

The red ring nematode undergoes four molts before reaching adulthood, as is typical for plant parasites. The entire life cycle takes about ten days. The dauer stage is the transmissible/survival stage. J_3 is the survival stage. This nematode's relationship with its vector determines how widely it spreads. This nematode causes chlorosis, which starts in the oldest leaves, and leaves a noticeable reddish-brownish ring around the tree trunk.

iv. Meloidogyne sp.

Root-knot nematodes, *Meloidogyne* sp. (Goldi 1889) is a combination of two Greek words that signify "female" and "apple-shaped". The female has two prodelphic ovaries and is swollen in adulthood. Six rectal glands secrete a matrix that contains the eggs, which are not retained in the female body. The excretory pore is typically found in front of the metacorpus. The female body doesn't develop cysts. The female's posterior cuticle striations create a perineal pattern resembling a fingerprint. The esophageal gland overlaps over the intestine. The male has one testis, though occasionally two. Males have a characteristic half-twist in their bodies and lack caudal alae.

In order to finish their life cycle, endoparasites such as root-knot nematodes must be introduced to a permanent feeding site. *M. incognita* juveniles, which are in their second stage, enter the vascular cylinder by penetrating the roots and moving through the intercellular space to create feeding sites. As the feeding site matures, the J_2 molt into third- and fourth-stage juveniles (J_3) and then become sedentary (J_4); then J4 develops to the egg-laying stage when they molt into females and begin producing eggs, which are deposited in a gelatinous matrix under favorable conditions and with enough nutrition. During this phase, the globose-shaped females have started laying eggs into a gelatinous matrix that is affixed to their posterior end. When susceptible plants become infected with RKN, their root systems typically develop galls or knots as a response. The ability of the nematode to reproduce in a particular host plant and the plant's capacity to support nematode reproduction define the parasitic relationship between nematodes and plants. One of the main factors determining a pathogen's pathogenicity and a key factor in the development of disease is its reproductive fitness.

22.3 NEMATODE-INFESTED PLANT SYMPTOMS

Coconut palms with burrowing nematodes present general decline symptoms like stunting, yellowing, decreased leaf quantity and size, delayed blooming, button shedding, and decreased production. Root symptoms are quite specific. On delicate, creamy white roots, *R. similis* infestation results in small, elongated, orange-colored lesions. These lesions expand and consolidate as a result of nematode parasitization and multiplication, which results in severe root rotting. On the epidermis of the semi-hard, orange-colored main roots, cracks appear as lesions merge. On the secondary and tertiary roots, lesions and rotting are only visible in the delicate region of the roots. On the old, hard, dark brown roots, lesions are rarely visible. Coconut seedlings with severe infestations develop soft, spongy roots. The plumule, leaf bases, and haustoria of seedlings are also attacked by the nematode. Plant growth is restricted by nematode parasitism which causes a sharp decrease in the quantity and mass of tertiary feeder roots (Koshy and Sosamma 1987). Since secondary and tertiary roots are small and decay quickly when infected, lesions are also difficult to see on them. Early in the infestation process, roots form distinct cavities that later converge as a result of nematode feeding and reproduction. The cavities could contain eggs and all nematode immature stages in longitudinal sections (Koshy and Sosamma 1987; Sosamma and Koshy 1998).

The red ring disease of coconut is caused by *R. cocophilus* and transmitted by the black palm weevil. This disease was initially reported from Trinidad (Hart 1905). Later, Grenada received news of it (Nowell 1918). Cobb (1919) first referred to *R. cocophilus* as *Aphlenchus cocophilus*, and Goodey (1960) later changed the name to *R. cocophilus*. It has been reported from the West Indies (Trinidad, Tobago, Grenada, and St. Vincent) and Latin America (Venezuela, Guyana, Surinam, French Guyana, Colombia, Ecuador, Peru, Mexico, Brazil, Panama, Nicaragua, Costa Rica, Honduras, Belize, and El Salvador), but it is not known to occur in India or South-East Asia. In Surinam, a small leaf disease of the coconut and oil palm is also linked to this worm. Up to 60% of crop loss was seen in coconut gardens that were severely damaged.

The emergence of internal lesions is the most significant defining symptom. Lesions appear as orange to brick-red colored rings in a cross-section of the stem (2–4 cm wide). Red ring disease is easily fatal to young palms (2–10 years old). The oldest leaves are first affected by chlorosis, which then spreads to their bases. The dark lower leaves may split across the petiole, or they may get partially detached and dangle down from the base. Nuts prematurely shed. In the first 4–6 weeks after the onset of symptoms, the crown frequently collapses. However, the trunk stays in the field for a few months before it starts to rot.

Management

Numerous management techniques for nematodes that parasitize plants have been reported (Rajkumar et al. 2016, 2019, 2021). The application of oilcake, the use of farmyard manure, and the growth of sun hemp in basins and interspaces and their integration as green manure are examples of cultural practices that exist in Kerala and Karnataka, India, and may lessen nematode multiplication. Additionally, cultivating plants like cocoa that enrich the soil with a significant amount of shed leaf aids in the development of advantageous organisms and antagonistic microorganisms that may prevent nematode expansion.

Intercrops vulnerable to root-knot and burrowing nematodes, such as ginger, turmeric, papaya, solanaceous vegetables, and elephant foot yam, may be avoided, and if planted, planting locations may be changed annually. Combinations of crops that are sensitive to the same kind of nematode should be avoided at all costs. Opening pits during dry weather to expose immature stages to sunlight and burning rubbish in planting pits are examples of sanitation techniques.

The sole known reservoir of inoculums was the diseased tree in which the vector palm weevils originated because the nematode neither multiplied in the insect nor survived for any meaningful length of time in the dead tree. As soon as red ring symptoms showed, the sick tree was destroyed by fire or poisoning, which dramatically reduced the amount of the pathogen that was present and also managed the vector population. When banana was utilized as a shade crop, *R. similis* prevalence in coconut nurseries increased. Over 50% of the seedlings grown in these nurseries did not survive to establish in the main field. Nematicide therapy was shown to be quite successful. Carbosulfan* was discovered to be efficient after carbofuran* and phorate* were phased out (*non-banned area).

Infested regions should be planted with tolerant or less sensitive varieties or hybrids. In India, *R. similis* was discovered to be susceptible to the majority of coconut cultivars. The least nematode multiplication and lesion indices were observed by the dwarf cultivars Kenthali and Klappawangi. Hybrids including Java Giant x Kulasekharan Dwarf Yellow, Kulasekharam Dwarf Yellow x Java

Giant, Java Tall x Malayan Yellow Dwarf, and San Ramon x Gangabondam all displayed similar reactions (Sosamma et al. 1980, 1988; Sosamma 1984).

The nematode population was reduced in coconut seedlings cultivated in polybags filled with potting soil enhanced with bio-agents such *Paeceilomyces lilacinus*, *Pasturia penetrans*, and mycorrhizae. Multiple endophytes, including *Acaulospora bireticulata*, *Glomus fasciculatum*, *Glomus macrocarpum*, *Glomus mosseae*, *Glomus versiforme*, *Sclerocystis rubiformis*, and *Scutellospora nigra* was found effective in improving the plant growth and reducing *R. similis* infestation on coconut seedlings.

To control weevils, sick palms' leaf axils should be sprayed with 0.1 percent Lannate (Methomyl). Frequently occurring disease outbreaks are guarded against using guard baskets constructed of 2 cm mesh wire. Freshly diseased tissue is placed in these baskets, and a 0.1 percent Lannate suspension is sprayed over them. The tissues in the basket are what draw the palm weevils in.

2. Areca nut crop

In India's western ghats, east coast, and northeastern regions, areca nuts, also known as betel nuts, are a significant cash crop. The areca plant is an upright palm with a long stem that can grow to a variety of heights depending on the surrounding conditions. In India, areca nut plays a significant role in religious, social, and cultural events as well as daily life. Ayurvedic and veterinary medications both use areca nut. The Indian subcontinent and its surroundings are known for their areca nut-chewing behavior. Although areca nut manufacturing is concentrated in a few states, the commercial product is sold all across the nation. The world's biggest producer and consumer of areca nut is India. The three states that cultivate the most areca nuts are Assam, Kerala, and Karnataka. In Tamil Nadu, Meghalaya, West Bengal, Maharashtra, and Goa, it is grown on a small basis.

It has been discovered that the migratory endoparasite burrowing nematode *R. similis* damages the areca nut. On delicate, white to orange-colored roots, the nematode was observed to generate characteristic root lesions and severe root rotting. Banana, cardamom, and pepper are the intercrops that have been discovered to be nematode-infested. The plant nematode infestation rises with the intercropping of these plants in areca nut gardens. Nematode infestation has not been found to significantly affect the economy of areca nut production, in contrast to other pests and insects.

Management

Nematode populations are kept in check by integrated nematode management techniques including the use of resistant planting materials, the manufacture of nematode-free seedlings in nurseries, and organic amendments such as oil cake/neem (*Azadirachta indica*) cake @ 2 kg/palm. To reduce the population of *R. similis*, soil treatment with carbosulfan 25 EC @ 0.025% (2 ml / 2 liters of water/ palm was found to be efficient.

3. Cocoa crop (*Theobroma cacao* L.)

One of the most important cash crops in the tropics, cocoa, is a native of the South American Amazon region and is grown in 60% of West Africa. Although numerous plant nematodes have been associated with the cocoa rhizosphere, only *M. incognita* pathogenicity is known with certainty. Despite being brought to India in the first part of the 20th century, commercial cultivation of cocoa didn't begin until 1970. In Kerala and Karnataka, cocoa is produced alongside coconut and areca nut trees as companion crops. There are also a few areas in Tamil Nadu and Andhra Pradesh where cocoa is farmed.

As a result of its extensive attack on the rootlets of new cocoa plantations all over the world, the root-knot nematode *Meloidogyne* sp. has a prominent spear with strong basal knobs causing rapid drying of the aerial portions and premature mortality of the tree. Morte subsite, sometimes known as "unexpected death", is a perplexing infection that sporadically affects most of Brazil's cocoa-growing regions. It causes yield loss, the sudden death of trees, and growth retardation. *Pratylenchus* spp., *Helicotylenchus* spp., *Scutellonema* spp., and *Xiphinema* spp. are some additional species that have been identified. The root-knot nematode on cocoa, a dominant species in India, was first observed by Rotzema Bos. Other nematodes include *Hoplolaimus* sp. and *Helicotylenchus* sp. Studies on the effects of *M. incognita* females with exposed egg masses on the root surface found that cocoa seedlings had significant root galling. On the other hand, the *R. similis* banana race found in coconuts was found to penetrate cocoa roots sparingly but not to reproduce.

Management

In the nursery, nematodes can stunt seedling growth or even kill them. Nematode-infested seedlings that have been transplanted may succumb to the nematodes on the plantations. In the nursery, protection is easily accomplished by growing seedlings in polythene bags with soil that has been fumigated with methyl bromide @ 500 g/100 cubic feet for 24–48 hours. It has been discovered that plants like Azadirachta and marigold, used as antagonistic crops, are beneficial in controlling plant nematodes in cocoa plantations. Herbal remedies are inexpensive, easily accessible, and safe for the environment.

4. Cashew crop (*Anacardium occidentale* L.)

One of the most delectable, beneficial, and significant nut crops in the world is the cashew. It is one of the most significant export crops for India in terms of earning foreign currency. Up to an elevation of 1500 MSL, it flourishes in regions with 1,000 mm to 4,000 mm of annual rainfall (Singh 1992). Kerala and mostly the coastal regions of Andhra Pradesh, Tamil Nadu, Karnataka, Orissa, West Bengal, Goa, Damon, and Diu are among the Indian states that produce cashews. The cashew has taken on a significant role in the Indian economy. The peninsular region of India is the only place where cashews are grown. Along the west coast are Karnataka, Kerala, and Maharashtra, while on the east coast are Odisha, Andhra Pradesh, and Tamil Nadu. Goa, the Andaman and Nicobar Islands, Madhya Pradesh, Manipur, Meghalaya, Assam, and Tripura also grow it to a lesser level. Due to the discovery of superior clones, standardization of vegetative propagation methods, and nearly self-sufficiency in high-quality planting material, the area, production, and productivity of cashews have been rising.

Brazilian untidy trees were found to have the plant nematodes *Criconemoides* sp., *Xiphinema* sp., and *Scutellonema* sp. in their rhizosphere (El-Borai and Duncan 2005). Additionally, the "Xiphinematose" disease, which is frequent in cashews in northeastern Brazil, is caused by the *Xiphinema index*. Other plant nematodes connected to cashews include *R. reniformis* in Costa Rica and *X. ifaculum* (Lamberti et al. 1992) in Libeia (Lopez and Azofeifa 1985). *Hemicriconemoides mangifera* (Siddiqi 1961), *Tylenchorhynchus mashhoodi* (Siddiqi and Basir 1959), *Caloosia longicaudata* (Loos 1948); (Siddiqi and Goodey 1963), and *Hemicycliophora attapadii* have all been recorded from Orissa (Rahaman et al. 1996).

Management

Before planting, all field-raised seedlings should have their external roots cut off, and nematode-infested intercrops should be eliminated. Plant nematodes cannot be easily controlled, and there are currently no proven methods to eradicate the nematodes that infest living palm trees. Control relies more on prevention than treatment, particularly when it comes to destroying infected palm material and catching and removing weevil vectors before they propagate the nematodes.

22.4 FUTURE THRUSTS

Evaluation of improved cropping techniques in coconut for the local control of *R. similis* and *R. cocophilus* constructing integrated management plans for the complex of nematode and fungal diseases that cause wilt. Finding genotypes with high levels of red ring and burrowing nematode resistance. Maximizing output with minimal effort and effective plant nematode management. Determine nematode hotspots using surveys in important plantation crop growing regions, and then make worm control knowledge available to farmers. For polyspecific nematode environments, an integrated nematode management strategy is being developed and to investigate and monitor the movement of important nematode pests.

REFERENCES

Cobb NA (1919) A newly discovered nematode *Aphelenchus cocophilus* nov. sp., connected with a serious disease of the coconut palm. *Wolndian Bull* 17: 203–210.

El-Borai FE, Duncan LW (2005) Nematode parasites of subtropical and tropical fruit trees. In: Luc M, Sikora RA, Bridge J (eds.), *Plant Parasitic Nematodes in Subtropical and Tropical Agriculture - Second Edition*, CABI Publishing, Wallingford, UK, pp. 467–492.

Goodey JN (I960) *Rhadinaphlenchus cocophilus* (Cobb 1919) nov. comb, the nematode associated with 'red ring' disease of coconuts. *Nematologica* 5: 98–102.

Hart JH (1905) Coconut diseases. *Bull Misc Information Roy Bot Eds Trinidad* 6: 241–243.

Koshy PK, Sosamma VK (1987) Pathogenicity of *Radopholus similis* on coconut (*Cocos nucifera* L.) seedlings under green house and field conditions. *Indian J Nematol* 17: 108–118.

Koshy PK, Sosamma VK, Banu JG (2002, February 21–22) *Milestones of Nematological Research in Kerala*, CPCRI, Kayangulam, Kerala, India, pp. 21–23.

Lamberti F, Boiboi JB, Ciancio A, Tuopay DK, Jimenez EA, Elia F (1992) Plant parasitic nematodes associated with tree crops in Liberia. *Nematol Mediterranea* 20: 79–85.

Lopez R, Azofeifa J (1985) Nematodos fitoparasitos asociados con frutales en algunos cantones de la provincia de Alajuela. *Agronomia Costaricense* 9: 193–196.

Nowell W (1918) Root disease of coconut palm in Grenada. *Agric News Bridgetown* 17(4): 398–399.

Rahaman PF, Ahmad I, Jairajpuri MS (1996) Three new species of Hemicycliophora de Man, 1921 from India. *Nematologica* 42: 24–34

Rajkumar, Maheswarappa HP, Subramanian P, Surekha R (2016, July 22–23) Management of plant parasitic nematodes in coconut based cropping system. In: Chowdappa, Hebbar and Samsudeen (eds.), *National Seminar on Plantation Based Cropping System for Improving Livelihood Security*, CPCRI, Kasaragod.

Rajkumar, Pervez R, Maheshwarappa HP, Santosh (2020) Occurrence of plant parasitic nematodes associated with Noni (*Morinda citrifolia* L.) intercrop with coconut. *Ann Plant Protect Sci* 28(1): 101–103.

Rajkumar, Pervez R, Surekha, Ravi B (2019) Weed as alternate host of root knot nematode, *Meloidogyne incognita* in coconut garden. *Ann Plant Protect Sci* 27(2): 285–288.

Rajkumar, Surekha R, Sujithra M, Pratibha VH (2021) Management of root-knot nematode infestation in coconut cropping system. *Indian Coconut J* 64 (2): 33–34.

Rama K, Dasgupta MK (2000) Population ecology and community structure of plant parasitic nematodes associated with coconut and arecanut in northern west Bengal. *Indian J Nematol* 30: 175–182.

Singh R (1992) *Fruits*. National Book Trust, India, p. 209.

Sosamma VK (1984) Studies on the burrowing nematode of coconut. *PhD thesis*, Kerala University, Trivandrum, Kerala, India.

Sosamma VK, Koshy PK (1998) Etiology – nematodes. In: Nampoothiri KUK, Koshy PK (eds.), *Coconut Root (Wilt) Disease*-Revised Edition, CPCRI, Kasaragod, pp. 21–27.

Sosamma VK, Koshy PK, Bhaskara Rao EVV (1980) Susceptibility of coconut cultivars and hybrids to *Radopholus similis* in field. *Indian J Nematol* 10: 250–252.

Sosamma VK, Koshy PK, Bhaskara Rao EVV (1988) Response of coconut cultivars to the burrowing nematode, *Radopholus similis*. *Indian J Nematol* 18: 136–137.

23 Innovative Plant Nematodes Management Strategies in Major Fruit Crops

Rashid Pervez and B.S. Chandrawat

23.1 INTRODUCTION

After China, India is the second-largest global producer of fruits like mango, banana, sapota, papaya, and acid lime. The most significant and priceless commodities that nature has given to humanity are fruits. In addition to being high in calcium, iron, and vitamins A and C, they also have a low calorie count and a low fat content. Plants that produce such crucial fruit crops are susceptible to diseases, parasitic nematodes, and insect pests that lower fruit quality and yield (Sikora et al. 2018; Chandrawat et al. 2022). Plant nematodes are the disease-causing organisms that harm horticulture crops the most (Lopes-Caitar et al. 2019).

Fruit crops encourage nematode population growth due to their perennial nature. The nematodes' damage to the roots causes them to use soil moisture and nutrients less effectively, which reduces their functional metabolism. In addition, numerous fungi and bacteria find it easy to feed on the weakened and injured roots, speeding up root deterioration. Microscopically small roundworms known as plant nematodes reduce crop yields and threaten global food security. Nematodes can be classified into two groups based on their feeding habits: above-ground parasites (which feed on roots), and below-ground parasites (which feed on foliage). Three types of nematodes attack plant roots: ectoparasitic, endoparasitic, and semi-endoparasitic. The anterior region of the nematode is inserted into plants while the posterior region remains outside the roots known as the semi-endoparasitic nematode. Ectoparasitic nematodes feed on plant tissues from the outside of the plant, while endoparasitic nematodes feed within the plant, further can also be divided into migratory and sedentary types (Pervez 2008). The nematode is considered migratory if it freely moves across the soil or plant tissues. Sedentary refers to an adult nematode that stays stationary in a certain region of the root.

The symptoms of plant nematodes feeding on below-ground plant parts, such as roots, rhizomes, tubers, and bulbs, and those that emerge on above-ground plant parts are frequently mistaken for those caused by abiotic stress, such as nutrition and water shortages. Some plant nematodes exhibit a high degree of polyphony (Leela et al. 2012; Pervez et al. 2012). Nowadays fruit crop production is declining due to several reasons; among them, plant nematode is one of the main factors. The major plant nematodes *Radopholus similis* cause toppling disease in bananas, *Tylenchulus semipenetrans* cause citrus/slow decline in citrus and *Meloidogyne* spp. causing root-knot disease in guava and pomegranate, and *Pratylenchus coffeae* causes economic losses in coffee (Table 23.1). Hence focus on the major plant nematodes infesting fruit crops and their innovative management strategies in this chapter.

DOI: 10.1201/9781032631578-23

TABLE 23.1
Plant nematodes associated with fruit crops

Crop	Common Name	Scientific Name
Apple	Apple cyst nematode	*Globodera mali*
	Apple root-knot nematode	*Meloidogyne mali*
	Apple lesion nematode	*Pratylenchus vulnus*
Banana	Burrowing / Banana-root nematode	*Radopholus similis*
	Banana cyst nematode	*Heterodera oryzicola*
	Banana nematode	*Pratylenchus musicola*
	Banana root-lesion / meadow nematode	*Pratylenchus coffeae*
	Banana spiral nematode	*Helicotylenchus multicinctus*
	Blueberry root-knot nematode	*Meloidogyne carolinensis*
Citrus	Citrus nematode	*Tylenchulus semipenetrans*
	Citrus ring nematode	*Criconemoides citri*
	Citrus root nematode	*Tylenchulus semipenetrans*
	Citrus root-knot nematode	*Meloidogyne indica*
	Citrus spine nematode	*Criconema civellae*
Fig	Fig cyst nematode	*Heterodera fici*
	Fig pin nematode	*Paratylenchus hamatus*
	Fig spine nematode	*Criconema decalineatum*
Grapevine	Reniform nematode	*Rotylenchulus reniformis*
	Grapevine root-knot nematode	*Meloidogyne incognita*
	Grapevine Dagger nematode	*Xiphinema index*
Guava	Root-knot nematode	*M. enterolobii, M. incognita* and *M. javanica*
Pomegranate	Root-knot nematode	*Meloidogyne* spp. *Meloidogyne incognita*
Papaya	Reniform nematode	*Rotylenchulus reniformis*
	Root-knot nematode	*Meloidogyne incognita*
Passion fruit	Reniform nematode	*Rotylenchulus reniformis*
	Root-knot nematode	*Meloidogyne incognita*
Peach and Plum		*Macroposthonia xenoplax*
Pineapple	Reniform nematode	*Rotylenchulus reniformis*
	Root-knot nematode	*Meloidogyne incognita*
Strawberry	Strawberry bud nematode	*Aphelenchoides besseyi* *Aphelenchoides fragariae*
	Strawberry lesion nematode	*Pratylenchus penetrans*
	Sheath nematode	*Hemicycliophoroidea arenaria*

Source: Chandrawat et al. (2022).

23.2 ECONOMIC LOSSES

The root-knot nematode (*Meloidogyne* spp.) tops the list of all of these in terms of host range, distribution, host damage, and disease complexes with a destructive nature. In some kinds of fruit crops, the nematodes may act as a breakdown of resistance to bacterial and fungal diseases. Some nematodes are also thought to behave as bacterial and viral vectors. Nematodes are therefore also accountable for indirect damage to fruit crops by escalating the interaction of disease complexes, in addition to direct damage. To create disease complexes, plant nematodes frequently collaborate with bacterial, viral, and fungal pathogens. Because of the vague symptoms that are brought on by nematodes that parasitize plants, the qualitative and quantitative losses in fruit harvests are

TABLE 23.2
Estimated losses by plant nematodes to major fruit crops in India

Crops	Plant nematodes	Yield loss (%)	Monetary loss (Rs. in million)
Banana	*M. incognita*	15	9710.46
Citrus	*T. semipenetrans*	27	9828.22
Grapes	*M. incognita*	30	3940.44
Papaya	*M. incognita* and *R. reniformis*	30	2516.64
Guava	*Meloidogyne* spp.	28	2350.88
Pomegranate	*Meloidogyne* spp.	23	3023.44

Source: Kumar et al. (2020).

frequently underestimated. Studies by several scientists demonstrate that plant nematodes in India cause annual productivity losses in important fruit crops (Table 23.2).

23.3 MAJOR PLANT NEMATODES

23.3.1 Lesion Nematode (*Pratylenchus* spp.)

Lesion nematodes *Pratylenchus* spp., are well acknowledged as a significant pest of valuable economic fruit crops, such as bananas, coffee, peanuts, and many other fruits. Due to their extensive host range and widespread in almost all tropical and subtropical countries, they play an important economic role in agriculture (Castillo and Vovlas 2007). This nematode has infectious J_2, J_3, J_4, and adult life stages and migratory endoparasites. Due to their unrestricted mobility through the root and eating in the plant cortex, *Pratylenchus* sp. attacks host tissues, resulting in dark patches or lesions. A significant issue in replanting an orchard arises from affected trees losing all feeder roots. Particularly in peach and cherry trees, the diseased tree frequently displays stunting, chlorosis, and twig dieback along with a reduction in vigor. This nematode is isolated from roots, plant tissues like tuber, rhizome, and corm, and, also from the soil. The nematode enters the plant root's tissues, migrates, and feeds there before leaving behind noticeable dark brown or black lesions on the root surface. However, most nematodes are found in tuber tissues, and sampling these tissues is the most appropriate means of assessing populations. Infected roots are chopped finely, teased apart, or, preferably, macerated before placing on a support tissue or sieve in water. Between 30 and 50% of *Pratylenchus* spp. emerges from roots and plant tissues in the first 3 days but they continue migrating for over 20 days.

23.3.2 Symptoms

Infected plants don't experience the same seasonal growth as a healthy plant would. The leaves of the affected plant could seem chlorotic or lighter green. Leaf size and leaf yield are reduced in plants that are extremely damaged. Fruit yields may be drastically lowered while having substantially lower quality or quantity. Extremely high nematode populations in a field may work in conjunction with other soil-borne calamities to kill originally sensitive plants, resulting in barren patches. The impacted areas may expand as the season progresses. Plants that are only marginally damaged commonly appear to outgrow initial harm because the root systems can reach deep into the earth. However, unsuitable growing circumstances can still cause damage, as seen by a little reduction in output in plants that are only weakly diseased. The number of infected plants gradually decreased as the nematodes multiplied over time. The plant may be more vulnerable to winter damage or other infectious diseases since the host's strength is reduced. Without treating the soil, the injured plant is replaced, and the new plant typically grows poorly and occasionally dies.

The earliest signs of infection on infected roots are tiny, light- to dark-brown lesions. These lesions usually spread and converge as the growing season progresses, giving the roots a general discolored appearance. Plants with severe infection may exhibit complete root damage, and such plants are simple to pull out of the ground. Plants with moderate disease have brownish, stubby roots. The majority of the remaining roots will be found near the soil's surface. The wounds/lesions left behind by the nematodes could serve as entry points for a wide variety of generally unspecialized soil microorganisms, including diseases that infect roots and may also contribute to lesion production and worsen the condition.

23.3.3 Management

Planting nematode-free suckers/planting materials is advised for fruit crops. To harvest new planting material, mostly sword suckers, some growers or groups maintain "disease-free" nurseries. It is more typical to use planting material from already-existing banana fields, which are more likely to have nematode infestations. Commercial practices have involved submerging banana suckers in water maintained at a constant 55°C for intervals of 15 to 25 minutes. Crop rotation with cotton, paddy, sugarcane, green gram, hemp, or sun hemp aids in nematode population reduction. The nematode population is drastically reduced when antagonistic plants like *Tagetes erecta*, *Tagetes patula*, and *Crotalaria juncea* are grown as intercrops in fields. Neem cake, applied @ 250 g per plant at planting and again after four months, aids in nematode population management. Planting materials that are nematode-free can be produced by cutting the corm surface of necrotic lesions and soaking it in hot water at 50–55°C for 30 minutes. This nematode control is advised using biological agents such *Trichoderma*, VA mycorrhiza, *Glomus fasciculatum*, and *Pasteuria penetrans*. It is advised to apply carbofuran* @ 20 g per plant during planting and twice @ 20 g per plant every three months recommended for the effective control of plant nematodes. Apply 10 g of carbofuran* at planting and @ 20 g per plant at the third and fifth months of tissue culture plant implantation (sources: https://nrcb.icar.gov.in/oldwebsite/documents/NRCB%20Publications/20%20Years%20of%20NRCB.pdf) (* non-banned area).

23.3.4 Citrus Nematode (*Tylenchulus semipenetrans*)

In California, the citrus nematode was first identified as a citrus disease (Thomas 1913). *Tylenchulus semipenetrans*, a new species reported by Nathan Cobb later that year, was later determined to be the nematode responsible for the steady decline disease in citrus. Since its discovery, *T. semipenetrans* has been identified throughout the world's citrus-growing regions (Duncan 2005). According to research by Duncan (2009), field infestations affect 50–90% of citrus orchards in Arizona, California, Florida, and Texas as well as a few isolated vineyards in California. Citrus orchards with fine-textured soils or sandy soils with high organic matter contents typically have higher populations (Timmer et al. 2003).

A dimorphic species, *T. semipenetrans*, exhibits sexual dimorphism both as juveniles and as adults. The life cycle of *T. semipenetrans* begins as an egg, which includes the first stage juvenile, and is characteristic of plant nematodes (J_1). The second-stage juvenile (J_2) that hatches and looks for host roots is the result of this stage molting inside the egg. The vermiform J_3 and J_4 molt from the vermiform J_2 female to become the sedentary adult in the final. Female juveniles must eat the epidermis and outermost layers of the cortical parenchyma of the roots to mature into adults. Without typically getting to the central cylinder, the immature female starts to pierce the root's outer surface and enter its deep cortical layers (or exceptionally the endodermis). It settles down and creates a long-term feeding site with cells (nurse cells) that serve as the primary source of nourishment. As it reaches maturity, its elongated neck and head stay lodged in the cortex while the posterior portion of its body swells and protrudes from the root surface. Eggs produced by mature females are encased in a gelatinous matrix that is released through the excretory hole. From egg to egg, the female life cycle lasts between four and eight weeks. *T. semipenetrans*, unlike many nematodes, has a limited host range that consists of citrus, trifoliate orange, grapevines, persimmon, lilac, and olive (Inserra et al. 1994).

23.3.5 Symptoms

The initial signs of the seedlings are frequently not noticeable due to the general loss in tree growth, leaf shedding and yellowing. The dieback of twigs and the tiny fruits match the signs of both drought and hunger. Soil particles frequently cling to the rootlets when a root is extensively contaminated before the cortex eventually easily separates from the vascular stele. Strongly diseased roots have deeper hues and fewer branch rootlets. Replanting weak rootstock in a heavily infested orchard without soil fumigation causes young tree roots to be extensively parasitized, slower tree growth, and decreased productivity. This problem is known as the citrus replants problem. Infected plants display symptoms of weakness, chlorosis, and leaf drop. Root damage leads to plant death over a 3–5 year period.

23.3.6 Management

In both replanting conditions and established orchards, nematicide treatments and soil conditioning were once used as *T. semipenetrans* management techniques. In the past, nematicide treatments were frequently applied to mature orchards. Although the use of chemical remedies has been severely curtailed, producers still have few choices for controlling *T. semipenetrans*. Currently, using certified propagative citrus plants free of citrus nematode and resistant (*Swingle citrumelo*) root-stocks are effective ways to stop the harm that *T. semipenetrans* does to citrus (Ling et al. 2000; Galeano et al. 2003). A great approach to sustainable agriculture that should be applied for other fruit crops prone to nematode infections is planting nematode-certified citrus and grape root-stocks. Put the paired set in a hot water bath to 55°C for bananas, 50°C for apples, 54.4°C for strawberries, 50°C for papayas, and 45°C for citrus for 25 minutes to disinfect it.

23.3.7 Burrowing Nematode (*Radopholus similis*)

Radopholus simils has been found in American Samoa, Australia, Asia, Africa, North (Hawaii, Florida), South, and Central America, the Caribbean, Europe, Oceania, Cook Islands, Fiji, French Polynesia, Guam, Federated States of Micronesia, New Caledonia, Niue, Palau, Papua New Guinea, Samoa, Solomon Islands, and Tonga. *R. similis* has a significant impact on bananas, black pepper, tea, coconut, areca nut, coffee, and cocoa. All of the states that produce bananas in India, including Tamil Nadu, Kerala, Karnataka, Goa, and Maharashtra, sell them. The burrowing nematode, an endoparasitic migratory nematode, completes its life cycle inside root tissue. Females and all juvenile stages can infect root tissue at any point along the length of the root. The egg-to-egg life cycle is completed in 20–25 days between 24–32°C. The production of bananas and plantains suffers losses of between 30 and 60% as a result of the borrowing nematode. The difference relies on the fertility of the soil, with "poor" soils, those devoid of organic material the biggest losses.

23.3.8 Symptoms

The symptoms they cause and the kind of fruit crops that the burrowing nematodes prefer vary depending on the population and location. Dwarfing, poor growth, smaller leaves, color changes, leaf chlorosis, weak pseudostems, small bunches, and early lodging were observed on the above-ground plant. Infected plants exhibit root symptoms in the form of dark red lesions on the cortical, or outside, section of the root. On severely infected roots, the strong infection results in open and depressed sores, which fracture the root. Because diseased plants with short necrotic roots are unable to handle the weight of the bunch or the stress of high winds, they are uprooted or destroyed, giving rise to the nickname "blackhead toppling disease" of bananas. The fungus *Fusarium oxysporum* f. sp. *cubense* cannot infect healthy tissue until worms have caused lesions. This nematode also aids the "Panama wilt" which is caused by the fungus. Citrus trees that are afflicted have fewer, smaller leaves and more dead twigs than healthy trees due to decreased intake of water and nutrients. Fruit production is insufficient, and seasonal growth flushes are meager.

23.3.9 Management

Neem cake, applied @ 500 g per plant at planting and again after four months, aids in nematode population control. Nematode-free planting materials can be produced by pairing the corm surface by clipping away necrotic lesions and soaking it in hot water at 50–55°C for 30 minutes.

Plant nematode control is advised using biological agents such *Trichoderma,* VA mycorrhiza, *Glomus fasciculatum,* and *Pasteuria penetrans.* For the successful control of nematodes, it is advised to use carbofuran* @ 30 g per plant and to repeat the application @ 30 g per plant if necessary, after three months of planting. Pairing is followed by a dip in a monocrotophos solution at a concentration of 0.5% (14 ml in 1 liter of water) for 30 minutes, followed by 72 hours of drying in the shade before planting. Field tolerance is demonstrated by varieties including Karpooravalli, Monthan Nattupoovan, Kunnan, Pey Kunnan, Pedali Moongil, Yelakki Bale, Pidi Monthan, and Pisang Jari Buaya (sources: https://nrcb.icar.gov.in/oldwebsite/documents/NRCB%20Publications/20%20Years%20of%20NRCB.pdf) (* non-banned area).

23.3.10 Root-knot Nematode (*Meloidogyne* spp.)

In 1983, a population of root-knot nematode (RKN) *Meloidogyne enterolobii* was spotted in China on the pacara earpod tree (*Enterolobium contortisiliquum* (Vell.) Morong). Florida was the first location in the continental USA to report it in 2001. Because it can reproduce on bell peppers that are resistant to RKN (genotypes that have the Mi-1 gene), among other commercially significant crops, *M. enterolobii* is now regarded as one of the most significant RKN species (Kiewnick 2009). A tropical or subtropical species known as *M. enterolobii* has been found in Brazil, Venezuela, China, Cuba, France, Guatemala, Puerto Rico, Martinique, Malawi, Senegal, South Africa, Switzerland, Trinidad and Tobago, United States, West Africa (Ivory Coast and Burkina Faso) and India.

After the fourth year of farming, the nematode *M. enterolobii* severely decreases crop output by weakening the plants, rendering planting impractical. Pereira (2009) investigated how *M. enterolobii* parasitism affected guava crops in five Brazilian states, including Ceará, and calculated the direct damage caused by this nematode to be $ 112.7 million. They also estimated that thousands of workers would lose their jobs as a result of the decline in guava production five years after the establishment of orchards. There have been reports of *M. enterolobii* (or *M. mayaguensis*) from guava in numerous countries (Singh 2020). It was initially found in the aubergine of Puerto Rico, and then it was found in Africa, Europe, the United States, Vietnam, China, and Thailand. *M. enterolobii* was the first to be reported in India by Poornima et al. (2016). The nematode was found in Tamil Nadu's suddenly falling guava (*Psidium guajava*) orchards.

Wherever bananas and planta are produced, RKN has been found to harm fruit crops, as they do many other crops in tropical regions. *M. incognita* and *M. javanica* are the most prevalent and widely distributed of the six identified species in the areas. The states of Kerala, Karnataka, Madhya Pradesh, Andhra Pradesh, Tamil Nadu, Gujarat, Bihar, and Assam are reported to have the widespread distribution of *M. incognita*, but *M. javanica* is restricted primarily to mid hills and some areas of colder plains where the dessert bananas are cultivated. *M. incognita* and *M. javanica* are the two most significant RKN species in the world (Singh et al. 2019). Ibrahim and Mokbel (2009) reported on northern Egypt, while Vega-Callo et al. (2020) reported on Peru. In Maharashtra, it detected a serious problem with pomegranates.

23.3.11 Symptoms

Usually, infected plants show stunted growth and discoloration. chlorotic vegetation galls develop on the subsurface parts. Sizes of root galls vary from crop to crop. The egg mass's gelatinous structure draws dirt, which causes the roots to grow galls. The roots are different sizes and appear galled. Galled roots frequently show signs of rooting. There is extensive galling throughout the root system, many of which were compound galls.

23.3.12 Management

The root-knot nematode population is drastically reduced when antagonistic plants like *Tagetes erecta, Tagetes patula*, and *Crotolaria juncea* are used as intercrop in fields. Root-knot nematodes can be eliminated from affected roots of the planting material and then submerged in hot water at 50–55°C for 30 minutes. The root-knot nematode's egg was parasitized by the biocontrol agent *Verticillium chlamydosporium*, which considerably delayed the egg's hatching.

Apply potential biological agents such as *Paecilomyces lilacinus* (*Purpureocillium lilacinum*), *Trichoderma viride*, VA mycorrhiza, *Glomus fasciculatum*, and *Pasteuria penetrans* to reduce RKN populations. It is known that applying carbofuran* @ 20 g per plant, Rugby @ 5 g per plant, or Caldan @ 5 g per plant twice, once at planting and again three months later, will greatly reduce the nematode population and boost output. Neem cake applied twice, once at planting and again after four months, @ 250 g per plant, aids in nematode population control (sources: https://nrcb.icar.gov.in/oldwebsite/documents/NRCB%20Publications/20%20Years%20of%20NRCB.pdf) (* non-banned area).

Nematicidal effects against *M. incognita* were present in aqueous leaf extracts of *Xanthium strumarium* L. (cocklebur), *Acalypha indica* (Indian Mercury), *Colocasia gigantean* (Indian taro), and *Argemone mexicana* (Mexican poppy) (Akhter and Khan 2018). According to Jindapunnapat et al. (2013), the number of nematodes in the soil and roots was reduced when *T. harzianum* was injected into the root zones of guava plants as compared to untreated plants According to research, carbofuran* 10G was the most effective method for lowering nematode numbers in guava and fig trees (Dawabah et al. 2019). The most efficient management combinations for lowering nematode densities in the rhizosphere of guava trees, after carbofuran * 10G, were *P. lilacinum* + *P. penetrans* + urea 46%, *P. lilacinum* + *P. penetrans* + chicken manure, and *T. harzianum* + *P. penetrans* + chicken manure (66.54–69.22% nematode reductions) (* non-banned area).

23.3.13 Spiral Nematode (*Helicotylenchus* spp.)

Helicotylenchus is a worldwide species that is most prevalent in tropical and warm-temperate climates. The *H. multicinctus* disease affects a variety of crops, including grapevines, bananas, citrus, cocoa, coffee, rubber, oil palm, and rubber. Ectoparasites and endoparasites are present in the root cortex. In the tropics and subtropics, all varieties of bananas are reported to be infested by the spiral nematode *H. multicinctus*. Where *R. similis* is absent, *H. multicinctus* may be the main nematode issue in subtropical areas like Israel and Taiwan. *H. multicinctus, H. dihystera, H. africanus*, and *H. erythriane* have been reported to be the most significant pests in inflicting substantial economic loss. There are seventeen *Helicotylenchus* sp. known to exist on the banana.

23.3.14 Symptoms

Infected plants persist to be stunted, mature more slowly, yield smaller bunches, and decline faster than uninfected plants. Large, conspicuous, and maybe coalescing brown lesions are present on the root's outer surface. The more severely the thin secondary and tertiary roots are attacked, where necrosis is more pronounced, the weaker the entire plant becomes. Compared to lesion or burrowing nematodes, spiral nematodes frequently feed on the roots' surface. Unlike the lesions caused by *R. similis* and *P. coffeae*, which enter the cortex, they are typically located just beneath the root's epidermis. It results in widespread root necrosis dieback, functional disruption, and final plant death. Where and when *R. similis* is not prevalent, this nematode favors colder temperatures and is most frequently found in association. It results in a major drop in banana production and is to blame for a yield loss of 34%, a fruit loss of 56%, and a 134-day flowering delay.

23.3.15 Management

Neem, Mahua, Castor, Karanji, and other oil cakes have demonstrated a specific potential for plant nematode reduction. Neem cake was applied twice, once at planting and again after four months @ 500 g/plant both times, it reduced the population of main nematodes and boosted bunch weight. The production is increased, and the plant nematode population is reduced by crop rotation with paddy, sugarcane, green gram, hemp, cotton, or intercropped with suppressed *Tagetes* species (sources: https://nrcb.icar.gov.in/oldwebsite/documents/NRCB%20Publications/20%20Years%20of%20NRCB.pdf).

23.3.16 Strawberry Nematode (*Aphelenchoides* spp.)

Significant strawberry nematodes include *Aphelenchoides* spp. (*A. fragariae, A. ritzemabosi* and *A. besseyi*), which cause the ailments dwarf crimp, spring dwarf of strawberry, and red plant of strawberry. However, it also eats strawberry leaves and flower buds. Endoparasite in strawberry leaves. Nematodes can enter leaves by the stomata or directly. Life cycles last between 10 and 13 days. Each female nematode produces about 30 eggs.

23.3.17 Symptoms

A. besseyi, which causes the summer crimp of strawberry, exhibits symptoms. The nematodes that attack strawberry plants usually congregate close to the plant's crown, where they spend a lot of time as migratory ectoparasites that live on the buds and leaflets of the growth points or in the leaf axils. The susceptibility of diseased leaves increases. If a plant's main bud is destroyed, it may not be able to produce more leaves and may regularly wither and die. Buds, leaves, and flowers are stunted and malformed. Symptoms are more obvious in the spring when the immature leaves are crimped. In extreme cases, the primary crown might be aborted, which would result in the growth of subsidiary crowns and a delay in flowering. The nematode requires cool, damp conditions to move on plant surfaces, where it feeds and reproduces. Poor yields are the result of the impact on fruit, blooming, and fruiting.

23.3.18 Management

The transportation of nematode-infected rootstock must be strictly regulated. Use nematode-free planting material or abstain from planting seedlings that have nematodes. Trees and planting materials with nematodes should be removed and destroyed. By dipping strawberry suckers in hot water at 43.3°C for 20 minutes or 46.5°C for 10 minutes before planting, healthy, vivacious strawberry plants free of nematode infestation are generated.

23.4 FUTURE THRUST

Develop more effective crop management strategies to control the infestation of fruit crops by *Pratylenchus* sp., *R. similis*, and *M. incognita*. Developing integrated management strategies to address the "blackhead toppling disease" caused by nematodes and fungi as well as the Panama wilt, progressive decline, root-knot, and guava disease. Identifying genotypes with high plant nematode resistance levels. Maximizing production with the least amount of work and efficient nematode management. Identify nematode hotspots in significant fruit-growing regions using surveys, and give farmers advice on how to control harmful nematodes. An integrated nematode management strategy is being created for polyspecific nematode habitats and to look into and keep track of the movement of significant nematode pests.

REFERENCES

Akhter G, Khan TA (2018) Evaluation of some plant extracts for nemato-toxic potential against juveniles of *Meloidogyne incognita in vitro. The J Phytopharmacol* 7(2): 141–145.

Castillo P, Vovlas N (2007) *Pratylenchus* (Nematoda: Pratylenchidae): diagnosis, biology, pathogenicity and management. In: Hunt DJ, Perry RN (eds.), *Nematology Monographs and Perspectives*, volume 6 , Brill Academic Publishers, Leiden, The Netherlands, p. 530.

Chandrawat BS, Kanwar H, Pervez R, Siddiqui AU (2022) Management of plant parasitic nematodes infesting pomegranate and guava. In: Gupta A, Verma N (eds.), *Soil Microbes and Industrial Application*, pp. 192–204. Bentham Science Publishers Pvt. Ltd., Singapore.

Dawabah AAM, Al-Yahya FA, Lafi HA (2019) Integrated management of plant-parasitic nematodes on guava and fig trees under tropical field conditions. *Egypt J Biol Pest Control* 29: 29.

Duncan LW (2005) Nematode parasites of citrus. In: Luc M, Sikora RA, Bridge J (eds.), *Plant Parasitic Nematodes in Subtropical and Tropical Agriculture*, CAB International, Wallingford, UK, pp. 437–466.

Duncan LW (2009) Managing nematodes in citrus orchards. In: Ciancio A, Mukerji KG (eds.), *Integrated Management of Fruit Crops and Forest Nematodes*, Springer Science and Business Media, New York, NY, pp. 135–173.

Galeano M, Verdejo-Lucans S, Sorribas FJ, Ornat C, Forner JB, Alcaide A (2003) New citrus selections from *Cleopatra mandarin* x *Poncirus trifoliata* with resistance to *Tylenchulus semipenetrans* Cobb. *Nematology* 5: 227–234.

Ibrahim IK, Mokbel AA (2009) Occurrence and distribution of the root-knot nematodes *Meloidogyne* spp. and their host plants in northern. *Egypt J Exp Biol (Bot)* 5: 125–129.

Inserra RN, Duncan LW, O'Bannon JH, Fuller SA (1994) Citrus nematode biotypes and resistant citrus rootstocks in Florida. *Nematology Circular No. 205*. Florida Department of Agriculture and Consumer Services. Division of Plant Industry. https://ccmedia.fdacs.gov/content/download/10992/file/nem205.pdf

Jindapunnapat K, Chinnasri B, Kwankuae S (2013) Biological control of root-knot nematodes (*Meloidogyne enterolobii*) in guava by the fungus *Trichoderma harzianum*. *J Develop Sustain Agric* 8: 110–118.

Kiewnick S, Dessimoz M, Franck L (2009) Effects of the *Mi-1* and the *N* root-knot, nematode-resistance gene on infection and reproduction of *Meloidogyne enterolobii* on tomato and pepper cultivars. *J Nematol* 41: 134–139.

Leela NK, Pervez R, Rammanna V, Rosana OB, Eapen S J (2012) Nematicidal activity of *Strychnos nuxvomica* leaf and its constituents against the root-knot nematode, *Meloidogyne incognita. Nematol Medit* 40(2): 116–123.

Ling P, Duncan LW, Deng Z, Dunn D, Hu X, Huang S, Gmitter FG, Jr (2000) Inheritance of citrus nematode resistance and its linkage with molecular markers. *Theor Appl Genet* 100: 1010–1017.

Lopes-Caitar VS, Pinheiro JB, Marcelino-Guimarães FC (2019) Nematodes in horticulture: An overview. *J Horti Sci Crop Res* 1(1): 106.

Pereira FOM (2009) Estimativa do impacto econômico e social direto de *Meloidogyne mayaguensis* na cultura da goiaba no Brasil. *Nematol Brasileira Piracicaba* 33(2): 176–218.

Pervez R, Leela NK, Eapen SJ (2012) Nematicidal activity of *Strychnos nuxvomica* leaf and its constituents against the *Radopholus similis* infesting black pepper (*Piper nigrum* L.). *Curr Nematol* 22 (1–2): 49–53.

Pervez R (2008) Chemosensory responses of *Aporcelaimellus nivalis* (Nematoda: Dorylaimida) towards kairomones emitted by prey nematodes belonging to different trophic groups. *Arch Phytopathol Plant Protect* 41(5): 318–327.

Poornima K, Suresh P, Kalaiarasan P, Subramanian S, Ramaraju K (2016) Root knot nematode, *Meloidogyne enterolobii*in guava (*Psidium guajava* L.) a new record from India. *Madras Agric J* 103: 359–365.

Sikora R, Coyne D, Hallmann J, Timper P (2018) *Plant Parasitic Nematodes in Subtropical and Tropical Agriculture*, CABI Publishing, Wallingford, pp. 234–242.

Singh N (2020) Emerging problem of guava decline caused by *Meloidogyne enterolobii* and *Fusarium oxysporum* f. sp. *psidii*. *Indian Phytopath* 73: 373–374.

Singh T, Prajapati, Anjana, Maru AK, Chaudhary R, Patel DJ (2019) Root-knot nematodes (*Meloidogyne* spp.) infecting pomegranate: a review. *Agricult Rev* 40(4): 309–313.

Thomas EE (1913) A preliminary report of a nematode observed on citrus roots and its possible relation with the mottled appearance of citrus trees. *California Agr Expt Sta Circ No. 85*. University of California, California, 14 pp.

Timmer LW, Garnsey SM, Broadbent P (2003) Diseases of citrus. In: Ploetz RC (ed.), *Diseases of Tropical Fruit Crops*, CAB International, Wallingford, UK, pp. 163–196.

Vinod Kumar V, Khan MR, Wali RK (2020) Crop loss estimations due to plant-parasitic nematodes in major crops in India. *Natl Acad Sci Lett*. 43(5):409–412https://doi.org/10.1007/s40009-020-00895-2

Vega-Callo RA, Mendoza-Lima MY, Mamani-Mendoza NR, Lozada-Villanueva LS, Tamo-Zegarra JJ, Casa-Ruiz TG, Bellé C (2020) First report of southern root-knot nematode, *Meloidogyne incognita*, infecting pomegranate, *Punica granatum*, in Peru. *J Nematol* 26(52): 2020–2026.

24 Biotic Stress Management Strategies in Guava

P. K. Shukla and Gundappa

24.1 INTRODUCTION

Guava is among the most important subtropical fruit crops after mango. It is a preferred fruit crop because of the beginning of production within 20–24 months after plantation, fruit production over half of the year, better price in comparison to input cost and high net profit. Its lovely taste, rich nutritional status, over 90 percent edible portion, digestion-friendly quality and low price in comparison to other fruits attract the consumers. Production of guava in India was estimated at 4053.51 thousand MT from 264.85 thousand hectare area during 2017–18 (NHB Database 2018). The highest area (49.53 thousand hectares) and maximum productivity (18.79 MT/ha) were recorded in Uttar Pradesh. Twenty-two states of our country have over 1000 hectares of area under guava, which indicates the wider adaptability of guava under different sets of climatic conditions. Almost all the states are recording continuous area expansion under guava.

24.2 IMPORTANT DISEASES AND THEIR MANAGEMENT

The significant reduction in guava productivity is caused by infection in roots or fruits. The wilt and decline are the most important diseases because their management is difficult. Diseases of fruits are mostly severe during the rainy season, which can be escaped by regulating flowering and fruiting for winter season crops. Leaf and stem infections are rarely severe. Guava crop is parasitized by over 177 species of pathogens, but only several of them have been identified as pests of regular occurrence and causing serious damage (Misra 2004). We are including only major diseases of economic importance, out of many reported, in this chapter.

24.2.1 Wilt

The wilt is the most important disease of guava throughout the world (Lim and Manicom 2003; Gupta et al. 2010; Misra 2017; Shukla et al. 2019). The disease was reported first in 1926 in Taiwan; later, in South Africa, Malaysia, India and many more countries (Lim and Manicom 2003). It was first observed in India in 1935, and since then a lot of research has been done on the subject. Heavy yield and tree losses were reported till 1980 (Shukla et al. 2019), but later disease management technology and farmers' efforts worked to minimize it.

Fusarium species (*F. chlamydosporum*, *F. oxysporum*, *F. proliferatum*, *F. solani*), *Macrophomina phaeseolina*, *Acremonium diospyri*, *Rhizoctonia bataticola*, *Gliocladium vermoesenii*, *Gliocladium roseum* syn. *Clonostachys rosea* f. sp. *rosea*, *Verticillium albo-atrum*, *Nalanthamala psidii*, *Clitocybe tabescens*, *Myxosporium psidii*, and species of bacteria like *Pseudomonas* and *Erwinia psidii* were found present in infected tissues of wilt-affected plants in different guava growing countries (Misra

DOI: 10.1201/9781032631578-24

2017). On the basis of the literature survey and research results at ICAR-CISH, the species of *Fusarium* (*F. oxysporum* and *F. solani*) had the most frequent association with guava wilt in India (Misra 2017; Shukla et al. 2019).

Wilting of guava trees has been recorded under every type of soil and climatic condition, however, different types of reports have been published at various locations. The age of guava plants does not determine the susceptibility to the infection by species of *Fusarium* and the period for development of symptoms after infection is highly variable. The soil conditions, plant health status and extent of infection are the factors affecting the progress of infection. High soil moisture, particularly during the monsoon period is considered conducive to rapid rotting of infected roots followed by wilting of trees during the rainy or post rainy season (Misra 2017). Recovery of infected trees is possible under proper care and treatment. The disease progression has been faster at high soil moisture and temperature.

The tree suffering from infection of wilt pathogen initially looks nutrient deficient. Later, symptoms of yellowing or browning and curling of leaves have also been recorded. Depending on the weather factors, leaves may drop down or stay on twigs for months. The expression of typical wilt symptoms has been recorded in the majority of cases after the age of 5 years. When a bearing tree gets wilted, the fruits become mummified and remain hanging on trees for several months (Gomes et al. 2011; Misra 2017). The roots of infected or wilted trees have dark brown or blackish discoloration. A similar discoloration has also been observed in the lower stem of the tree.

The guava wilt is a soil-borne disease. *Fusarium* spp., the major cause of the disease, is a facultative saprophyte, which can survive in cultivated soil for years in the absence of a suitable host. The fungus produces chlamydospores for long-term survival and numerous macro and microconidia for fast dissemination. The active growth and survival of the fungus depend upon the soil temperature and availability of nutrition and moisture in the soil. The invasion of roots is made by the active hyphae growing in soil followed by infection and extensive colonization of root tissues. The infection proceeds gradually in the roots and may progress into the stem till the tree is wilted. If wilted trees remain in orchards for long periods, the disintegrated infected tissue may retain inoculum.

Management: The management efforts made in the past and the success achieved have also been reviewed (Misra 2017; Shukla et al. 2019). The soil of the pit prepared for transplanting should be treated with 2 percent formalin solution, the soil surface should be concealed for a few days to avoid evaporation of formalin, and then transplanting should be done 25–27 days after removing the concealing sheet (Misra 2017). Several more treatments like neem cake + gypsum (Misra 2017); growing intercrops of marigold and turmeric (Prasad et al. 2003); application of *Trichoderma harzianum*, *T. viride* and *Aspergillus niger* AN17 strain in rhizosphere soil (Misra et al. 2000; Dwivedi and Shukla 2002) minimized the progress of infection. Rajan et al. (2005) reported resistance in a cross, *Psidium molle* × *Psidium guajava* root stock against wilt and utilized its root stocks for planting material production.

24.2.2 Decline

Guava decline disease, caused by *Meloidogyne enterolobii* has been recorded in the recent past (Kumar and Rawat 2018, Khan et al. 2019; Shukla et al. 2019; Singh 2020; Ghule et al. 2020). It has spread in most of the major guava-growing states of India during the last ten years and has become a major issue in guava production (Khan et al. 2019). Early age decline and wilting of guava plants have been caused due to severe infection of *M. enterolobii* in the presence as well as the absence of infection of *Fusarium* spp. (Khan et al. 2019; Shukla et al. 2019). The infection incidence by *Fusarium oxysporum* was higher in *M. enterolobii*-infected plants as compared with root-knot nematode-free orchards. Our findings are in general agreement with Gomes et al. (2011, 2012, 2014) for the decline of guava plants due to infection of *Meloidogyne mayaguensis*, *M. enterolobii* and *F. solani*. Gomes et al. (2010) reported a significant reduction in yield at 25 galls/10 g roots and 10 J2/ 100 cc soil. Studies carried out by Veloso et al. (2020) in Brazil, confirmed that the guava decline is caused by the complex of *M. enterolobii* and *F. solani*. They did the RNA sequencing of aggressive isolates of *F. solani* from guava and suggested that it is *Neocosmospora falciformis*.

The nematode-infected trees initially express symptoms similar to the deficiency of nutrients, and extensive rotting of infected roots occurs due to colonization by pathogenic fungi in fresh galls and by saprophytic fungi in old galls (Gomes et al. 2008). Infected trees may die in a shorter period and dry leaves stay attached to twigs if the infection is severe; otherwise, leaves may turn yellow or brown and shed gradually. The development of infected trees becomes slow, they bear fewer flowers and fruits, and the size of fruits is also reduced. Galls of 1 to 10 mm size are formed on roots and the whole root system may be converted into galls in severely infected plants.

The root-knot nematode has mostly spread long distances through planting material and localized spread took place by irrigation water and movement of infested soil by different means. The life cycle of *M. enterolobii* is completed within 27–30 days on guava. The juveniles existing in the soil penetrate the feeder roots, galls are formed, and egg laying begins from 15–21 days after infection. New generations of juveniles cause infection of roots and several generations are completed before plants die. The wilt fungi present in soil are promoted to cause infection in root-knot-infected roots. If both pathogens are involved, symptoms appear quickly.

24.2.2.1 Management

Soil-borne disease management in perennial crops is always a challenge due to competition among the pathogens, plants and the inputs applied against the disease. Guava decline management should be based on precautionary avoidance of the entry of the nematode in the nursery or orchard. If the infection has arrived in the plants, it should be managed by integration of chemicals, cultural practices and bio-control agent application. The efficacy of nematicides *viz.* carbofuran and fluopyram against plant parasitic nematodes is well established and applicable to guava. Application of *T. harzianum* and *T. virens* in root zone soil efficiently manages the progress of infection (Jindapunnapat et al. 2013; Shukla et al. 2018). Intercropping of root-knot-resistant cultivars of marigold, turmeric and garlic with guava was advocated for minimizing nematode infestation (Khan and Misra 2003). Experiments conducted at ICAR-CISH, Lucknow, indicated that a combination of neem cake + *T. harzianum* + carbofuran was effective against the disease complex, and application of *Purpuricillium lilacinus*, *Pochonia chlamydospora* and *Bacillus* sp. was effective against *M. enterolobii*. Alves et al. (2021) conducted a series of experiments and reported satisfactory efficacy of fluensulfone 40EC @ 2 liter per hectare and *Pochonia chlamydosporia; P. chlamydosporia* + fluensulfone; *Trichoderma harzianum*; *T. harzianum* + fluensulfone against *M. enterolobii*. Biocontrol agents *Pochonia chlamydosporia*, *Purpureocillium lilacinum* and *T. harzianum* effectively managed *M. enterolobii* under controlled conditions (Shukla et al. 2023).

24.2.3 Anthracnose

The anthracnose pathogen causes infection not only of fruits but also shoots. It is well known to cause pre- and postharvest losses to fruits. It is caused by *Colletotrichum gloeosporioides* (teleomorph: *Glomererella cingulata*) or *Gloeosporium psidii*, which mostly causes infection in each tender part of the plant (Misra 2004). The extent of disease severity depends on the number of rainy days and the amount of rainfall in a specified period. Since the fungus causes infection of every tender portion of the plant, it may cause dieback of growing tips and spots on leaves, flowers and fruits. The dieback symptoms develop if young shoots are infected (Misra 2004). Initial infection of the growing tip can be observed as tiny brown spots that grow fast under favorable conditions. The shape of the spots remains round till they fuse to make larger patches and turn black. The growing shoot may be killed in the process. Similar symptoms also develop on leaves and in severe cases, leaves may drop down (Misra 2004). Infection on fruit is very common during monsoon. Severely infected fruits do not grow further and become hard. If mature fruits are severely infected, they rot and drop down. Infection of flowers causes their blight and death.

24.2.3.1 Management

Pruning for canopy management followed by removal of all infected leaves, fruits and branches reduces the disease. Spray of copper oxychloride 0.3 percent or carbendazim + mancozeb @ 0.2 percent or propineb @ 0.2 percent just after initiation of disease or preventive spray of fungicide followed by bagging of fruits provide protection when done during early development stages of fruits (Shukla et al. 2022).

24.2.4 Stylar End Rot

The disease, caused by *Phomopsis psidii*, has been recorded for a long time but its increased severity has been recently recorded in normal as well as high density plantings, thus getting a higher economic significance. Initially light brown discoloration at the distal end of the fruit is seen, which gradually progresses and becomes dark in color. The infection also proceeds inside the fruit and rotting of pulp takes place. Cracking of fruits due to boron deficiency followed by infection of fungus has also been recorded (Shukla et al. 2022).

24.2.4.1 Management

Proper canopy management to harness maximum solar radiation, spray of solubore @ 2 g per liter of water followed by bagging of fruits for the rainy season crop provided success against the disease in farmers' fields. Repeated spray of zineb @ 2 g or guaiacol @ 0.025 g or α-naphthol @ 0.005 g per liter of water provides a satisfactory reduction in disease severity. Postharvest rot can be minimized by dipping fruits in neem leaf extract and keeping fruits at <10 °C at low relative humidity (El-Eryan et al. 2017).

24.2.5 Phytophthora Fruit Rot

The infection of *Phytophthora nicotianae* var. *parasitica* and *P. citricola* in fruits has been reported during the rainy season crop of guava. Although the disease has been of regular occurrence since 1971 we could not observe any significant losses in the past 10 years (Shukla et al. 2022).

The water-soaked lesions develop anywhere on the fruit surface during the monsoon period. The infection progresses very fast in mature fruits and the whole fruit surface may be covered within days. Superficial growth of fungus may appear on fruits in a damp micro-climate. Fruits covered with a canopy and near soil level are more severely affected as compared to fruits exposed to sunlight (Shukla et al. 2022).

24.2.5.1 Management

Sanitation is important for its management. Diseased fruits should be collected from orchards and be properly disposed of. Proper canopy management and drainage should be maintained during the rainy season. Disease incidence can be minimized by the spray of Dithane Z-78 (0.2 percent).

24.2.6 Fruit Canker

The disease is reported to cause severe losses throughout the world in heavy rainfall areas (Lim and Manicom 2003); however, it is confined to a few geographical locations and not found in severe form in central Uttar Pradesh. The infection of the causal organism *Pestalotiopsis psidii* is mostly observed during the rainy season (Lim and Manicom 2003). The maximum sporulation of the fungus occurs at 28–30 °C with relative humidity above 96 percent. The ascospores of *P. psidii* develop in perithecium, and conidia in black color circular pycnidia in pustules on fruits. The conidia are released at maturity to serve as the source of secondary inoculum.

Immature fruits are most commonly infected but also the leaves occasionally. The initial symptoms appear as tiny brownish or orange spots on fruits. The fully developed spots or cankers have a raised periphery and are depressed in the center. Cracking of fruits occurs in later stages, otherwise, fruits remain underdeveloped. The infestation of bugs enhances the infection and severity of the disease (Lim and Manicom 2003). The orange-brown lesions may develop on infected leaves.

24.2.6.1 Management

Disease can be managed by adopting measures similar to anthracnose and Phytophthora fruit rot management.

24.3 IMPORTANT INSECT PESTS OF GUAVA AND THEIR MANAGEMENT

24.3.1 Tea Mosquito Bug – *Helopeltis antonii*

24.3.1.1 Biology

It is a polyphagous pest found feeding on guava. The adult lays elongated and sausage-shaped eggs that possess two filamentous long processes that remain jutting out from the tender plant tissue in which the eggs are bedded by the female. Eggs are laid in buds, axils of leaves, tender stems, and developing fruits. Fecundity varies from 40 to 50 eggs per female. The incubation period varies from 5 to 7 days. The nymphs are greenish-yellow with elongated appendages and pass through 5 instars in 15–18 days. The adult is a slender insect 6–8 mm in length with a yellowish brown head and white abdomen, a dark red thorax and long dark appendages.

24.3.1.2 Damage

Nymphs and adults feed on tender leaves, shoots and fruits. The nymphs and adults cause scab formation on fruits and make them unsuitable for marketing. Young leaves bear brown patches and later get dried. Brown to black spots develops on the infested flower.

24.3.1.3 Natural Enemies

They include the parasites *Erythmeleus helopeltidis*, *Telenomus*, *Chaetostricha* & *Gonatocerus*, predators *Cremastrogaster wroughtonii*, *Sycannus collaris* and the entomogenous fungus, *Aspergillus flavus*.

24.3.1.4 Management

Spraying of malathion (0.10%) or L-cyhalothrin (0.003%, profenophos (0.05%) and triazophos (0.05%) or neem oil (2%) twice at fortnightly intervals at pea stage of fruits was found effective. Weekly spraying of biopesticide *Beauveria bassiana* @ $1x10^9$ (IIHR strain) with suitable adjuvant cause significant reduction in mirid damage on guava (Vishalakshy and Mani 2011).

24.3.2 Mealybugs – *Ferrisia virgata*, *Maconellicoccus hirsutus*, *Planococcus citri*, *Planococcus lilacinus* and *Rastrococcus iceryoides*

24.3.2.1 Biology

Mealybugs are reported highly damaging in peninsular India. The eggs of mealybugs are oval and can be seen with the naked eye. As many as 500 eggs are laid by a female. It has three larval instars for the female completing in 20 days and four instars in 18 days for the male. Females are wingless and as they mature, become more sessile, with an oval and flattened, segmented body. The adult male has a brown-colored body bearing two white, anal filaments. The life cycle is completed in 25–30 days.

24.3.2.2 Damage

Nymphs and adults suck the sap from leaves, stems and fruits. The sticky honeydew excreted by them serves as a substrate for the growth of sooty mold. In the case of *M. hisutus*, the mealybug injects toxic saliva into the plant while feeding which results in malformation of the leaf and shoot growth stunting (Mani and Krishnamoorthy 2001).

24.3.2.3 Natural Enemies

Natural enemies include *Scymnus coccivora*, *Aenasius advena* on *F. virgata*, *C. montrouzieri*, *Allotropa* sp., *C. perminutus*, *Spalgis epius* on *P. citri*, *Praleurocerus viridis* and *S. coccivora* on *R. iceryoides*.

24.3.2.4 Management

Release of Australian ladybird beetle *C. montrouzieri* @ 20/plant is recommended to control the mealybugs in guava. Releases of the exotic parasitoid *Leptomastix dactylopii* are found to be highly effective against *P. citri* in guava orchards. The encyrtid *Praleurocerus viridis* and *S. coccivora* are found very effective in reducing the population of *R. iceryoides* in guava.

24.3.3 Fruit Flies – *Bactrocerazonata*, *B. dorsalis*

Guava is the next preferred host of fruit flies after mango and is found active in rainy season fruits. Incidence of *B. dorsalis* on guava fruits goes up to 40% in Punjab. *B. zonata* may cause crop damage to 75 percent (Kapoor 2000).

24.3.3.1 Biology

Females lay eggs under the skin of fruits. The egg hatches in about two days. Maggots tunnel through the fruit feeding on the pulp, shed their skins twice, and emerge through exit holes in 8 to 17 days, depending on temperature. The larvae drop from the fruit and burrow into the soil to pupate. The pupal period varies from 7 to 18 days. The newly emerged adult females need 16 to 38 days to mature sexually (Kapoor 2000).

24.3.3.2 Damage

The damage starts when the female fruit fly punctures the fruit with its long and sharp ovipositor. Ovipositional damage in the form of minute depression with dark punctures may be seen on the fruits. The infested fruit softens at the site of larval feeding and later on, rots and drops down.

24.3.3.3 Management

It involves plowing around the tree basins exposing the pupae in early harvesting; early harvesting of fruits, collection and destruction of fallen and infested fruits; the male annihilation technique with methyl-eugenol traps to kill males along with need-based insecticidal cover sprays during fruit maturity, the stage most vulnerable to attack by ovipositing gravid females (Verghese and Jayanthi, 2001). Bait containing a mixture of protein hydrolysate and malathion (0.1-0.25 + 0.05%) as cover spray on host trees is useful in killing adult flies (Singh et al. 2003).

24.3.4 Pomegranate fruit borer – *Virachola (Deudorix) isocrates*

24.3.4.1 Biology

It is a polyphagous pest having a very wide range of host plants, including guava. The bluish-brown butterfly has an orange spot on each of the forewings and black spots on the hindwings. The eggs are laid singly on flowers and tender fruits. The female lays eggs singly on calyx of flowers or small fruits. Eggs are shiny white in color and oval in shape. The egg hatches out within 7–10 days. The caterpillar is dirty brown and has short hairs on its body. Larval period is 14–45 days. Larvae pupate therein or on the fruit stalk. The total life cycle may take 30–40 days depending upon the weather.

24.3.4.2 Damage

The caterpillars bore into ripening fruits and affected fruits rot and drop. Boreholes often plugged by the anal segment of the caterpillar or its excreta are seen on infested fruits. Average fruit loss in weight goes up to 20%. The pest incidence attains its peak in the months of August and November to December.

24.3.4.3 Management

Regular collection and destruction of infested fruits will reduce the infestation level. Spray application of fenvalerate 20% EC @ 0.01 % or triazophos 40% EC @ 0.05% at the fruit-setting stage is found effective against the pest.

24.3.5 Castor Capsule Borer – *Conogethes (Dichocrocis) punctiferalis*

It is an important pest of guava causing serious damage (Gupta and Arora 2001). The incidence is more common in the rainy season crop.

24.3.5.1 Biology

The moth is of medium size and yellow with numerous black spots on the body. Larvae of this moth mainly bore fruits, but it may also attack buds and tender shoots. The mature larva is pinkish in color and speckled with minute black spots. The larvae feed on pulp and seeds of fruits resulting in premature drop of fruits. The pupa is formed in a pupal chamber made at the basal end of the fruit. The pest completes its life cycle in 25–33 days. The incidence of *C. punctiferalis* may range up to 25 percent.

24.3.5.2 Damage

The affected fruits are generally deformed at the point of entry of larva. Such fruit becomes weak, rots and falls down.

24.3.5.3 Management

Spray application of fenvalerate 20% EC @ 0.01 % or triazophos 40% EC @ 0.05% at intervals of three weeks commencing at the initiation of fruit-setting is found effective against this pest.

24.3.6 Fruit Borer – *Rapala varuna*

Caterpillars are found feeding on flower buds and fruit. The adult is a metallic red-colored butterfly. Management involves the collection and destruction of fruits and spray application of insecticide malathion 50 EC @ 0.1% in two rounds, one at flower formation and the next at fruit set (Rai and Shankar 2012).

24.3.7 Bark Eating Caterpillars – *Indarbela quadrinotata* and *I. tetraonis*

24.3.7.1 Biology

The occurrence of bark-eating caterpillars is regular throughout the year. Eggs are laid in clusters of 15–20 directly on the bark of a branch or bole. The freshly laid eggs are pale colored and oval in shape and become creamy white after some time. The larva bores a short tunnel downwards into the wood, usually at the junction of a dead branch or snag and the hole. The full-grown larva is 1.5–2.0 inches long and smooth-skinned with dark chitinised patches on the segments. Pupation takes place in the tunnel in the wood and the pupal period lasts about 3 weeks. The life cycle of this pest is annual. The male adults are whitish grey, short, stout with brownish streaks on pale whitish forewings. The female moths are pale brownish with forewings having rows of dark rusty red spots.

24.3.7.2 Damage

The larva bores into the trunk or branches, usually at forks or angles, to a depth of 15–25 cm. The larvae emerge from the tunnel at night and eat the bark of the tree in the immediate vicinity of the hole. Large dark-brown webby masses, comprising chewed wooden particles and fecal matter, are conspicuously seen plastered loosely on tree trunks or main branches, especially near the forks. The incidence of this pest varies from 50 to 80% in orchards. Its initial incidence was observed during the second week of April and its peak incidence during the first week of October (Haseeb 2007).

24.3.7.3 Management

The pest could be checked by inserting cotton wool soaked in chlorpyriphos 20% EC or quinalphos 25% EC @0.05% in the borer hole and plugging it with the mud (Verghese and Jayanthi 2001). Application of conidial suspension of naturalis – L (0.40%) in the borer hole also proves effective in bringing down (up to 81.48%) the incidence of this pest. Chorpyriphos (0.5%) along with sealer cum healer (IIHR product) and copper oxychloride (3%) reduced the bark-eating caterpillar infestation significantly.

24.3.8 Stem Borer – *Aristobia testudo*

It has attained the status of a major pest in the NEH region for guava, causing 80 to 100 percent damage in Allahabad Safeda, L-49 and local varieties (Shylesha et al. 2000).

24.3.8.1 Biology

Eggs are laid in the bark by making inverted U-shaped punctures on the branches and main trunks. Grubs initially feed on the bark for about a month and later enter into the wood, making zig-zag and later single, long tunnels. The adults are large, long-horned beetles, black in color with yellow spots, measuring 2.5–3 × 8–1.2 cm. The adults after emergence (June–October) feed on the bark of these hosts (Shylesha et al. 2000).

24.3.8.2 Damage

Small exit holes are found at a distance of 20 to 30 cm to push away straw-colored fecal matter in the form of pellets seen accumulated at the base of the plant, indicating the infestation of this borer. Emergence holes of the beetle, measuring 1 to 1.5 cm diameter may also be seen on the stem indicating the presence of beetle. Severe infestation results in leaf fall and drying of twigs and branches. Young, infested plants are killed.

24.3.8.3 Management

Mechanical destruction of beetles; ovipositional sites can be easily identifiable because of U-shaped markings and eggs can be destroyed with the help of a sharp knife during September and October; removal of infested branches and its destruction by burring from December to February; injection of 5 ml malathion 50% EC (0.1%) is found effective against the pest in the case of severe infestation.

24.3.9 Shoot Borer – *Microcolona technographa*

Newly hatched larvae are found feeding mostly at the base of the leaf of the newly emerged twigs, flower buds (both auxiliary and apical buds), leaf petioles, etc. The presence of the larva within the tunnel could be detected with the presence of the products of the larvae *i.e.*, the fecal matter coming out from the entry. The infested twigs just above the bored hole dried up, affecting further growth of the plant, and as the pest also attacked flower buds and larvae fed indirection which resulted in a decrease in fruit production. The infestation of the shoot borer starts from the end of January to mid-February coinciding with the increase of temperature and emergence of new flush and with its peak activity from the middle of July to the second week of August (Rama Devi and Jha 2017; Sharma et al. 2004). Spray chloropyriphos 20% EC @ 4.5 ml/liter or Quinalphos 25% EC @ 4 ml/liter of water to manage the borer. Guava varieties Red flesh, Sardar Guava, Dharar and Annu Ishkwalaare are moderately resistant to the shoot borer.

24.3.10 Aphid – *Aphis gossypii*

It occurs on new flush, damaging leaves and growing shoots by sucking sap.

24.3.10.1 Damage

Colonies of nymphs and adult aphids infest tender shoots and leaves and suck the cell sap. They excrete honeydew resulting in the development of sooty mold on leaves and shoots.

24.3.10.2 Natural Enemies

Five coccinellids *Cheilomenes sexmaculata*, *Scymnus latemaculatus*, *Pseudaspidimerus trinotatus*, *Chilocorus nigrita*, *Cryptolaemus montrouzieri*; two syrphids *Ishiodon scutellaris* and *Paragus serratus* and the green lacewing *Chrysoper lacarnea* (Steph.) are known to prey on *A. gossypii* in guava orchards. The population of this pest is kept under check in nature within 50 days.

24.3.10.3 Management

In case of severe aphid incidence, application of any systemic insecticides like dimethoate 30% EC @ 0.05% is advocated.

24.3.11 Wax scale – *Drepanococcus chiton* Green (Hemiptera: Coccidae)

The scale insect infestation is confined to the lower surface of the leaves and shoots of guava plants. They suck the cell sap and excrete honeydew resulting in the development of sooty mold thereby interfering photosynthetic activity of guava plants.

24.3.11.1 Natural Enemies

Five parasitoids *Anicetus ceylonensis*, *Cephaleta brunniventris*, *Diversinervus elegans*, *Metaphycus helvolus*, *Philosindia longicornis*, and four predators namely *Chilocorus nigrita*, *Cryptolaemus montrouzieri*, *Cheilomenes sexmaculata* and *Scymnus* sp. are found to attack this scale.

24.3.11.2 Management

The scale insect is suppressed mainly due to the activity of locally occurring natural enemies chiefly *C. nigrita*, *A. ceylonensis* and *C. brunniventris* in guava orchards. The percentage of parasitism goes up to 98%.

24.3.12 Thrips – *Selenothrips rubrocinctus* and *Rhipiphorothrips cruentatus*

24.3.12.1 Biology

Eggs are minute and dirty white in color. Nymphs and adults are also tiny, slender, fragile, and yellowish straw in color; adults have heavily fringed wings that are uniformly grey in color. A single life cycle is completed in 23 to 35 days. Nymphs are yellow with a broad band around the base of the abdomen. Adults are dark brown, long (1–1.5 mm), fragile, slender insects with dark grey heavily fringed wings.

24.3.12.2 Damage

Thrips are the major pests causing severe damage to guava. *Selenothrips rubrocinctus* is found in large numbers on the leaves. Nymphs and adults suck the sap on the undersurface of the leaves. De-sapping of leaves by *S. rubrocinctus* results in their withering and shedding. Heavy incidence of *R. cruentatus* results in scar formation on fruits and their premature dropping. Severe infestation of pests may result in a reduction in the quality of fruits and their production.

24.3.12.3 Management

Clean cultivation and clipping off the infested plant parts at the initial stages of infestation helps in minimizing the thrips. Installation of yellow sticky traps or yellow pan water is to be considered for monitoring and management of thrips. Chemical insecticides, synthetic pyrethroids, carbamates and organophosphate compounds have proved effective in significant reduction of population of thrips in different crops.

REFERENCES

Alves CDS, Starling RZC, Rodrigues LL, et al. (2021) Greenhouse and field assessment of biological and chemical agents against guava decline. *Summa Phytopathol.* 47(2). https://doi.org/10.1590/0100-5405/225806

Dwivedi BP, Shukla DN (2002) Biocontrol of Fusarium wilt of guava (*Psidium guajava* L.) using *Trichoderma* and *Gliocladium* species. *Karnataka J Agric Sci* 15: 399–400.

El-Eryan EE, Tarabih ME, EL-Metwally MA (2017) Influence of natural extracts dipping to maintenance fruit quality and shelf life of Egyptian guava. *J Plant Product* 8(12): 1431–1438.

Ghule TM, Phani V, Somvanshi VS, Patil M, Bhattacharyya S, Khan MR (2020) Further observations on *Meloidogyne enterolobii* (Nematoda: Meloidogynidae) infecting guava (*Psidium guajava*) in India. *J Nematol.* 52(1). www.researchgate.net/publication/347303072

Gomes VM, Souza RM, Almeida AM, Dolinski C (2014) Relationship between *M. enterolobii* and *F. solani* spatial and temporal dynamics in the occurrence of guava decline. Nematoda. 2(1). http://dx.doi.org/10.4322/Nematoda.01014

Gomes VM, Souza RM, Correa FM, Dolinski C (2010) Management of *Meloidogyne mayaguensis* in commercial guava orchards with chemical fertilization and organic amendments. *Nematol Bros* 34: 139–142.

Gomes VM, Souza RM, Midorikawa G, Miller R, Almeida AM (2012) Guava decline: Evidence of nationwide incidence in Brazil. *Nematropica* 42: 153–62.

Gomes VM, Souza RM, Mussi-Dias VDA, Silveira SF, Dolinski C (2011) Guava decline: A complex disease involving *Meloidogyne mayaguensis* and *Fusarium solani. J Phytopath* 159: 45–50.

Gomes VM, Souza RM, Silva MM, Dolinski C (2008) Caracterizacao do estado nutricional de goiabeiras em declinio parasitadas por *Meloidogyne mayaguensis. Nematol Bros* 32: 154–160.

Gundappa B, Rajkumar MB, Singh S, Rajan S (2018) Pests of guava. In: Omkar (Ed.), *Pests and Their Management*. Springer, Singapore, pp. 491–516.

Gupta RK, Arora RK (2001) Lepidopteran fruit borers on guava in Jammu. *Insect Environ* 7: 83–84.

Gupta VK, Misra AK, Gaur RK, Jain PK, Gaur D, Sharma S (2010) Current status of *Fusarium* wilt disease of guava (*Psidium guajava* L.) in India. *Biotech* 9: 176–195.

Haseeb M (2007) Current status of insect pest problems in guava. *Acta Hortic* 735: 453–467

Jindapunnapat K, Chinnasri B, Kwankuae S (2013) Biological control of root-knot nematode (*Meloidogyne enterolobii*) in guava by the fungus (*Trichoderma harzianum*). *J Dav Sustainable Agric* 8: 110–118.

Kapoor VC (2000) Fruit flies (Diptera: Tephritidae): status, biology and management strategies. In: Upadhyay RK, Mukerji KG, Dubey OP (Eds.), *IPM System in Agriculture*. Aditya Books Private Ltd., New Delhi, pp. 123–142.

Khan RM, Misra AK (2003) Influence of co-cultivation of marigold, garlic and turmeric on nemic population in guava cropping system. In: *Proceedings of Zonal Conf of ISMPP on Plant Disease of National Importance with Special Reference to Guava Wilt and Mango Malformation*, 4–5 April. CISH and IISR, Lucknow, pp. 24–25.

Khan RM, Ahmad I, Kumar HK, Singh A, Shukla PK (2019) Infestation of root-knot nematode, *Meloidogyne enterolobii* in newly established/old orchards and nurseries in Madhya Pradesh, Rajasthan and Uttar Pradesh. *Ann Pl Prot Sci* 27(1): 170–171.

Kumar S, Rawat S (2018) First report on the root-knot nematode *Meloidogyne enterolobii* Yang and Eisenback, 1988 infecting guava (*Psidium guajava*) in Udham Singh Nagar of Uttarakhand, India. *IJCMAS* 7: 1720–1724.

Lim TK, Manicom BQ (2003) Diseases of guava. In: Ploetz RC (Ed.), *Diseases of Tropical Fruit Crops*. CABI Publication, UK, pp. 275–290.

Mani M, Krishnamoorthy A (2001) Suppression of *Maconellicoccus hirsutus* (Green) on guava. *Insect Environ* 6(4): 152.

Misra AK (2004) Guava diseases–their symptoms, causes and management. In: Naqvi SAMH (Ed.), *Diseases of Fruits and Vegetables, Volume II*. Kluwer Academic Publishers, The Netherlands, pp. 81–119.

Misra AK (2017) Progressive steps in understanding and solving guava wilt- a national problem. *Indian Phytopath* 70(1): 1–11.

Misra AK, Prakash O, Sen B (2000) Biological control of guava wilt by *Aspergillus niger* strain AN17 (Pusa Mrida). In: Chadha KL (Ed.), *National Seminar on Hi-Tech Hort,* 26–28 June, Bangalore, Horticulture Society of India, New Delhi, p. 149.

NHB Database (2018) *National Horticulture Board, Department of Horticulture and Cooperation, Government of India*. Aristo Printing Press, New Delhi, pp. 185–186.

Poornima K, Suresh P, Kalairasan P, Subramanian S, Ramaraju K (2016) Root knot nematode, *Meloidogyne enterolobii* in guava (*Psidium guajava* L.) a new record from India. *Madras Agri J* 103: 359–365.

Prasad D, Shukla SK, Prasad B, Misra AK (2003) Effect of intercrop and different doses of NPK on the incidence of guava wilt. In: *Proceeding of Seminar on Plant Diseases of National Importance with Special Reference to Guava Wilt and Mango Malformation*, April 4–5, CISH and IISR, Lucknow, p. 53.

Rai S, Shankar U (2012) Integrated pest management in dry land horticultural crops. In: Abrol DP, Shankar U (Eds.), *Ecologically Based Integrated Pest Management*, New India Publishing Agency, New Delhi, p. 681. www.nipabooks.com

Rajan S, Kumar Ram, Misra AK, Negi SS (2005) Hybrid guava, *Psidium molle* × *P guajava* (INGR No. 04116; IC427822), with resistance to guava wilt. *Indian J Plant Genetic Resour* 18(3): 301.

Ramadevi A, Jha S (2017) Incidence pattern of guava shoot borer, *Microcolona technographa* Meyrick (Lepidoptera: Agonoxidae) in relation to weather conditions in West Bengal. *India Pest Manage Horticult Ecosyst* 23(1): 1–6.

Sharma DR, Batra RC, Dhaliwal GS (2004) Seasonal incidence, varietal susceptibility and chemical control of shoot borer, *Microcolona technographa* Meyrick (Lepidoptera: Cosmopterygidae) on guava. *J Res Punjab Agri Univ* 41(1): 39–46.

Shukla PK, Rajan S, Fatima T (2018) *Root-Knot Disease of Guava: Identification and Management*. ICAR-CISH, Lucknow Publication, Lucknow, p. 4.

Shukla PK, Fatima T, Rajan S (2019) Research on Fusarium wilt disease of guava. *Indian Phytopathol*. 72:629–636. https://doi.org/10.1007/s42360-019-00167-0

Shukla PK, Kumari N, Fatima T (2022) Major diseases of guava, Bael and Aonla. In: Khan MR, Haque Z (Eds.), *Diseases of Fruits and Plantation Crops and Their Sustainable Management*. Nova Science Publ. Inc., New York, pp. 307–328.

Shukla PK, Kumari N, Fatima T, Singh H (2023) Evaluation of bio-control agents for the management of guava decline. *J Eco-Friendly Agri* 18(1): 193–199.

Shylesha AN, Thakur NSA, Ramchandra (2000) Incidence of litchi trunk borer, *Aristobia testudo* Voet (Coleoptera:Lamiidae) on guava in Meghalaya. *Pest Manage Horticult Ecosyst* 6: 56–157.

Singh G, Misra AK, Haseeb M, Tandon DK, Pathak RK (2003) *The* Guava. CISH, Lucknow, p. 38.

Singh N (2020) Emerging problem of guava decline caused by *Meloidogyne enterolobii* and *Fusarium oxysporum* f. sp. *psidii*. *Indian Phytopathol*. 73: 373–374. https://doi.org/10.1007/s42360-020-00198-y

Veloso JS, Câmara MPS, Souza RM (2020) Guava decline: updating its etiology from '*Fusarium solani*' to *Neocosmospora falciformis*. *Eur J Plant Pathol* 159: 455–460. https://doi.org/10.1007/s10658-020-02161-z

Verghese A, Jayanthi PDK (2001) Integrated pest management in fruits. In: Parvatha Reddy P, Verghese A, Krishna Kumar NK (Eds.), *Pest Management in Horticultural Ecosystems*. Capital Publishing Company, New Delhi, pp. 1–23.

Visalakshy PG, Mani M (2011) Record of *Beauveria bassiana* on tea mosquito bug, *Helopelti santonii* Signoret in guava ecosystem. *J Biol Control* 25(2): 150–151.

25 Biotic Stress Management of Makhana Crops

Santosh Kumar, Tribhuwan Kumar, Sah S. B., Tamoghna Saha, Sailabala Devi, Kishan Lal and Dhirendra Kumar Singh

25.1 INTRODUCTION

Makhana, generally known as 'fox nut/gorgon nut', is a floating annual aquatic cash crop. It has large leaves and prickly petioles and grows in standing water bodies (1.5 m deep) under a wetland ecosystem. The botanical name of makhana is *Euryale ferox* Salisb, which comes under the Nymphaeceae family of the plant kingdom. This crop is well adapted to tropical and subtropical climatic conditions. Bihar, Assam, West Bengal and Odisha are some Indian states where makhana is grown as a high-value commodity on a commercial scale (Shanker et al. 2010; Kumar et al. 2011). The estimated total acreage of makhana cultivation in India is 20,000 ha, which constitutes more than 80% of the total makhana crop in terms of production (3.18 lakh quintals) and productivity (21.25 q/ha) as per the report from the ICAR-National Research Center for Makhana, Darbhanga (Minten et al. 2014). The total yield of makhana seed is 1,200,000 MT which gives rise to 40,000 MT makhana pop after processing. The approximate production value of makhana at the farmers' field is Rs. 250 crore which generates revenue of Rs. 550 crores at the traders' level. Across the nation, only Bihar contributes 80% of acreage and more than 90% of makhana production, which supports the livelihood of more than 2 million poor farmers (Jana 2017; Vision 2050, 2015). It has also been observed that North Bihar ranks first not only in India but also in the world. Makhana is cultivated particularly in agro-climatic Zone II (Purnea, Madhepura, Saharsa and Supaul, Araria, Katihar districts) and Zone I (Darbhanga, Motihari, Madhubani, Samastipur, Sitamarhi districts) of Bihar where the pH of the soil is 6.5–7.8 and the average rainfall is about 1450.00 mm. Darbhanga and Madhubani are the traditional and conventional centers for makhana cultivation and processing, while Katihar and Purnea are the fast-emerging districts in terms of trading and cultivation of makhana in Bihar.

The magnitude of makhana in the Indian economy is rising day by day due to their nutraceutical and cultural values. But the productivity of makhana seems to be affected by several abiotic and biotic factors resulting in a lower crop yield than estimated. Therefore, understanding various biotic and abiotic stresses affecting makhana is essential for farmers and scientists to devise agronomical and pathological management systems leading to higher cultivation and processing of makhana at a commercial scale. However, limited sources and information are available regarding the biotic stress and management in makhana due to the poor attention of pathologists and entomologists towards makhana. However, we have tried our best to write a book chapter on "Biotic stress management on makhana crop" referring to different sources. This book chapter comprises different pieces of information pertaining to major diseases and insects of makhana along with their control and management strategies.

 DOI: 10.1201/9781032631578-25

25.2 BIOTIC STRESSES IN MAKHANA

The outbreak of diseases and infestation of insects in makhana were reflected during the surveillance program steered by the team of Bihar Agricultural University, Sabour. Yield losses due to biotic stress in makhana are 20–25% of the total yields (Kumar et al. 2021a). The most common diseases and insects in makhana are leaf blight, leaf spot, botrytis grey mold, tumor formation, aphid (*R. nymphaeae* L.), case worms (*E. depunctalis* G. & *E. crisonalis* W.), Singhara beetle (*Galerucella birmanica*), and gastropod (Mollusca's) in addition to many more which are evolving with the passage of time and changing climatic conditions and establishment of different ecosystems (Nath et al. 2017; Kumar et al. 2021a). Nath et al. (2018) reported that yield losses due to insects are up by 17.04 to 23.66 percent in makhana. Though there are insufficient investigations pertaining to occurrences and statuses of insects. However, four-dozen insects and gastropods have been found associated with the makhana crop (Mishra et al.1992).

25.3 MAJOR DISEASES OF MAKHANA

25.3.1 Alternara Leaf Blight and Leaf Spot

Alternara leaf blight and leaf spot caused by *Alternaria alternata* and *Alternaria tenuissima* respectively have emerged as the most dreadful diseases in makhana (Kumar et al. 2021 b, c). North Bihar, the major area of makhana cultivation is facing the threat of these diseases and is resulting in severe loss of crop yield.

The symptoms of both Alternaria leaf blight and spot disease are more or less similar except for the pattern of appearance. Typical symptoms of Alternaria leaf blight disease are large, light brown necrotic sunken with a circular ring, first visible on the margin sides of leaves (Figure 25.1a) to the middle sides of leaves. During favorable environmental conditions, severe infections are seen as coalesced lesions and blighted leaves. Such severe infections occur in the period when there is high humidity and warm temperatures. Symptoms of Alternaria leaf spot disease are characterized as lesions appearing on the leaves of infected plants. This lesion looks like a necrotic sunken, light tan to brown and usually has a concentric ring or target board pattern with a yellow halo (Figure 25.1b). Blighting begins with the expansion of the leaf spot lesion. Gradually spots coalesce to form large dark blighted areas and begin to cover the entire leaf causing premature defoliation (Figure 25.1c). Produced seeds from the infected plant are grey to black in color, wrinkled in shape, small in size and have low viability. As symptoms progress into the fruits, it is likely to be small, distorted and twisted with less seed.

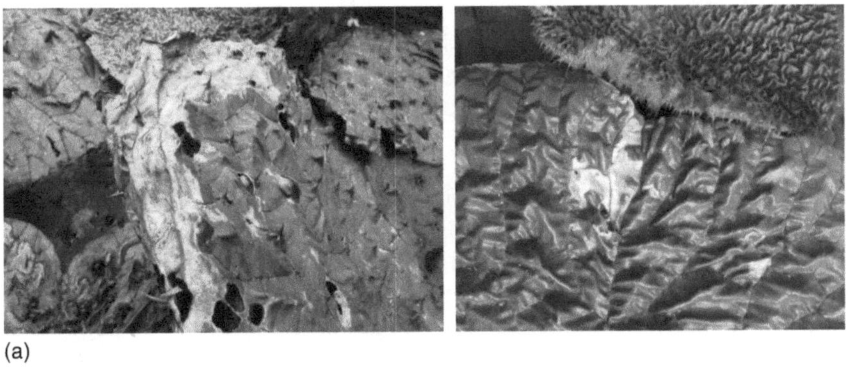

(a)

FIGURE 25.1 (a) Symptom of leaf blight appearing from margin side of the leaves. (b) A number of spots on a single leaf. (c) Symptoms of leaf blight in advance stage.

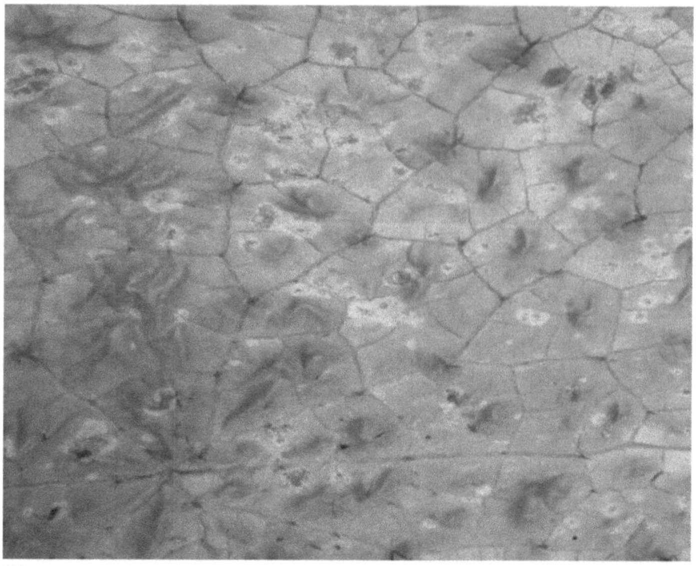

(b)

(c)

FIGURE 25.1 (Continued)

The development and progress of Alternaria leaf blight and leaf spot are influenced by weather parameters from 30 days of sowing to the harvesting stage. Alternaria leaf blight and leaf spot severity has been observed in the month of April when there is a gradual increase in temperature, relative humidity and rainfall, which continues up to the harvesting stage. It has been noted that the maximum temperatures and relative humidity and rainfall exhibit a strong positive linear relationship and influence the occurrence of leaf blight disease (Kumar et al. 2021, 2021b, 2021c). Selvamani et al. (2014) reported that the maximum and minimum temperatures positively correlated to disease development in leaf spots in crucifers caused by *Alternaria brassicae*, but the average temperature showed a higher degree of correlation than the minimum (1.5–15.3°C) and the maximum temperatures (10–28°C). In dry weather, the development of disease is minimal, whereas during moist weather, there is an increase in the progress of leaf blight causing rapid blighting and rotting of entire leaf/leaves. Maximum temperatures, relative humidity and rainfall show a strong positive linear relationship and significantly influence the occurrence of Alternaria leaf blight and leaf spot disease.

25.3.2 Tumors

Tumors or galls caused by a fungus known as *Doassansiopsis euryaleae* result in huge losses of makhana (Verma and Jha 1999). Leaves and other parts including the petiole, pedicel, etc. are the most affected parts. The infection usually extends from the leaf lamina to the petiole and from the pedicel to the basal part of the flower rendering great distortion in shape due to hypertrophy. Initially, small tumors or gall-like chlorotic growth appear on leaves, which are dark green to light yellow in color with pinkish leaf veins. It later attains a major dimension with a hard and dark grey color (Figure 25.2). Symptoms appear under the leaf surface, and the main leaf veins are a faded orange in color with swelling. It turns brown in color later. The internal cross-section of the swollen portion also shows white powdery growth. Infection on the basal part of the flower, including the ovary, greatly reduces the number of viable seeds and causes economic loss to the farmers (Verma et al. 2003). Sometimes it affects fruits and causes fruit galls. No control measure for this disease is available to date.

25.3.3 Root Rot

Root rot disease is assumed to be caused by the *Phytophthora* species (Mishra et al. 2003; Kumar et al. 2011). Apparently, healthy plants abruptly show yellowing, dying and finally wilting. The affected roots show dark brown or black, which become soft and rotten. The outer layer of cells strips off the roots easily leaving only the central strand of water-conducting tissue.

25.3.4 Botrytis Grey Mould

This is the prevalent disease caused by the fungus *Botrytis cinerea* (Mishra et al. 2003; Kumar et al. 2011). Fungus *Botrytis* is omnipresent and opportunistic in nature and infects only physiologically-weakened plants. Botrytis grey mold is common in seeds and seedlings. Initially, it appears as a white growth on the plant but rapidly darkens to a grey color. A smoky-grey "dusty" spore forms and spreads with the help of wind or water. Buds or small leaves infected with grey mold may affect growth and photosynthesis. The damage can be minimized by reducing pathogen populations and increasing plant vigor. Sanitation is the key step, but is not enough alone, to control this disease.

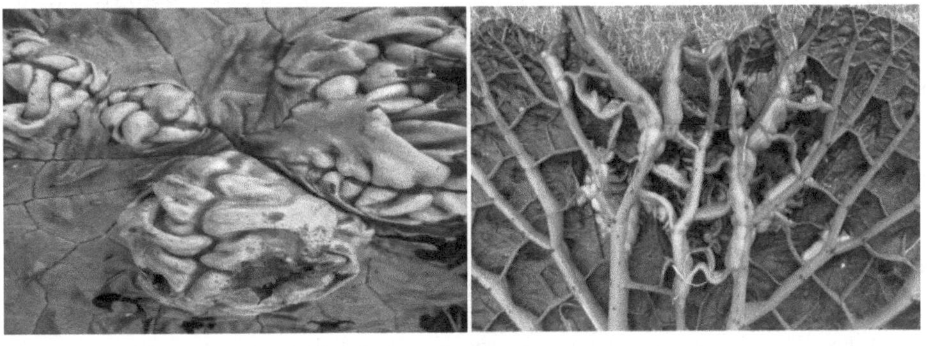

FIGURE 25.2 Symptoms of tumour/gall caused by *Doassansiopsis euryaleae* on leaves of makhana.

25.4 MANAGEMENT STRATEGIES FOR MAKHANA DISEASE

Management of diseases of makhana either in the pond or submerged field condition is not easy, as most of the farmers cultivate makhana along with fish culture. Hence, the application of bio-control agents is more advisable. Pond sanitation should be maintained by removal of infected leaves and weeds. Removal of dead or dying tissue from the plants and from the soil surface should be encouraged. Augmentation in the natural plant defense mechanism can be done by the application of Mycorrhiza, etc. along with farmyard manure/vermi-compost in the pond. Dipping of seedlings in *Pseudomonas fluorescens* @ 5 gm/l of water reduced the primary infection of the disease. Seeds should be soaked in hydrogen peroxide and water at a ratio of 1:10. Until the seed has swollen and is ready to split its skin and extend stems and roots, a fresh solution should be added every day. It's a time-consuming process, so patience must be exercised (Mishra et al. 2003; Kumar et al. 2011). Three-time foliar applications of *P.* fluorescens @ 5 ml/l of water at 15-day intervals not only manage the disease but also favor the growth and development of makhana without harming fish. However, in ponds, where there is no fish culture, foliar sprays with copper oxychloride or dithane M-45 @ 0.3% twice or thrice at fortnightly intervals have been found very effective in checking the disease (Kumar et al. 2011). Application should be done when symptoms first appear and repeated at fortnightly intervals as required (Mishra et al. 2003; Kumar et al. 2011). For the management of root rot, the pond should be cleaned thoroughly and treated with a solution of sodium hypochlorite or swirling a muslin bag filled with copper sulfate crystals through the water before being flushed out with fresh water. Once clear water has been run back in, new plants can be safely introduced. Fish must be saved alive before beginning any such treatment and not returned until the pool has been emptied, swilled out and refilled with fresh water.

25.5 MAJOR INSECT PESTS OF MAKHANA

Like other biotic factors, insect pests are major threats limiting the production and productivity of makhana. About fifty insect species are known to be associated with the makhana crop (Mishra et al. 1992). Out of which, aphids (*Rhopalosiphum nymphaeae* L.), case worms (*Elophila depunctalis* W. & *E. crisonalis* W.), rib borer (*Chironomous* sp.) and gastropods which have emerged as major insect pests (Nath et al. 2017) are discussed below.

25.5.1 Aphid (*Rhopalosiphum nymphaeae* L.)

Aphids are major insects infesting the makhana crop particularly during the initial period of crop growth (Mishra et al. 2003; Kumar et al. 2011). The aphid infestation on makhana was first observed by Saraswati et al. (1990). The peak activity of aphid is from February to March, and it sharply declines and disappears in the month of April. The aphids are brownish or more or less shiny reddish-brown to dark olive in colour, with a wingless form body length of about 1.6–2.6 mm. They colonize aquatic plants of several families including Nymphaeaceae. An adult female produces 2–6 young ones per day parthenogenetically and continues up to 10–12 days. In its life, it undergoes four nymphal instars before attaining the adult stage. Cloudy weather coupled with western winds favor aphid infestation. Both nymph and adult suck the sap from the upper surface of the leaves. Under heavy infestation, the affected parts of leaves turn yellow/rust red and decay quickly, resulting in a reduction in yield.

25.5.2 Case worms (*Elophila depunctalis* G. &*Elophila crisonalis* W.)

Case worm is reported to be the key insect of makhana as it causes heavy damage and a significant reduction in the yield of makhana. The case worms appear in February on the crop with peak activity from March to June, and their population declines in July (Mishra et al. 2003; Kumar et al. 2011).

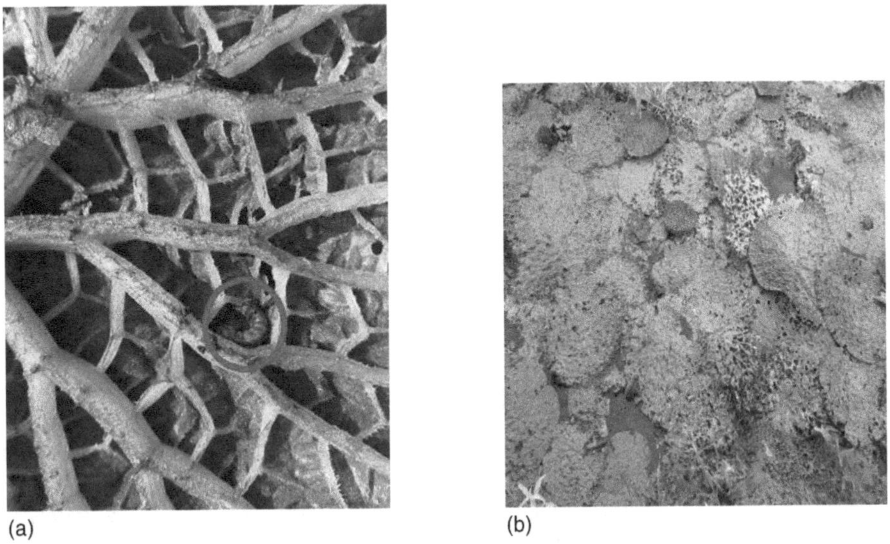

FIGURE 25.3 (a) Larva of case worm at lower surface of makhana leaf. (b) Damage of makhana leaf by case worm larva.

The young larva feeds on the makhana leaf either in the middle or at the margin by scraping. The larva makes an oval case out of the portion of the leaf cut out from it and remains inside it, moving with the case on the leaves (Figure 25.3a and 25.3b.). The larval stage of case worms is a major damaging phase, which causes a significant reduction in the yield of the makhana crop.

25.5.3 Singhara Beetle (*Galerucella birmanica*)

Singhara beetle affects the makhana crop particularly in the early crop growth, especially in the March month and shows soaring trends in April with extreme population in May. The pest shows falling trends from the end of May and vanishes in June. Grub and adult stages cause injuries in the younger and mature leaves, and even sometimes the petiole and integument of the makhana fruit. Grubs are more destructive than adults.

25.5.4 Gastropods

Gastropods, especially different species of Mollusca, are the most notorious incipient pests of the makhana crop causing heavy damage and are found in the entire crop growth period, leading to a substantial reduction in yield (Kumar 2020) (Figure 25.4). They lay eggs on fibrous root and leaves. The adults remain attached below the surface of the leaves, shoots and fibrous roots. It feeds on the plant tissue of the host plant and causes fast deterioration of affected parts of the makhana plants.

25.5.5 Sucking Bug (*Plea litruata* F.)

The sucking bug usually appears in tender seedlings of makhana and shows increasing trends for the next two-three months till April; however, its population declines thereafter. The population of bugs is low and is observed on leaves, tender shoots and roots of makhana. Both the nymphs and adults suck the cell sap from the lower surface of the tender leaves, shoots and roots of the crop.

FIGURE 25.4 Gastropods and damaged leaf of makhana by gastropods.

25.6 MANAGEMENT STRATEGIES OF MAKHANA INSECTS

Since makhana is a freshwater cash crop, therefore application of chemical pesticides should be avoided unless there is economically severe damage (Kumar et al. 2020). Organic pesticides are an approach to the control of insect pests. Therefore, the application of neem cake (@ 25 kg/ha) during land preparation or neem oil spray (0.3%) in standing crops has been recommended (Kumar et al. 2011). However, if chemical pesticides are necessary, then use imidacloprid 70WS or thiomethoxam 25WG @ 5 gm/kg seed and root dipping @ 5 gm/lt of water for half an hour along with three foliar sprays of NSKE @ 5% can be done (Nath et al. 2018). Removal of dead or dying tissue from the plants is advised to prevent further infestation. Seed treatment with thiomethoxam 25 WG (@ 5 g/kg seed) followed by root dip treatment with thiomethoxam 25 WG @ 5 g/litre of water prevents the infestation of soil-borne and other insects. Three times foliar spray of neem oil @ 5 ml/l of water at intervals of 25 days after 40 days of transplantation control aphids, case worms and other small insects. It also restricts the formation of wounds in the leaves of makhana and minimizes the chances of infection by the *Alternaria* pathogen. Gastropods may be controlled by holding leaves mechanically and hand-picking.

25.7 CONCLUSION

The productivity of makhana is affected by various biotic factors. Diseases and insect pests *viz*., Alternaria leaf blight, leaf spot, root rot, botrytis grey mold, tumor, aphid, case worms, rib borer and gastropods along with changing cropping patterns and the intensification of crops are some of the major biotic factors affecting makhana. The cultivation of makhana over a large area in Bihar especially in North Bihar, provides the pathogen and insects the opportunity to spread and could pose a serious threat to the cultivation resulting in a severe crop yield loss. Weather parameters such as temperatures, relative humidity, rainfall and eastern winds play a pivotal role in the infection processes and spread of pathogens. Since makhana is an aquatic cash crop, management of diseases and insects either in the pond or submerged field conditions is not easy, as most of the farmers cultivate makhana along with fish culture. Hence research and application of bio-control agents should be encouraged.

25.8 FURTHER STUDY

Makhana has received little attention from pathologists and entomologists; therefore, the biology of biotic stresses of makhana and their management is not deeply understood. Studies on insect pests and disease dynamics in makhana are further required to formulate management strategies against various diseases and pests of this aquatic crop encompassing huge future potential. The research work on the control of insect pests like aphids and some prominent diseases must be accelerated. The establishment of a correlation of disease incidence and insect infestation with weather parameters is essential for the prediction and forecasting of disease outbreaks. The development of a disease and pest-tolerant makhana variety using a collaborative approach is required. Moreover, being an aquatic crop, makhana is a high water-requiring crop; climate change impact on plant health should be monitored properly and minutely. Makhana diseases and pest management practices require the involvement of effective and novel bio-pesticides as makhana is an aquatic crop and chemical pesticides may be lethal to aquatic flora and fauna. Eventually, a detailed study of the interaction of insect pests and plant pathogens including their involvement in disease development and their role in crop loss is required to formulate potent management practices.

REFERENCES

Jana BR (2017) Impact of water depth on growth of gorgon nut and associated weed under wetland ecosystem of North Bihar province in India. *J Crop Weed* 13(1): 77–82.

Kumar L, Gupta VK, Jha BK, Singh IS, Bhatt BP, Singh, AK (2011) Status of makhana (*Euryale ferox* Salisb.) cultivation in India. *Technical Bulletin No. R-32/PAT-21*. ICAR Research Complex for Eastern Region, Patna.

Kumar M (2020) Makhana triumphs over flood: A success story from Darbhanga, Bihar. *Bio Res Today* 2(11): 1113–1116.

Kumar M, Padala VK, Ramya N, Chakraborty A, Dey JK (2021) Plant health issues in fox nut/ makhana (*Euryale ferox*): an agronomic perspective. *J Plant Health Issue* 2(1): 30–35.

Kumar S, Akhtar MN, Kumar S, Kumar M, Kumar T (2021a) Influence of weather parameters on alternaria leaf blight severity in makhana under the agro-climate of Koshi Region of Bihar. *J Agrometerol* 23(1): 54–59.

Kumar S, Kumar T, Singh M, Kumar M, Kumar M, Saha T (2021b) *Alternaria alternata* causes leaf and fruit blight in makhana. *Curr Sci* 121 (11): 1491–1495.

Kumar S, Akhtar MN, Kumar T, Kumar M (2021c) *Alternaria tenuissima* causes leaf spots in makhana. *Curr Sci* 120 (5): 749–750.

Minten B, Singh KM, Sutradhar R (2014) Branding in food retail of high value crops in Asia: Case of makhana from Bihar (India). *Food Policy* 38 (C): 23–34.

Mishra RK, Jha BP, Jha V, Singh SK, Mahto A (1992) Insect associations of *Euryale ferox* Salisb. in the ponds of Darbhanga, north Bihar. *J Freshwater Bio* 4(3): 199–208.

Mishra RK, Jha VN, Dehadrai PV (Eds.) (2003) *Makhana*. DIPA, ICAR, New Delhi.

Nath P, Kumar A, Yadav PK, Kumar R (2017) Emerging pests of makhana (*Euryale ferox* Salisb.) crop in koshi region of Bihar. *Int J Curr Microbiol App Sci (Special Issue)* 7: 4605–4609.

Nath P, Kumar A, Yadav PK, Kumar R (2018) Estimation of losses caused by aphid (*Rhopalosiphum nymphaeae* L.), case worms (*Elophila depunctalis* W. & *E. crisonalis* W.) and rib borer (*Chironomous* sp.) in makhana crop. *J Pharmacogn Phytochem (Special Issue)* 4: 199–201.

Saraswati KC, Mishra RK, Kumar R, Jha V (1990) *Rhopalosiphum nymphaeae* (L.) infestation on the leaves of *Euryale ferox*. *J Aphidol* 4: 89–92.

Selvamani R, Pandian RTP, Sharma P (2014) Role of weather on Alternaria leaf spots development in crucifers. *Indian Phytopath* 67(3): 285–290.

Shankar M, Chaudhary N, Singh D (2010) A review on Gorgon nut. *Int J Pharm Bio Arch* 1(2): 101–107.

Verma RAB, Jha V, Devi S (2003) Leaf and floral hypertrophy of makhana caused by *Doassansiopsis euryaleae*. In: Mishra RK, Jha V, Dharai PV (Eds.), *Makhana*. ICAR, New Delhi, pp. 163–168.

Verma, RAB, Jha V (1999) New *Doassansiopsis* associated with freshwater plant *Euryale ferox* Salisb. (*Makhana*) in north Bihar (India). *J Freshwater Biol* 11: 7–10.

Vision 2050 (2015) *Research complex for Eastern Region*, Indian Council of Agricultural Research, Patna, p. 33.

26 Disease Management Strategies in Medicinal and Aromatic Crops

Nidhi Didwania, Paridhi Saini, Nilanjana Saha, Vishal Kumar, Sneha Agarwal, Kritik Verma and Sasanka Kalita

26.1 INTRODUCTION

In India, approximately 8000 plant species are considered as medicinal plants. A large portion of the country's rural population uses medicinal plants to treat a variety of diseases. Approximately 1.5 million practitioners of Indian Systems of Medicine and Homeopathy (ISM&H) use medicinal plants for prevention, promotional, and therapeutic purposes (Mondal et al. 2018). Medicinal and aromatic crops play vital roles in disease prevention in plants because of the existence of phytochemicals, which are the key ingredients in the production of green pesticides as they are sustainable and safe (Gupta and Didwania 2021). Medicinal plants have been found to possess antibacterial, antifungal, anticancer, antidiuretic, anti-inflammatory, and anti-diabetic properties. Some of the examples are Kalmegh *(Andrographis paniculata)*, which is used to treat malarial fever and as a liver tonic; neem (*Azadirachta indica)* as seed oil is used in treating skin diseases; the leaves of tulsi *(Ocimum sanctum)* are used to cure coughs, colds, boils, and ulcers (Garg et al. 2021). A study proved that the plant extracts of *Azadirachta indica, Allium cepa, Vernonia anthelmentica, Tamarix aphylla, Tagetes erecta* and *Plumbago zelanicum,* possess strong antimicrobial properties in 50% concentration against plant pathogen *Xanthomonas campestris pv. campestris*, resulting in improved germination of seeds and plant vigour than streptomycin (Didwania et al. 2013). Aromatic plants are used as natural medicines as they have remedial and inherent pharmacological properties and therefore they are used for extraction of essential oils and various essences which find their applications in industries such as cosmetics, repellents, fragrance, herbal beverages, flavouring, pesticides, and spices (Samarth et al. 2017). Bacteria, viruses, nematodes, fungi, insects, arachnids, and weeds are examples of living organisms in existence that induce biotic stress in medicinal and aromatic plants. Biotic stress agents deprive their hosts of nutrients, which can result in plant death. Severe biotic stress can develop because of pre-harvest and post-harvest losses. Plants can cope with biotic stress despite the lack of an adaptive immune system by employing sophisticated strategies. A plant's genetic code controls the defence mechanisms that respond to these threats. Biologic stress is entirely distinct from abiotic stress, which harms plants due to non-living factors such as heat, cold, salinity, and drought. The type of biotic stress that plants can withstand depends on their environment, as does how well-adapted a given plant species is to that specific stress. Plant physiology, morphology, and agricultural productivity can all be impacted by biotic stress (Gull et al. 2019). Below listed are the various biotic stress causative organisms, followed by the biological, chemical, and cultural strategies to control them:

 DOI: 10.1201/9781032631578-26

26.2 FUNGAL DISEASES IN MEDICINAL AND AROMATIC CROPS AND THEIR MANAGEMENT

A variety of fungi present in the environment can cause diseases in crops that cause harmful deviations or alterations from the normal physiological processes of the plant cells. There are reports of more than 19,000 fungi causing diseases in crop plants across the globe (Jain et al. 2019). Fusarium wilt, fungal leaf spot, leaf blight, downy mildew, and powdery mildew are the primary types of fungal diseases that affect medicinal and aromatic plants.

26.2.1 Fungal Diseases

26.2.1.1 Fusarium Wilt

Fusarium wilt is a fungal plant disease caused by the fungus *Fusarium oxysporum* thriving in soil temperatures above 24 °C. Fusarium wilt is the most prevalent type of destructive disease of *Ocimum sanctum* L. (tulsi), causing a severe problem in the commercial production of tulsi (Prasad et al. 2019). Once infected, the plant loses turgidity, and the colour changes from a lighter green to a greenish-yellow, turning yellow and then brown, leading to wilting and plant death. Dark streaks appear in the xylem vascular tissue of the roots and the lower stem of the plant, resulting in root decay. Seedlings are also susceptible to infection and ultimately dies.

26.2.1.2 Fungal Leaf Spot

Fungal leaf spot is caused, by insects, environmental factors, toxicity, herbicides, or by fungal spores in the air which find a warm, wet, plant surface as their host, after which sporulation occurs and fungal leaf spot begins, resulting from necrosis (cell death). Medicinal plants such as aloe vera are infected by the fungus *Alternaria brassicae* (Ghosh et al. 2018), Oregano *(Origanum vulgare),* and sweet basil *(Ocimum basilicum)* are infected by the fungal pathogen *Alternaria alternata* (Avan 2021). The spots appear greyish-white towards the edges and are enclosed by reddish-brown, brownish, or yellowish borders. Newly emerging leaves are also susceptible to infection.

26.2.1.3 Leaf Blight

Leaf blight causes yellowing of leaf margins and tips (Jain et al. 2019). As the disease spreads throughout the plant, the spots turn reddish-purple or tan, forming large lesions. Eventually, the leaves wither and die. Early blight spreads through rain, irrigation, insects, and garden tools. The disease-causing pathogen in ashwagandha (*Withania somnifera* L. *Dunal)* is *Alternaria chlamydospora* Mouch. (Mondal et al. 2018). In lemon grass *(Cymbopogon citrates)*, the fungal pathogen is *Colletotrichum caudatum* sensu lato and *Curvularia andropogonis* (Zimm.) Boedijn (Meltem 2021). In oregano *(Origanum vulgare)* the fungal pathogen is *Rhizoctonia solani* causes irregular spots that spread inward from the leaf edges (Avan 2021).

26.2.1.4 Downy Mildew

Downy mildew *(Plasmopara viticola)* is a disease of the foliage causing destruction to the leaves, stems, flowers and fruits of plants. Downy mildew found on sweet basil (*Ocimum basilicum*), holy basil or tulsi *(Ocimum sanctum)* (Avan 2021) is caused by the fungus *nospora belbahrii* (Farahani-Kofoet et al. 2014). The disease first appears as yellow to white patches on the upper surfaces of older leaves. Eventually, the leaves turn brown and wither, even if the plant has sufficient water. The disease thrives in cold weather and almost immediately disappears after exposure to bright sunlight. It spreads through insects, wind, rain, and garden tools.

26.2.1.5 Powdery Mildew

There have been reports of powdery mildew in medicinal plants such as Chandan *(Santalum album* L.*)* and Amla *(Emblica officinalis* Gaertn.), caused by *Podosphaera xanthii* (Castagne) Braun &

Shishkoff. The fungal pathogen *Erysiphe graminis* DC infects lemon grass *(Cymbopogon citratus)* causing chlorotic and brownish spots that appear as a powdery coating on the leaf surface (Meltem 2021). As a result, the leaves curl and bend (Avan 2021). There appear to be white to yellow spots on the top faces of fully grown leaves which gradually turn brown. The infected leaves then fall off. The mycelium of infected *Santalum album* Linn. (Chandan) under the microscope was observed as amphiphyllous, thick coating of powdery masses on leaves, persistent, hyphae were septate thin-walled, smooth, hyaline, branched, appressorium, solitary, lobed shaped (Patel et al. 2015). The disease thrives in colder temperatures. The spores spread through insects, wind, rain, and garden tools (Jain et al. 2019).

26.2.2 Management Strategies for Fungal Diseases

26.2.2.1 Biological Control

Biofumigation (biological soil disinfection) utilizes specific plant species that produce toxic molecules, and biodisinfection employs organic materials to create anaerobic conditions that inactivate phytopathogenic fungi. (El Baky et al. 2021). Suppressive soil comprising antagonistic microorganisms can be added to natural pathogen-conductive soil to suppress plant fungal diseases by managing phytopathogenic fungi (Cha et al. 2016). To prevent downy mildew the infected crops are uprooted and burnt (Jain et al. 2019). Despite years of infection, plant fungal disease can be reduced by monocultures of the same crop in soil conducive to pathogens. (Kim et al. 2019). To combat the disease known as leaf spot caused by *Alternaria brassicae* (Berk.) Sacc in *Aloe vera*, seeds are primed with growth-promoting rhizobacteria (PGPR) such as *Pseudomonas aeruginosa*, *P. putida*, *Bacillus subtilis*, *B. cereus*, *Azotobacter chroococcum*. This technique has been proven to enhance seed germination, seedling, vigour, fruit size, chlorophyll content, indole-3-acetic acid (IAA) production, and rhizospheric phosphate solubilization, thereby improving plant growth. Fungal bio-control agents for example *Trichoderma* and *Beauveria* are frequently employed in the treatment of fungal diseases on crop plants. Formulations from *T. viride*, *T. harzianum*, and *T. virens* have been generated to regulate these diseases. (Ghosh et al. 2018).

26.2.2.2 Cultural Control

The cultural strategies to combat fungal diseases in medicinal plants include cleaning or disinfecting farming tools frequently regardless of there being the presence of diseases in plants. After handling infected plants, individuals should wash their hands, body, and clothes to prevent infecting healthy plants. Infected plants are to be eradicated when symptoms are seen. Some of the pathogens can be prevented by maintaining a lower garden temperature. Humidity among plants is to be reduced by adequately spacing them while planting and pruning lesser branches (Jain et al. 2019). Arid weather suitable for seed production should be maintained as seed-borne fungal diseases are thrive in a wet climate. Practicing crop rotation is an effective method to prevent fungal plant diseases (Singh et al. 2012).

26.2.2.3 Chemical Control

The chemical strategies to prevent fungal diseases in medicinal plants contaminated with fusarium wilt include killing fungicides like Terraclor, Mycostop, Cap Tan, and Thiram to ward off infections. Medicinal plants such as *Santalum album* Linn. (Chandan) and *Emblica officinalis* Gaertn. (Amla) infected with powdery mildew can be treated with sulphur. Copper-based fungicides should be applied at the early signs show for leaf blight. The disease of downy mildew in medicinal plants such as sweet basil (*Ocimum basilicum*) can be treated using Forestry-grade fungicides (Jain et al. 2019). To treat leaf blight carbendazim 50WP @ 0.1% can be applied three times at an interval of 15 days (Mondal et al. 2018). Plant extracts are used in the environmentally friendly synthesis of silver nanoparticles (AgNPs), which have fungicidal properties against a variety of phytopathogenic fungi. (Rabab et al. 2018). The addition of AgNPs to soil or coating the plant seeds with AgNPs can control phytopathogenic fungi, promoting plant growth (El-Baky et al. 2021). Zinc Oxide Nanoparticles (ZnO NPs) extracted from *Parthenium hysterophorus* L. leaves can inhibit fungal growth against *Aspergillus fumigatus*, *Aspergillus flavus* and *Aspergillus niger* (Rajiv et al. 2013). The green synthesis of Au NPs using extracts of fresh leaves and flowers proves to be antimicrobial and antifungal to many phytopathogenic fungi such as *A.flavus*. The vapour action of fungicide elemental sulphur can be employed to destroy fungal spores (El-Baky et al. 2021).

26.3 BACTERIAL DISEASES IN MEDICINAL AND AROMATIC CROPS AND THEIR MANAGEMENT

Bacterial pathogens that cause serious and frequently fatal infections in plants. Different groups of pathogenic bacteria are classified based on their cell walls, shape, and motility. The taxonomy of plant pathogenic bacteria is continuously evolving due to advancements in bacterial classification. The most prevalent plant pathogenic bacteria are *Erwinia*, *Pectobacterium*, *Pantoea*, *Agrobacterium*, *Pseudomonas*, *Ralstonia*, *Burkholderia*, *Acidovorax*, *Xanthomonas*, *Clavibacter*, *Streptomyces*, *Xylella*, *Spiroplasma*, and phytoplasma. The most common diseases in medicinal plants are bacterial spots and cankers. Bacterial galls, bacterial vascular wilts, bacterial soft rots, bacterial scabs, etc. (Mansfield et al. 2012).

26.3.1 Bacterial Diseases

26.3.1.1 Bacterial spot

Leaf spots are the most prevalent indication of bacterial disease. Spots occur on the leaves, flowers, fruits, and stems of plants. If the spots form and spread rapidly, the disease is classified as a blight. Bacterial spots are usually resulting from the genera *Pseudomonas* and *Xanthomonas*. The most prevalent type is turmeric bacterial leaf spot, a disease of turmeric (*Curcuma longa* L.). The symptoms shown in the turmeric plant were on the leaves, there are oblong brown dots with grey centres (Utpal 2013). Bacterial leaf spots, caused by *Pseudomonas viticola* and *Xanthomonas azadirachtin*, appear as black spots on the surface of neem and betel vine leaves. (Thyagharajan and Raji 2021). In tulsi, basil disease was caused by *Pseudomonas cichorii* (Swingle) Stapp, the symptoms of the bacterial leaf spot were black, and spots were observed on the leaves (Mondal et al. 2018)

26.3.1.2 Bacterial Canker

Bacterial canker primarily affects woody species of medicinal plants. The development of cankers and gum exudation is the most visible indication of bacterial canker disease in plants. Cankers can be deep, dark brown, and significantly longer than they are wide. Canker cortical tissue can range from orange-brown to dark brown. Citrus canker is caused by the bacterium pathogen *Xanthomonas citri* pv. *citri* (Hasse) Constantin et al., and the disease symptoms appear as lesions of the leaf, fruit and twig (Sharma and Sharma 2009).

26.3.1.3 Bacterial wilt

Ralstonia solanacearum (Smith) Yabuuchi et al., a soil-borne bacterium, is the primary pathogen causing bacterial wilt in plants. The causative pathogen enters the host plant's xylem vessels, multiplies there, and moves through them. It obstructs the movement of nutrients and water by creating gum. The infection frequently breaks down portions of the xylem vessels' cell walls, leaving behind pockets of bacteria, gum, and cell debris. The aboveground plant sections will wilt and eventually die due to bacterial wilt disease. The aboveground plant sections will wilt and pathogen known as *Xanthomonas campestris* pv. *turf*, the major symptoms of this blight disease were throughout the entire leaf, a water-soaked necrotic lesion was grown (Singh et al. 2016). eventually die due to bacterial wilt disease (Sarkar and Chaudhari 2016). The primary symptoms of bacterial wilt initially include dull green leaves that wilt during the day but recover at night. As the disease progresses, the edges of the leaves turn brown and yellow before completely withering and dying.

26.3.2 Management Strategies for Bacterial Diseases

26.3.2.1 Biological control

Biological control involves the employment of beneficial bacteria or their by-products, or extracts from plants or animals, to inhibit plant disease. Biological control agents (BCAs) can be managed in the field using two main strategies: inundative control, which involves introducing BCAs in sufficient quantity to suppress disease even in the absence of reproduction, or augmentative, which involves giving BCAs in sufficient quantity to create a stable and reproducing population in order to control the disease. For woody plant cuttings, the most notable example of biological control for bacterial infections in plants has been the use of *Agrobacterium radiobacter* strain K84 to protect exposed surfaces from crown gall caused by *Agrobacterium tumefaciens*. (Kerr 2016). The application of bacteriophages to the management of bacterial diseases is another biological control method that has been the subject of research activity on occasion over the past few decades. (Jones et al. 2007).

26.3.2.2 Cultural Control

The cultural practices used by the farmers to protect their harvest from bacterial diseases are evaluating that seeds or plant propagation materials are free of bacteria, cleaning, and especially clearing off pests from pruning tools, rotation of crops to lessen overwintering, avoiding superficial wounds that allow bacteria to enter the inner tissues, using only germ-free nursery stock for propagation. Sometimes, prolonged exposure to hot, dry air and sunlight will kill microorganisms in plant matter (Fry 2012).

26.3.2.3 Chemical Control

There are various pesticides or insecticides available that available to control plant diseases by either killing the disease-causing pathogens or reducing their growth. Chemicals used to control bacterial diseases in medicinal plants are bactericides such as copper-based bactericides which protect plants from both bacterial and fungal infections. These bactericides are usually sprayed on the leaves and stems. To combat bacterial resistance, oxytetracycline and streptomycin have been approved for use against bacterial blight disease (Lindsey et al. 2020). Additionally, fastidious vascular bacteria-related diseases are treated with antibiotics. Some medications, particularly tetracycline, are effective against *Candidatus* Phytoplasma and *Spiroplasma*. It's been proven that using copper-containing compounds or Bordeaux mixtures (copper sulphate and lime) to treat the diseases is successful (Garland et al. 2017).

26.3.2.4 Genetic Strategies

In order to treat bacterial diseases, the creation of transgenic plants with bacterial disease resistance holds great potential. Recent developments in CRISPR technology have enabled researchers to develop a range of CRISPR variants with diverse applications, including the creation of bacterial disease-resistant cultivars. Through the disruption of a plant susceptibility (S) gene, which alters the way pathogens and plants interact and reduces pathogen fitness in the host plant, this method has been applied to the design of disease resistance (Zaidi et al. 2018). By using genome editing technologies like the TALEN and CRISPR/Cas9 systems, plant breeders can create new sources of resistance to bacterial infections and other diseases on a much larger scale without introducing foreign DNA into the host genome. Additionally, compared to conventional breeding procedures, the greater precision of these new methodologies and the potential for resistance stacking present a very attractive scenario for the creation of multiple resistances (Sundin et al. 2016).

26.4 VIRAL DISEASES IN MEDICINAL PLANTS AND THEIR MANAGEMENT

26.4.1 Viral Diseases

One of the most serious diseases is plant virus infection, which leads to significant large global losses in the production of medicinal plants. Viral diseases are mosaic diseases such as tobacco mosaic virus (TMV), tomato spotted wilt virus (TSWV), tomato yellow leaf curl virus (TYLCV), and cucumber mosaic virus (CMV), (Scholthof et al. 2011). The tobacco mosaic virus typically spreads from plant to plant through mechanical wounds caused by contaminated hands, items of clothing, or implements. Viral diseases can affect medicinal plants by affecting their medicinal properties. It was seen that the medicinal plants associated with TMV infections show the following symptoms stunting, on the leaves, there is a light and dark green (or yellow and green) mosaic pattern, incorrectly shaped leaves or growing points, yellow spotting on leaves, distinct vein-specific yellowing, etc. Diseases like these can cause physiological changes in plants. For instance, it was found that plants contaminated with TMV in *Datura metel* produced less chlorophyll, which in turn reduced photosynthesis (Srivastava et al. 2016).

Tomato yellow leaf curl (TYLC) is one of the most destructive viral diseases affecting cultivated medicinal plants in subtropical and tropical regions worldwide. It is caused by a virus from the genus *Begomovirus*, the family *Geminiviridae,* and is usually transmitted by the whitefly *Bemisia tabaci.* The most prominent signs of TYLCV in medicinal plants were leaf curling, yellowing of veins in small leaves, wrinkling of leaves, etc. In *Ocimum basilicum,* it was observed that cultivars affected with tomato yellow leaf curl virus showed stunted growth, and excessive branching, reducing the size of leaves (Zaim et al. 2016). Cucumber Mosaic Virus belongs to the genus *Cucumovirus*, the family *Bromoviridae*. It affects more than 1200 species of plants. Aphids, are the primary vectors which feed on healthy plants while spreading the virus to uninfected ones, are the primary vectors for the advancement of MV. More than 80 species of aphids may spread CMV. The number of infected plants functioning as the virus reservoirs in an area, the size and activity of the aphid populations, and the severity of the cucumber mosaic are all frequently related. A plant infected with CMV infection can show symptoms like stunting, deformity of the developing points, leaves, leaf with yellow spots (Li et al. 2020).

26.4.2 Management Strategies for Viral Diseases

26.4.2.1 Cultural Control

The strategies that are preferably used by the farmers to prevent the spread of viral diseases in medicinal plants are identifying and eliminating symptomatic plants, well-known substitute weeds, or volunteer plants that may serve as a reservoir for a particular virus, after handling infected plants, it is crucial to wash their hands, and tools must be washed as the virus is commonly spread in plants by contact with contaminated tools geographic isolation of production facilities may also aid in avoiding plant virus losses. Isolating newly arrived plant material before introducing it into the remainder of a manufacturing system can also help to reduce the unintentional introduction of diseases (Luvisi et al. 2017).

26.4.2.2 Chemical Control

The chemical strategies have been proven to be challenging as the virus is highly unstable, although these strategies do not affect the quality of the plant, they impact its harvest. Synthetic pesticides are the conventional approach, and they have detrimental effects on both the environment and human health. The search for viral disease treatments continues, even though no virus-curative agrochemicals are currently available on the market, except the antiviral agent ningnanmycin (NNM), which appears to have cure rates of 30–60% (Ouyang et al. 2008). Plant breeders typically strive for total resistance, which prevents the virus from causing a systemic infection. Since a significant number of plants may be evaluated concurrently, ELISA and molecular hybridization have been employed to test germplasm and cultivars for resistance (Rubio et al. 2020). Applying insecticides to populations of vector-borne viruses, such as *B. tabaci* in tomato yellow leaf curl virus, in fields and greenhouses is the most widely utilized method of controlling viral diseases (Cathrin and Ghanin 2014).

26.4.2.3 Genetic Strategies

To understand the function of biotic stress, a gene-by-gene approach has traditionally been employed. Functional genomics enables large-scale gene function analysis using high throughput technology and incorporates interactions between gene products within cells and organisms. Significant alterations in gene expression, impact the transcriptome, proteome, and metabolome composition of plants, and mediate plant adaptation to biotic stress. Genome editing techniques like the CRISPR-Cas9 system may be employed to directly mutate host genes implicated in recessive resistance to avoid viral contact in crops and to adopt resistance genes generated through engineered evolution. (Haggag et al. 2015). RNA interference (RNAi) technology has been proven effective in conferring resistance to several viral infections in plants (Parmar et al. 2017).

26.5 NEMATODE DISEASES IN MEDICINAL AND AROMATIC PLANTS AND THEIR MANAGEMENT

26.5.1 Nematode Diseases

Plant parasitic nematodes are microscopic, filiform worms ranging from 0.5 to 4–5 mm. A wide range of nematodes impacts on medicinal and aromatic plants, and some of them have a significant economic impact. The most prevalent nematode species include *Meloidogyne javanica*, *Heterodera*, *Globodera*, *Pratylenchus,* and *Radopholus similis*, root knot, cyst, root lesion, and the most common nematode diseases in food and fragrance plants are burrowing ones.

26.5.1.1 Root-knot Nematode

Meloidogyne incognita is a nematode, a type of roundworm in the family Heteroderidae. It is commonly called the "southern root-knot nematode. Root knot is the most prevalent ailment affecting medicinal plants. A species of nematode in the *Heteroderidae* family called *Meloidogyne incognita* and *M. javanica* damages the roots of medicinal plants. It was discovered that root-knot nematodes, specifically *M. incognita* and *M. javanica*, are present in Aloe vera plants. By consuming the root cells with their needle-like mouthparts, they harm plants (stylets). Instead of the typical root-knot nematode galls, the infected plants showed irregular thickening of the roots and reddish-brownish discolouration of the leaves connected to inadequate plant growth (Palomares-Rius et al. 2015). The Solanaceae family includes the annual evergreen shrub called ashwagandha (*Withania somnifera* (L.)). *M. incognita* is a common root-knot nematode that causes significant plant damage and has an approximate $100 billion global economic impact. The root is the main point of infection. Infestations result in root galling and stunted growth, lowering the output of the ashwagandha plant (Pandey and Kalra 2003). Yellowing, stunted growth, wilting, tubers, bulbs, and galling of the roots are signs of an infected plant (Bakr et al. 2014).

26.5.2 Management Strategies of Nematode Diseases

The important medicinal plant *W. somnifera* (ashwagandha) is extremely susceptible to the root-knot nematode *Meloidogyne incognita.* Various fungi antagonistic to nematodes have been investigated as potential bio-control agents. It was determined whether the nematode *M. incognita* could be controlled by the fungi *Arthrobotrys oligospora, Trichoderma harzianum*, and *Paecilomyces lilacinus*, when combined with organic natural products (Neem compound mix). They were also compared to the chemical control agent carbofuran in terms of their effectiveness in controlling nematodes. The examined fungal agents significantly reduced nematode populations and increased plant growth. (Poornima and Pandey 2009). Nematodes that parasitize plants are a threat to the global agricultural industry's ability to produce various crops. New nematode control methods must be developed due to the high toxicity of chemical nematicides. Filamentous fungi can offer an intriguing bio-control alternative in this regard. The main filamentous fungi studied and used as biological control agents (BCAs) against nematodes as resistance inducers are those of the genus *Trichoderma*, mycorrhizal fungi, and endophytic fungi. Furthermore, by producing lytic enzymes, antibiosis, paralysis, and parasitism, they can directly lessen the harm done by plant parasitic nematodes. By increasing the plant's ability to absorb nutrients and water, or by changing the root morphology and/or rhizosphere interactions, which benefits the growth of plants, they also lessen the harm caused by competition for space and resources. Additionally, filamentous fungi have the capacity to elicit the defence mechanisms of plants that are hormone-mediated (e.g., strigolactones, salicylic and jasmonic acids). A plant's ability to defend itself can also be improved by changing the way that chemical defence components are transported through the plant or by synthesising secondary metabolites and other enzymes. Therefore, using filamentous fungi from the aforementioned groups as BCAs in medicinal plants is a promising, long-lasting bio-control method against nematodes that parasitize plants (Poveda et al. 2020).

26.6 PHYTOPLASMA DISEASES

The obligate parasite phytoplasmas lack a cell wall and are related to gram-positive bacteria. They are small ranging (0.3–0.5 M in size; 580–2200kb circular DNA genome) (Hogenhout et al. 2008). The numerous ailments caused by phytoplasma and plant viruses affect biomass yield and world production of medicinal and aromatic plants. Numerous medicinal and aromatic plants, such as Mentha species, *Andrographis paniculata*, *Hyoscyamus muticus*, *Catharanthus roseus*, *Rauwolfia serpentina*, *Hyoscyamus albus*, and *Papaver somniferum*, have been reported to have viral infections. The causal organisms for phytoplasma diseases in *Catharanthus roseus*, *Withania somnifera*, *Foeniculum vulgare*, *Phyllanthus amarus*, *Matricaria chamomilla*, and *Portulaca grandiflora* are yellow aster, witches' broom made of peanuts, coconut yellowing that is fatal, clover proliferation, Ash's yellows, Witches' broom made of pigeon peas, Bermuda leaves, and Group Stolbur Luffa (Ajaykumar et al. 2007).

26.7 CONCLUSIONS AND FUTURE PERSPECTIVES

Medicinal plants provide significant benefits to humans and the environment, and their cash value far outweighs that of any other crop plant in terms of functions and efficacy. However, medicinal management of plant diseases has been largely ignored, and thus there is little information regarding their genetic system, heterozygosity, growth cycle, and self-incompatibility are all factors to consider and have stifled the advancement of biotic stress-tolerant medicinal plants. A multifaceted analysis of the effects of management strategies on economics, sociology, and ecology to completely understand the mechanisms underlying the epidemics of aromatic and medicinal plants, the functioning of healthy agro-ecosystems, and the individual and collective roles of RAER (Resistance, Avoidance, Elimination, and Remedy) approaches to medicinal plant disease management (Pywell et al. 2015). The most sensible and efficient method for managing plant diseases is host resistance. To ensure a mismatch between the critical periods of plant and pathogen development, the disease should not be allowed to alter the geographic and temporal pattern of host plant cultivation, such as by changing planting time, location, or cultivation system. Elimination produces incredible plant disease management results with no adverse ecological effects by eliminating or lowering inoculum sources, some techniques that reduce overwintering locations (places and hosts) of diseases and their transmission vectors. Spraying pesticides to kill diseases and/or their insect vectors is a necessary part of plant disease control when other methods fall short of the required level of pathogen population density reduction and epidemic amelioration. To accomplish this, further studies on the management of aromatic and medicinal plant diseases should focus on the following topics: (i) the pattern of plant disease evolution and epidemic under shifting environmental conditions and agricultural production ideologies; (ii) the contribution of ecological considerations to agricultural productivity and plant health; (iii) social-economic analysis of plant disease epidemics and management; and (iv) technology development (He et al. 2016).

REFERENCES

Ajayakumar PV, Samad A, Shasany AK, Gupta MK, Alam M, Rastogi S (2007) First record of a *Candidatus* phytoplasma associated with little leaf disease of *Portulaca grandiflora*. *Austr Plant Dis Notes* 2(1): 67–69.

Avan M, Palacıoğlu G, Aksoy C, Kaya R, Bayraktar H, Katırcıoğlu YZ, Maden S (2021) Characterization and pathogenicity of *Rhizoctonia* species causing root rot and damping-off on sugar beet in Turkey. *Curr Microbiol 78*(5): 1939–1948.

Bakr RA, Mahdy ME, Mousa ESM (2014) Biological control of root-knot nematode *Meloidogyne incognita* by *Arthrobotrys oligospora*. *Egypt J Crop Protect* 9(1): 1 1–11.

Carris LM, Little CR, Stiles CM (2012) Introduction to fungi. *Plant Health Instructor* 48. https://doi.org/10.1094/PHI-I-2012-0426-01

Cathrin PB, Ghanim M (2014) Recent advances on interactions between the whitefly *Bemisia tabaci* and begomoviruses, with emphasis on Tomato yellow leaf curl virus. In: Gaur RK, Hohn T, Sharma P (Eds.), *Plant Virus–Host Interaction*. Academic Press, Cambridge, MA. pp. 79–103. https://doi.org/10.1016/B978-0-12-411584-2.00004-4

Cha JY, Han S, Hong HJ, Cho H, Kim D, Kwon Y, Kwon SK, Crüsemann M, Bok Lee Y, Kim JF, Giaever G (2016) Microbial and biochemical basis of a Fusarium wilt-suppressive soil. *The ISME J* 10(1): 119–129.

Dajic-Stevanovic Z, Pljevljakusic D (2015) Challenges and decision making in cultivation of medicinal and aromatic plants. In: Máthé Á (Ed.) *Medicinal and Aromatic Plants of the World*. Springer, Dordrecht, pp. 145–164.

Didwania N, Sadana D, Trivedi PC (2013) Antibacterial activity of a few medicinal plants against *Xanthomonas campestris* pv. *campestris*. *Int J Resn Pharmaceutical Sci* 4(2): 177–182.

El-Baky NA, Amara AAAF (2021) Recent approaches towards control of fungal diseases in plants: an updated review. *J Fungi* 7(11): 900.

Farahani-Kofoet RD, Römer P, Grosch R (2014) Selecting basil genotypes with resistance against downy mildew. *Sci Horti* 179: 248–255.

Fry WE (2012) *Principles of Plant Disease Management*. Academic Press, Cambridge, MA, USA. p. 378.

Garg AK, Singh S (2021) Role of medicinal plant in human health disease. *Asian J Plant Sci Res*. 2(11): 14919–14921. https://doi.org/10.24941/ijcr.40170.11.2020

Garland M, Loscher S, Bogyo M (2017) Chemical strategies to target bacterial virulence. *Chemical Rev* 117(5): 4422–4461.

Ghosh SK, Banerjee S, Pal S, Chakraborty N (2018) Encountering epidemic effects of leaf spot disease (*Alternaria brassicae*) on *Aloe vera* by fungal biocontrol agents in agrifields—an ecofriendly approach. *PloS One* 13(3): e0193720.

Gull A, Lone AA, Wani NUI (2019) Biotic and abiotic stresses in plants. *Abiotic Biotic Stress Plants* 7: 1–19.

Gupta S, Didwania N (2021) Role of medicinal plants as green pesticides against Alternaria blight. *Bulgarian J Agri Sci* 27(3): 562–568.

Gutteridge JM, Halliwell B (2010) Antioxidants: molecules, medicines, and myths. *Biochem Biophys Res Commun* 393(4): 561–564.

Haggag WM, Abouziena HF, Abd-El-Kreem F, El Habbasha S (2015) Agriculture biotechnology for management of multiple biotic and abiotic environmental stress in crops. *J Chem Pharm Res* 7(10): 882–889.

He DC, Zhan JS, Xie LH (2016) Problems, challenges and future of plant disease management: from an ecological point of view. *J Integrative Agri* 15(4): 705–715.

Hogenhout SA, Oshima K, Ammar ED, Kakizawa S, Kingdom HN, Namba S (2008) Phytoplasmas: Bacteria that manipulate plants and insects. *Mol Plant Pathol* 9(4): 403–423.

Idu M (2009) *The Plant Called Medicine: The 104th Inaugural Lecture Series of University of Benin City*. Calameo, Nigeria.

Inoue M, Hayashi S, Craker LE (2019) Role of medicinal and aromatic plants: past, present, and future. In: Perveen S Al-Taweel A (Eds.), *Pharmacognosy-Medicinal Plants*, pp. 1–13. IntechOpen. http://doi.org/10.5772/intechopen.82497

Jain A, Sarsaiya S, Wu Q, Lu Y, Shi J (2019) A review of plant leaf fungal diseases and its environment speciation. *Bioengineered* 10(1): 409–424.

Jones JB, Jackson LE, Balogh B, Obradovic A, Iriarte FB, Momol MT (2007) Bacteriophages for plant disease control. *Annu Rev Phytopathol* 45: 245–262.

Kerr A (2016) Biological control of crown gall. *Austra Plant Pathol* 45(1): 15–18.

Kim DR, Jeon CW, Shin JH, Weller DM, Thomashow L, Kwak YS (2019) Function and distribution of a lantipeptide in strawberry Fusarium wilt disease–suppressive soils. *Mol Plant-Microbe Interact* 32(3): 306–312.

Li N, Yu C, Yin Y, Gao S, Wang F, Jiao C, Yao M (2020) Pepper crop improvement against cucumber mosaic virus (CMV): a review. *Front Plant Sci* 11: 598798.

Lindsey APJ, Murugan S, Renitta RE (2020) Microbial disease management in agriculture: current status and future prospects. *Biocatalysis Agri Biotechnol* 23: 101468.

Luvisi A, Panattoni A, Materazzi A, De Bellis L (2017) Chemical outbreak for tobacco mosaic virus control. *Int J Agri Biol* 19(4):792–800.

Mansfield J, Genin S, Magori S, Citovsky V, Sriariyanum M, Ronald P, Dow MAX, Verdier V, Beer SV, Machado MA, Toth IAN (2012) Top 10 plant pathogenic bacteria in molecular plant pathology. *Mol Plant Pathol* 13(6): 614–629.

Meltem AVAN (2021) Important fungal diseases in medicinal and aromatic plants and their control. *Turkish J Agri Engin Res* 2(1): 239–259.

Mondal G, Dasgupta B, Sharma R (2018) Diseases of medicinal and aromatic plants and their management. *Recent Approach Manage Plant Dis* 251: 283.

Ndhlala AR, Moyo M, Van Staden J (2010) Natural antioxidants: fascinating or mythical biomolecules? *Molecules* 15(10): 6905–6930.

Ouyang G, Chen Z, Cai XJ, Song BA, Bhadury PS, Yang S, Jin LH, Xue W, Hu DY, Zeng S (2008) Synthesis and antiviral activity of novel pyrazole derivatives containing oxime esters group. *Bioorganic Med Chem* 16(22): 9699–9707.

Palomares-Rius JE, Castillo P, Rapoport H, Archidona-Yuste A, Tzortzakakis EA (2015) Host reaction of *Aloe vera* infected by *Meloidogyne incognita* and *M. javanica* in Crete Island (Greece). *Eur J Plant Pathol* 142(4): 887–892.

Pandey R, Kalra A (2003) Root-knot disease of ashwagandha *Withania somnifera* and its eco-friendly cost effective management. *J Mycol Plant Pathol (India)* 33(2): 240–245.

Parmar N, Singh KH, Sharma D, Singh L, Kumar P, Nanjundan J, Khan YJ, Chauhan DK, Thakur AK (2017) Genetic engineering strategies for biotic and abiotic stress tolerance and quality enhancement in horticultural crops: A comprehensive review. *3 Biotech* 7(4): 1–35.

Patel RP, Pandey GN, Tiwari G, Patidar H, Patidar DK (2015) First report of powdery mildew fungi on. *Int J Current Res* 7(6): 16705–16708.

Poornima S, Pandey R (2009) Biological control of root-knot nematode; *Meloidogyne incognita* in the medicinal plant; *Withania somnifera* and the effect of biocontrol agents on plant growth. *African J Agri Res* 4(6): 564–567.

Poveda J, Abril-Urias P, Escobar C (2020) Biological control of plant-parasitic nematodes by filamentous fungi inducers of resistance: *Trichoderma*, mycorrhizal and endophytic fungi. *Front Microbiol* 11: 992.

Prasad R, Pande SK, Chandra S, Kumar N, Maurya N (2019) *In vitro* evaluation of bioagents botanical and fungicides against *Fusarium oxysporum* f.sp. *baisilici*. *Int J Chem Stud* 7(6): 746–748.

Pywell RF, Heard M.S, Woodcock B.A, Hinsley S, Ridding L, Nowakowski M, Bullock JM (2015) Wildlife-friendly farming increases crop yield: evidence for ecological intensification. *Proc R Soc B Biol Sci* 282(1816): 20151740.

Rabab ME, Raida EAH, Awatif AH (2018) Biosynthesis and characterization of silver nanoparticles using *Trichoderm longibrachiatum* and their effect on phytopathogenic fungi. *Egypt J Biol Pest Control* 28: 28.

Rajiv P, Rajeshwari S, Venckatesh R (2013) Bio-Fabrication of zinc oxide nanoparticles using leaf extract of *Parthenium hysterophorus* L. and its size-dependent antifungal activity against plant fungal pathogens. *Spectrochim Acta Part A Mol Biomol Spectroscopy* 112: 384–387.

Reddy K (2001) Electron microscopy and molecular characterization of phytoplasmas associated with little leaf disease of brinjal (*Solanum melongena* L.) and periwinkle (*Catharanthus roseus*) in Bangladesh. *J Phytopathol* 149(5): 237–244.

Rubio L, Galipienso L, Ferriol I (2020) Detection of plant viruses and disease management: relevance of genetic diversity and evolution. *Front Plant Sci* 11: 1092.

Samad A, Shasany AK, Gupta S, Ajayakuar PV, Darokar M.P, Khanuja SPS (2006) First report of a 16SrVI group phytoplasma associated with witches'-broom disease on *Withania somnifera*. *Plant Dis* 90(2): 248–248.

Samarth RM, Samarth M, Matsumoto Y (2017) Medicinally important aromatic plants with radioprotective activity. *Future Sci OA* 3(4): FSO247. PMID: 29134131; PMCID: PMC5674267. https://doi.org/10.4155/fsoa-2017-0061

Sarkar S, Chaudhuri S (2016) Bacterial wilt and its management. *Curr Sci* 110(8): 1439–1445.

Scholthof KBG, Adkins S, Czosnek H, Palukaitis P, Jacquot E, Hohn T, Hohn B, Saunders K, Candresse T, Ahlquist P, Hemenway C (2011) Top 10 plant viruses in molecular plant pathology. *Mol Plant Pathol* 12(9): 938–954.

Sharma SK, Sharma RR (2009) Citrus canker approaching century: A review. *Tree Forestry Sci Biotechnol* 2: 54–56.

Singh VK, Chawla S (2012) *Cultural Practices: An Ecofriendly Innovative Approach in Plant Disease Management*. In: Singh VK, Singh Y, Singh A (Eds.), *Eco-friendly innovative approaches in plant disease management* (pp. 1–20). International Book Distributors and Publisher. https://doi.org/10.13140/RG.2.1.4216.2327

Singh A, Gupta R, Saikia SK, Pant A, Pandey R (2016) Diseases of medicinal and aromatic plants, their biological impact and management. *Plant Genet Res* 14: 370–373.

Srivastava D, Shukla K (2016) Viral diseases on medicinal plants of north-eastern Uttar Pradesh. In: Gaur RK, Petrov NM, Patil BL, Stoyanova MI (Eds.), *Plant Viruses: Evolution and Management*. Springer, Singapore, pp. 89–129.

Sundin GW, Castiblanco LF, Yuan X, Zeng Q, Yang CH (2016) Bacterial disease management: challenges, experience, innovation and future prospects: Challenges in bacterial molecular plant pathology. *Mol Plant Pathol*, *17*(9), 1506–1518. https://doi.org/10.1111/mpp.12436

Thyagharajan KK, Raji IK (2021) Diagnosis of neem leaf diseases using fuzzy-HOBINM and ANFIS algorithms. *CMC-Comput Mater Continua* 69(2): 2061–2076.

Utpal D (2013) Bioefficacy of fungicides, bio-control agents and botanicals against leaf spot of turmeric incited by *Colletortricum capsici*. *African J Microbiol Res* 7(18): 1865–1873.

Wachtel-Galor S, Benzie IF (2011) Herbal medicine. *Lester Packer,* Ph. D., p. 1.

Wang W, Xu J, Fang H, Li Z, Li M (2020) Advances and challenges in medicinal plant breeding. *Plant Sci* 298: 110573.

Zaidi SSEA, Mukhtar MS, Mansoor S (2018) Genome editing: Targeting susceptibility genes for plant disease resistance. *Trends Biotechnol* 36(9): 898–906.

Zaim M, Pandey R, Singh RP, Smita Singh, Patel AY, Lal RK (2016) Occurrence of yellow mosaic and leaf curl disease of *Ocimum* sp. and screening for disease management. *Indian Phytopath* 69: 415–418.

27 Biotic Stress Management Strategies for Soft Rot Disease of Vegetable Crops

Yenjerappa S. T., Pavitra, Ganesh H. S., Ranjana Joshi, Savitha A. S., Aswathanarayana D. S. and Ajithkumar K.

27.1 INTRODUCTION

Soft rot is a destructive disease infecting fleshy storage tissue of fruits, vegetables and ornamentals worldwide. The disease commonly infects the storage tissues of vegetables. Bacterial soft rots are known to cause maximum losses worldwide as compared to fungal soft rots. Although, infection initiates in the field, symptom expression will be more in storage. Soft rot of vegetables is caused by different species of *Bacillus* (*B. pumilus, B. cereus etc.*), *Pseudomonas* (*Pseudomonas aeruginosa, P. cichorii, P. marginalis* and *P. viridiflava*) and *Erwinia* (Agrios 2005). *Erwinia carotovora* subsp. *carotovora* is however recognized as one of the important soft rot-causing bacteria in carrot (Larka 2004). The soft rot *Erwinia* belongs to the family Enterobacteriaceae. *Dickeya* and *Pectobacterium* genera were classified previously under the genus *Erwinia*. But the subsequent revisions gave the name *Dickeya* and *Pectobacterium* as a separate genera based on biochemical, host range and molecular traits. Although the genus *Erwinia* is commonly used to narrate the group, but an alternative genus, *Pectobacterium* was proposed recently for the soft rot-causing species of *Erwinia* and the nomenclature has been done based on 16S rDNA sequence analysis (Portier et al. 2017). The pathogen *Pectobacterium carotovorum* was found to be the most vulnerable and common in occurrence worldwide. Losses due to post-harvest soft rots are estimated to be around 15–30 percent of the harvested crop (Agrios 2005).

Bacterial pathogens causing soft rot produce tissue-degrading enzymes which play an important role in causing soft rot of vegetables. Pectinases are the enzymes that hydrolyze the pectin material of the host tissue and thereby leading to soft rots. These enzymes are produced by soft rot-causing bacterial pathogens such as *Dickeya* and *Pectobacterium*, which are commonly referred to as Soft Rot Enterobacteriaceae (SRE). Members of the soft rot Enterobacteriaceae secrete large amounts of exoproteins encompassing plant cell wall-degrading enzymes (PCWDEs) such as polygalacturonases (*Pgl*), pectate lyase (*Pel*), proteases (*Prt*) and cellulases (*Cel*). The exoproteins released by SREs are secreted by type III secretion systems. Cellulolytic and pectinolytic enzymes such as cellulase and pectate lyase are secreted by type II secretion systems (T2SS), and proteases are secreted through type I secretion systems.

During infection, the enzymes disintegrate the plant cell wall structures resulting in the maceration and soft rotting of host tissue. Soft rot bacteria degrade pectate molecules that bind plant cells together causing the plant structure to eventually fall apart. The quantity of the enzymes produced is correlated positively with bacterial virulence (Islam et al. 2019). The secretion of plant cell wall-degrading enzymes (PCWDEs) is affected by various factors such as the presence of substrates and cell density, plant extracts, temperature, pH, oxygen, nitrogen starvation and osmolarity. Hence, the virulence of the pathogen is influenced by a combination of these factors. Rot can occur over a

DOI: 10.1201/9781032631578-27

wide range of temperatures with the worst decay between 30 and 40 °C, particularly when oxygen is limited (Matsumoto et al. 2013).

The bacterium chiefly attacks succulent tender tissues of storage organs such as fleshy tubers, fruits, roots, bulbs, corms and rhizomes as well as bud, stem, petiole and leafstalk tissues. The bacterial soft rot pathogens have a broad spectrum of hosts including genera from all the plant families of vegetables, fruits and ornamentals having fleshy tissues but not woody tissues. The host range includes carrot, potato, cabbage cauliflower, capsicum, egg plant, pepper, rhubarb, leafy greens, squash, other cucurbits, tomato, pea, chilli, brussel sprouts, turnip, radish, lettuce, green peppers, African violets and horse radish (Bhat et al. 2012).

27.2 SYMPTOMS

The symptoms of the soft rot are similar in most of the host plants. In field, above-ground symptoms are general yellowing of the foliage followed by brown lesions on leaves and stem, which later on become mushy and colorless, finally leading to wilting and collapse of the foliage (Figure 27.1A). Infection was which spread across the field. Symptoms on vegetables appear as soft, watery and slimy decay of the fruit found to spread rapidly to the core of the fruit leading to the formation of cavities often leaving the epidermis intact. A foul, unpleasant odor is associated with rotting caused by secondary invading bacteria that are growing in decomposed tissues and are often attracted by insects.

Symptoms differ from vegetable to vegetable *i.e.,* the appearance of water-soaked lesions in the beginning followed by gradual spread of the disease covering the entire vegetable surface with remittance of foul smell were the characteristic symptoms observed in tomato, cucumber, bitter gourd, pumpkin and bell pepper. Symptoms on potato tuber manifests as brown to black colored lesions at the point of infection in the beginning, which later on lead to complete rotting. Similarly, the pathogen produces brown to black-colored lesions on cabbage at the point of infection (Figure 27.1b). Leafy vegetables such as cabbage may turn into watery, green, slimy masses within a day or two at room temperature if stored in plastic containers or in warm humid storage areas.

27.3 CAUSAL ORGANISM

The bacterial pathogenic genera causing soft rot belong to the SRE family (Soft Rot Enterobacteriaceae) in many crops. It was earlier described as a *Bacillus* (Jones 1900), but later on moved to new genus *Erwinia*, derived from the name of the founder of Phytobacteriology, Erwin Frank Smith (Winslow et al. 1917). For many decades, the genus *Erwinia* was used for all the members of plant-pathogenic Enterobacteriaceae. Soft rot in vegetables is also caused by different species of *Bacillus*, *Pseudomonas* and *Erwinia* (Agrios 2005). *Erwinia carotovora* subsp. *carotovora* is

FIGURE 27.1A Foliar symptoms of soft rot on carrot.

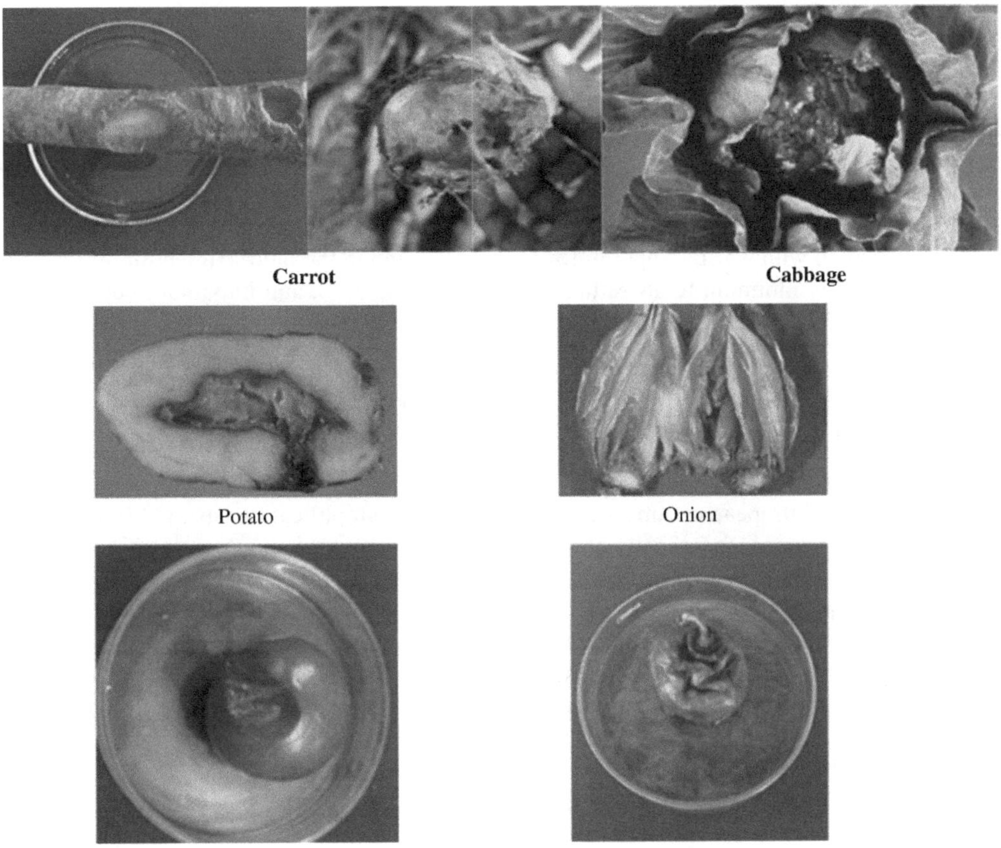

FIGURE 27.1B Soft rotting on different vegetables.

however recognized as one of the important soft rot-causing bacteria in carrot (Larka 2004). But subsequent revisions gave the name *Dickeya* and *Pectobacterium* as separate genera based on biochemical, host range and molecular traits. Although, the genus name *Erwinia* is the one most oftenly used to describe the group. A substitute genus named *Pectobacterium* was recently proposed for the soft rot-causing species and the nomenclature has been done based on 16S rDNA sequence analysis (Anonymous 2019).

Soft rot Enterobacteriaceae secretes large quantities of exoproteins including plant cell wall-degrading enzymes (PCWDEs) such as polygalacturonases (Peh), pectate lyases (Pel), cellulases (Cel) and proteases (Prt). The exoproteins released by SREs are secreted by type III secretion systems, cellulolytic and pectinolytic enzymes such as cellulase and pectate lyase are secreted by type II secretion systems (T2SS), whereas proteases are secreted by type I secretion systems. During infection, these enzymes disintegrate the plant cell wall structures which results in maceration and soft rotting of host tissues. The extent to which, these enzymes are produced correlates positively with bacterial virulence (Islam et al. 2019). Characterization of enzymes was carried out using both conventional (Turbidometric method, HPLC, Spectrophotometer, Colorimeter, Titration methods, *etc.*) and molecular methods (16S rDNA sequencing, AFLP, RAPD, RFLP, qPCR, RT-PCR, *etc.*). The secretion of plant cell wall-degrading enzymes (PCWDEs) is affected by various environmental factors such as the presence of substrates and cell density, plant extracts, temperature, pH, oxygen, nitrogen starvation and osmolarity, hence the virulence of *Pectobacterium* is influenced by a combination of these factors on bacteria (Matsumoto et al. 2003)

27.4 DISEASE CYCLE AND EPIDEMIOLOGY

Soft rot bacterium overwinters in an infected fleshy tissues in storage, field, garden or greenhouse, soil (especially in the rhizosphere around the roots of many plants) and on contaminated tools, equipment, containers and in certain insects. The bacteria enter primarily through wounds made during planting, cultivation, harvesting, grading, and packing and through freezing injuries, insects, hail wounds, growth cracks and sunscald. Uninjured tissues may be infected when the humidity reaches 100 percent or whenever free moisture is present. More rainfall, and poorly drained water-logged soils coupled with warm temperatures favors infection in the field. High humidity in storage or transit along with minimum temperatures between 18 and 35°C and maximum between 35 and 41°C favors the infection and spread depending on the species of the bacterium.

27.5 CHARACTERIZATION OF SOFT ROT PATHOGENS USING ADVANCED TECHNIQUES

Eisse et al. (2008) identified the bacterial strains that caused soft rot in temperate regions of Western and Central European countries on the basis of PCR amplifications using different primers. ECA1f and ECA1r primers specifically amplified *P. carotovorum* subsp. *atrosepticum (Pca)* strains and ADE1 and ADE2 primers specifically amplified strains of *Dickeya* spp. All *Pectobacterium carotovorum* subsp. *carotovorum* strains were positive for primers EXPCCF/EXPCCR. Fang et al. (2008) sequenced the *Pectobacterium* isolates SXR1 and ZJR1 by 16S rRNA sequencing. Phylogenetic analysis revealed that both the isolates shared 99.9 percent homology and they had which 97–99 percent similarity with *Pcc* strains. The DNA sequence similarity was analyzed by NCBI BLAST server. Isolates SXR1 and ZJR1 were grouped together under *P. carotovorum* subsp. *carotovorum* and *P. carotovorum* subsp. *odorifera* in one cluster. PCR amplification for the total genomic DNA of 53 isolates of *Pectobacterium* with two sets of primers Y1/ Y2 which yielded an amplified fragment of an expected size of 434 bp indicated that all the isolates that were collected and tested belonged to the genus *Pectobacterium* (Terta et al. 2010).

PCR amplification with specific primers PCCSSF/PCCSSR (5'ATAACTACTGGAAACGGTA-3') (5'-TTCTCTTTGTATACGCCATT-3') led to the identification of 20 isolates as *P. carotovorum* subsp. *carotovorum* which was extracted from infected potato, cabbage, bell pepper and carrot samples from Iran. Six isolates from infected bell peppers and five isolates from infected potatoes were identified as *P. wasabiae* and *P. atrosepticum* respectively. Molecular assays indicated the occurrence of *P. wasabiae* as a novel species in bell pepper in Iran (Rafiei et al. 2012).

Katarina et al. (2014) isolated two soft rot bacterial strains *i.e.*, KBI 1 and KBI 2 from broccoli. Molecular identification was carried out with PCR amplification of 16S rRNA and 23S rRNA intergenic transcribed spacer region (ITS) using L1 and G1 primers and validated that the strains isolated in the study belonged to *Pectobacterium* spp. Strains KBI 1 and KBI 2 were further characterized by partial sequencing of their 16S rRNA gene with 1350 bp aligned. The sequences of KBI 1 (GenBank acc. no. KC527051) and KBI 2 (GenBank acc. no. KC527052) showed 100 percent identity with the sequence of *P. carotovorum* subsp. *carotovorum*, which was deposited in GenBank. Dadasoglu and Kotan (2017) isolated four different *P. carotovorum* subsp. *carotovorum* strains, two from potato and other two from onion or pepper. The strains were identified at the genus level by PCR using Y1/Y2 primers. These four strains were recognized by specific PCR tests which found to generate 434 bp long bands. This was the first study to reveal that these four strains are causative agents of soft rot disease in potato, onion and pepper in East Anatolia and also in Turkey.

Pectobacterium spp. was isolated from potato plants showing typical black leg symptoms using CVP medium. Sequences and phylogenetic analysis of partial 16S and 23S rDNA intergenic spacer region indicated that the isolates were *P. carotovorum* subsp. *brasiliensis*. Comparison of PCR-RFLP patterns of the 16S and 23S rDNA of isolates to the reference cultures validated the similarity of South African blackleg strains as *P. carotovorum* subsp. *brasiliensis* which was identical to strain 8 and was isolated in Brazil. This was the first report of *P. carotovorum* subsp. *brasiliensis* in South Africa and it appeared to be the major causal agent of blackleg in South Africa (Johanna et al. 2019). Using 16S rDNA sequencing and genus-specific PCR, Li et al. (2020) identified 72 *Pectobacterium carotovorum* strains from Chinese cabbage, bok choy and celery and then assessed their pathogenicity on Chinese cabbage petioles and potato tubers. Based on phylogenetic analysis of *pmr*A sequences and confirmation by subspecies-specific PCR, the strains were divided into 18 *P. carotovorum* subsp. *carotovorum*, 29 were *P. carotovorum* subsp. *odoriferum* and 25 were *P. carotovorum* subsp. *brasiliensis*.

27.6 MANAGEMENT STRATEGIES FOR SOFT ROT

Pectobacterium carotovorum subsp *carotovorum* causing soft rot in vegetables is so versatile and ubiquitous and is regarded as a major soft rot-causing pathogen infecting a wide variety of vegetables. In light of present-day constraints with the use of chemical pesticides for plant disease management, biological control using botanicals and biocontrol agents is becoming popular and occupying the minds of people all over the world as evidenced that biopesticides are eco-friendly and cost-effective. The mechanism of inhibition of pathogens by biocontrol agents inhibiting the colonization of host plant root, secretion of toxic microbial metabolites and competition with pathogen for food and space are worth mentioning. Hence, management involving eco-friendly approaches with non-toxic residue-free chemicals finds a better place to address the disease effectively. In this regard, efforts were made by several workers to screen the botanicals, biocontrol agents, edible salts, antibacterial chemicals and bactericides by *in-vitro* and *in vivo* methods.

27.7 *IN-VITRO* EFFECT OF BOTANICALS, BIOAGENTS AND ANTIBACTERIAL CHEMICALS AND BACTERICIDES AGAINST THE PATHOGEN

The impact of various phytobiocides against the pathogen has been tested and found that turmeric followed by neem is effective in inhibiting the growth of the pathogen. Among the 14 aqueous extracts of different plant species used to inhibit the growth of *P. carotovorum*, the clove (*Syzygium aromaticum*) and white willow (*Salix alba*) were found promising against the pathogen (Viswanath et al. 2019). Pavitra (2022) made an attempt to evaluate different botanicals and found that Nilgiri plant extract can be used effectively to suppress the soft rotting bacterium.

Viswanath et al. (2019) tried to use aqueous plant extracts for the management of soft rot in carrot. Aqueous extracts of 14 plant species, most of them belonging to the temperate ecology were evaluated against *P. carotovorum* subsp. *carotovorum* under *in vitro*. Among them, clove and *Salix alba* (white willow) proved highly effective against the pathogen. *Azadirachta indica, Urtica dioica, Xanthium* sp., *Cupressus torulosa* and *Morus alba* also inhibited the growth of the pathogen. *Pinus* sp., *Allium sativum, Artemisia annua, Juglans regia* showed moderate inhibition while aqueous plant extracts of *Plantago* spp. and *Mentha arvensis* were least effective. *S. aromaticum*, *Salix alba* and *A. indica* which proved effective against the growth of test pathogen *in-vitro*. Among the various biocontrol agents under evaluation, *Streptomyces* spp. was observed very effective against two strains Ecc-1 and Ecc-2 of *E. carotovora* subsp *carotovora* with considerable inhibition zone (Salem and Shafea 2018). Pavitra (2022) suggested the use of *Pseudomonas fluorescens* and *Bacillus subtilis* for the effective control of soft rotting pathogen *P. carotovorum* (Table 27.1).

Different antibiotics under evaluation by Thiyagarajan (2016) against the pathogen *E. carotovora* subsp *carotovora* revealed that streptocycline could significantly inhibit the growth of the pathogen followed by K cycline (Table 27.2). Pavitra (2022) assessed the efficacy of different antibacterial chemicals and observed that Tagmycin exhibited significantly superior efficacy over other antibiotics with a mean inhibition zone of 14.33 mm. Bacterianashak (10.92 mm) and plantomycine (8.00 mm) were found moderately effective.

TABLE 27.1
Bioefficacy of bioagents against *Pectobacterium carotovorum* subsp. *carotovorum*

Sl. No.	Bioagents	Mean inhibition zone (mm)
1	*Pseudomonas fluorescens*	12.66 (3.62)*
2	*Bacillus subtilis* (SE76)	9.00 (3.08)
3	*Bacillus amyloliquefaciens*	4.16 (2.15)
4	*Bacillus thuringenesis*	0.00 (0.70)
5	*Trichoderma asperillum*	5.09 (2.36)
6	*Lactobacillus* sp. (MYSVA 20)	0.00 (0.70)
7	*Lactobacillus* sp. (MYSVA 21)	0.00 (0.70)
8	*Lactobacillus* sp. (MYSVA 23)	4.13 (2.15)
9	Control	0.00 (0.70)
	S Em ±	0.13
	CD at 1 %	**0.45**

* Figures in the parenthesis are square root transformed values.

TABLE 27.2
Efficacy of antibiotics against the growth of *E. carotovara* sub sp. *carotovora* under *in vitro*

		Inhibition zone (mm) at different concentrations			
Sl. No.	**Antibiotics**	**100 ppm**	**250 ppmn**	**500 ppm**	**Mean (mm)**
1	Streptocycline	44.30 (41.71)	51.10(45.61)	66.30(54.49)	53.90(47.27)
2	K-cycline	41.00(39.79)	49.80(44.86)	56.60(48.77)	49.10(44.48)
3	Bionol-100	32.60(34.80)	47.30(43.26)	63.50(52.81)	47.70(43.63)
4	Bacterinashak	28.50(32.25)	44.00(41.53)	54.60(47.62)	42.30(40.55)
5	Control	0.00	0.00	0.00	0.00
Mean		29.20(32.69)	38.40(38.20)	48.20(43.90)	38.60(38.39)
Source		S.Em. ±		C.D. at 1%	
Treatments (A)		0.04		0.15	
Concentration (B)		0.03		0.12	
A × B		0.07		0.27	

** The values in the parentheses are arc sine values.

27.8 MANAGEMENT OF SOFT ROT DISEASE UNDER GREENHOUSE/ FIELD AND STORAGE CONDITIONS

Several measures are adopted to manage the disease in greenhouse, field and storage conditions. *Bacillus velezensis* strains A-6 and P42 were found effective for the control of soft rot in carrot under glasshouse conditions as evidenced by minimum disease incidence of 20 percent in A-6 strain treated pots. The field performance of *B. velezensis* strain A-6 was also good in reducing the soft rot incidence (4.33 percent) as demonstrated by Chandrashekar (2017). *Bacillus* spp. is known to produce AHL Lactonase which is found to inactivate AHL signaling thereby disrupting the quorum-sensing phenomenon of the pathogen (Sneha and Anuradha 2017).

Potato tuber infection in the field caused by two pathogenic isolates of *P. carotovorum* subsp *carotovorum* was successfully reduced with the treatment of *Streptomyces* spp. (Salem and Shafea 2018). Muttanna et al. (2021) conducted a field experiment on the integrated management of tip over disease of banana caused by *P. carotovorum* subsp. *carotovorum*. Results indicated that, no disease incidence was recorded in plots treated with streptocycline (500 ppm) + COC (3000 ppm) and K-cycline (500 ppm) + COC (3000 ppm) after fourth drenching (120 DAP).

A field management study on soft rot of carrot was conducted by Pavitra (2022) and the disease was effectively controlled through drenching of copper oxychloride @ 3 g/lit. The disease was also considerably reduced through soil application of *P. fluorescens* @ 2.5 kg mixed with 100 kg FYM added with neem cake at the time of sowing followed by drenching with *Pseudomonas fluorescens* @ 5 g/lit soon after the notice of foliar symptoms.

The efficacy of biocontrol agents observed under field conditions was also reflected in storage; carrot roots treated with *Pseudomonas fluorescens* and *Bacillus subtilis* along with FYM and neem cake are well protected without any rotting in storage. Biological control of soil-borne diseases using strains of *Pseudomonas* has been thoroughly studied by Cook et al (1995). *Pseudomonas* spp. has been identified as a potential biocontrol agent and is an aerobic and gram-negative bacterium. It is ubiquitous in agricultural lands and well adapted to grow in rhizosphere. Fluorescent *Pseudomonads* are aggressive rhizosphere colonizers and produce a wide variety of antimicrobial compounds (Kloepper 1983). Most of the biocontrol agents belonging to the genus *Pseudomonas* produce metabolites such as phenazines, pyrrolnitrin, pyoluteorin, which play a role in antimicrobial activity in the suppression of plant pathogens.

Protection during storage is another important step to prevent the damage from soft rot as there are enormous chances for the pathogen to be carried from field to storage and to cause infection in storage. Chlorination by using sodium or calcium hypochlorite was effective in removing dirt and debris from the carrot root as suggested by Galati et al. (2005). Chlorination acts as a preventive measure against pre-harvest infection, decay or injury. Chlorine helps to limit the spread of soft rot in storage. Efforts were made to prevent the infection using various salt compounds. As a preventive measure potato tuber treated with aluminum acetate, alum and sodium bicarbonate found less soft rot incidence than the untreated control under normal storage conditions. These compounds can be the potential replacements for commercially available pesticides for post-harvest disease control (Mills et al. 2006).

Pavitra (2022) assessed the effect of various salts against the invasion of soft rot under storage conditions. Among the different salt compounds, calcium hypochlorite inhibited the pathogen completely at all the concentrations without any rotting incidence, whereas a moderate disease incidence of 48.46 percent was recorded in copper sulfate pentahydrate treated samples. Severe incidence of soft rot was observed in calcium propanate (73.17 %), sodium chloride (77.56 %), alum (78.87 %) and potassium sorbate (80.00 %) treated samples where they are on par with each other, whereas 100 percent of disease incidence was reported in untreated samples. Between the concentrations of each salt, efficacy was significant from lower to higher concentrations. The effect among the salts and concentrations indicated that, the performance of calcium hypochlorite completely inhibited the

TABLE 27.3
Efficacy of different salts against *Pectobacterium carotovorum* subsp. *carotovorum* under storage condition

		Percent Incidence			
		Concentration (M)			
Sl. No.	**Salt**	**0.002**	**0.02**	**0.2**	**Mean**
1	Alum	86.64	76.65	73.33	78.87
		(68.56)	(61.10)	(58.90)*	(62.63)
2	Calcium hypochlorite	0.00	0.00	0.00	0.00
		(0.00)	(0.00)	(0.00)	(0.00)
3	Calcium propionate	86.03	76.66	56.84	73.17
		(68.05)	(61.11)	(48.93)	(58.80)
4	Copper sulfate pentahydrate	53.33	55.55	36.52	48.46
		(46.90)	(48.18)	(37.17)	(44.12)
5	Potassium sorbate	86.68	80.00	73.33	80.00
		(68.59)	(63.43)	(58.90)	(63.43)
6	Sodium bicarbonate	73.33	63.52	46.68	61.17
		(58.90)	(52.84)	(43.09)	(51.45)
7	Sodium chloride	96.01	80.00	56.67	77.56
		(79.47)	(63.43)	(48.83)	(61.72)
8	Control	100.00	100.00	100.00	100.00
		(90.00)	(90.00)	(90.00)	(90.00)
	Factors	**Salts (A)**	**Concentration (M)**	**Ax M**	
	S Em ±	1.34	1.01	1.70	
	CD at 1%	4.05	3.11	6.95	

* Figures in the parenthesis are arcsine transformed values.

pathogen at all the concentrations, significantly followed by copper sulfate pentahydrate exhibited 36.52 percent disease incidence at 0.2 M concentration, whereas, a maximum disease incidence of 96.01 percent was recorded from sodium chloride-treated samples at 0.002 M concentration (Table 27.3).

27.9 PREVENTIVE MEASURES

- Careful eradication and destruction of infected plants, rotted bulbs, fruits, corns, rhizomes, tubers and any other infected plant parts in field and storage.
- Planting of disease-free seeds and other planting materials.
- Planting of vegetables in well-drained soils accompanied by optimum and adequate irrigation.
- Thick planting and dense population to be avoided, soil fertility status needs to be optimum.
- Soil drenching with copper oxychloride @ 3 g/lit of water or soil application of *Pseudomonas fluorescens* @ 2.5 kg along with 100 kg FYM added with neem cake, soon after the notice of disease symptoms can effectively control the soft rot.

27.10 CONCLUSION

Soft rot is the most destructive disease infecting fleshy storage tissues of fruits, vegetables and ornamentals worldwide. The rotting incited by bacterial pathogens is most common during transit and in storage. The tissue-degrading enzymes produced by the pathogens play a key role in soft rotting of vegetables. The disease aggravates with injury caused to the fruits and vegetables, accompanied by low temperature and high humidity in storage. Hence care is to be taken to manage the disease effectively in field and storage with the application of recommended botanicals, biocontrol agents, suitable bactericides accompanied with maintenance of appropriate storage conditions.

REFERENCES

Agrios GN (2005) Bacterial soft rots. In: *Plant Pathology*, 5th edn. Elsevier Academic Press, Amsterdam, p. 656.

Bhat KA, Bhat NA, Mohiddin FA, Sheikh PA, Wani AH (2012) Studies on pectinase activities of isolates of *Erwinia carotovora* and *Rhizopus* sp. causing soft rot in cabbage (*Brassica oleracea* var. *capitata* L.). *African J Agric Res* 7(45): 6062–6067.

Chandrashekar BS (2017) Studies on etiology and management of bacterial soft rot of carrot. *M Sc Thesis*, University of Agricultural Sciences, Bangalore, p. 90.

Cook RJ, Thomashow LS, Weller DM, Mazzola M, Bangera G, Kim DS (1995) Molecular mechanisms of defense by rhizobacteria against root disease. *Proc Natl Acad Sci* 92: 4197–4201.

Dadasoglu F, Kotan R (2017) Identification and characterization of *Pectobacterium carotovorum*. *J Animal Plant Sci* 27(2): 647–654.

Eisse GD, Toos CN, Ge WB, Arjen GS, Patricia SZ, Jan MVDW (2008) *Pectobacterium carotovorum* subsp. *carotovorum* can causes potato blackleg in temperate climates. *European J Plant Pathol* 122: 561–569.

Fang XH, Xiang Y, Bo YH, Yuan YG, Hai MC, Ji-Shuang C (2008) Characterization of *Pectobacterium carotovorum* subsp. *carotovorum* causing soft rot disease on Pinellia ternate in China. *European J Plant Pathol* 120: 305–310.

Galati A, McKay A, Tan SC (2005) *Minimizing Post Harvest Losses of Carrots*. Department of Agricultural Food, Govt Western Australia, South Perth.

Islam RY, Shyretha B, Ali T, Korsi CD (2019) The gene encoding NAD dependent epimerase/dehydratase, wcaG, cell surface properties, virulence, and extracellular enzyme production in the soft rot phytopathogen, *Pectobacterium carotovorum*. *Microorganism* 17(7): 3–13.

Johanna J, Vander M, Teresa A, Coutinho S, Korsten L, Jacqueline E, Van W (2019) *Pectobacterium carotovorum* subsp. *brasiliensis* causing blackleg on potatoes in South Africa. *European J Plant Pathol* 30(5): 1–2.

Jones LR (1900) A soft rot of carrot and other vegetables caused by *Bacillus caratovorus* Jones. *Vt Agric Exp Sta Rpt* 13: 299–332.

Katarina G, Veljko G, Nenad D, Nenad T, Svetlana Z, Danijela R, Aleksa O (2014) *Pectobacterium carotovorum* subsp. *carotovorum*- the causal agent of broccoli soft rot in Serbia. *Pestic Phytomed* 29(4): 249–255.

Kloepper JW (1983) Effect of seed piece inoculation with plant promoting rhizobacteria on populations of *E. carotovora* on potato roots and daughter tubers. *Phytopathology* 73: 217–219.

Larka BS (2004) Integrated approach for the management of soft rot (*Pectobacterium carotovorum* subsp. *carotovorum*) of radish (*Raphanus sativus*) seed crop. *Haryana J Agron* 20: 128–129.

Li X, Lu F, Chen C, Wangwang S, Yu T, Xie, H (2020) Characteristics and rapiddiagnosis of *Pectobacterium carotovorum* subsp. associated with bacterial soft rot of vegetables in China. *Plant Dis* 104: 1158–1166.

Matsumoto H, Jitareerat P, Baba Y, Shinji T (2003) Comparative study of regulatory mechanisms for pectinase production by *Erwinia carotovora* subsp. *carotovora* and *Erwinia chrysanthemi*. *Am Phytopathol Soc* 16(3): 226–237.

Mills AS, Platt HW, Hurta RA (2006) Sensitivity of *Erwinia* spp. to salt compounds *in vitro* and their effect on the development of soft rot in potato tubers in storage. *Postharvest Biol Technol* 41: 208–214.

Muttanna, Dinesh K, Ravikumar MR, Shamarao J, Ganapathi T (2021) Survey on incidence of banana rhizome rot disease caused by *Pectobacterium carotovorum* subsp. *carotovorum* in major banana growing parts of Karnataka. *Pharma Innov J* 10(8): 177–180.

Pavitra (2022) Investigations on bacterial soft rot of vegetables with special emphasis on soft rot of carrot and its management. *Ph.D Thesis*, University of Agricultural Sciences, Raichur, p. 250.

Portier P, Taghouti G, Briand M, Dutrieux C, Lathus A, Fischer-Le Saux M (2017) Protocols for strains identification at CIRM-CFBP.

Rafiei GH, Khodakaramian, Baghaee RS (2012) Characterization of *Pectobacterium* species isolated from vegetable crops in North- West of Iran. *African J Agric Res* 10(46): 4258–4267.

Salem EA, Shafea YM (2018) Biological control of potato soft rot caused by *Erwinia carotovora subsp. carotovora*. *Egyptian J Biol Pest Control* 28: 94–98.

Sneha SG, Anuradha SN (2017) Evaluation of quorum quenching *Bacillus* spp. for their biocontrol traits against *Pectobacterium carotovorum* subsp. *carotovorum* causing soft rot. *Biocatalysis Agric Biotechnol* 9: 48–57.

Terta M, El- Karkouri A, Ait MA, Achbani E, Barakatte M, Amdan M, Annajar B, El Hassouni M, Val F, Bouteau F, Ennaji M (2010) Occurrence of *Pectobcterium carotovorum* strains isolated from potato soft rot in Morocco. *Cell Mol Biol* 6: 1324– 1333.

Thiyagarajan V (2016) Investigation on tip-over disease of banana caused by *Erwinia carotovora* subsp. *carotovora* (Jones) Holland. *M Sc (Agri) Thesis*, University of Agricultural Sciences, Raichur, p. 120.

Viswanath HS, Bhat KA Illiger Ks, Stanzin D (2019) Ecofriendly management of post harvest bacterial soft rot of potato caused by *Pectobacterium carotovorum* using different botanicals. *J Pharma Phytochem* 8(6): 236–243.

Winslow C, Broadhurst J, Buchanan RE, Krumwiede CJ, Rogers LA, Smith GH (1917) The families and genera of the bacteria, preliminary report of the committee of the society of American bacteriologists on characterization and classification of bacterial types. *J Bacteriol* 2: 505–566.

28 Integrated Pest Management and Good Agricultural Practices for Sustainable Solanaceous Vegetable Crop Production

Raghavendra K.V., Ramesh K.B., Singh S.K., Sehgal M. and Chander S.

28.1 INTRODUCTION

To maintain sustainable agricultural productivity and soil health farming practices should be more eco-friendly, and importance should be given to developing production technologies that are sustainable in the long run. Hence a "resource protective" Integrated Pest Management (IPM) approach is the need of the hour. In a real sense, IPM refers to a comprehensive approach towards the improvement of both the health of the underlying productivity of the soil and plants, leading to the enrichment of the surrounding ecology, which is a pre-requisite criterion for sustainable agriculture.

IPM, especially in vegetables, is becoming popular across the world as pesticide-residue-free vegetables are preferred to a greater extent in developed countries and in major urban centers within the country. India has a strategic geographic structure, varied climate and variety of soils, which makes the country extremely potential for vegetable production. Among vegetable crops, solanaceous vegetables *i.e.,* tomato, brinjal, potato and chillies are served on each and every plate of India, either cooked or raw, regularly because of the availability of a wide range of solanaceous vegetables year round. The share of tomato (*Solanum lycopersicum* L.), brinjal (*Solanum melongena* L.), chilli (*Capsicum annuum* L.) and potato (*Solanum tuberosum* L.) is 21% of total vegetable production in India (Tiwari 2018).

Insects and other vertebrate pests are the predominant biotic obstacles to the production of solanaceous vegetables and can affect any part of the plant (Table 28.1). The occurrence of these pests and diseases, which depend on the prevailing weather conditions, significantly influences the quantity and quality of the harvest. In the recent past, with shifts in cropping patterns, ecosystems and habitat, climate and wider use of high-input-intensive vegetable varieties/hybrids there has been a paradigm shift in insect pest infestations, both in terms of timing and location. Depending on the extent of the infestation over time and space, these insect pests can cause enormous losses.

Management of insects and other vertebrate pests needs a comprehensive approach by integrating all ecologically feasible methods in crop protection. Farmers use pesticides as their first line of defense and frequently resort to indiscriminate and non-judicious use of pesticides. Most of the plant protection recommendations for vegetables so far indicate the calendar-based application of pesticides. This has become a common practice over time by the majority of the farmers growing vegetables (Juraske et al. 2008). Also, due to scarcity of information, farmers tend to rely upon chemical pesticides to manage the pests. But, indiscreet use of chemical pesticides leads to several

DOI: 10.1201/9781032631578-28

TABLE 28.1
Impact of key insect pests on solanaceous vegetable crop yields in India

Crop	Insect pest	Yield loss (%)	Reference
Tomato (*Lycopersicon esculentum* L.)	South American tomato pinworm/tomato leaf miner/ tomato pinworm, *Phthorimaea* (*Tuta*) *absoluta* (Meyrick) (Lepidoptera: Gelechiidae)	Upto 100	Buragohain et al. 2021
	Fruit borer, *Helicoverpa armigera* (Hübner) (Lepidoptera: Noctuidae)	50.00 – 80.00	Tewari and Krishnamoorthy 1984
Brinjal (*Solanum melongena* L.)	Fruit and shoot borer, *Leucinodes orbonalis* Guenée (Lepidoptera: Crambidae)	85.00 – 90.00	Patnaik 2000
Potato (*Solanum tuberosum* L.)	White grub/Root grub, *Holotrichia longipennis* (Blanchard) (Coleoptera: Scarabaeidae)	Upto 62.92	Singh et al. 2004
Chilli (*Capsicum annuum* L.)	Thrips, *Scirtothrips dorsalis* Hood (Thysanoptera: Thripidae)	12.00 – 90.00	Rai et al. 2014

problems such as environmental pollution, pesticide residue in the harvested products (Arora and Gopal, 2002), development of resistance/resurgence in pests, emergence of new pests, destruction of natural enemies and pollinators, and increased cost of production. In this circumstance, a paradigm shift in pest management approaches is the current necessity and this can be effectively achieved through rationalization of pesticide use and integration of several eco-friendly approaches to minimize over-reliance on pesticides.

Hence, it is imperative to develop, validate and implement multipronged approaches consisting of cultural practices, mechanical practices, resistant / tolerant varieties / hybrids / genotypes, biocontrol agents, botanicals, chemical pesticides (judicious and need-based use with appropriate dose, and time of application) in an integrated manner to tackle the management of pests and diseases (Rai et al. 2014). This chapter specifically provides detailed information on integrated management techniques to tackle both insect and non-insect pests in various solanaceous vegetable crops, organized under distinct headings.

28.1.1 Solanaceous Vegetable Crops and Their Insect and Non-insect Pest Management

28.1.1.1 Tomato

Tomato holds significant importance as a vegetable crop in India. Tomato is grown in a wide range of climatic conditions across different states of India. India's diverse climatic conditions support tomato cultivation across various states, collectively contributing around 90% of the nation's total tomato production (DAC&FW data; Anonymous, 2020).

In tomato, farmers generally practice calendar-based application of pesticides for management of economically important insect and non insect stresses (Table 28.2). This has become a common practice over the years by most of the farmers growing tomatoes. It is in this context that Integrated Pest Management (IPM) plays a major role in the judicious use of pesticides, reducing costs of production and increased crop yield.

28.1.1.1.1 IPM Practices in Tomato for Insect and Non-insect Pest Management

* Employing crop rotation involving non-host crops like cereals and legumes and avoiding growing solanaceous crops such as chilli and brinjal, after tomato or tomato after tomato to avoid the carry-over of pests from one season to the other.
* Deep plowing during the summer to expose insect pupae to sunlight and natural predators.

TABLE 28.2
Important insect and non-insect pests of tomato

Sl.No.	Common name	Scientific nomenclature	Taxonomic details
		Insect Pests	
1.	South American tomato pinworm / leaf miner	*Phthorimaea* (*Tuta*) *absoluta* (Meyrick)	Lepidoptera: Gelechiidae
2.	Tomato fruit borer	*Helicoverpa armigera* (Hübner)	Lepidoptera: Noctuidae
3.	Leaf-eating caterpillar	*Spodoptera litura* Fabricius	Lepidoptera: Noctuidae
4.	Serpentine leaf miner	*Liriomyza trifolii* (Burgess)	Diptera: Agromyzidae
5.	Melon fruit fly	*Zeugodacus cucurbitae* (*Bactrocerca cucurbitae*) (Coquillett)	Diptera: Tephritidae
6.	Whitefly	*Bemisia tabaci* (Gennadius)	Hemiptera: Aleyrodidae
7.	Thrips	*Thrips tabaci* Lindeman, *Frankliniella schultzei* (Trybom)	Thysanoptera: Thripidae
8.	Aphids	*Myzus persicae* Sulzer, *Aphis gossypii* (Glover)	Hemiptera: Aphididae
9.	Mealybugs	*Maconellicoccus hirsutus*(Green), *Phenacoccus solenopsis* Tinsley, *Paracoccus marginatus* Williams and Granara de Willink	Hemiptera: Pseudococcidae
		Non-insect pests	
1.	Red spider mite	*Tetranychus* spp.	Acarina: Tetranychidae
2.	Root knot nematode	*Meloidogyne* spp.	Tylenchida: Heteroderidae

* Seed treatment or seedlings treatment with Imidacloprid 17.80% SL against sucking pests.
* Treatment of nursery beds and bedding materials with Dazomet 98% G against root knot nematode.
* Use of 100-mesh nylon net in nursery beds to avoid entry of whitefly and transmission of leaf curl disease in tomato.
* Application of well-decomposed enriched Farm yard manure @ 2 t/ha, enriched with *Pseudomonas fluorescens* @ 2 kg/t of well-decomposed FYM against nematodes.
* Application of 500 kg of neem cake to the soil to combat serpentine leaf miner and root knot nematodes and 1 tonne of vermin-compost per hectare to the main field before transplanting (induced resistance against biotic stress – pest incidence).
* Plastic mulching reduces pupation in the soil, repels whiteflies in the case of silver-coloured sheets and also avoids weed growth. Polythene sheets of 25–30 micron thickness (black or silver colored) should be spread on the raised beds as mulch before transplanting. Round holes should be made according to the recommended spacing (4 ft x 1.5 ft or 120 cm x 45 cm) using a punch or a large-diameter pipe and a hammer and seedlings should be inserted into the holes.
* Intercropping a tall variety of marigold as a trap crop @ 16:1 ratio. Plant 45-day-old marigold seedlings and 25-day-old tomato seedlings to synchronize flowering in both crops. Eggs and larvae are manually collected and destructed from the field along with the flowers. This trap cropping system also helps in reducing the root knot nematode infestation.
* Growing 5–6 rows of sorghum or maize all around the tomato field at least a month before transplanting as a barrier crop against sucking pests like whiteflies, thrips and aphids.
* Staking tomato plants improves fruit quality by keeping plants and fruits off the ground and by providing better spray coverage. Staking tomatoes and using a system of drip irrigation will reduce leaf dampness, along with allowing enough space for aeration.
* For monitoring purposes, 10 yellow sticky traps per hectare is suggested to be installed at a height of 15 cm above the canopy to monitor whiteflies, and 10 blue sticky traps per hectare to monitor thrips.

* Installation of pheromone traps designed for particular host species like South American tomato pinworm, fruit borer and tobacco/leaf-eating caterpillar @ 10–12 traps/ha, with lures replaced every 2–3 weeks. Traps should be positioned in the field such that the lure is consistently 1 to 1.50 feet above the crop canopy. The trapped moths should be periodically collected and killed.
* Installation of indigenously prepared light trap by hanging one incandescent bulb @ one bulb on a tub of water. Such traps may be installed @ 2–3 traps/ha for monitoring and mass trapping of adult moths of South American tomato pinworm.
* Releasing of egg parasitoids *Trichogramma chilonis* at weekly intervals @ 50,000/ha for 6 times at an interval of 7 days against fruit borer/ pin worm/ leaf eating caterpillar.
* Spraying of Entomopathogens like *Beauveria bassiana/Lecanicillium lecanii/Metarhizium anisopliae* @ 1 x 10^8 cfu/ml against sucking pests
* Spraying of *Bacillus thuringiensis* var. *kurstaki* @ 500 g/ha against fruit borer/ pin worm/leaf-eating caterpillar.
* Spraying of Ha/Sl NPV 250 LE with 1% jaggery at 28, 35 and 42 days after transplanting in the evening hours against fruit borers, leaf-eating caterpillar.
* Spraying of neem oil 1% (10 ml/lit) or Pongamia oil 1% (10 ml/L) or commercial formulation of neem-based insecticide (Azadirachtin 1500 PPM) @ 2 ml/L is effective against fruit borer, defoliators and sucking pests.
* Mechanical collection and destruction of lepidopteron larvae at periodic intervals (3–4 times) for management of fruit borer and defoliator. Gathering and disposing of diseased leaves, shoots, and fruits to minimize disease transmission.
* Spraying of label claim pesticides *viz.,* Indoxacarb 14.50% SC @ 0.80 ml/L (against South American pin worm, fruit borer), Chlorantraniliprole 18.50% SC @ 0.25 ml/L (against fruit borer), Thiomethoxam 25% WP @ 0.40 g/L or Imidacloprid 17.80% SL @ 0.50 ml/L (against whitefly), Cyantraniliprole 10.26% OD @ 1.20 ml/L (against South American pin worm, leaf miner, aphids, thrips, whitefly, fruit borer) and Fenazaquin 10% EC @ 2.50 ml/L or Spiromesifen 22.90% SC @ 1.50 ml/L (against red spider mite).
* Application of Dazomet 98% G @ 300 kg/ha against nematodes (however does not have label claim).

28.1.1.2 Potato

The potato has always been a 'poor man's friend'. Potato (*Solanum tuberosum* L.) ranks third among the most important food crops in the world. The potato is one of the most widely grown vegetables in India and has been farmed for more than 300 years. Potato is considered an economical wholesome food. Insect pests cause major problems for potato farmers in India (Table 28.3). Around 60 billion rupees are reported as lost due to pest damage annually, which accounts for 10–20% of total potato production (Chandel et al. 2012).

28.1.1.2.1 IPM Practices in Potato for Insect and Non-insect Pest Management

* **Cultural Method** – The greatest preventive measure in insect pest management is to use resistant cultivars or disease-free seed potato tubers. Effective methods to tackle the population of pests include maintaining biodiversity, using trap crops, planting geometry, planting time (seed plot technique for aphid management), multiple cropping, effective crop rotation (against *Globodera rostochinensis*, a cyst nematode), and habitat manipulation.
* **Mechanical Method** – Collecting and disposing of diseased plants and plant parts, along with removing egg masses and immature stages of insect pests, establishing bird perches, deploying light traps @ 03 traps/ha, placing yellow sticky traps @ 01 trap per 900 square meters, and employing sex pheromone traps are effective mechanical methods for managing insect pests.

TABLE 28.3
Important insect and non-insect pests of potato

Sl.No.	Common name	Scientific nomenclature	Taxonomic details
		Insect Pests	
1.	White grubs / root grubs	*Brahmina coriacea* (Hope), *Holotrichia longipennis* (Blanchard), *Anomala dimidiate* (Hope), *Melolontha indica* Hope, *Lepidiota mansueta* (Burmeister) and *Holotrichia serrate* (Fabricius)	Coleoptera: Scarabaeidae
2.	Cutworms	*Agrotis ipsilon* Hufnagel, *Agrotis interacta* Walker,	Lepidoptera: Noctuidae
3.	Potato tuber moth	*Phthorimaea operculella* (Zeller)	Lepidoptera: Gelechiidae
4.	Termites	*Microtermes obesi* (Holmgren) and *Odontotermes obesus* (Rambur)	Isoptera: Termitidae
5.	Aphids	*Myzus persicae* (Sulzer) and *Aphis gossypii* Glover	Hemiptera: Aphididae
6.	Leafhoppers	*Empoasca fabae* Harris and *Amrasca devastans* Dist.	Hemiptera: Cicadellidae
7.	Thrips	*Thrips palmi* Karny	Thysanoptera: Thripidae
8.	Whiteflies	*Bemisia tabaci* Gennadius	Hemiptera: Aleyrodidae
9.	Leaf defoliator	*Helicoverpa armigera* (Hübner)	Lepidoptera: Noctuidae
10.	Leaf-eating caterpillar	*Spodoptera litura* Fabricius	Lepidoptera: Noctuidae
11.	Bihar hairy caterpillar	*Spilosoma oblique* (Walker)	Lepidoptera: Erebidae
12.	Hadda beetle	*Henosepilachna vigintioctopunctata* (Fabricius)	Coleoptera: Coccinellidae
		Non-insect pests	
1.	Yellow mite	*Polyphagotarsonemus latus* (Banks)	Trombidiformes: Tarsonemidae
2.	Root knot nematode	*Meloidogyne* spp.	Tylenchida: Heteroderidae
3.	Potato cyst nematode / golden nematode of potato	*Globodera rostochiensis* (Wollenweber), *Globodera pallida* (Stone)	Tylenchida: Heteroderidae

* **Biological Method** – Employing parasitoids, entomo-pathogens, and insect predators has also shown to be a successful pest management strategy. Mass release of parasitoids like *Trichogramma sp*. (Egg parasitoid) @ 40,000 to 50,000 eggs/ ha, *Chelonus blackburni* (Egg-larval parasitoid) @ 15,000 to 20,000/ha, *Apanteles* sp. (Larval parasitoid) @ 15,000 to 20,000/ha for the management of lepidopteran pests like potato tuber moth (*Phthorimaea operculella*) – both in field and storage conditions, leaf-eating caterpiller (*Helicoverpa armigera*), tobacco caterpiller (*Spodoptera litura*), cut worm (*Agrotis segetum*) and semilooper (*Trichoplusia ni*) after 30–40 days of planting and predators like *Chrysoperla* sp. @ 5,000/ha after 15–20 days after sprouting will effectively manage the pest problem in potato.

Use of bio-pesticides like *Beauvaria bassiana* / *Verticillium* sp. (against aphids, white flies, leaf hoppers and mites), *Nomuraea rileyi* (against lepidopteran pests), *Metarizium anisopliae* (against root grubs) and *Pseudomonas fluorescens* + *Trichoderma harzianum* (against nematodes) @ 4.00 g or ml/L (spore load – $1x10^8$ cfu/g or ml) which are available in the market can manage the specific host insect and non-insect pests. *Bacillus thurengensis* formulation is recommended @ 0.5–1.0 kg/ha for the management of lepidopteran insect pests. Viral biopesticides of baculovirus group *viz.* Granulosis Viruses (GV) and Nuclear Polyhedrosis Viruses (NPV) can also be used in potato insect pest management. Spray of NPV of *Helicoverpa armigera* (Ha NPV) or *Spodoptera litura* (Sl NPV) and GV of potato tuber moth (PTM) @ 250 larval equivalents are very effective tools to manage the *Helicoverpa* / *Spodoptera* / PTM, respectively.

* **Botanical Method** – The utilization of pesticidal properties found in different plants, either in the form of extracts or refined derivatives, can effectively control pest populations. The application of neem oil at concentrations of 0.50–1.00% (5–10 ml/L) or Neem Seed Kernel Extract (NSKE) at 5% (50 ml/L) is commonly practiced to manage insect pests such as leafhoppers, aphids, whiteflies, spider mites, and larval lepidopterans in potato crops.
* **Other mineral origin compounds** – Spraying of Horticultural Mineral oil @ 50.00 ml/10 L of water against sucking insect pests is practicable in organic potato cultivation.
* **Chemical methods** – Application of label claim pesticides like Imidacloprid 70.00 % WG @ 5.00 g/kg (seed tuber treatment against aphids and white fly), Imidacloprid 48.00 % FS @ 5.00–10.00 g/kg (seed tuber treatment against against aphids and jassids), Thiamethoxam 25.00 % WG @ 0.25 g/L (against aphids).

28.1.1.3 Chilli

Chilli (*Capsicum annuum* L.) also known as 'red pepper' is most valuably used as spice, condiment, vegetable in our Indian cuisine, while also offering potential benefits in pharmaceuticals, cosmetics, and beverages. India stands as the leading producer, exporter and consumer of chillies, globally. Like other crops, chilli also faces challenge of various pests (insects, plant pathogens, nematodes, etc.).

These insect and non-insect pests (Table 28.4) accounts for up to 50–90 % yield loss under favorable conditions. In recent past, chilli farmers of Andhra Pradesh, Telangana, Karnataka, Tamil Nadu and Maharashtra has faced the serious problem due to an invasive insect pest known as South East Asian thrips (*Thrips parvispinus*) resulted in serious crop loss (50–80 %) irrespective of the cultivars grown across the chilli belt.

To tackle these insect and non-insect pests, farmers spray up to 25–30 pesticide sprays in a season comprising of various pesticides solely or in combination. But still farmers are unable to manage these pests and instead, farmers lose a huge amount of money. As an alternative, IPM would be the better option for the farmers producing chillies of export standard and also high-grade chillies for domestic consumption.

TABLE 28.4
Important insect and non-insect pests of chilli

Sl.No.	Common name	Scientific nomenclature	Taxonomic details
		Insect Pests	
1.	Thrips	*Scirtothrips dorsalis* Hood, *Frankliniella schultzei* (Trybom), *Thrips florum* Schmutz, *Thrips hawaiiensis* (Morgan), *Thrips palmi* Karny and *Thrips parvispinus* (Karny).	Thysanoptera: Thripidae
2.	Fruit borer	*Helicoverpa armigera* (Hübner)	Lepidoptera: Noctuidae
3.	Aphids	*Myzus persicae* (Sulzer), *Aphis gossypii* Glover	Hemiptera: Aphididae
4.	Tobacco caterpillar	*Spodoptera litura* Fabricius	Lepidoptera: Noctuidae
		Non-insect pests	
1.	Yellow mite	*Polyphagotarsonemus latus* (Banks)	Trombidiformes: Tarsonemidae
2.	Red spider mite,	*Tetranychus* spp.	Acarina: Tetranychidae
3.	Root knot nematode	*Meloidogyne* spp.	Tylenchida: Heteroderidae

28.1.1.3.1 IPM Practices in Chilli for Insect and Non-insect Pest Management

* Deep plowing during the summer to eliminate both the dormant and lingering stages of thrips.
* Cultivating resistant or early/short-duration varieties, if accessible, to evade the peak occurrence of thrips and mites
* Practicing clean cultivation by keeping weed-free boundaries and borders around crop fields, as several weeds such as *Cleome viscosa*, *Prosopis* sp., *Lantana camara*, *Calotropis* sp., *Tecoma* sp., *Abutilon* sp., *Parthenium hysterophorus*, and wild *Solanum* sp. serve as alternative hosts for thrips, mites, and aphids.
* Seed treatment with label claims systemic insecticides.
* Application of well-decomposed farmyard manure or compost @ 2 t/ha, enriched with *Metarhizium anisopliae* or *Pseudomonas fluorescens* @ 2 kg/t in addition to the 25 to 30 t/ha of FYM.
* Applying 500 kilograms of neem cake and 1 tonne of vermicompost per hectare to promote resistance against thrips and mites through soil application.
* Application of balanced fertilization with increased potassium application, along with nitrogen and phosphorus fertilizers, to stimulate plant resistance to pests.
* Using silver-colored polyethylene sheets of 25–30 micron thickness for mulching to decrease thrips pupation in the soil. The silver color reflects pests, preventing them from anchoring onto plants.
* Growing border crops with 2–3 rows of tall crops such as sorghum/maize/bajra/ fodder grasses, to serve as a barrier hindering the movement of aphids, thrips, and mites.
* Intercropping chilli with a tall variety of yellow-colored marigold as a trap crop and cow pea as banker crop @ 16:1:1 ratio against fruit borer and other sucking pests.
* Regular intercultural practices (soil racking and awakening) effectively eliminate thrips and other insect pest pupae that reside in the soil.
* Plants with severe infestations from thrips and mites can be mechanically destroyed by uprooting, burying, or burning them.
* Installing blue and yellow sticky traps @ 60–75 traps/ha at 0.50 feet above the crop canopy to mass trap thrips and aphids, respectively.
* Installation of host-specific pheromone traps for fruit borer and tobacco/leaf-eating caterpillar @ 10-12 traps/ha (replace the lures with fresh ones every 2–3 weeks).
* Releasing of egg parasitoids *Trichogramma chilonis* at weekly intervals @ 50,000/ha for 6 times at an interval of 7 days against fruit borer and tobacco/leaf-eating caterpillar.
* Using sprinkler irrigation in place of flood irrigation, since the water jet spray from the sprinklers can hinder the growth and spread of aphids, mites, and thrips.
* Conservation of natural enemies by minimal application of chemical pesticides. Instead, can spray botanical-based pesticides like Neem Seed Kernel Extract (NSKE) 5% (50 ml/L) or neem oil 0.50–1.00% (5–10 ml/L), Pongamia oil 0.50–1.00% (5–10 ml/L), *Vitex negundo* extract @ 50–80 ml/L, *etc.* or microbial-based insecticides like *Beauveria bassiana* @ 4.00 g or ml/L (spore load – 1x10^8 cfu/g or ml), *Pseudomonas fluorescence* – NBAIRPFDWD @ 20 g/L or *Bacillus albus* – NBAIR-BATP @ 20 g/L uniformly covering whole plant against insect pests and mites. Spraying of *Bacillus thuringiensis* var. *kurstaki* @ 500 g/ha, Ha/Sl NPV 250 LE with 1% jaggery in the evening hours against fruit borer and leaf-eating caterpillar
* The last recourse is to spray label claim insecticides according to need.

28.1.2 Examples of Label Claim Insecticides in Chilli against Insect Pests and Mites

Acetamiprid 20% SP @ 0.10 g/L, Chlorantraniliprole 18.50 % SC @ 0.25 ml/L, Cyantraniliprole 10.26% OD @ 1–1.20 ml/L, Emamectin benzoate 1.90% EC @ 0.75 ml/L, Fipronil 5% SC @ 1.50 ml/L, Imidacloprid 70% WS @ 3–5.00 g/Kg of seeds, Imidacloprid 17.80% SL @ 0.25–0.50 ml/

L, Spinetoram 11.70% SC @ 1.00 ml/L, Spinosad 45% SC @ 0.30 ml/L, Spirotetramat 15.31% w/w OD @ 0.80 ml/L, Thiacloprid 21.70% SC @ 0.50 ml/L, Tolfenpyrad 15% EC @ 2.00 ml/L, Emamectin Benzoate 1.50% + Fipronil 3.50% SC @ 1–1.50 ml/L, Indoxacarb 14.50% + Acetamiprid 7.70% w/w SC @ 1.00 ml/L, Fenazaquin 10 % EC @ 1.50–2.00 ml/L, Propargite 57 % EC @ 1.50-2.00 ml/L and Spiromesifen 22.90 % SC @ 0.80–1.00 ml/L.

28.1.1.4 Brinjal

Among various vegetables, brinjal (*Solanum melongena* L.) is the most common and extensively grown solanaceous vegetable crop all over the country (0.39 m ha) and occupies an important place in the food basket of Indian consumers. Despite covering an area of 0.39 million hectares in the country, brinjal productivity remains low, primarily due to the rising occurrence of insect pests, diseases, and nematodes, often leading to significant yield reductions. Because of its tender and delicate characteristics, along with cultivation under high moisture and input conditions, brinjal is highly susceptible to pest attacks, resulting in an estimated loss of about 35–40% (Sardana et al. 2006).

To mitigate the losses due to these pests (Table 28.5), a huge quantity of pesticides is used in brinjal, and it is not unusual for the vegetable growers to give 10–12 sprays in a season and thus the fruits, which are harvested at short intervals, are likely to retain an unavoidably high level of pesticide residues which may be highly hazardous to consumers. The IPM in this context would be the better option for the farmers to produce residue-free brinjal for the domestic and international markets.

28.1.1.4.1 IPM Practices in Brinjal for Insect and Non-insect Pest Management

* Deep summer plowing and destruction of debris, crop residues, weeds and other alternate hosts to destroy the resting as well as residual stages of insect pests and nematodes.
* Adoption of proper crop rotation and avoiding growing of solanaceous crops in sequence

TABLE 28.5
Important insect and non-insect pests of brinjal

Sl.No.	Common name	Scientific nomenclature	Taxonomic details
		Insect Pests	
1.	Fruit and shoot borer	*Leucinodes orbonalis* Guenée	Lepidoptera: Crambidae
2.	Jassids	*Amrasca biguttula biguttula* Ishida	Hemiptera: Cicadellidae
3.	Whiteflies	*Bemisia tabaci* Gennadius	Hemiptera: Aleyrodidae
4.	Aphids	*Myzus persicae* Sulzer	Hemiptera: Aphididae
5.	Thrips	*Thrips palmi* Karny	Thysanoptera: Thripidae
6.	Mealybugs	*Phenacoccus solenopsis* Tinsley, *Coccidohystrix insolitus* Green and *Paracoccus marginatus* Williams and Granara de Willink	Hemiptera:Pseudococcidae
7.	Lace bug	*Urentius echinus* Distant and *Urentius hystricellus* Richter	Hemiptera: Tingidae
8.	Hadda beetle	*Henosepilachna vigintioctopunctata* Fabricius	Coleoptera: Coccinellidae
9.	Ash weevil	*Myllocerus subfasciatus* Guerin, *M. discolor* Schoenherr *M. viridanus* Fabricius	Coleoptera: Curculionidae
		Non-insect pests	
1.	Yellow mite	*Polyphagotarsonemus latus* Banks	Trombidiformes: Tarsonemidae
2.	Red spider mite	*Tetranychus* spp.	Acarina: Tetranychidae
3.	Root knot nematode	*Meloidogyne* spp.	Tylenchida: Heteroderidae
4.	Reniform nematode	*Rotylenchulus reniformis* Linford and Oliveira	Tylenchida: Hoplolaimidae

* Use of resistant and tolerant varieties recommended by the State Agricultural Universities of the region.
* Covering the nursery beds with a polythene sheet of 45 gauge (0.45 mm) thickness for three weeks during May–June for soil solarization which will help in reducing the soil-borne insects and nematodes. However, care should be taken that sufficient moisture is present in the soil for its solarization
* Application of well-decomposed FYM @ 20–25 t/ha or vermicompost @ 12 t/ha enriched with *Pseudomonas* sp. and *Metarhizium anisopliae* @ 2 kg/t (against insect pests) and *Paecilomyces lilacinus* 1.15% WP @ 3 kg per 500 kg organic manure (against nematodes) for nursery treatment and soil application.
* Soil application of 250 kg/ha of neem cake for nematode management
* Planting 2–3 rows of tall crops such as sorghum/maize/bajra/fodder grasses densely as border crops acts as a barrier, hindering the movement of whiteflies, aphids, thrips, mealybugs, and mites.
* Installation of bird perches @ 25/ha for facilitating predatory birds to feed on insect pests, installation of yellow/blue sticky traps @ 50–60 traps /ha for aphids, hoppers, white fly, and thrips and Delta pheromone traps @ 10–12 traps /ha for shoot and fruit borer. The lures in the pheromone traps are to be changed at 15–20 day intervals.
* Releasing of egg parasitoids, *Trichogramma brasiliensis* (Ashmead) @ 1.0–1.50 lakh/ ha for shoot and fruit borer management for 4–5 times at weekly intervals.
* Clipping of borer-damaged shoots and collection & destruction of damaged fruits *i.e.,* clean cultivation helps in the management of shoot and fruit borer effectively. Periodical collection and destruction of egg masses, larvae and adults of hadda beetle will help in the effective management of the pest.
* Spraying of botanical pesticides such as Neem Seed Kernel Extract (NSKE) 5% (50 ml/L) or neem oil 2% (20 ml/L), Pongamia oil 2% (20 ml/L), *Vitex negundo* extract @ 50–80 ml/L, etc. against sucking and chewing insect pests and mites.
* Applying microbial-based insecticides such as *Beauveria bassiana* @ 4.00 g or ml/L (for sucking insect pests and mites), *Metarizium anisopliae* (for hadda beetle and ash weevil) (spore load – $1x10^8$ cfu/g or ml) and *Bacillus thuringiensis* var. *kurstaki* @ 500 g/ha (for shoot and fruit borer) evenly covering the entire plant
* As a last resort, need-based and judicious application of label claim pesticides like Chlorantraniliprole 18.50 SC @ 0.25 ml/L, Emamectin benzoate 5 SG @ 0.40 g/L, Flubendiamide 39.35 SC @ 0.25 ml/L (against shoot and fruit borer), Cypermethrin 25 % EC @ 1.00–2.00 ml/L (against shoot and fruit borer, jassids and grubs of hadda beetle), Fenazaquin 10 EC @ 1.50 ml/L (against spider mites), Diafenthiuron 50 WP @ 0.80–1.00 ml/L (against whitefly), Fenpropathrin 30 EC 0.25–0.30 ml/L (against whitefly, shoot and fruit borer and mites), Fenvalerate 20 EC @ 0.50 ml/L (against shoot and fruit borer and aphids).

28.1.3 Good Agricultural Practices in Sustainable Solanaceous Vegetable Production

Good agricultural practices (GAP) is the application of available knowledge to address environmental, economic and social sustainability for on-farm production and post-production processes resulting in safe and healthy food products. The concept of GAP has evolved in recent years in the context of a rapidly changing and globalization of the food economy and due to the concerns and commitments of a vast range of stakeholders about food production and security, food safety and quality, and the environmental sustainability of agriculture. Many farmers in developed and developing countries are applying GAP through sustainable agricultural methods such as integrated nutrient management, IPM and conservation agriculture.

28.1.3.1 Integrated Nutrient Management (INM)

In sustainable crop production, nutrient management is very important as soil fertility has to be not only maintained but also improved. It has been estimated that in India, every year 280 mt of cattle dung, 273 mt of crop residues, 285 mt of compost and 6351 Mm^3 of domestic wastes are produced which can be recycled and reused effectively in order to promote sustainable agriculture like organic farming in India (Singh *et al.*, 2017). Sole use of synthetic chemical fertilizers is prohibited in sustainable agriculture and therefore application of an appropriate quantity of organic origin plant nutrients along with synthetic chemical fertilizers in an integrated approach will not only sustain the yield of crops but can also enhance it. To fulfill this objective, it is recommended to follow Integrated Nutrient Management (INM) practices. INM is the process of maintaining optimal plant nutrient supply and soil fertility for maintaining targeted productivity by optimizing the benefits of all available sources of organic, inorganic, and biological components in a coordinated way. Application of well-decomposed biogas slurry, farmyard manure, compost, vermi-compost, crop residue recycling and any safe organic source is beneficial for maintaining the sustainability of the soil.

28.1.3.1.1 Plant Nutrients Derived from Organic Sources

a. Crop rotation and crop diversification: In general, vegetable crops improve the organic load of the soil. Crop rotations and taking up multiple crops on a given piece of land ensures better utilization of resources. Legume crops like beans, peas, cowpea, groundnut, *etc.,* should be included in the crop rotation to improve the soil fertility as these leguminous crops fix atmospheric nitrogen into the soil.

b. Green manures: Green manuring is simply the incorporation of live biomass into the soil in order to supply plant nutrients. The crops grown to be incorporated are termed green manuring crops. In general, Dhaincha (*Sesbania esculenta, S. rostrata*), a rigorously growing nitrogen-fixing crop (fix atmospheric nitrogen to the extent of 60–100 kg/ha), should be plowed into the soil after about 6 to 8 weeks of sowing, once sufficient vegetative growth has been achieved. Green manures not only contribute nutrients to the soil but also enhance its physical and microbial properties. In addition, *ex-situ* addition of green leaves from Pongamia (*Pongamia pinnata*) or Gliricidia (*Gliricidia sepium*) trees and other plants grown on bunds is highly recommended. Green manuring practice supplements the nutrient requirement of the crop by 20–30 percent and improves vegetable yield.

c. Farm yard manure / Enriched compost: Any manure prepared in the backyard using farm waste, cattle urine and dung is referred to as farmyard manure (N – 0.95%, P_2O_2 – 0.62% and K_2O – 2.20%). Compost (rural compost – N – 0.75%, P_2O_2 – 0.63% and K_2O – 1.05% and urban compost – N – 1.35%, P_2O_2 – 0.62% and K_2O – 1.43%) is organic waste such as food or plant material biologically decomposed by bacteria, fungi and other decomposing organisms under controlled aerobic (occurring in the presence of oxygen) conditions. Compost enriched by the incorporation of biofertilizers, microbial inoculants, *etc.,* is known as enriched compost. 5–6 t/0.4 ha of farmyard manure or enriched compost is recommended to incorporate into the soil during land preparation, preferably a fortnight before planting. After one to one and a half months of sowing, another 2–3 t/ 0.4 ha of farmyard manure or enriched compost may be applied near the root zone.

d. Vermicompost: It is an organic manure where certain species of earthworms are used to enhance the process of organic waste conversion and produce a nutrient-rich end product *i.e.,* vermicompost (N –1.80%, P_2O_2 – 0.22% and K_2O – 0.40%). It is prepared by large-scale rearing of earthworms in an artificial or natural pit. Potato generates a large amount of crop residues during dehaulming (cutting of haulms/aerial parts mechanically) of the crop and the nutrients embedded in these residues is a potential source of organic nutrition (contains recyclable nutrients like N, P, K, S, Zn, Mn, Fe and Cu) and therefore effectively used in the production of valuable vermicompost. 4–6 tonnes of vermicompost is recommended to incorporate into the soil before planting of solanaceous vegetables.

e. Oil cakes: The non-edible oil cakes (neem cake – N – 3.90%, P_2O_5-1.80% and K_2O – 1.60%, castor cake – N – 4.30%, P_2O_5 – 1.80% and K_2O – 1.30% and linseed cake – N – 4.90%, P_2O_5 – 1.40% and K_2O – 1.30%) @ 80 kg/0.4 ha and oil-bearing trees (*Terminalia chebula, Madhuca indica, Jatropha cucras*) should be applied to the soil before planting of potato. The application of oil cakes like neem cake, mustard cake, *etc.*, will not only act as a source of nutrients but also help in the management of soil-borne insect pests, pathogens and nematodes.

f. Biofertilizers: For 0.4 ha of land, 2–4 kg of microbial agents Azotobacter/Azospirillum and 2–4 kg of Phosphorus Solublising Bacteria (PSB) are required. These two different kinds of biofertilizer should be mixed in 2–4 liters of water separately and sprinkled on two separate heaps of 50–100 kg of compost/farm yard manure. Later on, the two heaps should be mixed separately and left for incubation overnight. Next morning, the two heaps may be mixed together and the biofertilizer-manures mixture applied in furrows immediately before planting or the same biofertilizer-manures mixture can be prepared and applied in furrows near the root zone after 20 days of planting or at the time of earthing-up operations. These microbial agents will not only act as biofertilizers but also have antifungal activities without any residual effects, resulting in sustainable quality vegetable production.

28.1.3.2 Integrated Pest Management (IPM)

It is always recommended to follow IPM practices mandatorily in all the predominantly grown solanaceous vegetable crops throughout the country as furnished above under the insect and non-insect pest management section.

28.1.3.3 Conservation Agriculture (CA)

Conservation agriculture (CA) is a sustainable approach in crop production that aims to protect soil from degradation and erosion, improve its quality and biodiversity and contribute to the conservation of natural resources, water and air, whilst optimizing yields. CA is based on three core principles: minimum soil disturbance, maintenance of permanent soil cover and use of crop rotations with a diversity of crop species (FAO, 2015). The benefits arising from the ease of crop management, energy/cost/time savings, and soil and water conservation have influenced a few farmers for widespread adoption of CA in the recent past. It is therefore suggested to adopt conservation agriculture practices in order to maintain sustainability in crop production.

REFERENCES

Anonymous (2020) *Monthly Reports on Onion, Potato and Tomato for the month of January, 2020.* Horticulture Statistics Division, Department of Agriculture, Cooperation & Farmers' Welfare Ministry of Agriculture & Farmers' Welfare Government of India. www.agricoop.nic.in

Arora S, Gopal M (2002) Status of pesticide residues in brinjal (*Solanum melongena* L.): Indian Scenario. *Ann Agric Res* 23(3): 352–354.

Buragohain P, Saikia DK, Sotelo-Cardona P, Srinivasan R (2021) Development and validation of an integrated pest management strategy against the invasive South American tomato leaf miner, *Tutaabsoluta* in South India. *Crop Prot.* 139: 105348. https://doi.org/10.1016/j.cropro.2020.105348

Chandel RS, Chandla VK, Verma KS, Pathania M (2012) Insect pest of potato in India: Biology and management. In: Alyokhin AC, Vincent, Giordanengo P (eds.), *Insect Pests of Potato: Global Perspectives on Biology and Management*. Academic Press, USA, p. 227.

FAO (2015) *Conservation Agriculture.* www.fao.org/ag/ca/index.html (Accessed March 29, 2022).

Juraske R, Anto´n A, Castells, F (2008) Estimating half-lives of pesticides in/on vegetation for use in multimedia fate and exposure models. *Chemosphere* 70: 1748–1755.

Patnaik HP (2000) Flower and fruit infestation by brinjal fruit and shoot borer, *Leucinodesorbonalis*Guenee damage potential vs. weather. *Veg Sci* 27(1): 82–83.

Rai AB, Loganathan M, Halder J, Venkataravanappa V, Naik PS (2014) Eco friendly approaches for sustainable management of vegetable pests. *IIVR Technical Bulletin No. 53*, Indian Institute of Vegetable Research, Varanasi, p. 104.

Sardana HR, Bambawale OM, Singh A (2006) *Integrated Pest Management Strategies for Okra and Brinjal.* ICAR- National Centre for Integrated Pest Management, New Delhi, India. https://ncipm.icar.gov.in/NCIPMPDFs/Publication/okrabrinjalExtensionFolder.pdf

Singh SK, Yadav RB, Singh J, Singh B (2017) Organic farming in vegetables. *IIVR Technical Bulletin no.77.* ICAR-IIVR, Varanasi, p. 47.

Singh MP, Mishra PN, Bisht RS (2004) Nature and extent of damage of white grub *Lachnosternalongipennis* Blanchard (*Holotrichia longipennis* (Blanchard)) under various farming situations of Uttaranchal hills. *Indian J Entomol* 66(3): 277–280.

Tewari GC, Krishnamoorthy PN (1984) Yield loss in tomato caused by fruit borer. *Indian J Agric Sci* 54: 341–343.

Tiwari AK (2018) Solanaceous vegetable in temperate region of Arunachal Pradesh–a production cost analysis. *Indian J Hill Farming* 31(1): 57–60.

29 Integrated Pest Management Technology in Oleiferous Brassicas

Yadav M. S., Singh S. K., Rekha Balodi, Raghavendra K. V. and Neelam Mehta

29.1 INTRODUCTION

Indian vegetable oil economy is the world's fifth largest after the US, Brazil, Argentina and China (Nanjundan et al. 2020). Oleiferous brassicas (= Rapeseed-mustard) is the world's third leading provenance of vegetable oil following soybean (*Glycine max*) and palm oil (*Elaeis guaneensis*) and the second most preferred oil for consumption in India. Although rapeseed-mustard accounts for 25% of the total oilseeds produced (compared to 38% for soybean), it accounts for 31% of the edible oil produced in India (Jat et al. 2019). In Asia, India contributes >40% to the total rapeseed-mustard production. Globally, India is the fourth-largest rapeseed-mustard producer after Canada, Germany and China (ASG 2022). In spite of being the fourth-biggest producer (10.01%) of rapeseed-mustard after Canada, Germany and China in the world, India achieves 55% of the domestic edible oil necessities via imports (ASG 2022). The losses in oilseed crops due to biotic stresses are about 1/3 in *brassicas* in several parts of India (Jat et al. 2019). Of the anticipated requirement of 82–101 mt of oilseeds by 2030, the contribution of brassicas is projected to be 16.4–20.5 mt, considering its share of 20%–25% of production (Chauhan et al. 2020).

The crops thrive in both tropical and temperate regions, requiring cool, moist conditions for growth and dry weather for harvesting. The rapeseed-mustard group of crops comprises Indian mustard/raya (*Brassica juncea* L.), toria (*Brassica rapa* var. *dichotoma*), brown sarson (*Brassica rapa* subsp. *sarson*), yellow sarson (*Brassica rapa* var. *triculuris*), gobhi sarson (*Brassica napus* subsp. *oleifera* var. *annua*), black mustard (*B. nigra*) and taramira, *Eruca sativa* (Chauhan et al. 2020). These crops were cultivated at a density of 8.06 million ha and a yield of 11.75 million tonnes (mt), with average productivity of 1,458 kg/ha during 2021–22 (ASG 2022). The cultivation of these crops is primarily carried out during the *rabi* season across India. Nevertheless, it holds a significant presence in four states, namely Rajasthan, Madhya Pradesh, Haryana, and Uttar Pradesh, which collectively account for 83% of the total cultivated area and 86% of the overall production. Among these states, Rajasthan takes the lead with the highest proportion of both cultivated area (46%) and production (46%) (Jat et al. 2019). Previously, brown mustard (*B. rapa* var. brown mustard) was the primary rapeseed-mustard crop in India, but now Indian mustard, *B. juncea*, holds the majority of the acreage (90%). The role of irrigation, seeds, efficient crop management and technological interventions act as the key sources of growth of oilseed production and yield.

Among the different management strategies that exist, predominantly, only chemical pesticides continue to act as the main element, causing harm to natural predators and disrupting their food sources, thus creating conditions for the proliferation of diseases such as Alternaria blight, white rust and Sclerotinia stem rot (SSR) (Yadav et al. 2023). The important factors causing low and fluctuating production of these crops in India are low or non-adoption of a package of improved

DOI: 10.1201/9781032631578-29

production, protection technology, and the susceptibility of rapeseed-mustard varieties to insect pests and diseases. The loss in yield may depend on the nature of the pest and the severity of the attack.

Considerable advancement in research has led to an increase in the productivity of the rapeseed-mustard, particularly in China and India. Despite the collective efforts of numerous plant protection researchers, the global adoption of IPM in rapeseed-mustard has not reached the intended level, for a variety of reasons. Earlier studies, in general, were established on rapeseed which is widely grown in the US, Germany and Canada, but in India, Indian mustard is predominately grown in which IPM is required. Integrated Pest Management is a sustainable method used in mustard farming to ensure crop protection and enhance production. By integrating various management strategies and practices, IPM promotes the growth of robust crops while reducing the reliance on pesticides. Mustard IPM strategies involve utilizing pest-resistant crop varieties, adjusting agronomic practices, implementing biological control methods, exploring innovative approaches, and employing chemical pesticides in a targeted and responsible manner to minimize pest damage (Yadav 2018).

29.1.1 Pest Scenario

Holo-parasitic weed broomrapes (*Phelipanche=Orobanche* spp.) have appeared as a severe threat to oilseed brassicas cultivation particularly in the semi-arid tracts of the world (Jat and Singh 2018; Das et al. 2020). Aphid (*Lipaphis erysimi* Kaltenbach), painted bug [*Bagrada hillaris* Burmeister (*cruciferarum*)], mustard sawfly (*Athialia lugens proxima* Klug) and leaf miner (*Chromatomyia horticola* Goureau) are the predominant insect- pests of rapeseed-mustard along with the regional pests prevailing at the specific locations (Patel et al. 2019). Among diseases, white rust (*Albugo candida* (Pers.) Kuntze), *Sclerotinia* stem rot [*Sclerotinia sclerotiorum* (Lib.) de Bary], Alternaria blight [*Alternaria brassicae* (Berk.) Sacc.], downy mildew [*Hyaloperonospora parasitica* (Pers.) Constant.], powdery mildew (*Erysiphe cruciferarum*, Opiz ex L. Junell), black rot [*Xanthomonas campestris* pv. *campestris* (Pammel 1895) Dowson], and phyllody *Candidatus* Phytoplasma are the major diseases, which reduced the yield potential of rapeseed-mustard substantially (Chattopadhyay et al. 2015). Key pests of rapeseed mustard are presented in Figure 29.1 and their brief descriptions are as follows:

29.1.1.1 Broomrape (*Phelipanche = Orobanche aegyptica*)

In India, *Phelipanche* spp. has emerged as a major threat to rapeseed-mustard production in northern Rajasthan, Haryana, Punjab, and north-east Madhya Pradesh. Broomrape weed is one of the important pests of rapeseed-mustard in India, which has threatened crop cultivation over large areas in the state of Haryana and Rajasthan (Sharma et al. 2019). Broomrape is a phenerogamic holo-parasitic weed, which grows on the plant root in response to germination stimulants. When affected host plants are uprooted, the parasitic roots are seen intertwined with the host root system. The seed of the parasite survives many years in the soil or as seed contaminants along with mustard seed. As infestation of this weed starts 7–10 days after the sowing of the crop, control measures should be applied in the initial growth of the crop (Punia 2014).

29.1.1.2 Mustard Aphid (*Lipaphis erysimi*)

It is a widely distributed, most destructive and severely disparaging major pest of rapeseed-mustard (Dhillon et al. 2022). It solely feeds on the crucifer plants. The aphid attacks generally during 50–51 SMW and persists up to the 13th SMW. It is recognized as the most serious pest of rapeseed-mustard as it regularly havocs the same with colossal yield losses year after year. If not managed timely, it may lead to complete crop failure. In case of drastic infestation, the yield of the crop may be reduced by even 80% or more (Yadav and Rathee 2020). Nymphs and adults suck sap from leaves, stems,

FIGURE 29.1 (1) Broomrape (*Orobanche aegyptica*) infestation in mustard crop field; (2) Mustard aphid (*Lipaphis erysimi*) on mustard inflorescence (3) Painted Bug [*Bagrada hillaris* (*cruciferarum*)] on mustard (4) Sclerotinia rot (*Sclerotinia sclerotiorum*); on mustard stem with sclerotia (5) (a) Alternaria blight of mustard (*Alternaria brassicae*) on leaf (b) Alternaria blight of mustard (*Alternaria brassicae*) on pod (6) White rust (*Albugo candida*) of mustard (7) Powdery mildew (*Erysiphe cruciferum*).

inflorescence and developing pods. As a result, plants remain undersized with reduced pod and grain number, pods shrivel up and seeds do not develop. The economic threshold level of this pest occurs when the population reaches 23–25 aphids/10 cm length of central shoot in at least 10% of the plant population (Nayak et al. 2014).

29.1.1.3 Painted Bug [*Bagrada hillaris* (*cruciferarum*)]

The painted bug is a highly polyphagous pest and is more serious on young / seedling stage. This insect often infests rapeseed-mustard and is more common in the drier areas of Rajasthan, Madhya Pradesh, Haryana, Uttar Pradesh & Punjab where the soil is lighter. (Yadav and Rathee 2020). Due to the frequent appearance of the *Bagrada Hillaris* on mustard at the seedling stage as a severe pest, farmers are forced to rely solely on the use of extremely toxic insecticides which can lead to severe environmental and health risks. The nymphs and adults suck cell sap from leaves and pods, which eventually wilt and dry up. This insect pest also reduces the oil quantity by sucking pods; pods shrivel up and seeds do not develop. In both cases, re-sowing of the field becomes necessary (Yadav and Rathee 2020).

29.1.1.4 Sclerotinia Stem Rot (*Sclerotinia sclerotiorum*)

Sclerotinia stem rot of rapeseed-mustard is recognized as one of the most significant diseases in the world due to its broad host range, worldwide occurrence and lack of effective management practices (Saharan and Mehta 2008; Chattopadhyay et al. 2015; Yadav 2021). The explosive pathogenicity has made it very difficult to safeguard the crop from its infection in India (Chattopadhyay et al. 2015; Singh et al. 2020). Its infection at the initial stages of plant proliferation results in complete failure of the plant whereas late infection lowers the yield quantity and quality as well. Disease increases with mono-cropping of rapeseed-mustard (Yadav et al. 2013). In individual affected plants, sometimes no grain is formed. The disease appears as elongated, buff to light brown water-soaked lesions which later rot and are covered with white, cottony mycelial growth of the fungus (Yadav 2021). All the affected parts of the plants rot in cool and wet weather. The affected plants show stunting and premature ripening, shredding of stem, wilting and drying. A large number of black sclerotia appear in fungal growth around the rotted stem (Saharan and Mehta 2008; Yadav 2021; Yadav et al. 2012b).

29.1.1.5 Alternaria blight (*Alternaria brassicae*)

In rapeseed-mustard crops, this disease is mainly caused by *Alternaria brassicae*. However, other species of *Alternaria* like *A. brassicicola*, *A. raphani* and *A. alternata* have also been believed to parasitize on these crops in India (Chattopadhyay et al. 2015). This is a widely distributed, more destructive and most damaging disease under epiphytotic conditions causing yield loss of up to 47% (Saharan et al. 2016). Significant reductions in crop yield can occur when ideal conditions, such as an average temperature of 18°C coupled with a relative humidity of 80% or higher, coincide with stormy weather, particularly during the flowering and pod-development phases (Chattopadhyay et al. 2015). The shriveled and discolored seeds fetch a lower market price (Saharan et al. 2016). The disease is described by the formation of brown-colored concentric rings on leaves, stems, siliquae and pods. *Alternaria brassicae* survives on diseased plant remains in the soil and on many alternate cruciferous hosts like, cauliflower (*Brassica oleracea* var. *botrytis*), cabbage (*B. oleracea* var. *capitata)*, turnip (*B. rapa* var. *rapa*), and radish, *Raphanus raphanistrum* subsp. *sativus* (Saharan et al. 2016).

29.1.1.6 White Rust (*Albugo candida*)

This disease of rapeseed-mustard attacks all the plant parts except roots. Prominent white creamy scattered raised and roundish pustules appear on the bottom surface of lower leaves. Large regions of several pustules combine to cover the entire undersurface of the leaf (Chattopadhyay et al. 2015). The dangerous stage is floral infection in which the floral parts are malformed and become thick leathery green. This disease, when it appears on the stag head phase distorts inflorescence, where it causes hypertrophy and hyperplasia causing a heavy loss of 17–34% in yield. White and creamy pustules also appear on the hypertrophied parts of inflorescence. Late sown crop (November) suffers more from this disease.

29.1.1.7 Powdery Mildew (*Erysiphe cruciferarum*)

It is a disease of warmer and drier tracts, where rapeseed-mustard is grown. The disease is gradually becoming quite severe in Rajasthan, Madhya Pradesh, Haryana, Uttar Pradesh and Gujarat with the shortening of winter and climate change, resulting in a considerable loss in yield (Nanjundan et al. 2020; Yadav 2018). The disease usually arrives in the later part of the crop season. However, it has been observed during vegetative stages in recent times, whereby the pathogen is causing significant loss to the crop (Meena et al. 2018).

29.1.2 Integrated Pest Management Strategy

29.1.2.1 Prevention

Many aspects of field and crop management reduce sources of primary inoculum and prevent the initial build-up of pest population (Yadav 2018). Crop hygiene and various agronomic practices such as deep plowing in summer, solarization of soil, cleaning of crop refuse, elimination of weed hosts, regional recommended crop rotation, water management, intercropping / mixed cropping, nutrition management, selection of regional specific recommended variety / hybrid, seed treatment, soil incorporation of green manuring, bio-agent, etc. along with genetic resistance to pests are major tools of prevention.

29.1.2.2 Monitoring

Regular monitoring is crucial to pest management and is an important decision-making tool. An efficient programme can pay big dividends in lowering pest management costs. Agro-ecosystems based on crop health at different stages of growth, population dynamics of pests and natural enemies, soil conditions, climate factors and farmers' past experiences are considered for decision-making (Yadav 2018). Regular surveys should be conducted throughout the cropping or pest-breeding season to predict future populations and forewarn the cultivators to initiate the control measures or not. Yellow sticky or water traps may be used to monitor population fluctuations of *L. erysimi*.

29.1.3 Interventions in Rapeseed-Mustard at Different Growth Stages

To minimize the economic losses brought on by pests, integrated pest management is a systems approach that integrates a variety of crop production and protection techniques. Hence, the use of low or judicious doses of chemical fertilizers, and pesticides, integrated with other means like growing pest-tolerant cultivars, sanitation, crop rotation, use of bio-agents and plant extracts seems to be the best method of pest management without environmental degradation. The ICAR-NCIPM, New Delhi, conducted multilocational field trials of IPM technology of Indian mustard in Rajasthan, Haryana, and Madhya Pradesh on farmers' fields and based on them IPM interventions at various stages of growth of rapeseed-mustard were developed, as summarized in Table 29.1.

29.1.4 Integrated Pest Management in Rapeseed-Mustard in Farmers' Participatory Mode: A Success Story

The ICAR-NCIPM has been working on developing, validating and promoting IPM in rapeseed-mustard for more than two decades. The scientific knowledge has been profitably exhibited at the village level in Haryana, Rajasthan and Madhya Pradesh in farmers' participatory modes. Since IPM is location-specific and dynamic; it needs regular updating because of changes in pest scenarios due to the introduction of new hybrids and monoculture in large tracts of irrigated and water-logged soil, where close spacing and heavy fertilization were practiced. After the introduction of high-yielding hybrids by the ICAR-AICRP on rapeseed-mustard, *Sclerotinia* stem rot disease (*Sclerotinia sclerotiorum*) posed a significant threat in mustard because a few hybrids are highly susceptible to this disease (Yadav et al. 2012b). A mustard-growing farmer of Siyali Khurd village (district Alwar of Rajasthan) approached the ICAR-NCIPM because of the severe incidence of *Sclerotinia* stem rot in 2011. Siyali Khurd Village (Latitude N27°54′23.2″ Longitude E76°36′27.7″) is about 150 km away from the ICAR-NCIPM, Rajpur Khurd, New Delhi, in a flood-prone eastern plain in zone III b of Rajasthan. Baseline information of the village indicated that the farmers of the area, prior to IPM implementation, were taking monocultures of mustard without crop rotation and deep summer plowing, and no priming of seeds with *Trichoderma* was undertaken. The recommended dose of fertilizer along with gypsum @ 250 kg/ha and Potash @ 40 kg/ha, soil incorporation of *Trichoderma* @ 2.5 kg/ha pre-incubated in 50 kg well-rotten FYM and seed treatment with *Trichoderma* @10g/kg seed were followed. By destroying the previous crop residue in IPM, the crop avoided painted bug infestation and loss due to disease-causing soil-borne pathogens. In IPM practices, an appropriate seed rate of 4 kg/ha was followed along with region-specific optimum sowing time, which escape the incidence of aphids, *Alternaria* blight, white rust and *Sclerotinia* stem rot. Farmers were not aware of the IPM concept and were not able to identify insect pests, diseases and beneficial organisms. Based on baseline information collected at Siyali Khurd in 2011, the IPM module was fine-tuned and implemented in 40 ha in 2011–12 in Haryana and Rajasthan, which was subsequently implemented in 40 ha during 2012–13 and 2013–14, as more farmers became part of IPM programme. Implementation of IPM in this village resulted in a significant reduction in the severity

TABLE 29.1
IPM interventions in rapeseed-mustard crop

Crop stage	IPM interventions
Pre-sowing stage	• Conduct deep summer plowing to eliminate fungal spores and reduce the population of pests; ensure adequate water drainage, preparation of level field (Yadav et al. 2012a) • Rotating crops and applying an appropriate amount of fertilizers and seeds as per location-specific recommendation for the region, for example: apply 15 kg zinc sulphate/ha (Kumar et al. 2015); apply gypsum, boron, zinc sulphate, etc. to the soil. (Yadav et al. 2023); *Sesbania* application as green manuring in addition to the assimilation of soil of mustard crop residue @2.5 t/ha in rainy season (Regar et al. 2009); planting at the appropriate time (1–31 Oct.) to avoid the intensity of aphids and diseases; use of disease-resistant hybrids and cultivars approved for the region (Yadav et al. 2012b); Refrain from using a high seed rate or narrow spacing while managing Sclerotinia stem rot (Yadav et al. 2012a) • Elimination of pest remains and crop leftover to avoid infestation of painted bug and infection of pathogens (Yadav 2018) • Soil incorporation of *Trichoderma* spp. @ 2.5 kg/ha in 50 kg FYM in order to manage soil-borne diseases (Yadav et al. 2023); seed priming with Metalaxyl-M 31.8 % ES @ 3.5 ml/kg seed and Imidacloprid 70 WS @ 7 g/kg for the management of seed-borne diseases and painted bug, respectively; application of freshly made aqueous garlic bulb extract (2% W/V) for seed treatment or *Trichoderma*-based product @ 10 g/kg (Yadav et al. 2019)
Seedling and vegetative stage	• Maintenance of optimum plant population with wide spacing (Yadav et al. 2012a) • Irrigation of crop at seeding stage to protect against painted bug (Ahuja et al. 2008) • Maintenance of weed-free crop by clean cultivation, which act as collateral hosts for pathogens (Kumar et al. 2021) • Regular monitoring of crop and destroying of pest-infested/ infected plants; for example, hand-picking of aphid-infested twigs in the initial attack and apply Azadirachtin 300 PPM @ 5 ml/l followed by its second application after fifteen days (Nayak et al. 2014); Release *Chrysoperla* for checking aphid (Singh and Kumar 2000) • Spray application of micro-nutrients like boron and zinc are also very useful in pest management (Kumar et al. 2015) • Safeguarding of natural enemies of aphids, namely *Coccinella septempunctata, Chrysoperla carnea, Syrphid* fly, etc. (Dwivedi et al. 2018) • Orobanche panicle initiation starts early, so post-emergence application of glyphosate at 25–50 g/ha at 30 and 50 DAS can reduce 60–80% of broomrape (Singh et al. 2020) • Spray of freshly prepared aqueous garlic bulb extract @ 2% (W/V) for diseases management at early bloom stage (Mahapatra and Das 2016; Yadav 2009, 2018; Yadav et al. 2019, 2023) • Need-based spray of Hexaconazole 5% EC @ 0.005% for management of powdery mildew
Flowering and pod formation stage	• Regular scouting of field for pest monitoring (Yadav 2018) • Use of 4 (four) bee colonies/ha at early bloom stage • For environment-friendly management of mustard aphid, spray of Azadirachtin 300 ppm @ 5 ml/l of water at 40 and 55 days after sowing and installation of yellow sticky traps @ 25/ha (Nayak et al. 2014) • If mustard crop is fertilized excessively with nitrogen, the crop tends to get affected more severely by white rust but can be protected from major diseases by spraying the crop at flowering to early pod formation stage with a mixture of Metalaxyl 14% and mancozeb 64 (Yadav et al. 2019) • Thresh the crop as early as possible to avoid the further losses due to painted bug and dispose of infected plant debris immediately (Yadav and Rathee 2020)

of *Sclerotinia* stem rot with higher yield as well as higher benefit-to-cost ratio in IPM as compared to farmers' practices. Work on the impact of input costs in IPM mustard was undertaken during 2013–2017 (Yadav et al. 2019). The developed IPM package enables the growers to decide the input based on the predominance of biotic stress(es). In subsequent years, IPM in rapeseed-mustard was validated and promoted in 180 ha in Madhya Pradesh, Haryana and Rajasthan. Horizontal spread of IPM mustard on large areas in Haryana, Madhya Pradesh and Rajasthan is in progress at the ICAR-NCIPM, New Delhi.

29.2 FUTURE THRUST

India is heavily dependent on imported edible oils to meet its domestic consumption, necessitating their increase in acreage and production. During 2021–22 edible oils demand was 25.76 mt It is projected that 39.2 mt of edible oil will be needed for the 1.68 billion population in India by 2050. In order to meet the demand, 20.5 mt of oilseed brassicas will need to be produced by the year 2050, assuming a 25% contribution to total oilseed production. This is a significant challenge that requires the cooperation and efforts of scientists and policymakers. However, this task is further complicated by biotic stresses such as insect pests, diseases and weeds – particularly *Phelipanche* (*P. aegyptica*) – which can result in a yield loss of up to one-third in various regions across the country. A comprehensive management approach and efforts to develop resistant / tolerant cultivars are necessary to minimize losses due to biotic stresses. Efforts must be exerted to disseminate potential technologies / products and identify the low-cost or no-cost technologies that can help to manage crops effectively. Enhancing the connections between farmers, institutions, and industries is crucial in addition to the current formal delivery systems, in order to close the gap between achievable and actual crop yields. A better oilseed economy can only be achieved through various risk mitigation strategies such as timely inputs and credit, MSP procurement with crop insurance, and establishing a connection between farmers and the market. The policy alternatives should also consider market entry not just for edible oil, but also for value-added products, animal feed, biodiesel, medications, and fortified foods. Adaptive strategies that are tailored to the agro-ecological, socio-economic and epidemiological conditions of oilseed brassica should be given significant attention to achieve the targets for oleiferous crops.

REFRENCES

Ahuja DB, Kalyan RK, Ahuja U, Singh S, Sundaria M, Dhandapani A (2008) Integrated management strategy for painted bug, *Bagrada hillaris* (Burm.) inflicting injury at seedling stage of mustard (*Brassica juncea*) in arid western Rajasthan. *Pesticide Res J* 20(1): 48–51.

ASG (2022) *Agriculture statistics at a glance-2022*. DES, DAFW, Ministry of Agriculture and Farmers Welfare, GOI, New Delhi.

Chauhan JS, Choudhury PR, Pal S, Singh KH (2020) Analysis of seed chain and its implication in rapeseed-mustard (*Brassica* spp.) production in India. *J Oilseeds Res* 37(2): 71–84.

Chattopadhyay C, Kolte SJ, Waliyar F (2015) *Diseases of edible oilseed crops*. CRS Press Taylor & Francis Group, Boca Raton, pp. xxvii + 456.

Das TK, Ghosh S, Gupta K, Sen S, Behera B, Raj R (2020) The weed *Orobanche*: species distribution, diversity, biology and management. *J Res Weed Sci* 3(2): 162–180.

Dhillon M, Singh N, Yadava DK (2022) Preventable yield losses and management of mustard aphid, *Lipaphis erysimi* (Kalt.) in different cultivars of *Brassica juncea* (L.) Czern & Coss. *Crop Prot* 161: 1060–1070.

Dwivedi SA, Singh RS, Gharde SK (2018) Population buildup of mustard aphid and their natural enemies in relation to biotic and abiotic factors. *Plant Arch* 18(2): 2495–2500.

Jat RS, Singh D (2018) Broomrape intimidation to rapeseed-mustard production in semi-arid regions. *Adv Plants Agric Res* 8(6): 560–562.

Jat RS, Singh VV, Sharma P, Rai PK (2019) Oilseed Brassica in India: Demand, supply, policy perspective and future potential. *Oilseeds Fats Crops Lipids* 26: 8. https://doi.org/10.1051/OCL/2019005

Kumar S, Malay K, Bhowmick S, Ray P (2021) Weeds as alternate and alternative hosts of crop pests. *Indian J Weed Sci* 53(1): 14–29.

Kumar S, Singh D, Yadav SP, Prasad R (2015) Studies on powdery mildew of rapeseed-mustard *(Brassica juncea* L.) caused by *Erysiphe cruciferarum* and its management. *J Pure Appl Microbiol* 9(2): 1–6.

Mahapatra S, Das S (2016) Spraying schedule of garlic bulb extract, mancozeb and salicylic acid against *Alternaria* blight of mustard. *Indian Phytopath* 69: 413–416.

Meena PD, Mehta N, Rai PK, Saharan GS (2018) Geographical distribution of rapeseed-mustard powdery mildew disease in India. *J Mycol Pl Pathol* 48(3): 284–302.

Nanjundan J, Manjunatha C, Radhamani J, Thakur AK, Yadav R, Kumar A, Meena ML, Tyagi RK, Yadava DK, Singh D (2020) Identification of new sources of resistance to powdery mildew of Indian mustard and studying its inheritance. *Plant Pathol J* 36(2): 111–120.

Nayak US, Mishra I, Mishra BK (2014) Field evaluation of bio-intensive IPM modules against important insect-pests of mustard under North Coastal Plain Zone of Odisha. *J Biol Contrl* 28(4): 192–196.

Pammel LH (1895) Bacteriosis of Rutabaga (*Bacillus campestris* n.sp.). Iowa State Coll Agric Exp Sta, 1942-43, Part-II: pp 52–57.

Patel S, Singh CP, Yadav SK (2019) Monitoring of insect-pest complex on rapeseed-mustard at Pantanagar. *J Ent Res* 43 (1): 73–76.

Punia SS (2014) Biology and control measures of *Orobanche*. *Indian J Weed Sci* 46(1): 36–51.

Regar PL, Rao SS, Vyas SP (2009) Crop residue management for sustainable production of Indian mustard (*Brassica juncea*) in arid and semi-arid region. *Indian J Soil Cons* 37(2):118–I22.

Saharan GS, Mehta N (2008) *Sclerotinia diseases of crop plants: biology, ecology and disease management*. Springer Science, The Netherlands, p. 485.

Saharan GS, Mehta N, Meena PD (2016) *Alternaria blight of crucifers: biology, ecology and management*. Springer Verlag, Singapore, p. 326.

Sharma R, Amarjeet A, Punia SS (2019) Response of various chemicals, neem cake and hand pulling on growth and development of Egyptian broomrape (*Phelipanche aegyptiaca)* in Indian mustard. *J Crop Weed* 15(2): 126–131.

Singh NN, Kumar M (2000) Potentiality of *Chrysoperla carnea* (Stephens) in suppression of mustard aphid population. *Indian J Entomol* 62(4): 323–326.

Singh M, Avatar R, Pal A, Punia R, Singh VK, Bishnoi M, Singh A, Chaudhary RR, Manduani S (2020) Genotype-specific antioxidant responses and assessment of resistance against *Sclerotinia sclerotiorum* causing *Sclerotinia* rot in Indian mustard. *Pathogens* 9: 892.

Singh SP, Yadav RS, Godara AS, Bairwa RC (2020) Screening of herbicides for broomrape (*Orobanche*) control in mustard. *Indian J Weed Sci* 52(1): 99–101.

Yadav MS (2009) Biopesticidal effect of botanicals on the management of mustard diseases. *Indian Phytopath* 62(4): 488–492.

Yadav MS (2018) Integrated pest management - rapeseed-mustard. In: Chattopadhyay C, Tanwar RK, Sehgal M, Birah A, Bhagat S, Ahmad N, Mehta N (eds.), *Handbook of integrated pest management*, ICAR, pp. 543–548.

Yadav MS (2021) *Sclerotinia* stem rot: A significant agricultural problem of rapeseed-mustard crop production. *Pl Dis Res* 31(1): 18–26.

Yadav MS, Ahmad N (2020) *Sclerotinia* rot: A major constraint of mustard cultivation in India. In: Dubey SC, Mondal KK, Bashyal BM, Das A, Akhtar J, Gupta K, Gogoi R, Gurjar MS, Singh D (eds.) *7th international conference on "Phytopathology in achieving UN sustainable development goals"*. ICAR-IARI, New Delhi, pp. 178–179.

Yadav MS, Ahmad N, Singh S, Yadava DK, Godika S, Gaur RB (2012a) Multilocational validation of integrated management practices for Sclerotinia rot Indian mustard. *Indian J Agric Sci* 82: 972–77.

Yadav MS, Godika S, Yadava, DK, Ahmad N, Mehta N, Bhatnagar K, Agrawal, VK, Kumar, A, Thomas L, Chattopadhyay C (2019) Prioritizing components of package of integrated pest management in Indian mustard (*Brassica juncea*) in India for better economic benefit. *Crop Protect* 120: 21–29.

Yadav MS, Singh N, Ahmad N, Yadava DK, Singh S, Bambawale OM (2012b) Sources of resistance to *Sclerotinia sclerotiorum* in Indian varieties of rapeseed-mustard. *Plant Dis Res* 27: 108–110.

Yadav MS, Singh N, Singh S, Ahmad N, Godika S (2013) Assessment of prevalence and severity of Sclerotinia rot of Indian mustard in Rajasthan and Haryana. *Indian J Plant Protect* 41: 249–252.

Yadav MS, Yadav NS, Yadava DK, Singh SK, Mehta N (2023) Effect of bio-intensive strategy on disease management in mustard (*Brassica juncea*). *Indian Phytopath* 76: 637–640.https://doi.org/10.1007/s42360-023-00611-2

Yadav S, Rathee M (2020) Sucking pests of rapeseed-mustard. In: Omkar (ed.), *Sucking pests of crops*. Springer Nature Singapore Pvt. Ltd, Singapore, pp. 187–232.

Index

Note: Page numbers in *italic* refers to Figures

C

D

G

M

U

V

W

X

Y

Z